SOLID STATE NMR OF POLYMERS

SOLID STATE NMR OF POLYMERS

Edited by

Lon J. Mathias

University of Southern Mississippi
Hattiesburg, Mississippi

PLENUM PRESS • NEW YORK AND LONDON

Library of Congress Cataloging-in-Publication Data

Chemistry Conference of North America on Solid State NMR of Polymers
(3rd : 1988 : Toronto, Ont.)
Solid state NMR of polymers / edited by Lon J. Mathias.
p. cm.
"Proceedings of the Third Annual Chemistry Conference of North America on Solid State NMR of Polymers, held June 5-10, 1988, in Toronto, Canada"--T.p. verso.
Includes bibliographical references and index.
ISBN 0-306-44015-6
1. Polymers--Analysis--Congresses. 2. Nuclear magnetic resonance spectroscopy--Congresses. I. Mathias, Lon J. II. Title.
QD139.P6C48 1988
547.7046--dc20 91-23687
CIP

Proceedings of the Third Annual Chemistry Conference of
North America on Solid State NMR of Polymers,
held June 5–10, 1988, in Toronto, Canada

ISBN 0-306-44015-6

A Division of Plenum Publishing Corporation
233 Spring Street, New York, N.Y. 10013

Printed in the United States of America

PREFACE

The chapters in this collection are from papers which were presented at a symposium on solid-state NMR of polymers. A two-part program on available NMR techniques applicable to solid polymer analysis was presented at the 3rd Chemical Congress of North American held in Toronto, Ontario, June 5-10, 1988. The program was sponsored by the Division of Polymer Chemistry with support provided by the Division, its Industrial Sponsors, and the Donors of the Petroleum Research Fund administered by the American Chemical Society. Co-organizers included Professor Colin Fyfe of the University of British Columbia (Vancouver, Canada), Professor Hans Spiess of the Max Planck Institut fur Polymerforschung (Mainz, West Germany), and myself.

The full-day tutorial, which was free to registered attendees, covered the range of topics. The purpose of the tutorial was to provide a basic introduction to the field so that newcomers to its present and future applications could develop sufficient understanding to learn effectively from the subsequent symposium. The first talk attempted to give listeners a feel for the way a novice spectroscopist can learn to use the various NMR techniques to explore his own areas of interest. Simple experiments can provide unique information about solid polymers that can be useful in interpreting synthetic results and in relating solid-state conformation, morphology and molecular motion to physical properties.

The remaining five talks by internationally recognized experts in their areas covered the most commonly used techniques available today. These range from proven methods such as cross-polarization/magic angle spinning (CP/MAS) to magnetic resonance imaging (MRI) of polymers that is just developing as a useful tool for rigid polymer samples. The former is the first and most widely used method, and is "routine" on most instruments available today. It has been applied to nuclei such as ^{15}N, ^{29}Si and ^{31}P as well as ^{13}C, and can provide information on chemical composition, molecular motion and relaxation processes, and polymer morphology. Advanced applications such as those employing variable temperature studies and two-dimensional NMR pulse programs provide a wealth of information not available with any other single solid-state method.

Multipulse NMR of nuclei like ^{1}H and ^{19}F is well-suited for evaluating the details of molecular mobility and relaxation in solids. When combined with basic and advanced ^{2}H NMR experiments of specifically labelled polymers, it is possible to probe molecular behavior throughout a polymer sample at relaxation times covering ca. 10 orders of magnitude. Not only is it possible to examine amorphous and crystalline or highly ordered regions of a sample independent of each other, it is now feasible to evaluate the types and rates of motions available to individual segments of polymer backbones and pendent groups. The contents of this talk are summarized in the chapter by Cecil Dybowski. Examples of ^{2}H NMR applications are given in the overview paper by Ron Colletti and myself.

One of the most exciting new areas of polymer characterization is a result of the enormous effort successfully devoted to the development of MRI for medical applications and biological research. Compared to CAT scans that are based on X-ray imaging of the hard or X-ray opaque regions of the body, MRI is most sensitive to the soft tissue. Skin, nerves, brain tissue and the circulatory system are examples of body components that can currently be resolved down to the 1 mm level. These methods, especially when coupled with the higher field strengths that can be used on inanimate objects, have been shown to provide 10 to 20 micron resolution images of soft, flexible polymers. New pulse programs and imaging techniques are also beginning to provide information on rigid polymers that are difficult to examine using traditional imaging methodology. This area is summarized by Jack Koenig in Chapter 3.

Over 50 talks were given in the four-day symposium that followed the tutorial (see **Polym. Preprints,** 1988, **29(1)**, 3-102). These were organized under the same general topics as the tutorial with roughly equal coverage of the five main areas. This provided the 80-100 attendees at each session with a timely and comprehensive survey of current research efforts around the world in solid-state NMR analysis of polymers. In addition to speakers from the US and Canada, there were 7 speakers from West Germany, 3 from Japan, 2 from Great Britain, and 1 each from France, the Netherlands, and New Zealand. Papers based on many of these talks are included in Chapters 4-25 and are arranged according to general subject areas.

This book should provide the polymer community with an extensive introduction to the many aspects of solid-state NMR analysis of polymers. Learn to apply these methods to your own projects and you will find insight and understanding of molecular and microscopic behavior that can open up new vistas of research. I think I speak for all the authors in hoping that you find this book informative and useful. Let me also thank all of the contributors for their hard work and excellent contributions, and especially for their patience.

Lon J. Mathias

CONTENTS

I. Overviews

1. Multi-Pulse ^{1}H and ^{19}F Techniques — 1
Cecil Dybowski

2. Solid State ^{2}H NMR: Overview with Specific Examples — 23
Ronald F. Colletti and Lon J. Mathias

3. Application of NMR Imaging to Polymers: A Tutorial — 61
Jack L. Koenig

II. ^{13}C CP/MAS Applications

4. Solid State ^{13}C NMR Studies of the Structures, Conformations, and Dynamics of Semi-Crystalline Polymers — 81
A.E. Tonelli, M.A. Gomez, Hajime Tanaka, and M.H. Cozine

5. CP/MAS ^{13}C NMR Analysis of Poly(Arylene Methylene)s — 107
Michael L. Froehlich and Martin B. Jones

6. Radiation Induced Morphological Changes in a Semi-Crystalline Polymer: A ^{13}C NMR Study — 117
A.L. Cholli, F.C. Schilling, and A.E. Tonelli

7. ^{13}C NMR Studies of Poly(Vinylidene Fluoride) Blends — 131
R.A. Grinsted and J.L. Koenig

8. Solid State NMR Detection of Molecular-Level Mixing Phenomena in Strongly Interacting Polymer Blends and Phase-Separated Copolymers — 145
Laurence A. Belfiore, Thomas J. Lutz, and Chihmin Cheng

9. CP/MAS NMR Spectra of Poly(N-Vinylcarbazole) and of its Charge Transfer Complexes: Preliminary Investigation — 167
Almera Natansohn

10. ^{13}C NMR of Crosslinked Poly(Methacrylic Anhydride) — 179
D. Granger

11. Application of Solid State ^{13}C NMR Spectroscopy to Sulfur Vulcanized Natural Rubber — 201
Jack L. Koenig and Mladen Andreis

12. A Solid State ^{13}C NMR Study of Polysiloxane Side Chain Liquid Crystalline Polymers Containing 2-[4-(2S)-Methyl-Butoxy)-Phenyl]-5-(ω-Alkyl)-1,3,2-Dioxaborinane with Alkyl Being Undecanyl and Octyl — 215
B.C. Perry and J.L. Koenig

13. VT/MAS NMR as a Probe of Molecular Mobility and Local Composition: Solidified Acrylic Dispersions — 233
R. Voelkel

III. Deuterium NMR

14. Molecular Motions in Crystalline and Mesomorphous Phases of Large Size Cycloalkanes as a Model for Polyethylene — 245
Martin Moller, Gerhard Kogler, Dieter Oelfin, and Hansotto Drotloff

15. Solid State NMR Relaxation Study of Liquid Crystal Polymers Employing a Two-Dimensional Technique — 261
A. Schleicher, K. Müller, and G. Kothe

16. Dynamics of Surface Bound Polymers and Coupling Agents — 271
Frank D. Blum, Robert B. Funchess, and Wiriya Meesiri

17. Solid-State ^{2}H and ^{13}C NMR Study of the Structure of Polyanilines — 283
S. Kaplan, E.M. Conwell, A.F. Richter, and A.G. MacDiarmid

IV. Multinuclear CP/MAS

18. The Solid-State ^{29}Si and ^{13}C NMR of Poly(Di-*n*-Alkylsilanes) — 295
F.A. Bovey and F.C. Schilling

19. ^{29}Si and ^{13}C Magic Angle Sample Spinning Nuclear Magnetic Resonance Spectroscopy of Ceramic Fibers Prepared by Pyrolysis of Organosilicon Polymers — 305
Jonathan Lipowitz and Gary L. Turner

20. Variable Temperature Solid-State NMR of Polyphosphazenes — 321
James F. Haw and Richard C. Crosby

21. Hydrogen Cyanide Polymers: Solid State NMR (^{15}N and ^{13}C) Investigations 331
C.N. Matthews and R. Ludicky

V. Imaging of Polymers

22. ^{1}H NMR Imaging of Solids with Magic Angle Spinning 343
D.G. Cory, A.M. Reichwein, J.C. De Boer, J.W.M. van Os, and W.S. Veeman

23. NMR Imaging: Application to Elastomeric Materials 363
C.C. Chang and R.A. Komoroski

24. NMR Imaging Investigations of Case II Diffusion in Polymers 377
L.A. Weisenberger and J.L. Koenig

25. Application of One-Dimensional NMR Imaging to Solvent Diffusion Measurements in Polymeric Samples 387
Stephen B. Pickup and Frank D. Blum

Contributors 399

Index 403

Chapter 1

MULTIPLE-PULSE ^{1}H AND ^{19}F NMR TECHNIQUES

CECIL DYBOWSKI

1. Introduction

The NMR spectroscopy of solids has developed into a useful tool for chemists in the last decade and a half because the control of the time evolution of the nuclear systems allows one to choose the effects one wishes to emphasize in a particular experiment. This remarkable concept -- that one could analyze only parts of the contributions to the time evolution of a nuclear magnetic system -grew out of an understanding of the physics of radiofrequency excitation of nuclear spins, as developed in several research laboratories in the middle-to-late 1960's [1]. In particular, the early experiments concentrated on the dipole-dipole interactions among protons or fluorines in strongly coupled solid materials. Thus, the first experiments in this class were ones that subverted the dipole-dipole coupling among the abundant spins of a system to reveal subtler interactions such as the chemical shift.

Rapid spinning of a solid had been shown, as early as 1958, to overcome the effects of dipole-dipole couplings and the anisotropic contributions to the chemical shift to yield a spectrum that depends only on the isotropic chemical shift -- essentially a high-resolution spectrum [2]. Thus, a short time after the introduction of radiofrequency line-narrowing techniques, the coupling of these two technologies permitted a further improvement of the resolution of spectra of abundant spins [3]. We briefly describe the developments in spin dynamics and how they are applied to polymer systems.

CECIL DYBOWSKI - Department of Chemistry and Biochemistry, University of Delaware, Newark, Delaware 19716.

Solid State NMR of Polymers, Edited by L. Mathias
Plenum Press, New York, 1991

2. Technology of Spin Control

NMR spectroscopy is a dynamic tool. The spectrum is obtained nowadays by the direct Fourier transformation of a time response [4]:

$$S(\omega) = \int_0^\infty e^{-i\omega t} F(t)\, dt \tag{1}$$

In the usual NMR experiment, F(t) is a *free* decay, *i.e.* the nuclear spins respond without any interference from external stimuli during the time evolution (after the initial excitation). Thus, the spectrum reflects *all* the information on all the internal forces which affect the time evolution of the spins. A typical spectrum (that of polyethylene at room temperature) is shown in Fig. 1a. Although all the information on the various interactions is present in this spectrum, it is not possible to access it all because the amount of information being received simultaneously is enormous, containing, as it does, information on all the various proton-proton dipolar couplings!

In such cases one needs a means to make F(t) reflect only a subset of the gamut of interactions a spin experiences. If that were possible, the Fourier transformation would produce a spectrum, S(ω), that reflects that subset of interactions only. NMR spectroscopists have been doing experiments of this kind on liquids and solutions for some time [5]. In double-resonance experiments, irradiating one of two coupled nuclei yields a spectrum of the region of the other nuclear transition that does not show the effects of the coupling under appropriate low-pass filtering. In this way, an NMR spectroscopist can perform a series of experiments to determine how nuclei are coupled, and thereby make inferences about the chemical structure of the material under study.

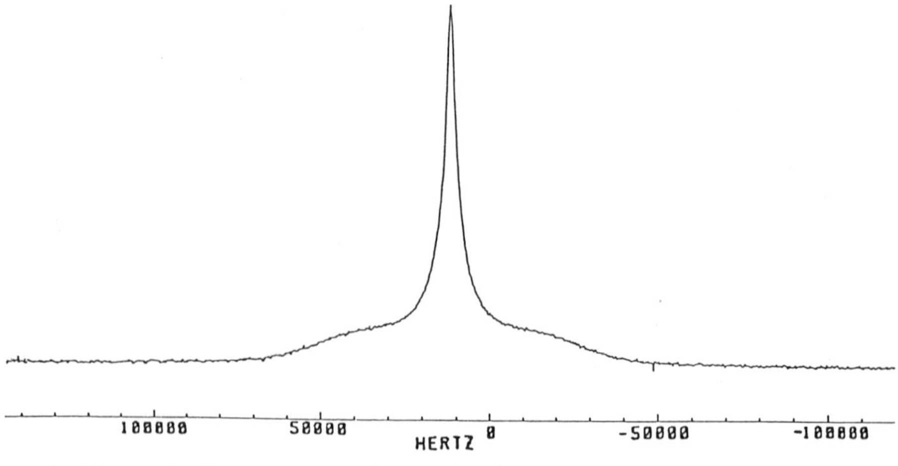

Figure 1. The proton NMR spectrum of a sample of poly(ethylene) at ambient temperature. The resonance is a composite one, showing the narrow resonance of protons in mobile regions and the broad resonance of protons in less mobile regions.

Decoupling abundant nuclei in solids to simplify the spectrum is much more difficult than decoupling in solutions because the strong coupling does not allow one to irradiate nuclei independently; this is the classic case of a homogeneously broadened spectrum. In the 1960s, several investigators had been studying the effects of high-power pulse irradiation on the magnetization [1]. Among other things, it was observed that the application of strings of pulses made the magnetization persist for a time long compared to the time it would have lasted had it evolved under the "internal" interactions only. Something that persists in time is something which has a narrow frequency spectrum. This was the beginning of a revolution that has substantially altered the meaning of "NMR spectroscopy of solids."

As I said, it had been shown that rapid spinning about an axis oriented at 54.7° to the magnetic field suppresses the effects of some dipolar couplings. However, this technique would have limited utility in tightly coupled systems, such as protons and fluorines, because of the requirement of using high spinning speed. That may no longer be a limitation since several research groups apparently can routinely spin samples at speeds in excess of 20 kHz, which approaches the speeds needed to suppress proton-proton dipolar couplings [6]. Thus, historically the suppression of dipolar couplings among abundant spins with the use of strings of radiofrequency pulses was accomplished in the late 1960's and was viewed as a means of analyzing materials containing protons and fluorine, including polymers.

2.1. Theory of Radiofrequency Pulse Averaging

The induction decay, F(t), whether free or driven, can be calculated theoretically from the density operator, $\rho(t)$, which specifies the time development of the nuclear spin system in the interaction frame of the Zeeman interaction by equation (2):

$$\langle M_x(t)\rangle = \mathrm{Tr}\, \rho(t)\, M_x \tag{2}$$

The problem of predicting the behavior of a spin system reduces to solution of the Liouville-von Neumann equation for $\rho(t)$ due to the modulated radiofrequency field and any internal interactions specified by an internal Hamiltonian, H_{int}. This latter Hamiltonian contains all the information on the chemistry available from the magnetic resonance experiment and is composed of several sub-Hamiltonians for the various interactions such as the chemical-shift and dipolar couplings. Because the radiofrequency excitation may be made sufficiently strong and fast-acting to dominate the internal Hamiltonians, the solution of the Liouville-von Neumann equation is generally given as an expansion in ever-higher orders of terms that account for the effects of the internal Hamiltonian, as modulated by the radiofrequency excitation.[1] Thus, the calculation of the density operator for the spin system is actually carried out in the interaction representation of the radiofrequency excitation. This frame is a peculiar frame of reference that changes orientation relative to the magnetic field depending on the sequence of pulses and has been called the toggling reference frame.

The transformation of the density operator between this frame and the Zeeman interaction frame is given by an operator, T, through the relation:

$$\rho(t) = T(t)\rho'(t)T^{-1}(t) \tag{3}$$

where $\rho'(t)$ is the density operator described in the toggling frame of reference. Of course, in this frame the density operator $\rho'(t)$ evolves under an effective internal Hamiltonian:

$$H_{int}'(t) = T(t)H_{int}T^{-1}(t) \tag{4}$$

From this Hamiltonian one can determine an effective time-evolution operator, U(t), such that

$$\rho'(t) = U(t)\rho'(0)U^{-1}(t) \tag{5}$$

The detected signal, F(t), is given by equation (6):

$$F(t) = \mathrm{Tr}\ T(t)U(t)\rho'(0)U^{-1}(t)T^{-1}(t)M_x \tag{6}$$

This is quite an imposing equation, but it is not difficult to comprehend the result. From it one finds that the observed behavior of the spin system is governed by both the radiofrequency excitation and the internal Hamiltonian, but the latter's effect is changed by the radiofrequency excitation.

For certain radiofrequency excitation schemes, the irradiation is cyclic and periodic. Then for times, Nt_c, $T(Nt_c) = T^1(Nt_c) = 1$, and for those times, the observed behavior is

$$F(Nt_c) = \mathrm{Tr}\ U(Nt_c)\rho'(0)U^{-1}(Nt_c)M_x \tag{7}$$

At these times the system appears to have evolved without the explicit effects of the radiofrequency excitation. The apparent evolution is determined by the internal interactions modulated by the radiofrequency excitation. Experimentally one can arrange to examine the system stroboscopically only at those times such that the detected signal gives the appearance of evolution under an "average Hamiltonian" as represented by equation (7).

2.2. The WAHUHA Sequence

The first excitation scheme proposed to suppress the action of the internuclear dipole-dipole couplings among abundant spins while retaining the effect of the chemical shift was a sequence of four precisely timed and phased 90° pulses that were repeated serially with stroboscopic detection. (Fig. 2a) It is based on a symmetry relation for the secular dipolar Hamiltonian, given in equation (8).

$$Z + Y + X = 0 \tag{8}$$

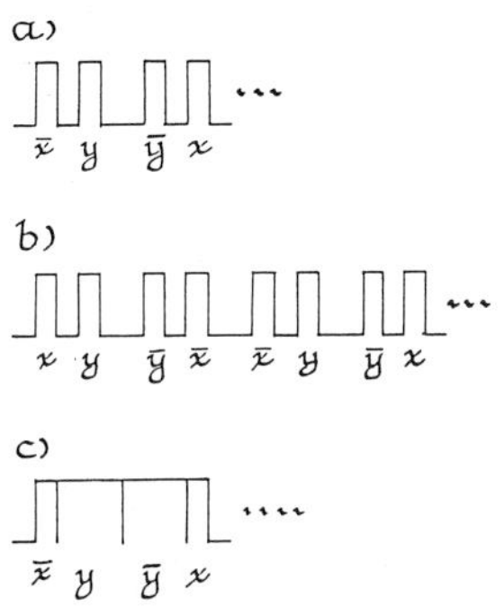

Figure 2. Three sequences for the decoupling of abundant nuclei: a) the WAHUHA-4 sequence; b) the MREV-8 sequence; c) the TREV-4 sequence, showing the spin-lock version of the WAHUHA-4.

In this equation $Z = [3I_Z^2 - I^2]$ and the other two symbols are similarly defined in terms of the components of the angular momentum. In the toggling frame, the operator for the dipolar interaction is rapidly switched by the action of the radiofrequency pulses through conditions in which it takes on each of these forms for one-third of the time. If the switching is rapid compared to the inverse of the magnitude of the dipolar coupling, one may base the average time-evolution operator on the time average of the total internal Hamiltonian:

$$U(Nt_c) = \exp(-iH_{int}^{(0)} Nt_c) \tag{9}$$

The average Hamiltonian is determined by the time average of the internal Hamiltonian *in the toggling reference frame.* It is composed, in the simplest model, of two parts:

$$H_{int}^{(0)} = H_c^{(0)} + H_D^{(0)} \tag{10}$$

and the time average is defined as follows:

$$H_i^{(0)} = t_c^{-1} \int^{tc} H_i'(t)\, dt \tag{11}$$

The calculation of the time average for the dipolar Hamiltonian gives an average value of zero.

$$H_D^{(0)} = (\omega_D/3)\ (X + Y + Z) \tag{12}$$

Thus, on average, the dipolar Hamiltonian appears to have no effect on the time evolution of the system. On the other hand, a similar calculation for the chemical-shift Hamiltonian gives a nonzero result:

$$H_c^{(0)} = (\sigma_{zz}{}^{\omega}{}_o/3)\ [\ I_x + I_y + I_z\ ] \tag{13}$$

because there is no similar symmetry relation among the components of the angular momentum in its switched forms. The apparent time evolution in such an experiment, at this level of approximation, then yields a spectrum upon Fourier transformation that is determined by the chemical shift only [7].

2.3. Other Sequences for Suppressing Dipole-Dipole Interactions

There are many effects that determine the nature of a spectrum. While the four-pulse sequence does produce a spectrum that is not dominated by the dipole-dipole interactions in this order of approximation, the resulting spectrum can still show effects of this substantial interaction. The fact that the system is not usually subject to ideal, instantaneous radiofrequency pulses, but rather to pulses of finite duration, can substantially affect the efficacy of the dipole-dipole suppression. In fact, the WAHUHA-4 sequence has been shown to reintroduce the dipolar interaction at this level of approximation when the finite width of the pulses is taken into account [9]. In addition, an improved approximation of the time evolution operator shows that it can have deleterious effects on the resolution of the spectrum. With these problems and others in mind, spectroscopists have developed more complex sequences tailored to minimize effects that can hamper getting the chemical-shift spectrum free of higher-order effects of the dipolar and other error Hamiltonians. One that has worked well is the MREV-8 sequence shown in Fig. 2b [10]. It consists of two WAHUHA-4 sequences of different phase sandwiched to produce a sequence twice as long as a single WAHUHA-4 sequence. It too suppresses the effects of dipolar coupling; however it also causes other effects to be minimized to improve resolution in strongly coupled systems. The resulting average chemical-shift Hamiltonian for this sequence has the form:

$$H_c^{(0)} = (\sigma_{zz}\omega_o/3)\ [\ I_x + I_z\ ] \tag{14}$$

Many versions of the dipole-dipole decoupling scheme have been created to enhance resolution by minimizing various subtle effects [11-13]. For example, the BR-24 sequence [11] consists of twenty-four pulses in a period three times that of a typical eight-pulse sequence that improves resolution further. The development of windowless and semi-windowless sequences have provided other improvements [12,13]. In every case, the motivation behind the development of a sequence of more complexity is sophisticated control of ever-smaller effects that influence time evolution, but every one must begin with the fundamental property of the WAHUHA-4 and the solid-echo sequence on which it is based -- the time-averaged refocussing of the effects of the dipolar coupling.

2.4. Chemical-Shift Scaling

Although the WAHUHA-4 sequence does not suppress the action of the chemical-shift Hamiltonian totally, it does change the effect it would have in the absence of such a radiofrequency Hamiltonian. The net effect is a retardation of the time evolution due to this interaction [7,9]. This effect produces a scaling of the frequency range of the chemical-shift spectrum obtained by Fourier transformation.

The ratio of the known frequency difference between two chemically shifted resonances to the observed frequency difference is the *chemical-shift scale factor*, SF [9]:

$$SF = (\omega_{1actual} - \omega_{2actual})/(\omega_{1observed} - \omega_{2observed}) \qquad (15)$$

This factor is predicted from theoretical arguments and is a characteristic of the sequence of radiofrequency pulses used. For a WAHUHA-4 sequence with infinitely sharp radiofrequency pulses [7]

$$SF_{WAHUHA} = 3^{1/2} \sim 1.732 \qquad (16)$$

For the MREV-8 pulse sequence with infinitely sharp radiofrequency pulses, [9]

$$SF_{MREV} = 3/2^{1/2} \sim 2.121 \qquad (17)$$

The fact that pulses are not infinitely sharp has a noticeable effect on the spectrum, in that the chemical-shift scaling factor depends on the ratio of the pulse width to the cycle time. Theoretically this gives a change in the scaling factor [9]:

$$SF_{WAHUHA} = 3^{1/2}(1 + a)^{-1} \qquad (18a)$$

$$SF_{MREV} = 3(2)^{-1/2} (1 + 2a)^{-1} \qquad (18b)$$

where $a = 3(t_w/t_c)(4/\pi - 1)$ and t_w is the 90° pulse length. For typical values of t_w (1.5 microseconds) and t_c (24 microseconds for WAHUHA-4 and 48 microseconds for MREV-8), one predicts the following values of SF:

$$SF_{WAHUHA} = 1.648$$

and

$$SF_{MREV} = 2.018$$

These values differ substantially from the "ideal" scaling factors and indicate a practical aspect of using line-narrowing sequences -- one must calibrate the spectrometer carefully to obtain reliable results with the techniques. In particular, the comparison of an experimentally determined chemical-shift scaling factor with that determined by equations [18] should yield substantial agreement before one uses the technique. With many variables (only a few of which I have touched on) that affect the efficacy and resolution of these complex sequences, caution should be used if such checks as agreement of the theoretical and experimental scaling factors do not indicate that all of these factors are adequately controlled [4].

3. Combined Rotation and Multiple-Pulse Spectroscopy [3b]

While the suppression of the action of the dipolar interaction is not a trivial feat, the resulting spectra would have limited utility for analysis of chemical systems. The reason for this limitation originates in the dependence of the chemical-shift interaction on orientation of the magnetic field relative to a frame fixed in the molecule. For a single-crystal sample, there can be as many resonances as there are geometrically unique positions in the unit cell, referred to the co-ordinate system of the magnetic field. Each resonance may depend on how the single crystal's axes are oriented relative to the magnetic field [4]. Thus, a single value is insufficient to describe the resonance position for even a single chemical structure. To describe this situation, one defines a co-ordinate system attached to the molecule. The axes of this co-ordinate system represent those axes that align with the magnetic field for the conditions that the chemical shift has its lowest value, its highest value, and a third axis that is mutually orthogonal to the others. For any other orientation of the molecular axes (called the chemical-shift *principal axes*), the observed chemical shift is a function of these three values and the direction cosines of the magnetic field in the molecular frame:

$$\sigma = \sigma_a l_a^2 + \sigma_b l_b^2 + \sigma_c l_c^2 \tag{19}$$

σ_a, σ_b, σ_c are the *principal values* of the chemical-shift tensor and l_i is the direction cosine of the magnetic field relative to axis i.

For single crystals, the resonances are relatively sharp. The positions vary with the orientation of the axes relative to the field and σ_a, σ_b, and σ_c may be determined from the variation of the chemical shift with orientation of the single crystal axes in the magnetic field. However, single crystals are not always available. The chemical-shift spectrum of a random powder of small microcrystallites is much different from that of a single crystal of the same material. Since every orientation of the magnetic field relative to the principal axes is present in the sample, every chemical shift is simultaneously seen in the spectrum. The result is a *band* spectrum (Fig. 3) for every chemically unique nuclear site. There are three possible shapes for these bands for a random powder (called *powder patterns*) depending on the relative equality of the principal values:

a) $\sigma_a = \sigma_b = \sigma_c$, spherically symmetric;

b) $\sigma_a = \sigma_b$ or $\sigma_b = \sigma_c$, axially symmetric;

c) σ_a, σ_b and σ_c all unique, totally asymmetric.

The shapes for each of the cases are given in Fig. 3.

The dispersion of the resonance intensity would not hamper the spectroscopy if the range, $\sigma_c - \sigma_a$, were small compared to the average isotropic shift difference for

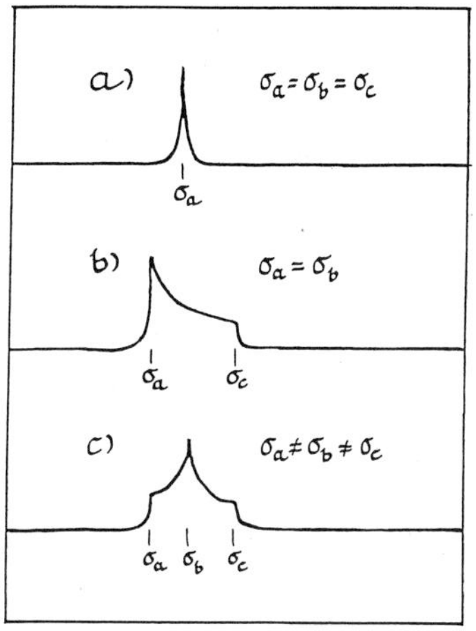

Figure 3. Powder patterns for a random distribution of microcrystallites. a) All principal elements of the tensor equal; b) the case of axial symmetry in which two of the three elements are equal; c) the general case in which all three elements have different values.

nuclei in different environments. Were it small, the lines for nuclei in different chemical environments in powders would then be well separated and easily identifiable in the spectrum. This is not generally the case. The range of typical principal values for a single environment is often comparable to the range of isotropic shifts for a nuclear species [14]. Hence, the spectrum of a powder of a material containing nuclei in several chemical environments is a complicated structure having substantial overlap of bands of the type shown in Fig. 3. The situation is somewhat analogous to that discussed above for the case of many simultaneously acting dipolar interactions -- there is too much information in the spectrum to allow complete, unambiguous deconvolution of it all.

While the situation is somewhat analogous to the multi-dipole case, there is an important difference. The dispersion of chemical shift produces an inhomogeneously broadened resonance [4]. Coherent modulation of an inhomogeneous resonance results in a spectrum that has a centerband flanked by a series of sidebands separated from the centerband by $\pm m\omega_m$, where m is a positive integer [15]. The shape of the sideband spectrum depends on the details of the modulation and on the spectral parameters. Modulation of the chemical-shift Hamiltonian is usually produced by a rotation of the powder sample about an axis inclined at some angle, Θ, relative to the magnetic field [3]. In particular, for one value of the angle Θ, the sidebands and centerband become symmetric. This angle is called the *magic angle* and has the value,
$\arccos(3^{-1/2})$ [3].

Schematically the effect is shown in Fig. 4. The sideband intensities are complicated functions of the spinning speed, the chemical-shift-tensor principal elements, and the magnetic field [16]. A general trend is, however, that sidebands which fall outside the range of chemical shifts, $\sigma_c - \sigma_a$, are strongly attenuated. Thus,

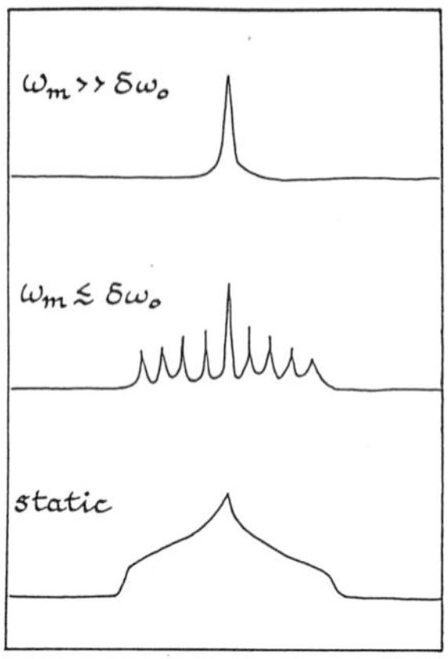

Figure 4. Sideband spectra for a solid being rotated about an axis inclined to the magnetic field at the magic angle. a) rapid spinning; b) intermediate spinning showing the appearance of sidebands; c) static spectrum.

if the spinning speed is high enough, effectively only the centerband of each resonance is detected -- a situation that gives a spectrum very much like a solution-state spectrum of the same substance. Obviously, such spectra are extremely useful as tools in analysis of intractable materials and materials whose character depends on the molecule being held rigidly relative to the magnetic field.

For abundant-spin NMR spectroscopy, one must carry out the suppression of the dipole-dipole couplings concomitantly with rapid spinning of the sample about an axis inclined at the magic angle to the magnetic field. The technique has been demonstrated and is called combined rotation and multiple-pulse spectroscopy, the acronym of which is CRAMPS [3b]. It appears that a critical consideration in carrying out a successful CRAMPS experiment is to ensure that the time scales of the rotation and the decoupling are much different [17]. In particular, if the spinning leaves the molecular orientation relative to the field essentially fixed during one decoupling cycle, the movement of the molecule is slow enough that the rotation will not significantly affect the efficacy of decoupling. From a practical point of view, the cycle frequency ($1/t_c$) is of the order of 25 kHz and the typical rotation frequency may be of the order of 2.5 kHz. This disparity in time scale means that one can neglect to the lowest order any synergy or interference between the two processes of radiofrequency decoupling and mechanical rotation.

4. Two-Dimensional Experiments

In abundant spin-1/2 systems, the two effects in which a spectroscopist is most interested are the chemical shift and the dipole-dipole coupling. The first contains primarily information on chemical identity, the latter information on structure. In many cases, the dipolar couplings at a nuclear site are dominated by only a few of the many interactions possible. For example, George Pake investigated the dihydrate,

$CaSO_4{\bullet}H_2O$, in a classic study [18]. In this system, the proton spectral features are determined essentially by a single coupling between the two closest protons. The other couplings, while extant, are weaker and do have an effect on the spectrum, but the basic doublet feature of the resonance is attributable to this single interaction.

When a spin system is subject to a composite Hamiltonian, it has become fashionable to carry out an experiment in which one determines how the action of one part of the Hamiltonian is correlated with another -- the quintessential two-dimensional experiment [19]. Such a two-dimensional experiment on abundant spins in the solid state is possible if the dipolar interaction is dominated by one or a few dipolar couplings. Figure 5 gives several possible variants of the procedure to determine the correlation between dipolar interactions and chemical shift interactions. In a) one creates a transverse magnetization by a pulse, which is the preparation. This magnetization then freely precesses for a time t_1, at which point multiple-pulse decoupling and detection begins. This provides the effective separation required in a two-dimensional experiment. Of course, during t_1 both the dipolar and chemical-shift interactions are active, but for strongly dominant dipolar couplings, the effective evolution is controlled by this interaction. In the t_2 dimension it evolves only under a sequence-averaged chemical-shift interaction. Thus such an experiment gives a correlation between the chemical shift and the dipole-dipole coupling. In b), a 180° pulse is inserted midway through the t_1 period to refocus the effects of chemical shifts during this period. In c) the chemical-shift evolution occurs in the t_1 dimension, followed by evolution under the combined Hamiltonian (assumed to by dominated by the dipolar interaction) during t_2. Other variants of these basic schemes are possible by adding spin-locking, refocussing, or Zeeman-storage sequences to the fundamental

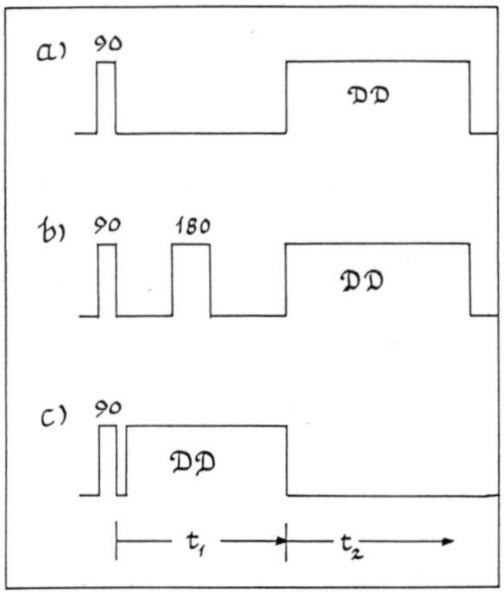

Figure 5. Sequences for carrying out a two-dimensional experiment on abundant spins in the solid state to correlate the chemical-shift and dipolar interactions. a) chemical-shift evolution in t_1 and t_2, dipolar interactions extant only in t_1. (The dipolar interaction is assumed to be much stronger than the chemical shift.) b) refocussing of the chemical shift in t_1 and refocussing of the dipolar interaction in t_2. c) chemical-shift evolution in t_1 , followed by evolution in t_2 under both the chemical-shift and dipolar interactions.

units. The goal is to separate the action of the two Hamiltonians in which one has an interest. Indeed, in early single-crystal chemical-shift studies with sequences such as the WAHUHA-4 and the MREV-8, one often calibrated the orientation of the crystal relative to the field by measuring the dipole splitting first, after which one would determine the chemical shift with a multiple-pulse experiment [20]. The two-dimensional experiment combined these two logical steps into a single unit.

5. Applications to Polymers

5.1. Fluorinated Polymers

Soon after its development, the multiple-pulse line-narrowing technique was applied to a polymeric material [21]. The sample was a copolymer of tetrafluoroethylene and perfluoromethylvinyl ether and the original spectrum is shown in Fig. 6. Disregarding the zero-frequency rise which is an artifact of spectra from this era, the spectrum consists of two resonances, a relatively sharp line for fluorine nuclei of the trifluoromethyl group and an axially symmetric powder pattern for the resonance of the backbone fluorines. The comparison of these two bands shows that the motion of the trifluoromethyl sidechain results in a smaller dispersion of the resonance than is seen for the fluorines on the backbone of the polymer. A resonance attributable to the methine fluorine was not detected, probably because the line broadening is significantly larger than in solution. Rough calculations of areas of the resonances were in agreement with the known composition of the copolymer.

Another early investigation centered on the ^{19}F resonance of poly(tetrafluoroethylene) [PTFE] powder and highly oriented fibers. The use of fibers allowed the extraction of the principal values of the chemical-shift tensor of a CF_2 group [22]. In Fig. 7a is shown the chemical-shift spectrum of the powder at 77K and in Fig. 7b is the chemical shift as a function of the orientation of the axis of a bundle of PTFE fibers relative to the magnetic field. A careful analysis of these data gave a value for the anisotropy, $\sigma_3 - (\sigma_1 + \sigma_2)/2$, of 104 ppm. Spectra taken above the 19°C transition (where rapid rotation about the chain axis causes an effective axial symmetry to be observed) showed axial symmetry of thc chcmical-shift powder pattern. At lower temperatures, the full asymmetry is observed (Fig. 7a) and the asymmetry, $(\sigma_2-\sigma_1)/(\sigma_3-(\sigma_1 + \sigma_2)/2)$, is found to be 0.85. The authors concluded that the most shielded component lies about 20° off the chain axis and is dominated by *gauche* interactions with other fluorine atoms.

Perhaps the most thorough investigation of a polymer with multiple-pulse NMR techniques has been carried out by Vega and English [23]. PTFE and a copolymer of tetrafluoroethylene and hexafluoropropylene were examined with the techniques over a wide range of temperature and crystallinity. For example, Fig. 8 [23] shows how the ^{19}F multiple-pulse spectrum of PTFE changes systematically as the crystalline content is varied. The lineshape also changes substantially over the range of temperature from -128°C to 344°C because of changes in molecular motion with temperature. A careful analysis of the details of these changes allowed an

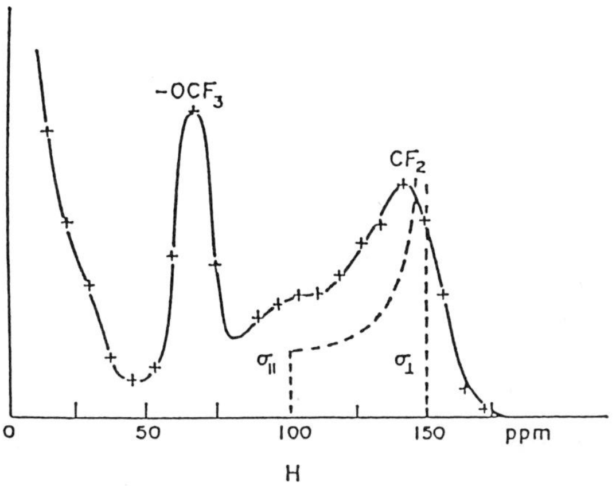

Figure 6. The fluorine-19 multiple-pulse spectrum of a copolymer of tetrafluoroethylene and perfluoromethylvinyl ether -- the first multiple-pulse spectrum of a polymer [21].

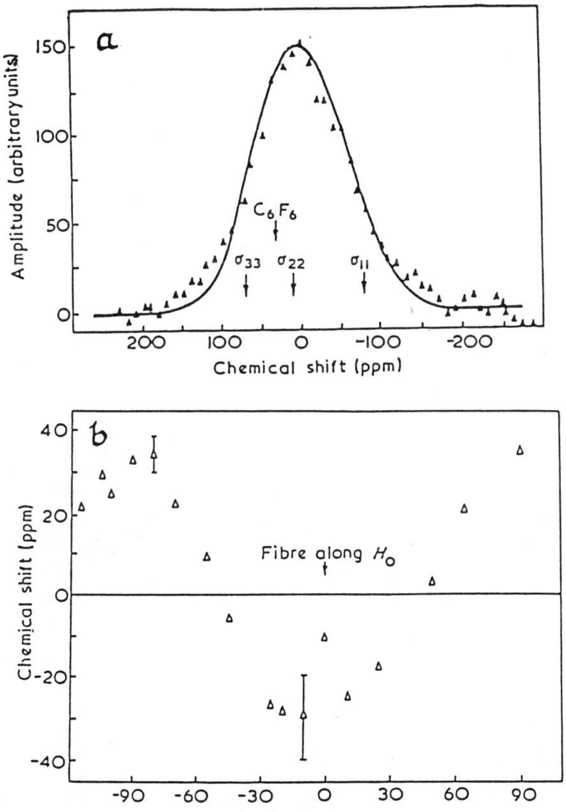

Figure 7. Multiple-pulse F-19 NMR spectrum of poly(tetrafluoroethylene). a) the fully asymmetric spectrum taken at 77K. b) the rotation plot for an oriented PTFE fiber. Chemical shift is defined relative to the average chemical shift, and the orientation refers to the angle between the fiber axis and the field [22].

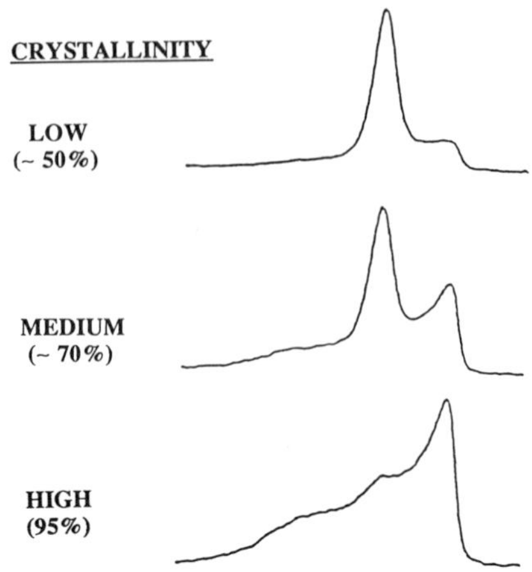

Figure 8. The effect of changes in crystallinity on the F-19 spectrum of a semicrystalline PTFE sample. The spectrum was taken with the temperature above the 30°C transition, so the crystalline resonance shows the effective axial symmetry [23].

association of the lineshape changes with dynamic transitions observed by other techniques and to specify the component most involved in a particular transition. For example, the γ transition was confirmed to take place in the amorphous regions. The addition of CF_3 groups was shown to facilitate chain-axis reorientation in the crystalline regions but to inhibit reorientation of the chain axes in the amorphous regions.

Once the association is made between chemical shift and orientation of some molecular axis in the magnetic field, one may look upon the resonance line as a map of the orientations of these molecular axes relative to the field for a sample. In particular, for PTFE the mapping of orientation on chemical shift is homomorphic when there is only one angular variable on which the chemical shift depends. This is the case for an axially symmetric tensor like the one for PTFE above the transition at 30°C. Thus, one can investigate the geometric effects of sample conditioning by multiple-pulse NMR spectroscopy [22-25].

PTFE can be deformed by a uniaxial stress to produce a material that has some chain alignment. Since the distribution of orientations of chain axes is not random for such a sample, the NMR spectrum can be used to investigate the average orientation in such a sample. Figure 9 shows how the ^{19}F multiple-pulse spectrum of a semicrystalline PTFE sample varies as a function of the angle, β, between the direction of stress and the magnetic field [24]. By the variation of the moments of these spectra about the isotropic chemical shift for a random powder, one can determine an approximate orientational probability distribution of the chain axes about the direction of stress. NMR-derived distributions for three uniaxially stressed PTFE samples are shown in Fig. 10 as a function of residual extension ratio. These may be compared to theoretical models. Figure 11, for example, shows how several of the moments of the distribution for PTFE are predicted to vary with extension ratio if

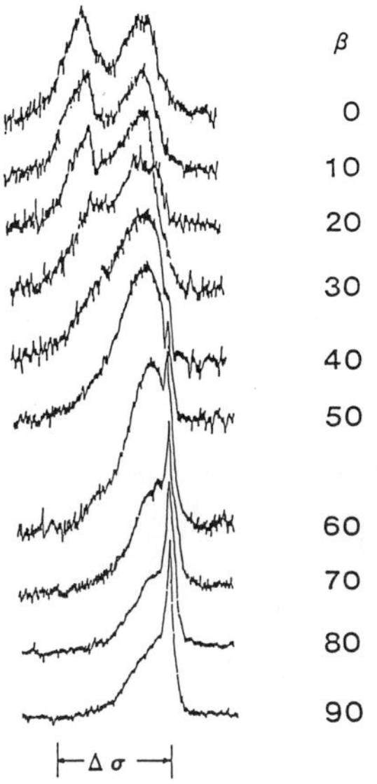

Figure 9. F-19 multiple-pulse spectra of a partially oriented PTFE sample that is semicrystalline. Spectra are shown as a function of the angle between the magnetic field and the direction of alignment [24].

PTFE undergoes a pseudoaffine deformation. The NMR-derived points follow the trends for pseudoaffine deformation, but there are systematic deviations of the NMR-derived points from the considering PTFE to deformed pseudoaffinely. The polymer's behavior is more complex than such a simple model can predict, and the reasonably close agreement is actually quite gratifying.

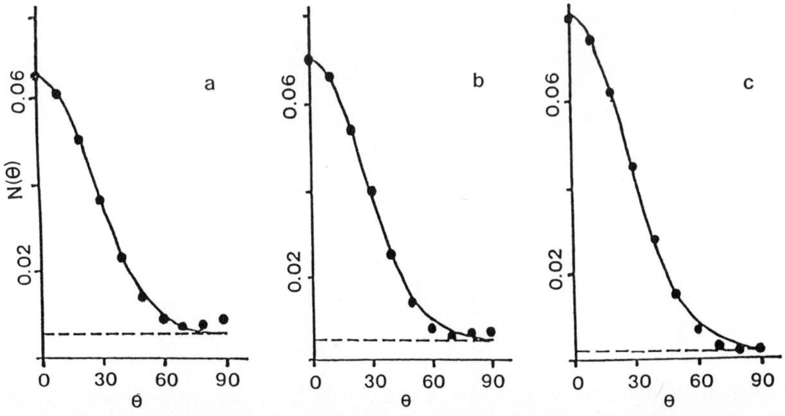

Figure 10. Orientation distribution functions for the crystalline parts of semicrystalline PTFE samples derived from the angular dependence of spectra. Extension ratios of a) 1.75; b) 2.10; c) 2.40 [24].

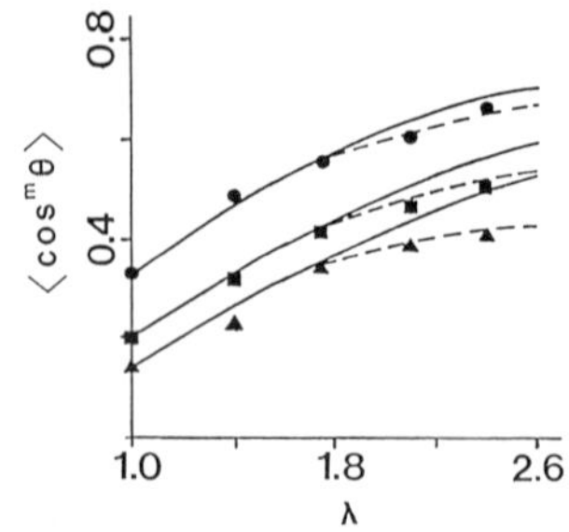

Figure 11. Comparison of the moments of the distribution function for the semicrystalline PTFE samples as a function of extension ratio. The solid lines are predicted theoretically assuming that PTFE deforms pseudoaffinely. ○ m=2; ■ m=4; ▲ m=6 [24].

5.2. Protonated Polymers

Protonated materials usually have a smaller range of chemical-shift elements than do similar fluorinated materials. Thus, the protons of poly(ethylene) have a much smaller anisotropy than do the fluorines of PTFE. This smaller anisotropy makes proton lineshapes more difficult to observe than those of fluorines. However, a number of proton studies have been carried out [26-31, 3b].

Perhaps the proton-containing material with the spectrum that is most difficult to obtain is solid poly(ethylene), excluding, of course, solid hydrogen. This is a result of the strong dipole-dipole coupling. Figure 12 shows the multiple-pulse spectrum of a poly(ethylene) sample [27] using the eight-pulse sequence. The spectrum shows two components, corresponding to amorphous and crystalline regions. Because the static

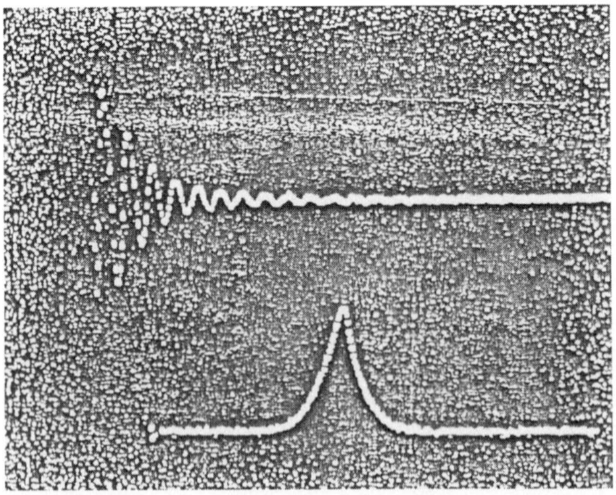

Figure 12. The multiple-pulse proton spectrum of poly(ethylene) at room temperature. The upper trace is the time-dependent response to the MREV-8 spectrum, detected stroboscopically. The bottom trace is the Fourier transform of the upper trace. The full width at half height of the resonance is only 150 Hz [27].

dipolar couplings in the two regions are of different magnitudes, the lineshape depends strongly on the cycle time [26]. For long cycle times, the decoupling of the most strongly coupled material is not very efficacious. Thus, a plot of detected magnetization (by comparison to an external water sample) extrapolated to an infinitely short cycle time gives the total number of protons in all phases of the sample. On the other hand, at long cycle times, only the spectrum of the protons in amorphous regions will be present in the spectrum. By a simple calculation, Gerstein *et al.* [27] estimated the crystalline fraction in a sample of poly(ethylene) to be 0.60 ± 0.02, in good agreement with density and heat-of-fusion measurements on the same sample. Thus, the spectra are quantitative and can be used to determine the number of spins having a specific value or range of values of NMR parameters. In this case the parameter is the dipolar coupling [27].

One can easily see that a theme of multiple-pulse NMR studies has been the characterization of not just chemical structure, but of the physical structure and molecular dynamics. English and Dybowski [30] used the comparison of multiple-pulse and free-induction-decay spectra to specify the rate and amplitude of molecular motion in *cis*-1,4-poly(butadiene) (PB). The line-narrowing of the resonance demonstrated that, on the time scale of the sequence ($\sim 10^{-4}$ s), there exists residual static dipolar couplings among the spins (Fig. 13). These static dipolar couplings are much smaller than the couplings in PB if it were rigid (below the glass transition). In fact, the couplings are weak enough that magic-angle spinning at 3.8 kHz suppresses the action of them and yields a sharp resonance. Hence, the results indicate that on the time scale of the sequence, the motion of PB is more complex than predicted by the simple isotropic theories often used to interpret NMR results. Both the amplitude and frequency of the motions affect the lineshape, and the temperature dependence of both must be considered when rationalizing the temperature dependence of NMR parameters such as the lineshape or the relaxation times.

5.3 Combined Rotation and Multiple-Pulse Spectroscopy

The multiple-pulse spectra shown in the last sections were all measured on static materials. The bandshapes therefore reflect orientational distributions in the magnetic field. Rotation of the sample about an axis inclined at the magic angle has been used for quite some time to investigate protonated polymers above the glass transition temperature or in gels, where the static dipolar couplings are weak [30,32]. Here I restrict the discussion to cases in which strong dipolar couplings are suppressed by CRAMPS [3b]. The first reported CRAMP spectrum was one of the ^{19}F nuclei in poly(chlorotrifluoroethylene). The sample was the commercial product, KEL-F™, a material of which many rotors for ^{13}C magic-angle spinning are made nowadays. The spectrum of fluorine in this material is shown in Fig. 14 [3b]. It consists of two resonances separated by 23 ppm. Also seen in the spectrum are two sidebands the authors do not mention. These result from the fact that the 2.5 kHz spinning rate was not sufficiently high to be much larger than the chemical shift anisotropy of the CF_2 resonance. The lines are not completely separated and the authors argue that a dispersion of isotropic chemical shifts is a factor in the broadening of the lines.

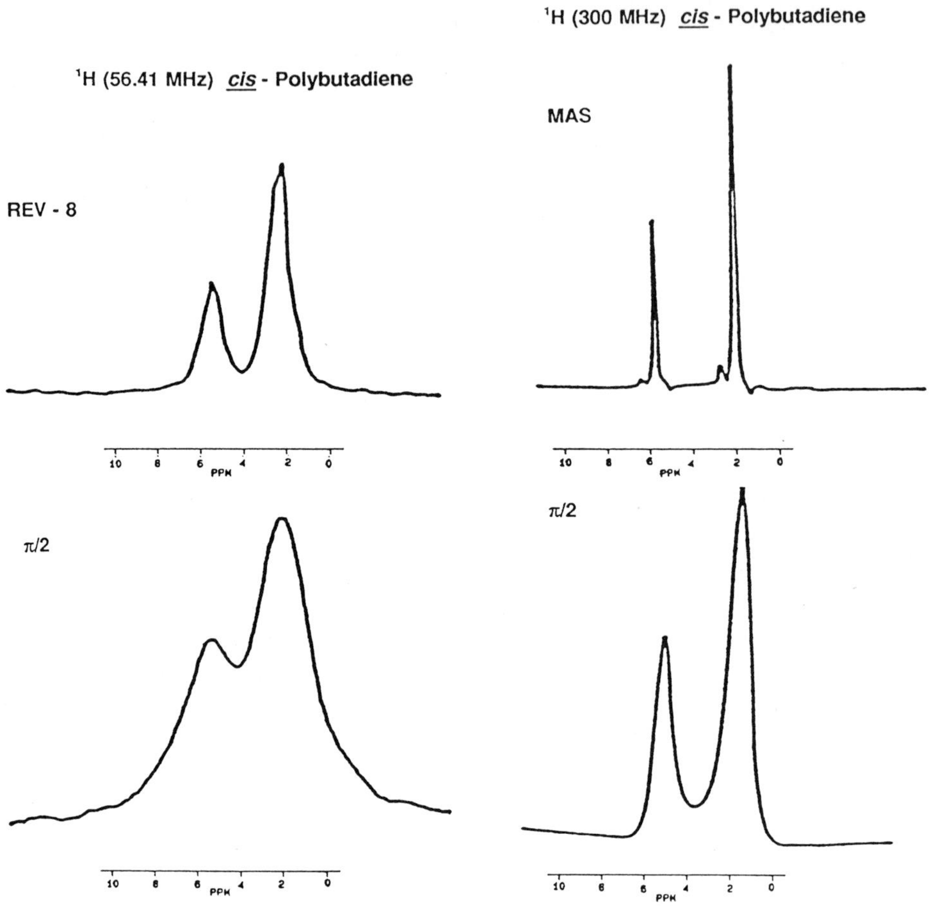

Figure 13. The effects of multiple-pulse H-1 NMR line-narrowing in a glassy polymer -- *cis*-poly(butadiene). The MREV-8 spectrum is compared to the FID spectrum at the left, showing the improved resolution. On the right is a comparison of the FID spectrum to a spectrum taken with magic-angle spinning at higher field. (υ = 300 MHz) Comparison of the FID spectra shows that the line broadening is field-independent [30].

CRAMPS may be applied to protons as easily as to fluorines [28]. For example, Fig. 15 shows the CRAMP spectra of the protons in isotactic poly(styrene) and poly(methylmethacrylate) [28]. These spectra give a feeling for the resolution attainable with this technique at this magnetic field strength. The aromatic resonance is resolved from the aliphatic one, however, the various aliphatic resonances are overlapping. In the poly(styrene) spectrum, the methine and methylene resonances are just barely discernible as a flattening of the resonance line. Thus, CRAMPS can distinguish chemical type or corroborate the results of other techniques, particularly at low field (proton frequency = 55.4 MHz). However, subtle differences are difficult to detect with it. It has been graphically demonstrated that performing CRAMPS at a higher field (270 MHZ provides substantially better resolution than is obtainable at lower field for crystalline materials. Perhaps the same may be true for polymers, but only if the broadening is *not* due to isotropic chemical-shift dispersion.

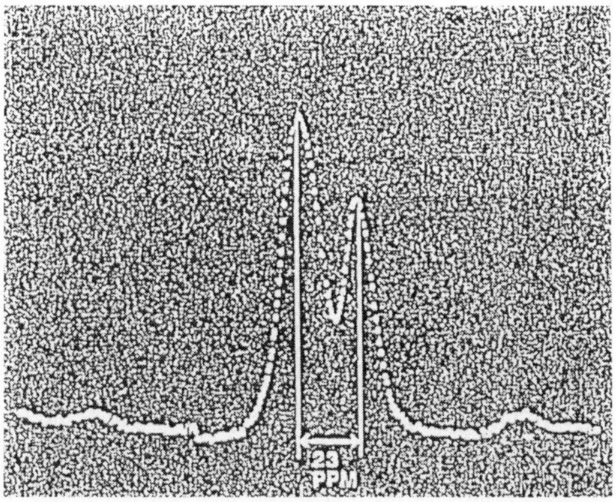

Figure 14. The CRAMP spectrum of poly(chlorotrifluoroethylene) (F-19 NMR) showing the different fluorine resonances for this material. Note the presence of spinning sidebands [3b].

6. Summary

Multiple-pulse spectroscopy of abundant spin-1/2 nuclei like ^{19}F and ^{1}H dramatically narrows the resonance lines by suppressing the action of the dipolar couplings, thus improving resolution. The resulting spectrum can be used in some cases for identification of chemical groups via chemical shifts. The resolution (at least with current technology) is not as good as that for carbon spectroscopy of solids. However, improved resolution can allow careful study of important physical properties of independent resonances of nuclei in various regions of a polymer, properties such as chain motion and orientation. It is reasonably quantitative, as was seen in the study of poly(ethylene). Thus, multiple-pulse spectroscopy is likely to facilitate understanding physical processes and their effects in various parts of a complex mixture or copolymer by allowing a measure of control of the spin interactions that affect the spins in each region, rather than be a routine tool for chemical group

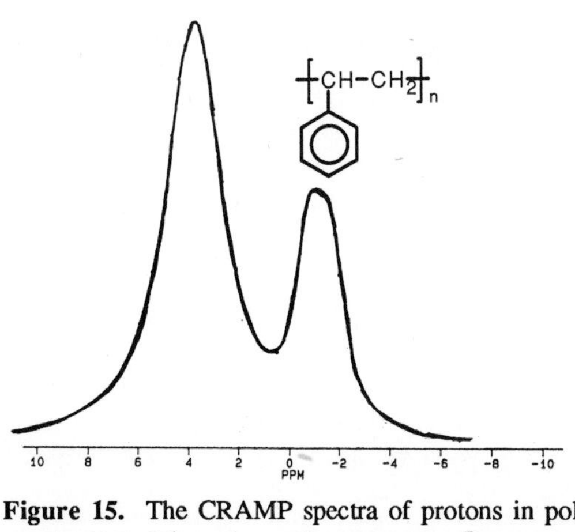

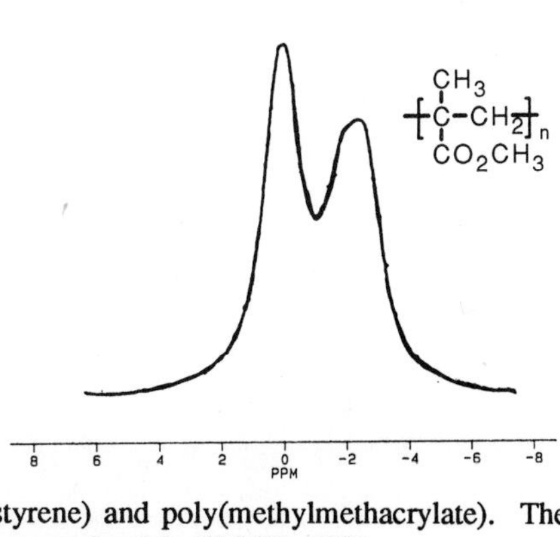

Figure 15. The CRAMP spectra of protons in poly(styrene) and poly(methylmethacrylate). These spectra were taken at a proton resonance frequency of approximately 55 MHz [28].

analysis of polymer, as is carbon NMR spectroscopy of solids. Thus, its role is complementary to this more well known technique and will find extended use in the studies of physical properties of polymers.

7. Acknowledgments

Over the years, I have been fortunate to receive funding for the work of our laboratory from a number of sources: the National Science Foundation, the Research Corporation, the Petroleum Research Fund of the American Chemical Society, the U. S. Department of Energy, the University of Delaware Research Foundation, the Sun Company, Hercules, Inc., and the Sponsors of the Center for Catalytic Science and Technology. I am deeply indebted to them for allowing me to do what I enjoy. I also owe gratitude to the graduate students who have continually refreshed me with their work and ideas: John Ballinger, Tom Apple, Anita Brandolini, Steve DeCanio, Neal Dando, Joel Miller, Frank DeThomas, David Mesaros, Charles Tirendi, Lori Hasselbring, Dan Barabino, Ron Colletti, Larry Kasuboski, Navin Bansal, Jon Kauffman, and Chris Tsiao. Of the many fine undergraduate research students I have had, I point out three who have had a significant impact on our research in polymer science: Bill Epstein, Mark Alvey, and Karen Rocco. I am indeed fortunate to have had the pleasure of working with them all.

8. References

1. a) Ellett, J.; Gibby, M.; Haeberlen, U.; Huber, L.; Mehring, M.; Pines, A. and Waugh, J., *Adv. Magn. Reson.*, *5*, 197.
   b) Mansfield, P., *Progr. Nucl. Magn. Reson. Spectrosc.*, *8*, 1971, 41.
2. a) Andrew, E. R.; Bradbury, A. and Eades, R., *Arch. Sci.*, *11*, 1958, 223.
   b) Lowe, I., *Phys. Rev. Lett.*, *2*, 1959, 285.
3. a) Schaefer, J.; Stejskal, E. and Buchdahl, R., *Macromolecules*, *10*, 1977, 384.
   b) Gerstein, B.; Pembleton, R.; Wilson, R. and Ryan, L., *J. Chem. Phys.*, *66*, 1977, 361.
4. Gerstein, B. and Dybowski, C., *Transient Techniques in the NMR of Solids*, Academic Press, New York, 1985.
5. Sarles, L. and Cotts, R., *Phys. Rev.*, *111*, 1958, 853.
6. a) Dec, S.; Wind, R.; Maciel, G. and Anthonio, F., *J. Magn. Reson.*, *70*, 1986, 356.
   b) Langer, V.; Daugaard, P. and Jakobsen, H., *J. Magn. Reson.*, *70*, 1986, 472.
7. Waugh, J.; Huber, L. and Haeberlen, U., *Phys. Rev. Lett.*, *20*, 1968, 180.
8. Mansfield, P., *J. Phys. Chem.*, *4*, 1971, 1444.

9. Rhim, W. K.; Elleman, D. and Vaughan, R., *J. Chem. Phys.*, *59*, 1973, 3740.
10. Rhim, W. K.; Elleman, D. and Vaughan, R., *J. Chem. Phys.*, *58*, 1973, 1772.
11. Burum, D. and Rhim, W. K., *J. Magn. Reson.*, *34*, 1979, 241.
12. Burum, D.; Linder, M. and Ernst, R., *J. Magn. Reson.*, *44*, 1981, 173.
13. Takegoshi, K. and McDowell, C., *Chem. Phys. Lett.*, *116*, 1985, 100.
14. Duncan, T., *J. Phys. Chem. Ref. Data*, *16*, 1987, 125.
15. Schaefer, J.; Stejskal, E. and McKay, R., *J. Magn. Reson.*, *25*, 1977, 569.
16. Herzfeld, J. and Berger, A., *J. Chem. Phys.*, *73*, 1980, 6021.
17. Pembleton, R., Doctoral Dissertation, Iowa State University, 1980.
18. Pake, G., *J. Chem. Phys.*, *16*, 1948, 327.
19. Bax, A., *Two-Dimensional Nuclear Magnetic Resonance in Liquids*, D. Riedel, Dordrecht, 1982.
20. McKnett, C.; Dybowski, C. and Vaughan, R., *J. Chem. Phys.*, *63*, 1975, 4578.
21. Ellett, J.; Haeberlen, U. and Waugh, J., *Polym. Lett.*, *7*, 1969, 71.
22. Garroway, A.; Stalker, D. and Mansfield, P., *Polymer*, *16*, 1975, 161.
23. Vega, A. and English, A., *Macromolecules*, *13*, 1980, 1635.
24. Brandolini, A.; Alvey, M. and Dybowski, C., *J Polym. Sci.*, 21, 1983, 2511.
25. McBrierty, V. and Douglass, D., *Phys. Rep.*, *63*, 1980, 61.
26. Dybowski, C. and Vaughan, R., *Macromolecules*, *8*, 1975, 50.
27. Pembleton, R.; Wilson, R. and Gerstein, B., *J. Chem. Phys.*, *66*, 1977, 5133.
28. Ryan, L.; Taylor, R.; Paff, A. and Gerstein, B., *J. Chem. Phys.*, *72*, 1980, 508.
29. Cheung, T.; Gerstein, B.; Ryan, L.; Taylor, R. and Dybowski, C., *J. Chem. Phys.*, *73*, 1980, 6059.
30. English, A. and Dybowski, C., *Macromolecules*, *17*, 1984, 446.
31. Havens, J. and vander Hart, D., *Macromolecules*, *18*, 1663.
32. a) Schneider, B.; Picova, H. and Doskocilova, D., *Macromolecules*, *5*, 1972, 120.
    b) Schneider, B. and Doskocilova, D., *Pure and Appl. Chem.*, *54*, 1982, 575.

Chapter 2

SOLID-STATE ^{2}H NMR: OVERVIEW WITH SPECIFIC EXAMPLES

RONALD F. COLLETTI AND LON J. MATHIAS

The value of high resolution (solution) NMR spectroscopy for investigating polymeric materials can be seen by the hundreds of applications and publications in this area each year. Owing to the sensitivity of chemical shifts to molecular bonding and stereochemistry, NMR can be used to look at polymer structure and tacticity, copolymer sequence distribution, and degree of branching.[1] Reaction mechanisms can be probed using isotopically labelled or natural abundance samples while NMR relaxation data can be related to polymer segmental motions.

Continued advances in instrument hardware and software allow extension of many techniques to examining solids. The combination of dipolar decoupling to remove ^{1}H-^{13}C dipole-dipole interactions, cross-polarization to increase sensitivity of low abundance nuclei, and magic angle spinning to narrow spectral peaks yields high-resolution solid-state NMR spectra analogous to solution spectra.[2] These methods are crucial for analysis of insoluble samples and polymeric materials in their end-use state.

Solid-state experiments can also be used to examine morphology, solid-solid transitions and molecular dynamics. Although it is not a high-resolution technique, wideline deuterium NMR has enormous capability for examining the physical behavior of both small molecules and polymers. In addition, it illustrates the extension of both solution and solid-state spectroscopy from ^{1}H and ^{13}C to other

---

RONALD F. COLLETTI - Monsanto Chemical, Pensacola, Florida 32575; and LON J. MATHIAS - University of Southern Mississippi, Hattiesburg, Mississippi 39406.

Solid State NMR of Polymers, Edited by L. Mathias
Plenum Press, New York, 1991

nuclei. For example, it is now routine to obtain high resolution spectra on ^{2}H, ^{3}H, ^{7}Li, ^{14}N, ^{15}N, ^{17}O, ^{19}F, ^{23}Na, ^{29}Si, ^{31}P. Even for solids, many of these nuclei can now be observed to provide unique information on composition, conformation, crystal packing, and molecular motion.

1. Polymer Dynamics

The dynamic behavior of polymers in both the solid and solution state is of significant practical importance, with much interest being focused on the relationships between structure and molecular motions, and the mechanical performance of these materials. Polymer molecular motions are a result of the interplay of thermal energy and cohesive forces between segments of the polymer chain. The structure of the polymer regulates interactions both along chains and between adjacent chains. Specific characteristics such as segment size and geometry, rotational flexibility and rigidity, and the types of intermolecular forces available are all important.[3] For example, studies have indicated that rapid phenyl ring mobility of polycarbonate below its glass transition is related to its high impact resistance.[4]

Dilute solution studies of polymer systems have resulted in numerous theoretical models to account for single chain polymer dynamics. In bulk polymers, new constraints are encountered dealing with restricted *inter-* as well as *intra-* chain associations. Techniques used to characterize molecular dynamics include variable temperature and/or frequency dielectric measurement and mechanical spectroscopy, electron spin resonance (ESR), fluorescence spectroscopy, and NMR (Table I).[5] NMR has the advantage of being more structurally specific than other methods. Choices of nuclei to be observed gives the advantage of observing specific atoms of a repeat unit or segment. For example, carbon and proton relaxation times and lineshapes provide much information about local polymer motions. Whether the motion is isotropic or anisotropic can be determined along with its timescale and temperature dependence. Spin-lattice (T_1) relaxation times can be correlated with segmental motion on, and local environments of, the polymer backbone, while spin-spin (T_2) relaxation data are more closely related to rotational and librational behavior.

2. Solid-State Deuterium NMR

Solid-state deuterium NMR is a unique probe for investigating molecular motions, and the alignment or orientation of chains.[6] Molecular geometry and type of motion are explicitly affected by a single interaction (the quadrupolar coupling) which is directly related to the chemical structure involving the C-^{2}H bond. The advantages deuterium NMR offers over other nuclei include the following:

- The NMR parameters of deuterium are well understood theoretically and experimentally, and the relaxation mechanisms involved are straightforward so that complications of data interpretation are minimized compared to other nuclei.

- Clear information is obtained about the type of motion, its timescale and frequency, and the degree of motional heterogeneity.

- By combining different deuterium NMR techniques, an enormous dynamic range can be covered from ca. 1 Hz to 10 MHz (Figure 1).

- Molecular order and orientation can be detected based on line-shape and quadrupole splitting.

- The main liability of deuterium NMR is the fact that labelled materials must be used, but this is also an advantage in that selectivity of observation is achieved. Use of specifically labelled samples can give detailed information regarding individual groups, or segments of microscopic domains of the sample.

Examples of polymer applications of solid-state deuterium NMR include studies of polymer chain orientation,[6] backbone and pendant group motion,[7] polymer interactions with plasticizers and water,[8] the influence of blending on

Table I. Relaxation techniques associated with molecular motion.

Technique	Constraints varied	Property observed	Motion involved
Dynamic mechanical relaxation	Stress, strain, temperature	Strain, stress, modulus	Translation and rotation of molecules and chain segments
Viscoelastic relaxation	Shear stress, shear rate, temperature	Dynamic viscosity, shear modulus	Translation and rotation of molecules and chain segments
Ultrasonic relaxation	Pressure, temperature	Acoustic absorption, velocity	Any molecular change for which either $\Delta V^\circ \neq 0$ or $\Delta H^\circ \neq 0$ (conformation change)
Dielectric relaxation	Electric field, temperature	Electric polarization, capacitance loss	Limited translation of charges, rotation of dipoles
Fluorescence depolarization	Polarized electromagnetic field, temperature	Polarized fluorescence	Rotation and libration of electronic transition dipole moment
Nuclear magnetic and electron spin relaxation	Magnetic field, temperature	Nuclear and electron spin magnetic polarization	Rotation of particle spin transition moment

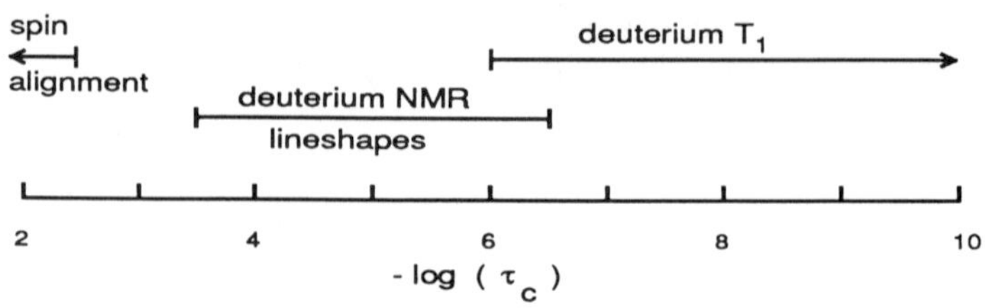

Figure 1. Deuterium NMR timescale.

changes in motion,[9] and motional identification of hard and soft phases of segmented copolymers.[10] The potential also exists for examining unique relaxation characteristics of crosslinks in polymer systems.[11,12] In the case of partially oriented systems, information can be gained by swelling with deuterated solvents and then making the NMR observations on these probe molecules.[13] Overall, ^2H NMR can provide a clear understanding of the type and timescale of a wide variety of molecular motions, and the degree and direction of segmental or molecular orientation. For many samples, this information is much more difficult or impossible to obtain by any other method.

For example, in contrast to other labelling methods for allowing ESR or fluorescence observation, selective ^2H labelling has a small to negligible effect on the properties of the sample. Labelling only specific portions of the compound and with an isotope of hydrogen minimizes perturbations and allows observation of localized motions. Molecular order and mobility can be determined directly by observing the orientation of the C-^2H bond within the magnetic field. The wide-range of frequencies of the various deuterium NMR techniques (Figure 1) is ideally suited for looking at polymers whose motions vary greatly with structure and temperature. In this chapter, we concentrate on the middle range of this timescale which uses lineshapes and quadrupole coupling constants to provide specific information.

3. Physical Background

There are several interactions which can cause experimental difficulties in obtaining NMR spectra of solids. These include direct dipole-dipole coupling, magnetic shielding effects, spin-spin coupling and quadrupolar interactions. A general Hamiltonian for the interactions on the nucleus of interest is given in equation (1).

$$H = H_Z + H_D + H_{SC} + H_{CS} + H_Q \tag{1}$$

The Zeeman interaction H_z is observed for all nuclei with odd atomic mass or odd atomic number when subjected to an applied magnetic field, H_o. This is the interaction which makes NMR possible. The magnetic moment of the nucleus also determines the size of this effect.

Dipole-dipole effects (H_D) are due to interaction between nuclear dipoles of directly bonded or spatially close nuclei. This includes coupling to both similar and dissimilar nuclei and depends on their total abundance. This interaction is also dependent on the magnitude of the magnetic moment of the nuclei involved and decreases rapidly with increasing internuclear distance of the nuclei. It is independent of the applied magnetic field and can range up to 10^5 Hz.

Scaler coupling (H_{sc}) is an interaction between nuclear spins (spin-spin coupling) and is usually smaller than the other interactions. It can involve similar or dissimilar nuclei, is also field independent, and ranges up to several kHz.

The chemical shift effect (H_{cs}) is a result of the magnetic shielding caused by through-bond or through-space interaction with electrons and gives rise to distinct chemical shifts for specific nuclei of individual functional groups or molecular segments. It is linear with the applied field and can range up to 10^5 Hz.

The quadrupolar interaction (H_Q) occurs *only* with nuclei that have a spin of greater than 1/2 and is due to the interaction of the nuclear electric quadrupole moment, eQ, with the non-spherically symmetrical field gradient around the nucleus. It is field independent and is usually very large, ranging up to 10^9 Hz.

A feature of all these interactions is their three-dimensional or anisotropic nature. In solution, rapid tumbling motions involving segmental and molecular reorientation reduce these interactions to zero or to a single non-zero value due to motional averaging. The chemical shift and spin-spin couplings are averaged, while the dipolar and quadrupolar interactions usually reduce to zero. NMR spectra of solids are complicated by the inherent anisotropy and orientational dependence of these interactions. This complexity, however, yields more information than obtainable from solution spectra. Depending on what type of information is desired, appropriate techniques can decrease all or some of these interactions to obtain solution-like or simplified spectra.

The deuterium nucleus has a spin of 1, resulting in three energy levels in a magnetic field and a non-zero quadrupole moment. The nuclear properties of deuterium are summarized in Table II along with those of other commonly studied nuclei. Deuterium NMR spectroscopy of solutions, where the quadrupole interaction is "zeroed-out", yields spectra that are similar to proton NMR spectra except that the deuterium peaks are slightly broader due to quadrupolar interactions. Solution-state deuterium NMR has been used extensively to monitor deuterium labelled species in kinetic and mechanistic studies.

Solid-state NMR of deuterium nuclei is quite different than that of other commonly observed nuclei. Even though the quadrupole moment for deuterium is relatively small (2.73×10^{-21} cm^{-1}), it dominates the spectrum rendering the other interactions negligible (except for Zeeman interactions). Equation (2) shows the general Hamiltonian as applied to solid-state deuterium spectra. Deuterium powder pattern spectra for unoriented samples are quite broad and can be up to 250 kHz in width for an aliphatic C-^{2}H.

$$H = H_Z + H_Q \tag{2}$$

Table II. NMR parameters and relative detectability of some common nuclei.

Isotope	$\gamma/2\pi$[a]	Natural Abundance	Relative Sensitivity[b]	Magnetic Moment (μ)[c]	Spin (I)
^{1}H	42.57	99.9844%	1.0000	2.7927	1/2
$^{2}H(D)$	6.54	0.0156	0.0096	0.8574	1
^{13}C	10.71	1.108	0.0159	0.7022	1/2
^{14}N	3.08	99.635	0.0010	0.4036	1
^{15}N	4.31	0.365	0.0010	-0.02830	1/2
^{17}O	5.77	0.037	0.029	-1.8930	5/2
^{19}F	40.06	100.00	0.834	2.6273	1/2
^{27}Al	11.09	100.00	0.207	3.6385	5/2
^{29}Si	8.46	4.70	0.0785	-0.5548	1/2
^{31}P	17.24	100.00	0.0664	1.1305	1/2

[a] gyromagnetic ratio in MHz/Tesla.
[b] compared to ^{1}H and based on $\gamma/2\pi$ and natural abundance at constant field.
[c] units of Bohr nuclear magnetons.

The quadrupolar nature of deuterium has several other implications. Due to the added neutron (compared with a proton), there is an asymmetric distribution of electric charge in the nucleus. This asymmetry results in an electric quadrupole moment which is important when the nucleus is in the presence of an electric field gradient (any environment with*out* cubic symmetry can exhibit such an EFG). The electric field gradient originates from the electrons in the C-^{2}H bond and is *intra*molecular in nature. The *intra*molecular interaction of the deuterium quadrupolar moment with the electric field gradient (EFG) of the C-^{2}H bond dominates ^{2}H NMR spectra and leads to characteristic solid-state spectral shapes.[14]

Another effect of the quadrupole moment is that there is an additional perturbation in the energy levels of the deuterium nucleus in the magnetic field. Figure 2 shows that the (m=-1) and (m=+1) energy levels of the deuterium nucleus are affected equally and in the same direction. The allowed transitions are (m=-1) ↔ (m=0) and (m=0) ↔ (m=+1). Thus, the spectrum for a single deuteron in a crystalline material would consist of a doublet with a peak separation indicating the angle of orientation of the C-^{2}H bond vector to the magnetic field (Figure 3). From observation of the orientation of individual C-^{2}H bond directions, the molecular order and mobility can be calculated.

The peak separation is related to the frequency of the NMR transition as given by equation (3).[12]

$$\Delta\nu = 3/4\ \delta\ (3\cos^2\theta - 1 - \eta\sin^2\theta\cos^2\Phi) \tag{3}$$

The quadrupole coupling constant δ is defined as e^2qQ/h. As stated previously, the quadrupole coupling constant is a product of the deuterium quadrupole

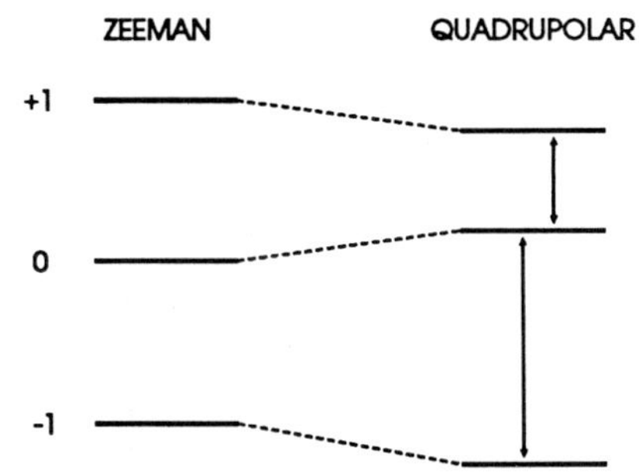

Figure 2. Energy diagram for the deuterium nucleus with spin = 1.

moment (eQ) and the electric field gradient (EFG). By definition of the quadrupolar Hamiltonian, the quadrupolar interaction is a 3 x 3 matrix or second rank tensor describing the three-dimensional nature of the interaction as in equation (4).

$$\hat{Q} = \frac{eQ}{2h} \cdot \hat{V} \tag{4}$$

The electric field gradient tensor, $\hat{V}$, is a function of the orientation and magnitude of the EFG. To visualize the geometry of the EFG around the C-^{2}H bond, it is broken down into three principle elements V_{11}, V_{22} and V_{33} which describe the size and direction of the field gradient. The V_{33} axis is typically taken to be along the C-^{2}H bond. Within this co-ordinate system, θ and Φ are polar angles specifying the orientation of the magnetic field with the principal axis system of the electric field gradient tensor (Figure 4). The three-dimensional nature of this interaction is related to the asymmetry parameter η as defined in equation (5).

$$\eta = \frac{V_{11} - V_{22}}{V_{33}} \tag{5}$$

In the case of the EFG being axially symmetric around the C-^{2}H bond (V_{33} principle axis), V_{11} and V_{22} will be equal and η will be zero. This situation is true

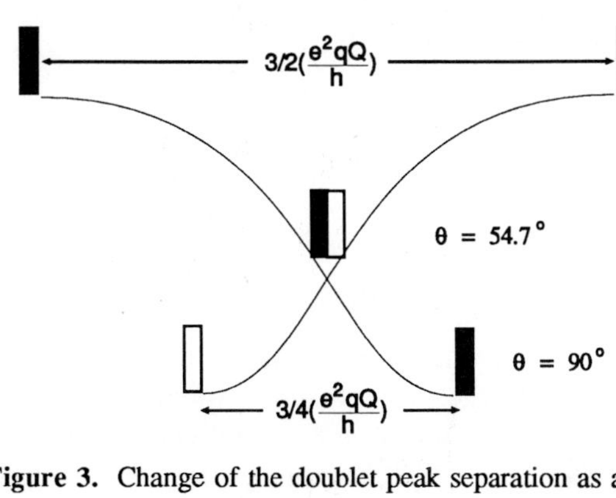

Figure 3. Change of the doublet peak separation as a function of θ.

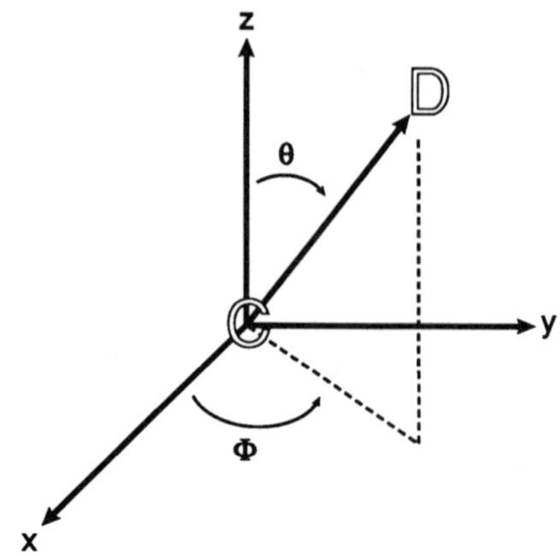

Figure 4. Coordinate system showing the interaction of the electric field gradient with the magnetic field.

in most aliphatic compounds, and to a good approximation, is usually true also for aromatic compounds. For the case of axial symmetry, equation (3) is reduced to give the quadrupole splitting given in equation (6).

$$\Delta\nu = 3/4\ \delta\ (3\cos^2\theta - 1) \qquad (6)$$

In the absence of motion, the deuterium lineshape is a result only of the angle the C-^{2}H bond forms with respect to the magnetic field and the two allowed transitions (Figure 5). For a single crystal or an oriented sample, the result would be a doublet with a splitting frequency as discussed above. Acquisition of spectra as a function of the sample orientation (with respect to the magnetic field) will give information relating to the geometry of the C-^{2}H bond to the sample.

Molecular motion, which changes the C-^{2}H bond orientation, is accounted for by the use of averaged terms for δ and η. Motion introduces a significant and predictable change in the width and shape of the deuterium NMR spectrum (Figure 6).

4. Experimental Considerations

As discussed above, the quadrupolar interaction dominates the deuterium spectrum. Because of the large spectral width of deuterium spectra an extremely short, high-power pulse is needed. The quadrupolar interaction also provides a very efficient mechanism for spin-lattice relaxation. This results in a very rapid decay of the magnetization signal after the high-power rf pulse. In fact, the decay is so rapid that much of the signal is lost due to the power of the pulse and the receiver dead time (the time needed for the sample coil to stop "ringing" from an rf pulse) associated with the spectrometer probe. As a result, a special quadrupole echo pulse sequence is used that is similar to the spin-echo technique used to measure T_2 relaxation times in solution NMR.[15] This sequence consists of two 90° rf pulses, the first to induce a net magnetization and a second 90° phase-shifted pulse given after a delay time τ applied to refocus the spin vectors at 2τ to form an "echo" of the FID (Figure 7). A Fourier transform is performed starting at the top of the echo to give the deuterium spectrum.

Probe and spectrometer tuning have a profound affect on the appearance of the spectrum. For solid-state deuterium NMR, spectra should be symmetrically centered about the carrier frequency. A mistuned spectrometer will lead to asymmetrical spectra which usually cannot be corrected by software phasing routines.[16] Ideally, symmetrical spectra can be obtained if the NMR signal is acquired entirely in the real channel of the receiver with only noise in the imaginary channel. Phase cycling routines are also employed. Currently, the preferred tune-up method is the same procedure which is used for multi-pulse NMR experiments.[17]

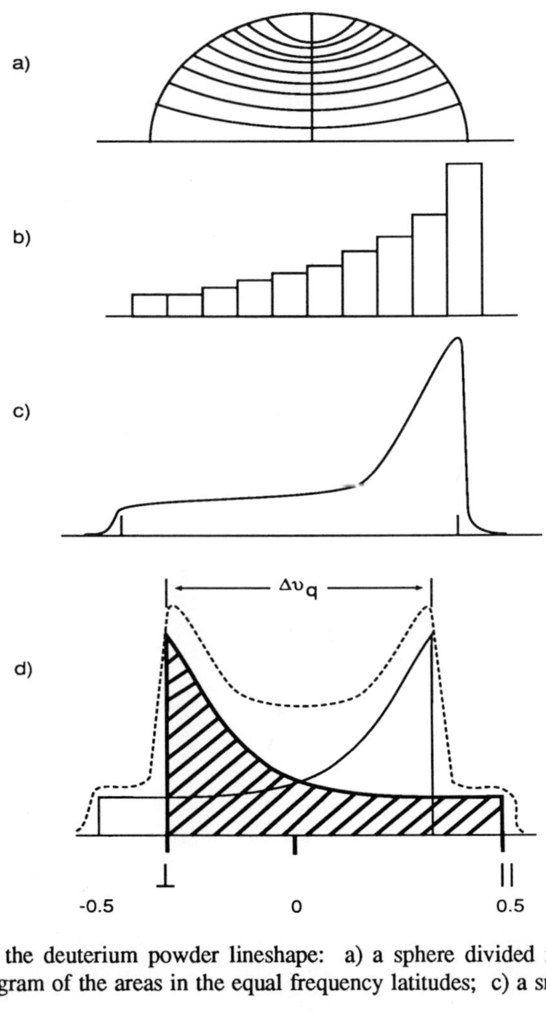

Figure 5. Origin of the deuterium powder lineshape: a) a sphere divided into latitudes of equal frequency; b) a histogram of the areas in the equal frequency latitudes; c) a smoothed version of b); d) both transitions.

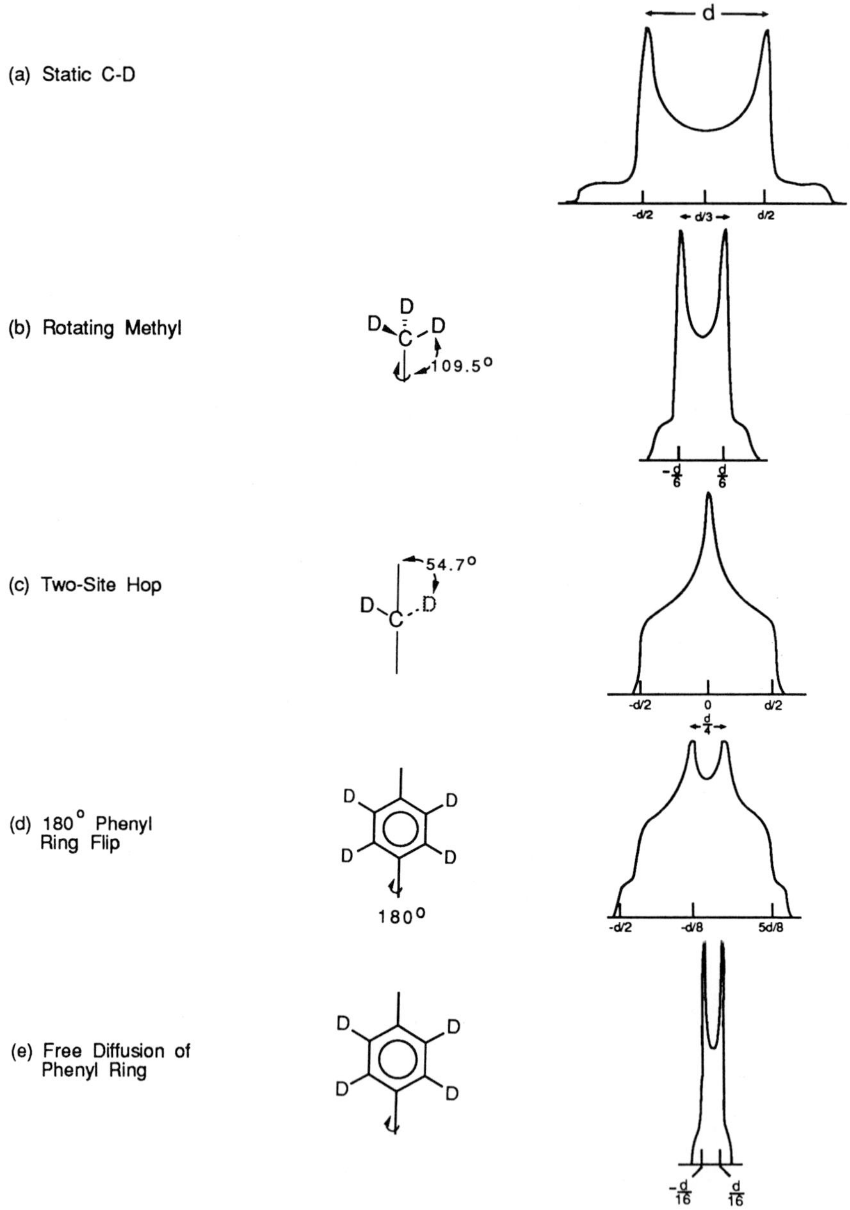

Figure 6. Theoretical deuterium NMR lineshapes for various types of anisotropic motion.

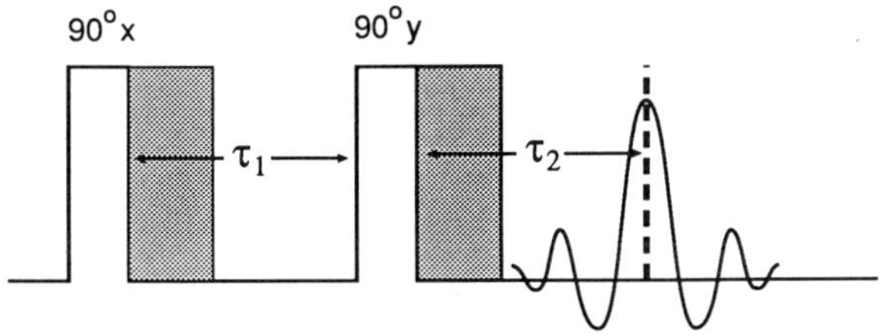

Figure 7. Quadrupole echo pulse sequence used to acquired deuterium spectra.

5. Experimental Variations

The quadrupole echo pulse program can be modified to gain additional NMR data such as deuterium spin-lattice and spin-spin relaxation times. Depending on the sample being examined, it is usually possible to selectively observe only mobile or rigid regions. Specific experimental setups that have been used in this research, as well as in the processing techniques, are detailed elsewhere.[18]

The spin alignment technique is a relatively new deuterium NMR method developed to study very slow motions.[19] This technique uses a Jeener-Broekaert pulse sequence to overcome the T_2^* limitation of the normal NMR experiment. The experiment is limited only by the T_1 relaxation time which is in the range of 10 ms to 10 s. By creating a period of spin alignment (a long lived state of order of the spin system), longer waiting periods can be used between the initial pulses. If motion takes place within this waiting period there will be a reduction of the spin alignment echo intensity and a characteristic lineshape will result. Therefore very slow motions can be directly observed.

2-Dimensional solid-state deuterium NMR spectroscopy is another experimental technique developed to give direct information regarding motional mechanisms.[20,21] As in traditional 2D solution NMR, additional information is derived from the cross peaks. Solution and single-crystal 2D exchange NMR methods give cross peaks which indicate the number of exchanging sites and the angles of exchange. 2D NMR spectra of solids and powders show the Pake pattern in the diagonal direction with cross ridges that form characteristic geometric figures, and molecular reorientational angles of jump-like motions can be determined directly by measuring the ridge patterns. Although valuable information is gained from this technique, the experiment is generally very time intensive and the quality of the results may vary from sample to sample.

6. Current Limitations

While the quadrupole echo pulse sequence can give much information and a clear description of the molecular motion, other deuterium techniques (spin alignment and relaxation studies) may be needed. These experiments may be very time consuming depending on the type of sample, the ^2H content, and the strength of the magnetic field.

To obtain quantitative information, computer simulation programs must be used along with detailed evaluation of several sample variables. Although it is straightforward to gain information on the molecular motion of a specific group, motional contributions from the rest of the system may be more difficult to determine. In general, many additional samples need to be evaluated by deuterium NMR to help clarify structure-motion-property relationships and develop an extensive data-base of peak shapes and relaxation times.

7. Summary of Background Information

Solid-state deuterium NMR has become a dominant technique for investigating molecular motions in small compounds as well as in polymers. The quadrupole echo pulse sequence provides an easy method of obtaining deuterium NMR spectra. Qualitative information can be readily determined directly from the spectrum while relaxation measurements and experimental variations can give additional insight into polymer dynamics at the segmental and molecular levels.

The following sections give specific applications of ^{2}H NMR in polymer and small molecule analysis that illustrate the type of informaiton available.

8. Deuterated Acrylates and Acrylonitrile

Selectively deuterated acrylates were prepared using a previously reported synthetic technique.[22] These monomers were polymerized to yield polymers with deuterium on every other carbon of the backbone. This deuteration method was also applied to diacrylates to produce selectively labelled crosslinking agents. Deuterium incorporation at specific sites in monomers, and therefore in their polymers, allows separate determination of backbone and crosslink site behavior.

Solid-state deuterium NMR spectra were acquired over a wide temperature range for the following deuterated polymers: poly(methyl acrylate-d_α); poly(ethyl acrylate-d_α); poly(butyl acrylate-d_α); polyacrylonitrile-d_α; and the crosslinked polymers, poly(ethyl acrylate-d_α-*cross*-hexanediol diacrylate), poly(ethyl acrylate-*cross*-hexanediol diacrylate-$d_{\alpha,\alpha}$) and poly(ethyl acrylate-*cross*-ethylene diacrylate-$d_{\alpha,\alpha}$). In all cases, the vinyl position next to the ester or nitrile group is the position of deuteration. In addition, polyacrylonitrile-d_α was studied further using the deuterium spin-alignment and 2D NMR techniques. Glass transition temperatures of the polymers were determined by differential scanning calorimetry (DSC) with no noticeable difference from those of the undeuterated species (Table III).

For methyl, ethyl and butyl acrylate polymers, the overall main chain behavior was similar as determined using the quadrupole echo pulse sequence: below T_g the spectrum was a rigid Pake pattern while at temperatures 25-50° above T_g, only a single sharp peak was present. On a molecular level this behavior implies that the

Table III. Glass transition temperature comparison.

Polymer	T_g K			
	Ref[a]	DSC[b]	^{2}H NMR[c]	^{2}H NMR[d]
Poly(methyl acrylate-d_α)	283	278	305	305
Poly(ethyl acrylate-d_α)	249	247	265	---
Poly(butyl acrylate-d_α)	219	218	260	260
Polyacrylonitrile-d_α	370	373	400	---
Poly(ethyl acrylate-d_α-*cross*-hexanediol diacrylate), 7.8 mol-%	---	--[e]	290	290
Poly(ethyl acrylate-*cross*-hexanediol diacrylate-$d_{\alpha,\alpha}$), 5.1 mol-%	---	--[e]	270	---
Poly(ethyl acrylate-*cross*-hexanediol diacrylate-$d_{\alpha,\alpha}$), 9.2 mol-%	---	--[e]	280	---
Poly(ethyl acrylate-*cross*-ethylene diacrylate-$d_{\alpha,\alpha}$), 13.7 mol-%	---	272	290	295

[a] Values for non-deuterated polymers listed in Polymer Handbook. [37]
[b] Determined by DSC at 10°C/min heating rate.
[c] Determined from ^{2}H NMR lineshape changes.
[d] Determined from ^{2}H T_2 curves.
[e] Not able to be determined from the DSC trace.

C-^{2}H bonds are rigid (on the quadrupole echo time scale) below T_g or that motion is ultraslow. Near the T_g the signal intensity decreased and a small isotropic peak was superimposed on the Pake pattern. At ca. 40° above T_g (as seen by DSC), polymer segments move in a random and isotropic manner with backbone motions that are fast on the deuterium NMR timescale ($\tau_c \leq 10^{-7}$s). This is expected due to the difference in frequency of observation and time scale of the ^{2}H NMR experiment compared to that of DSC. The ^{2}H NMR experiment is also an isothermal observation whereas the DSC measurement is influenced by the heating (or cooling) rate. This overall behavior is similar to that observed for other backbone labelled polymers.[23] Figure 8 illustrates how the NMR T_g's were taken to be as the temperature at which the sharp isotropic peak just becomes visible; see the 305K spectrum.

The quadrupole echo deuterium spectra for 7.8 mol-% crosslinked poly(ethyl acrylate-d_α-*cross*-hexanediol diacrylate), 5.1 and 9.2 mol% crosslinked poly(ethyl acrylate-*cross*-hexanediol diacrylate-$d_{\alpha,\alpha}$) and 13.7 mol-% crosslinked poly(ethyl acrylate-*cross*-ethylene diacrylate-$d_{\alpha\text{-}\alpha}$) showed similar behavior. For the crosslinked systems, a ca. 20° increase in T_g (onset of isotropic motion) was seen for the main-chain acrylate units. However, the crosslinks themselves seemed to be in a *less* constrained environment, showing T_g's that were *below* that of the uncrosslinked material at the lowest crosslink density. As the crosslink density increased, T_g increased (at least *at* the crosslink site) indicating increased rigidity and reduced segmental motion.

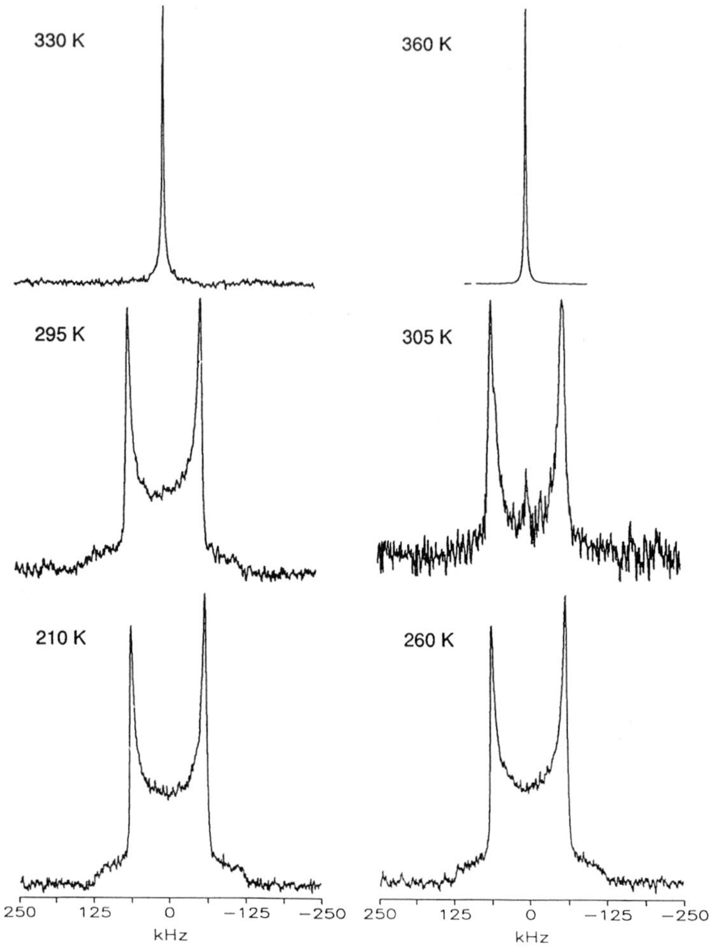

Figure 8. Solid-state deuterium NMR of poly(methyl acrylate-d_α).

For the acrylate polymers, reaching the T_2 minimum (by temperature variation) produced lower intensity lineshapes, and lower signal-to-noise. This was clearly evident in the crosslinked polymers which had a lower deuterium content. The result was the disappearance of the outer edges of the Pake pattern, while the horns of the pattern remained visible as two single peaks.

9. Polyacrylonitrile

Polyacrylonitrile-d_α gave very different results (Figure 9). Below T_g, the quadrupole echo lineshape consisted of a rigid Pake pattern. Above T_g, a non-isotropic pattern developed corresponding to restricted motion involving a multi-site jump process. There was a very slight decrease in the overall width of the pattern. The sample was taken through 4-5 heating cycles up to 435K (162°C) to examine motions around and above T_g (these results will be discussed below). All experiments performed after heat treatments displayed an additional "isotropic" peak in the center of the Pake pattern whose cause is unknown.

Deuteron spin alignment NMR was performed to examine very-slow molecular motions at 300K.[19,24] For polyacrylonitrile (with a T_g of 370K (97°C)) at 300K (27°C), the alignment echo decay curves are dependent on τ_1, which indicates the polymer backbone is undergoing diffusive motions over small but ill-defined angles. Initial spin alignment lineshapes also indicate that the motion is a restricted type of diffusion.

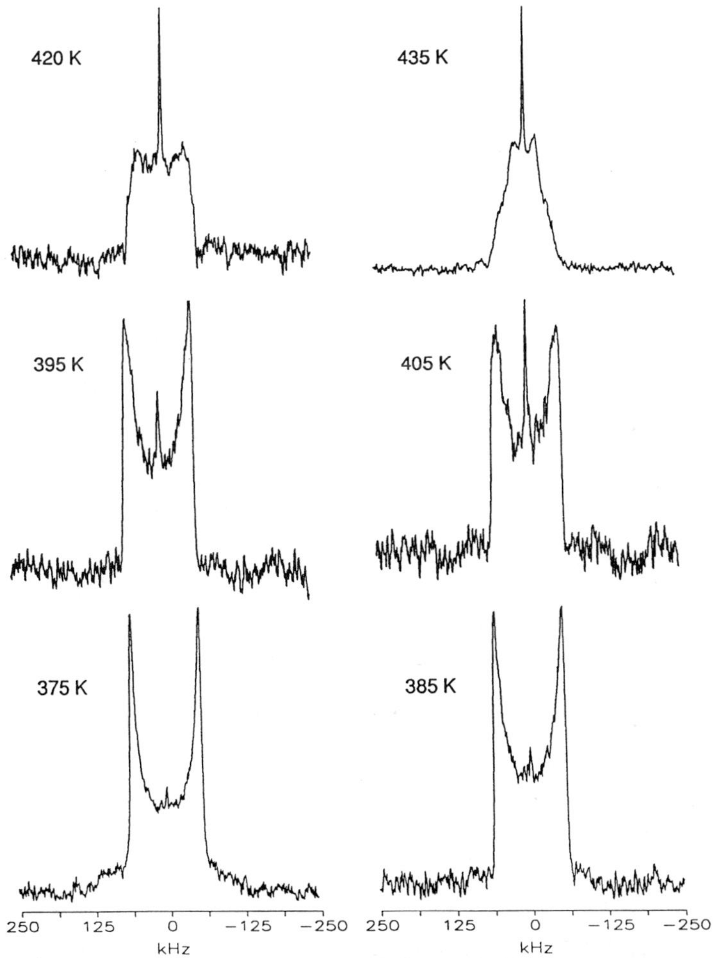

Figure 9. Solid-state deuterium NMR of polyacrylonitrile-d_α.

10. Deuterium Relaxation Experiments

Deuterium spin-spin relaxation (T_2) data for poly(methyl acrylate-d_α), poly(butyl acrylate-d_α) and the crosslinked polymer labelled on the mainchain [poly(ethyl acrylate-d_α-*cross*-hexanediol diacrylate)] showed a minimum at approximately 30° above the DSC T_g values (Figures 10 - 16). This is expected due to the difference in frequency of observation and time scale of the ^{2}H NMR experiment compared to that of DSC. The T_2 data for the crosslinked labelled polymer, poly(ethyl acrylate-*cross*-ethylene diacrylate-$d_{\alpha,\alpha}$) with 13.7 mol-% crosslinker showed a similar minimum (Figure 13). The greater scatter of data points may be due to the greater inhomogeneity associated with crosslinked systems resulting from a distribution of crosslink lengths. Overall, however, values of the T_2 minima correspond closely to the T_g's estimated from the onset of the isotropic peak for all of the acrylate polymers (Table III).

Polyacrylonitrile did not display a T_2 minimum for the temperature range employed (Figure 14). The minimum was most likely shifted to a higher temperature due to the specific jump motion as indicated from the spectrum for polyacrylonitrile above 400K (127°C).

Deuterium spin-lattice relaxation (T_1) data for poly(methyl acrylate) and polyacrylonitrile are shown in Figure 15 and Figure 16. The data of both samples give a much better fit using a two-component analysis rather than single component. Relaxation data for bulk polymers can be difficult to interpret due to the inherent heterogeneity of polymers.[25,26] For poly(methyl acrylate), the short T_1

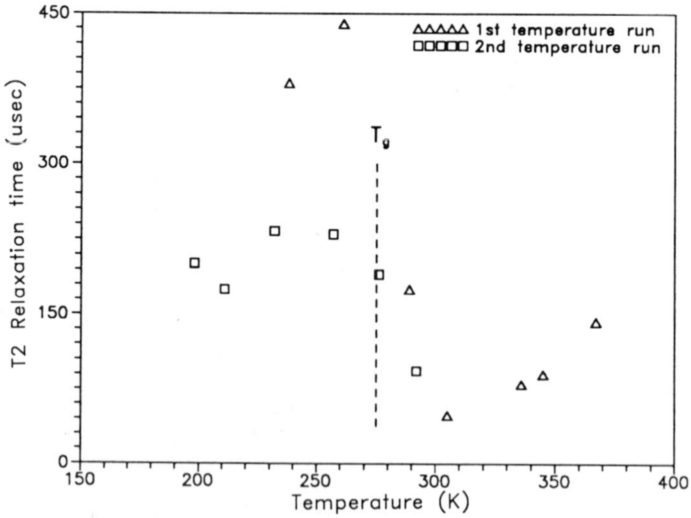

Figure 10. Spin-spin relaxation graph for poly(methyl acrylate-d_α).

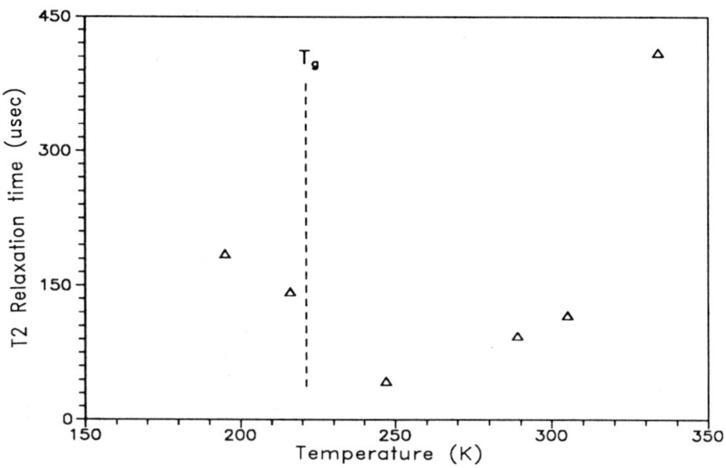

Figure 11. Spin-spin relaxation graph for poly(butyl acrylate-d_α).

component decreased with increasing temperature, while the long T_1 component went through a minimum analogous to the T_2 relaxation but at a temperature (280K) much closer to the T_g measured by DSC. The short T_1 component of polyacrylonitrile decreased almost linearly as the temperature increased. However, the long T_1 component went through a minimum at almost the same value as the DSC T_g (370 - 380K). After T_1 increased beyond this minimum, it abruptly dropped again at a temperature ca. 30° above the T_g. Such behavior was not seen for the acrylates and may be related to the onset of restricted motion indicated by the change in peak shape between 405 and 420K (Figure 9). Torchia and Szabo have shown that there may not be a unique T_1 for each frequency associated with a specific C-^{2}H bond

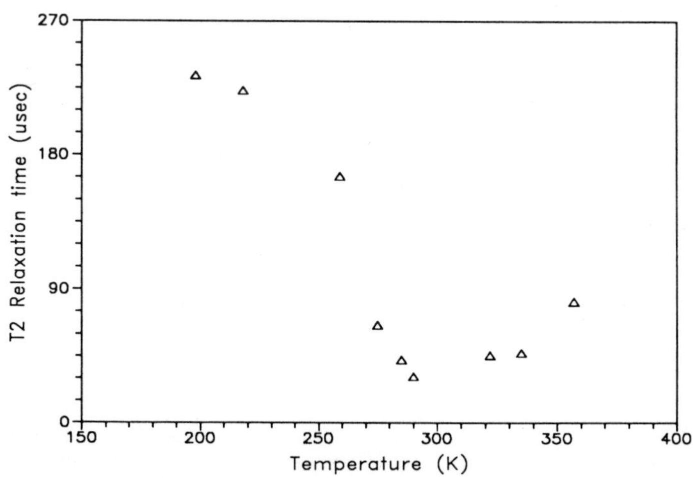

Figure 12. Spin-spin relaxation graph for poly(ethyl acrylate-d_α-*cross*-hexanediol diacrylate).

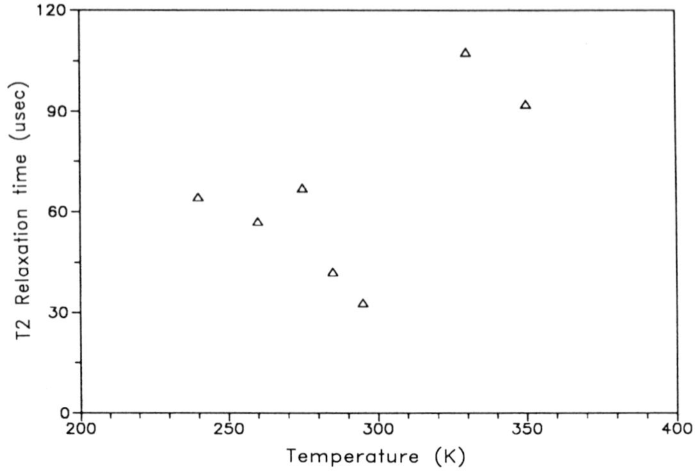

Figure 13. Spin-spin relaxation graph for poly(ethyl acrylate-*cross*-ethylene diacrylate-$d_{\alpha,\alpha'}$).

orientation.[27] Thus, the two component T_1's may be the result of morphological differences in chain segments in amorphous polymers.

11. 2D Deuterium NMR of Polyacrylonitrile-d_α

Polyacrylonitrile-d_α was also examined with a new 2D solid-state deuterium NMR technique developed by Spiess and coworkers as a direct method of determining rotational motions as compared to the indirect methods using the quadrupole echo procedure.[20] The spectrum produced from this experiment did not give clear results concerning the mechanism of motion due to the limited amount of time available to

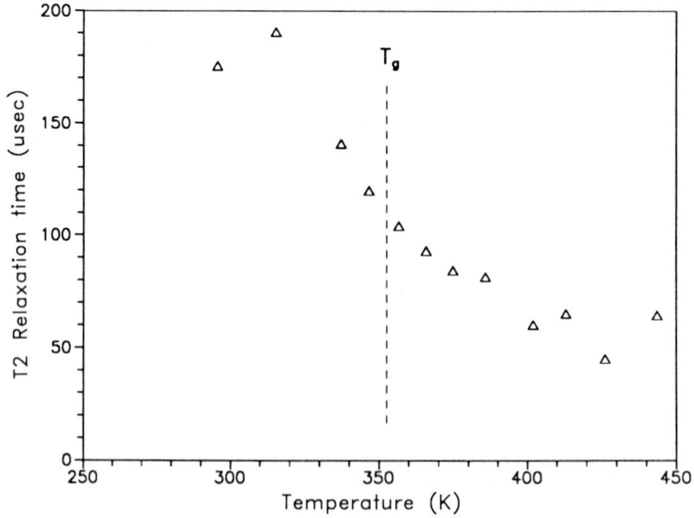

Figure 14. Spin-spin relaxation graph for polyacrylonitrile-d_α.

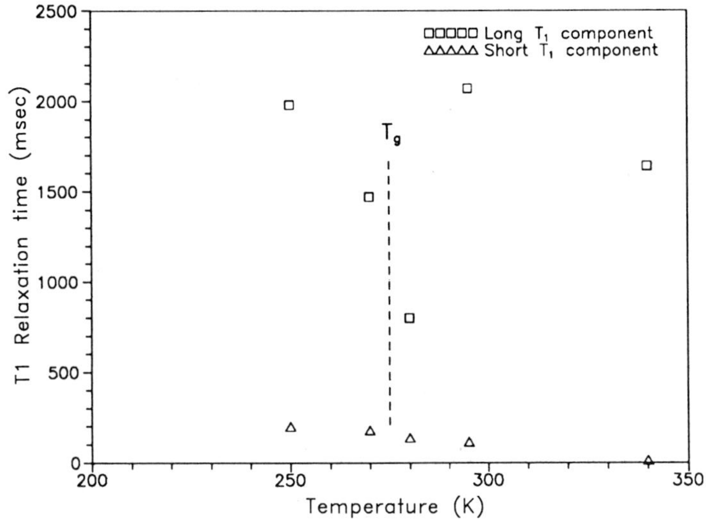

Figure 15. Spin-lattice relaxation graph for poly(methyl acrylate-d_α).

obtain the data (the spectrum obtained took over 48 hours of NMR acquisition time). The 2D spectrum does indicate a diffusive type of motion occurs at 300K (27°C) from the low intensity cross ridges (Figure 17). These are seen best to the right and center of the left-most main horn.

12. Summary of Sections 8-11

For polyacrylates, ^{2}H lineshapes and T_2 minima showed that molecular motions go from rigid to random and isotropic at temperatures ca. 30° above the glass

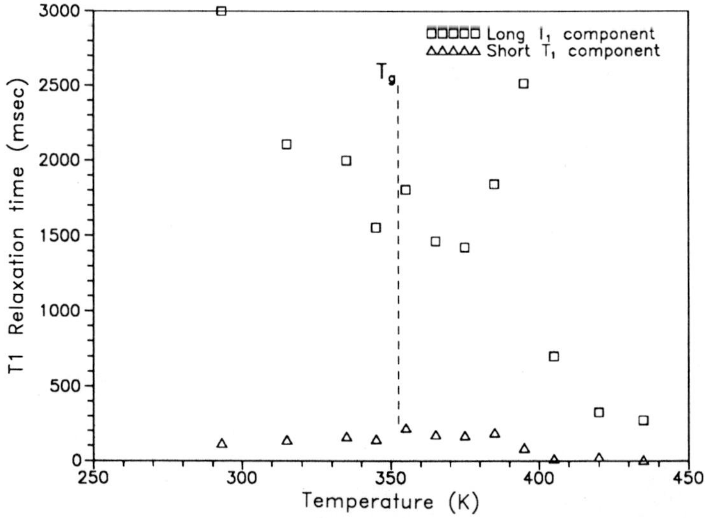

Figure 16. Spin-lattice relaxation graph for polyacrylonitrile-d_α).

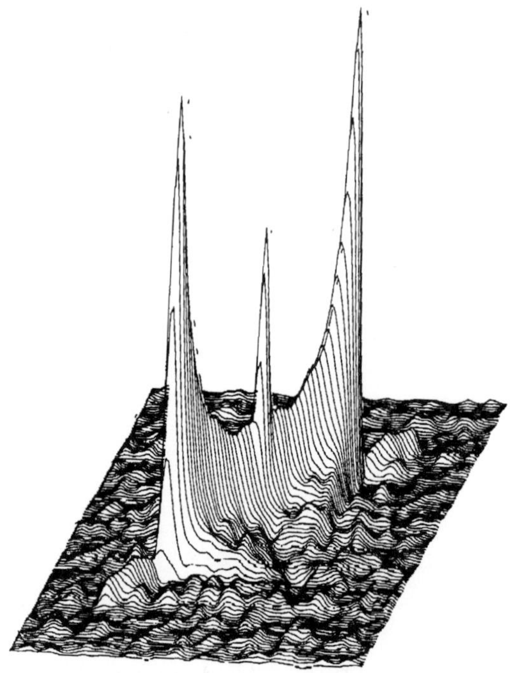

Figure 17. 2D solid-state deuterium NMR of polyacrylonitrile-d_α.

transition while polyacrylonitrile undergoes a transition to a multisite jump motion. T_1 data, however, indicated minima at about the same temperatures as the DSC T_g's. Crosslinking increased the T_g of the acrylate chains between crosslinks although the crosslink sites themselves showed *reduced* T_g's (compared to non-crosslinked material) that gradually increased with increasing crosslinker content.

13. ^{2}H NMR of N-^{2}H Nylon 6 and Nylon 11

The early observation of the Brill transition[28] in nylons has caused much speculation about the associated molecular dynamics. The crystal-crystal transition for nylon 11 involves a change from triclinic to pseudohexagonal unit cell (Figure 18). Previous x-ray and thermal results[29,30] were coupled with detailed ^{15}N CP/MAS NMR analysis of nylon 11[31] and nylon 6[32] and ^{13}C CP/MAS analysis of nylon 6[33,34] to understand this process. While earlier models[28] postulated random amide group rotational flips of 60° to explain the hexagonal lattice that develops, increased libration of the planar amide moieties, and especially of the intervening methylene units, is sufficient to account for the transition observed for nylon 11.

Extensive evidence for nylon 66 libration from ^{2}H NMR studies of C-^{2}H and N-^{2}H labelled samples has been reported to rule out amide flipping as a relaxation

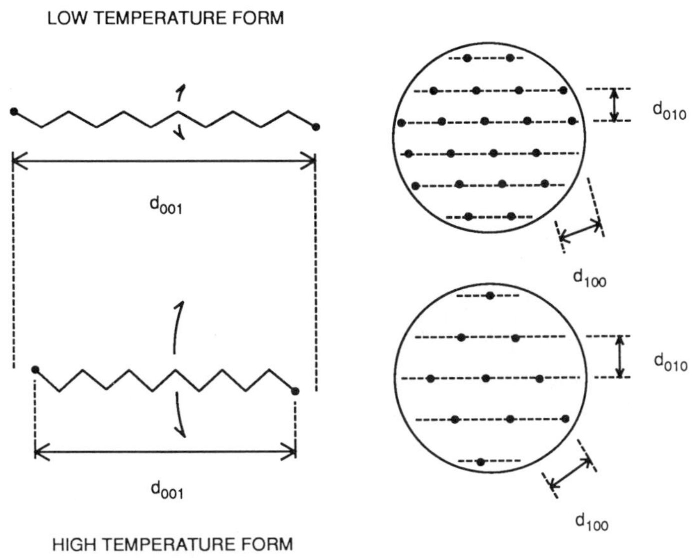

Figure 18. The proposed model for the crystal-crystal transition of nylon 11 showing the low temperature triclinic unit cell and the high temperature pseudohexagonal unit cell.

mechanism for this polymer on going through the Brill transition.[35] This work prompted us to use solid-state deuterium NMR to examine the transitions for nylon 11, nylon 6 and the model compound N-*n*-propyl stearamide, all synthesized with deuterium on the nitrogen.

Nylon 6 and 11 were selectively deuterated on the nitrogen using the method of Schuttenberg and Schulz.[36] The polymers were converted first to the N-trifluoroacylated derivatives which were then treated with D_2O to cleave the acetyl group and give the N-deuterated nylons (Figure 19). Samples were then dried and annealed just below T_m to induce maximum formation of the thermodynamically most stable crystal forms, the α-form in both cases.

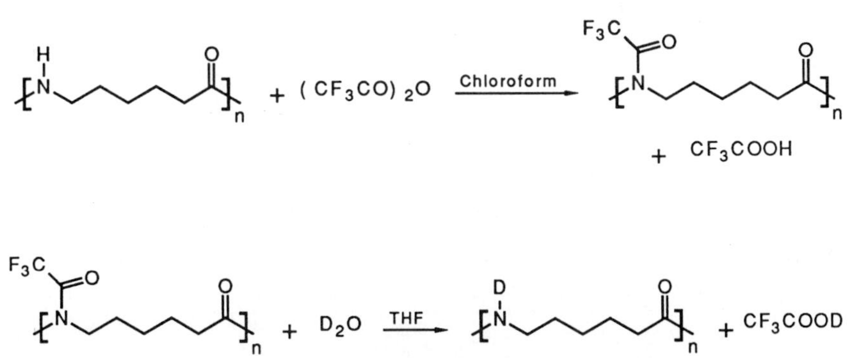

Figure 19. Formation of N-deuterated nylon.

14. ^{2}H NMR Results for Nylons

Solid-state CP/MAS NMR was used to determine the crystal contents of the deuterated, annealed nylons. The ^{13}C and ^{15}N spectra of nylon 6 (N-^{2}H) are shown in Figure 20 and Figure 21, respectively, with the corresponding spectra of nylon 11 (N-^{2}H) shown in Figure 22 and Figure 23. Figure 24 shows the ^{13}C CP/MAS spectrum of trifluoroacylated nylon 6 for comparison. Both ^{13}C and ^{15}N spectral data confirm that the crystalline domains of these deuterated samples were almost exclusively in the preferred α-form. The ^{15}N NMR data was especially clear with the strong, sharp peak of the α-form at 84-85 ppm overshadowing the broad, amorphous band at 85-90 ppm. No evidence of a sharp δ-form peak at 89 ppm was observed for either sample. It should be noted that only about 20 - 40% total crystallinity is usually attained for bulk nylons annealed under these conditions.[32,33]

The solid-state deuterium spectra of nylon 6 (N-^{2}H) are shown in Figure 25 and those of nylon 11 (N-^{2}H) in Figure 26. Spectra were acquired at temperatures below T_g, between T_g and the Brill transition, and between the Brill transition and T_m. For comparison, Table IV lists the reported transition temperatures for nylon 6 and 11,[37] and the thermal transitions observed by DSC for the model amide compound.

The deuterium spectra for both nylon 6 and 11 below their glass transitions were typical broad patterns ($\Delta\upsilon$ = 123 kHz) indicating that both crystalline and amorphous regions are rigid. At temperatures above T_g but below T_m, the spectra consisted of a narrow isotropic peak (rapidly moving amorphous component) superimposed on a rigid Pake pattern (rigid crystalline regions). The narrow peak gradually increased in intensity as the temperature was raised due to an increase in the fraction of amorphous component capable of rapid, isotropic-like motion. This increase is consistent with a gradient of mobility extending from rigid interphase near the surface of the crystallites out to highly flexible amorphous regions well-separated

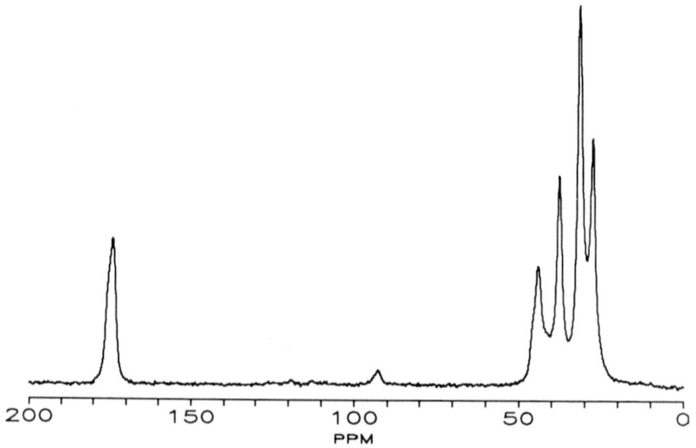

Figure 20. ^{13}C CP/MAS NMR spectrum of nylon 6 (N-^{2}H).

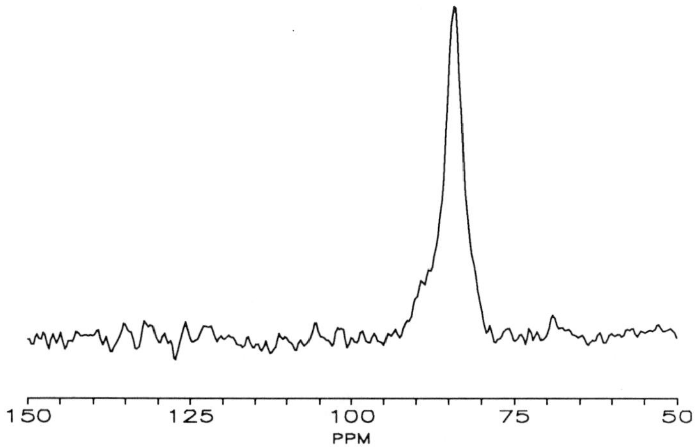

Figure 21. Natural abundance ^{15}N CP/MAS NMR spectrum of nylon 6 (N-^{2}H).

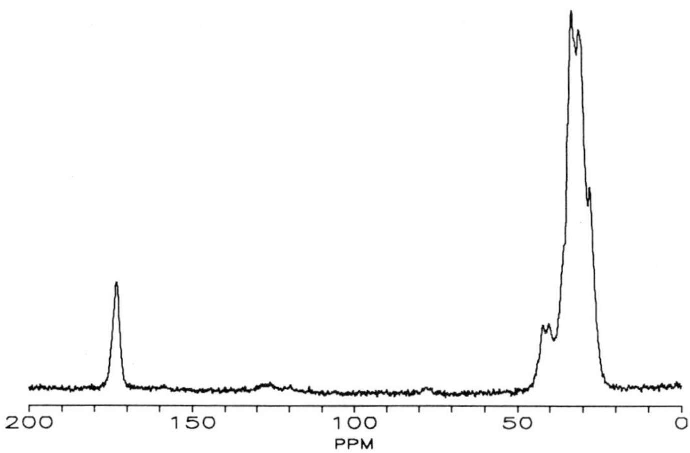

Figure 22. ^{13}C CP/MAS NMR spectrum of nylon 11 (N-^{2}H).

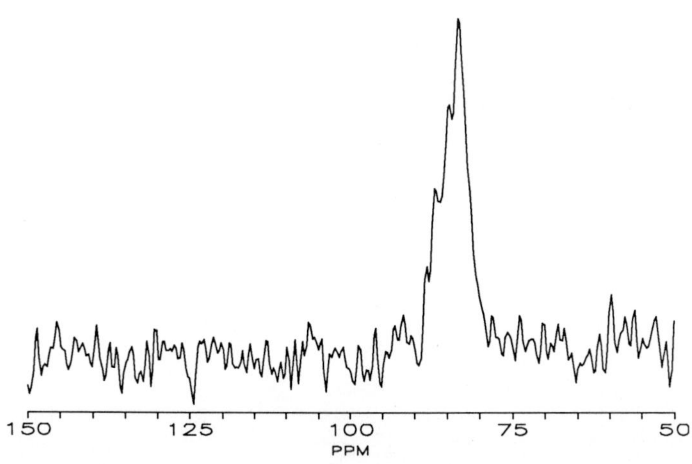

Figure 23. Natural abundance ^{15}N CP/MAS NMR spectrum of nylon 11 (N-^{2}H).

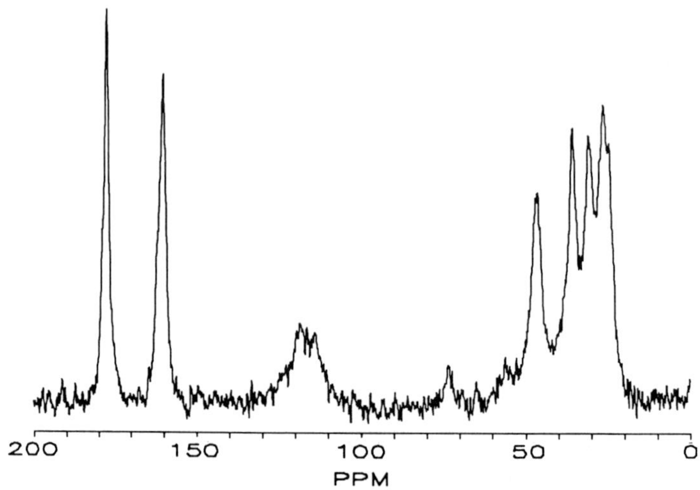

Figure 24. ^{13}C CP/MAS NMR spectrum of trifluoroacylated nylon 6.

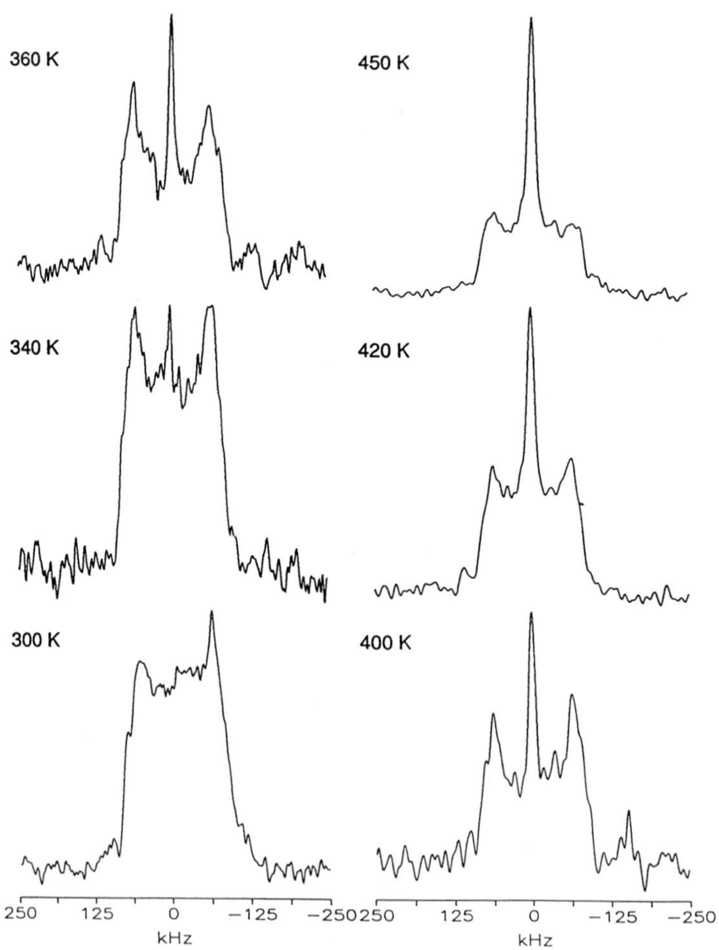

Figure 25. Solid-state deuterium NMR spectra of nylon 6 (N-^{2}H).

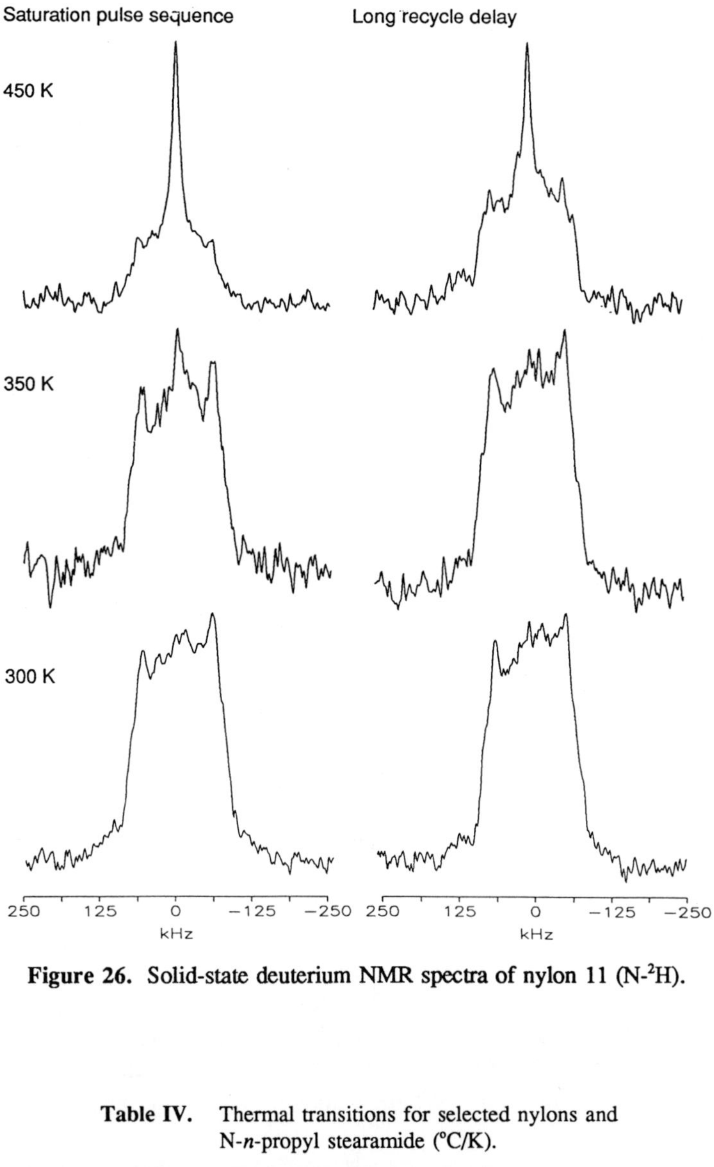

Figure 26. Solid-state deuterium NMR spectra of nylon 11 (N-^{2}H).

Table IV. Thermal transitions for selected nylons and N-*n*-propyl stearamide (°C/K).

Polymer	T_g °C/K	Brill transition	T_m °C/K
Nylon 6	55/328	---[a]	220/493
Nylon 11	50/323	95/368	195/468
N-*n*-propyl stearamide[b]	29/302	76/349	82/355

[a] temperature unreported
[b] temperatures listed for solid-solid transitions observed by DSC

from the crystallites. Similar multicomponent behavior was observed for relaxation times measured for the amorphous domains of ^{15}N-labelled nylon 6 and 11.[32,33]

For nylon 11 (N-^{2}H) above the Brill transition, the crystalline peak was of low intensity but of the same width as the crystalline component below the Brill transition. The fact that the spectral width did not change indicates that the previously postulated rotational motion involving 60° flips does *not* occur since this would involve motion of the N-^{2}H bonds leading to significant narrowing of the rigid pattern. A computer simulation of the nylon 11 (N-^{2}H) overlapping peaks by simple addition of two distinct patterns for the crystalline and amorphous regions at 385K (112°C) indicated ca. 24% crystallinity. This is comparable to the value determined by x-ray analysis on a similarly annealed sample.[32] These results are analogous to those found by English and coworkers[38] for nylon 66 and agree with the nylon 6 and 11 data above.

15. N-*n*-propyl Stearamide ^{2}H NMR Results

For comparison with the nylons, the model compound N-*n*-propyl stearamide was synthesized with deuterium on the nitrogen (analogous to the nylon samples) and studied by solid-state deuterium NMR (Figure 27). The behavior of N-*n*-propyl stearamide (N-^{2}H) observed by solid-state deuterium NMR was compared to the DSC transitions of this compound at 29°, 76° and 82°C. Spectra below 350K (77°C) showed a broad pattern ($\Delta\upsilon$ = 128 kHz) indicative of rigid N-^{2}H bonds despite thermal transitions observed below this temperature that we assign to rotator phase behavior of the propyl and stearamide alkane segments. As the temperature was increased, the center of the pattern filled in somewhat indicating the onset of librational motion. At 357K (82°C), a sharp transformation from a broad to a narrow peak was observed due to the onset of rapid ($\upsilon_c >> 10^5$ Hz), nearly isotropic motion in the melt. It is not surprising that this compound displays no dual-peak behavior like the nylons, since there is no amorphous component and the hydrogen bonded lattice of the amide groups maintains extended order right up to the melting point. While lower temperature transitions observed by DSC indicate increased cooperative motion of the two alkyl groups, the rapid amide group motion which involves breaking and making of transient hydrogen bonds only occurs above the melt.

16. Summary of Nylon Results

The deuterium NMR studies clearly indicated that, below the crystalline melting points of nylon 6, nylon 11, and the amide model compound, the extended hydrogen bonded arrays of the amide groups are rigid despite extensive motion of amorphous components of the nylons and the alkyl groups of N-*n*-propyl stearamide. Furthermore, the amide groups in the nylon amorphous regions were rigid below the glass transitions, but capable of isotropic motion above T_g. The relative increase in

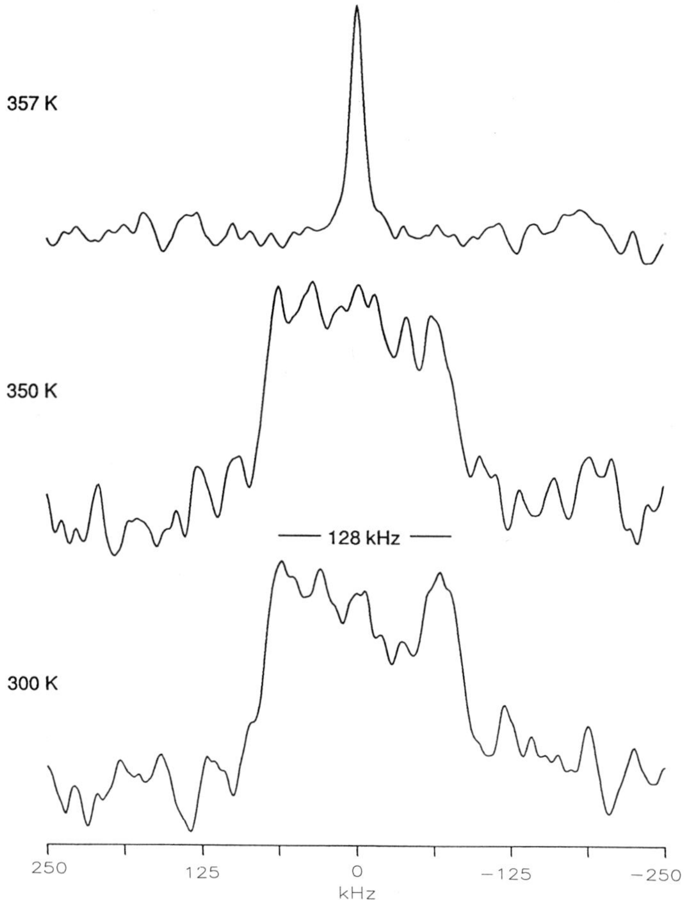

Figure 27. Solid-state deuterium NMR spectra of N-*n*-propyl stearamide.

mobile amorphous content with increasing temperature was consistent with an interphase gradient containing a distribution of non-crystalline material mobilieties which only gradually increased with increasing energy.

17. ^2H-Labelled Perfluorodecyldecane

Fluorocarbon polymers have unique thermal, chemical and mechanical properties but are difficult to process. As a result, studies are being done on oligomers and polymers which contain both fluorocarbon and hydrocarbon segments. Chemically-linked copolymers should have properties between poly(tetrafluoroethylene) and polyethylene. Recent research has focused on semifluorinated alkanes of the general structure **1** as models for potential polymers.[39,40] Semifluorinated *n*-alkanes exhibit unusual solid and liquid crystalline states consisting of lamellar bilayers. For the n = m = 9 compound (**1**) a reversible transition from LC1 to LC2 is observed with LC1 50°C - 61°C and LC2 below 50°C. Viney and coworkers have postulated a change in the bilayer tilt angle and degree of interdigitation between adjacent hydrocarbon and fluorocarbon chains on going from LC1 to LC2 (Figure 28).[41] Another group has used ^{13}C CP/MAS and MAS results to conclude that the transition involves not only a change in packing

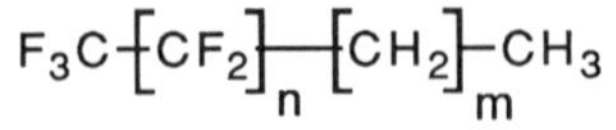

1

but also a conformational disordering.[42] The disorder arises from increased gauche conformations in an essentially liquid-like state. Despite additional data from calorimetric, Raman LAM, X-ray diffraction, and ^{13}C NMR, studies regarding these phases, there are still questions about molecular-level behavior.

Solid-state deuterium NMR is ideally suited for observing molecular dynamics,[6,12] and evaluation of a selectively deuterated material using this technique should give detailed information regarding the transitions described above. Perfluorodecyl decane was labelled as in Figure 29 and examined by ^2H NMR.

18. Results and Discussion

Figure 30 shows the solid-state deuterium spectra for perfluorodecyl decane deuterated in the β-position from the fluorinated carbon segment. At 200K (-73°C)

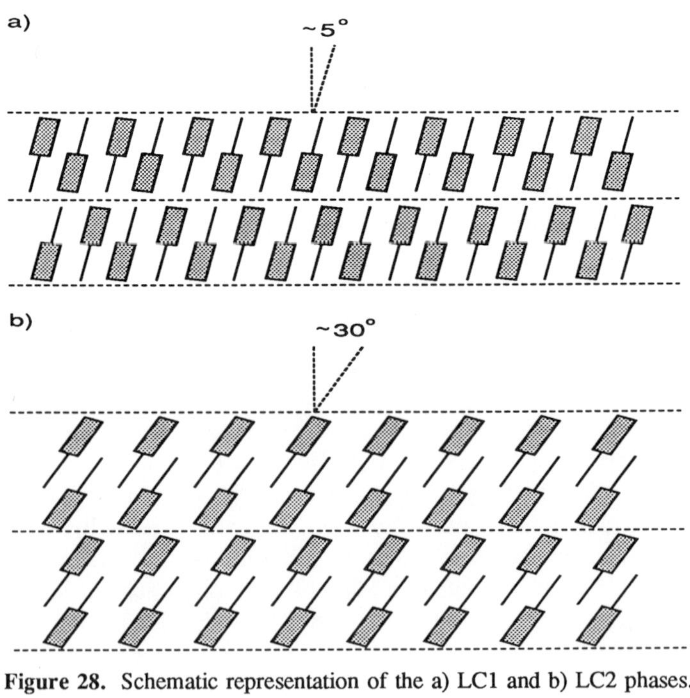

Figure 28. Schematic representation of the a) LC1 and b) LC2 phases.

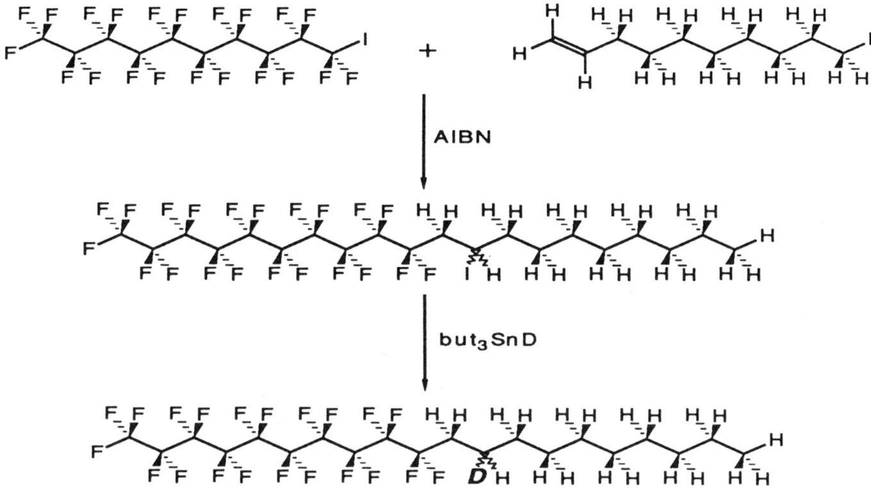

Figure 29. Synthesis of perfluorodecyl decane-d_β.

the spectrum consisted of a rigid Pake pattern with a contribution from motional averaging causing the center of the spectrum to be rounded upward. This is consistent with the low temperature motional behavior of simple alkanes which typically are static below -90 to -100 °C.[43] As the temperature was raised to the LC2 phase (303K, 30°C), a totally asymmetric spectrum was observed (η = 1) with a spectral width of ~122 kHz (base of the pattern). At the temperature corresponding to the LC1 phase (> 323K, 50°C), the spectrum changed to being axially symmetric with a quadrupole splitting ($\Delta\upsilon_{Q\perp}$) of ~51 kHz. This pattern can be accounted for by rapid diffusive motion ($\tau_c \leq 10^{-7}$s) with complete averaging due to rotation about the long axis.[44] At 334K (61°C) the expected isotropic spectrum is observed indicating essentially complete melting of the alkane segments. Spin-lattice T_1 measurements were made in each LC phase. A summary of the deuterium NMR characteristics and T_1 relaxation times is given in Table V.

After heating to the isotopic state the sample was slowly cooled inside the magnet. When the spectrum was acquired again in the LC1 phase at 324K (51°C) (Figure 30, cooling cycle), two superimposed deuterium patterns were observed. The relative amounts of each were found to depend on the cooling procedure. In one example, the outer pair of horns was more intense (Figure 31, 0 degree orientation) while another experiment resulted in the inner pair being more intense (Figure 32, 0 degree orientation). Spectra were acquired for each of these cases in the LC1 phase (sample vial initially at "0" degree orientation) and with the sample vial then rotated 45 and 90 degrees.[45] Figure 31 and Figure 32 show spectra obtained at each orientation. The stability of the oriented phase was shown by the fact that when the sample was cooled to room temperature and removed from the NMR, and allowed to stand for 5 days, the same spectrum was obtained at the same LC1 temperature. Examination of this sample with a hot-stage polarizing microscope showed macroscopic differences compared to material which was cooled from the isotropic state *outside* of the magnet.

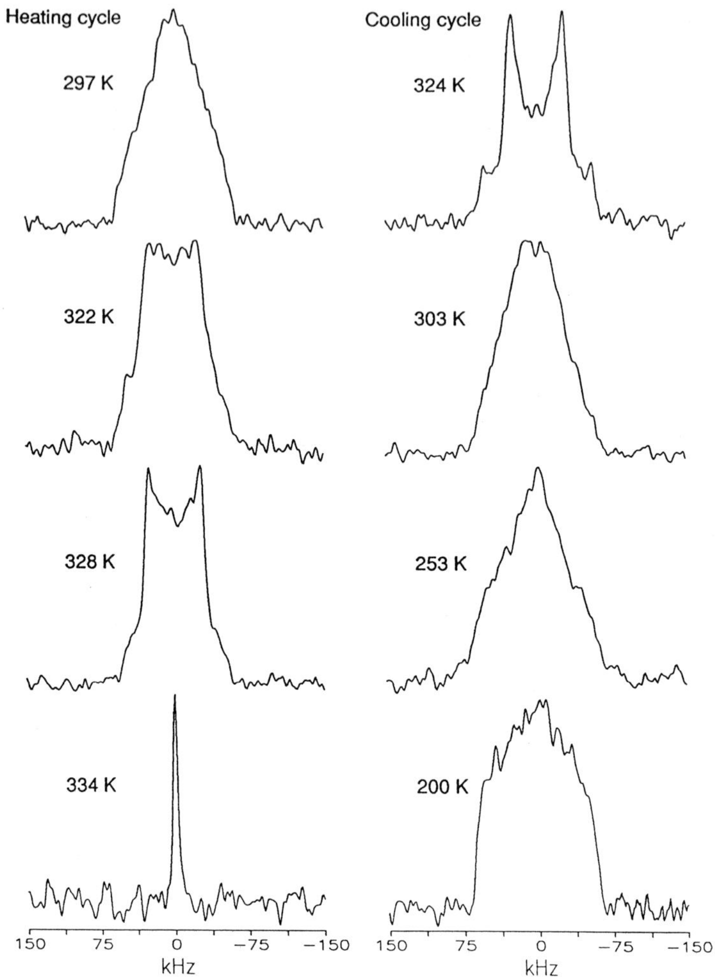

Figure 30. Solid-state deuterium NMR spectra of perfluorodecyl decane-d_β.

Table V. ^{2}H NMR characteristics for H10F10.

temperature	-73°C crystalline	-75°C - 50°C (LC2)	50°C - 61°C (LC1)	> 61°C isotropic
spectral breath	130 kHz	122 kHz	109 kHz	6 kHz*
quadrupole splitting	---	---	51 and/or 113 kHz	---
spin-lattice time	---	0.3 s	0.3 s	---

* peak width at half-height

The NMR probe containing one of the magnetically aligned samples was then taken out of the magnet, the temperature increased to the isotropic state and then slowly brought back to the LC1 state (no magnetic field interaction). The resulting spectrum (328K, 55°C) displayed only a Pake pattern with 51 kHz splitting which was similar to that in Figure 30 at 328K. From these results it is clear that the magnetic field aligned this material as it cooled from the isotropic state.

It was suspected that the two doublet patterns were due to two different regions of the sample, one being more ordered than the other or at a different orientation. To induce greater orientation and determine its direction, sample was spread between stacked glass slides as thin layers of molten material which were then placed in the NMR probe. This type of sample preparation should order the bilayers to give only a single orientation with respect to the glass surfaces. Due to the small amount of sample within the glass slides, the spectra were acquired on a 400 MHz NMR spectrometer at Bruker Instruments. The spectra of the semifluorinated material

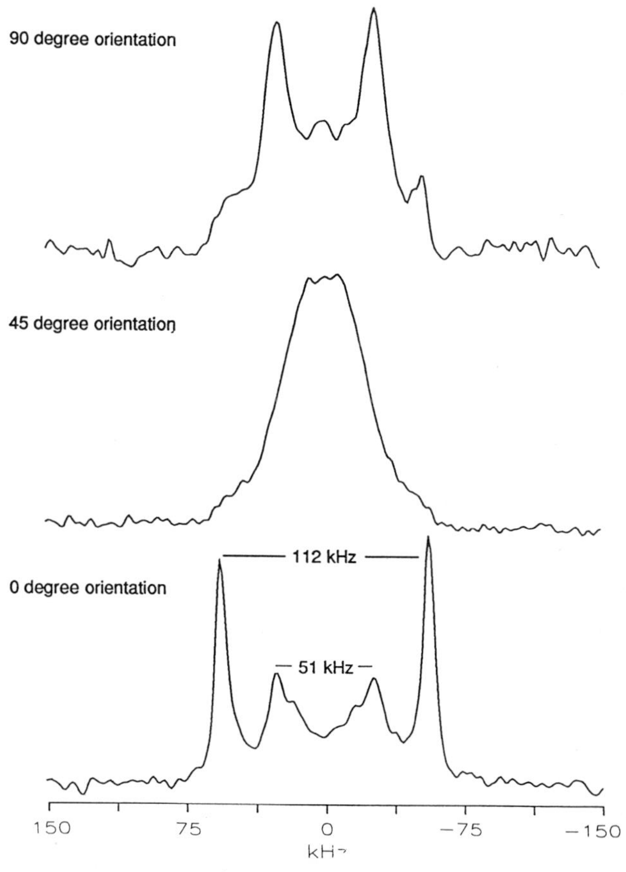

Figure 31. Solid-state ^{2}H NMR spectra of bulk perfluorodecyl decane-d_β in the LC1 phase with outer pattern more intense (cooled from isotropic state).

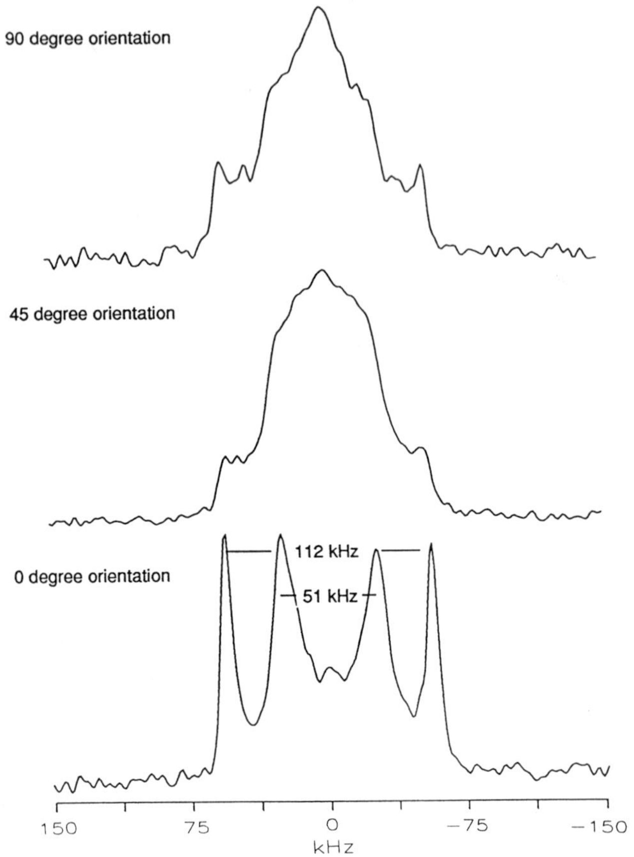

Figure 32. Solid-state deuterium NMR spectra of bulk perfluorodecyl decane-d_β in the LC1 phase when the inner pattern was more intense (after cooling from the isotropic state).

in the LC1 phase at 328K (55°C) clearly show that there was increased ordering of the material with this method (Figure 33). Even in the LC2 phase at 300K (27°C), the spectra (Figure 34) show ordering, although the spectral intensity was less due to peak broadening.

The question arises of how the sample orients in the magnetic field. Orientation is possible when molecules have distinctly different molecular dimensions which can result in diamagnetic anisotropy. Under appropriate conditions, the major axis of neighboring molecules can align as a result of cumulative effects of this weak interaction with the magnetic field. The direction of this orientation defines the director ***n***. Alignment in a magnetic field due to diamagnetic anisotropy is characteristic of the molecular structure and is defined by equation (7).[46]

$$\Delta\chi = (\chi_{\parallel} - \chi_{\perp}) \tag{7}$$

$\chi_{\parallel}$ and $\chi_{\perp}$ are the diamagnetic susceptibilities parallel and perpendicular to ***n***, respectively. For small molecules, the degree of order β (proportional to the diamagnetic susceptibility) is usually small, but for molecules which can arrange parallel to one another in a liquid crystal structure, the degree of order can be substantial as given by equation (8). N is the number of molecules grouped together, H_o is the magnetic field, k_B is Boltzmann's constant and T is the absolute temperature.[47]

$$\beta = \frac{N \Delta\chi H_o^2}{k_B T} \tag{8}$$

Depending on the value of the diamagnetic anisotropy, the sample may align with the director either parallel or perpendicular to the magnetic field. It has been reported that samples containing groups of alkyl chains generally align with the chain axis perpendicular to the magnetic field (negative value of $\Delta\chi$).[48,49] Although no previous reports have been found on the orientation of fluorinated materials (which

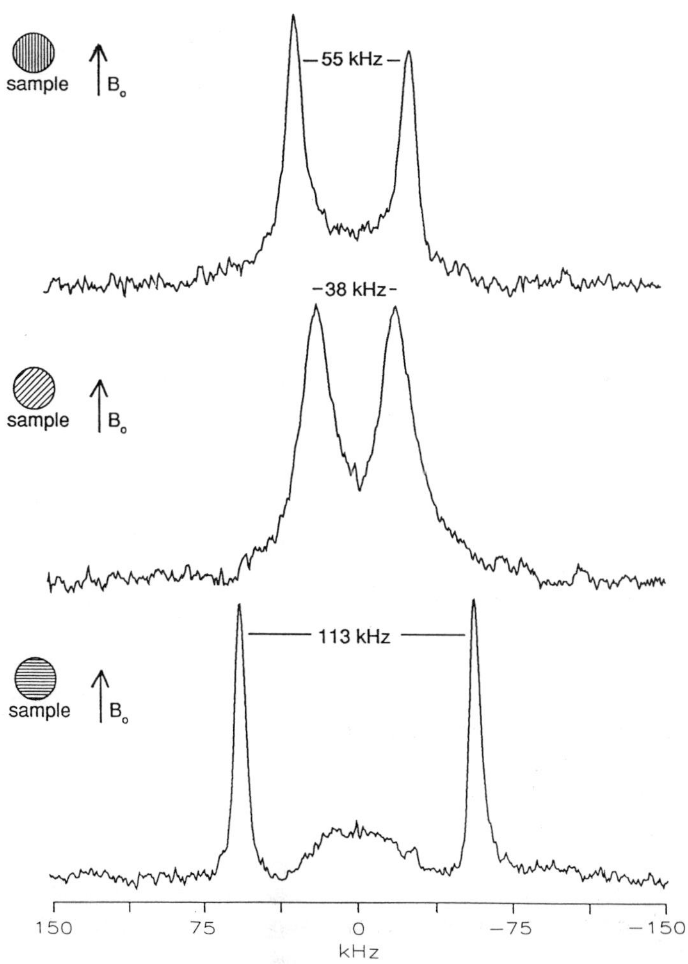

Figure 33. Deuterium NMR spectra of perfluorodecyl decane-d_β in the LC1 phase and oriented between glass slides.

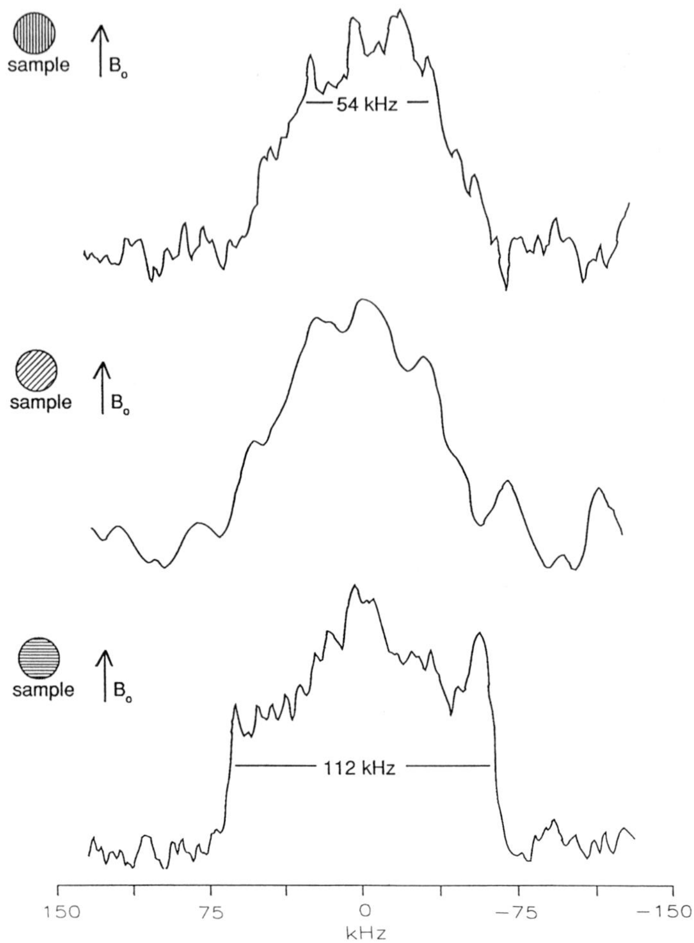

Figure 34. Deuterium NMR spectra of perfluorodecyl decane-$d_β$ in the LC2 phase and oriented between glass slides.

adopt helical conformations), it has been reported that α-helix polypeptides orient with the helix axis parallel to the magnetic field.[50] This would be perpendicular to what is expected based on alkane behavior. The spectrum obtained from the glass slide oriented sample has a maximum peak separation (113 kHz) when the glass slides are perpendicular to the magnetic field. Generally, the maximum doublet separation occurs when the major axis of the electric field gradient (the C-^{2}H bond in the absence of motion) is parallel to the magnetic field. Since the C-^{2}H bond is rotating around the chain axis for this compound, the major axis of the electric field gradient is averaged along the chain axis. This is consistent with the molecular axis being perpendicular to the plane of the glass slides and the lamellar sheets orienting parallel to the glass surface (Figure 35). This is also supported by the argument below.

In the case of ordered samples, the equation defining the deuterium doublet separation also contains an order parameter as given in equation (9).

$$\Delta\nu = 3/2\ \delta\ \left(\frac{3\cos^2\Omega - 1}{2}\right)\left(\frac{3\cos^2\gamma - 1}{2}\right) S_{CD} \qquad (9)$$

δ is the quadrupole coupling constant, Ω is the angle between the director and the magnetic field direction, γ is the angle of the C-^{2}H bond with respect to the molecular axis, and S_{CD} is the order parameter.[51,52] The order parameter is defined in equation (10).

$$S_{CD} = \left(\frac{3\cos^2\theta - 1}{2}\right) \qquad \mathbf{(10)}$$

θ is the angle between the motional axis of symmetry and the bilayer normal director. If Ω is taken to be 0 degrees (director parallel to the magnetic field), and the C-^{2}H bond angle with respect to the molecular chain axis is taken to be 90 degrees, the S_{CD} for this sample is 0.9. This indicates a very high degree of order compared to lipid membranes which show order parameters of up to 0.6. From the observed quadrupole splitting and the above equations, θ is estimated to be 18 degrees. When the sample vial containing the glass slides is rotated 90°, the quadrupole splitting of the resuling spectrum is 55 kHz. Using $\gamma = 90°$ and $\Omega = 90°$, the tilt angle is calculated from this splitting to be 19°. The tilt angle should not change with sample vial rotation since the lamella are oriented by the glass slides. Slight differences can be accounted for by the error involved with visual orientation of the sample vial. This data clearly indicate that the molecules in the magnetically oriented bilayer are "tilted" (Figure 35) in agreement with the model of Viney and coworkers who estimated a 5 - 10 degree tilt based on X-ray diffraction results.

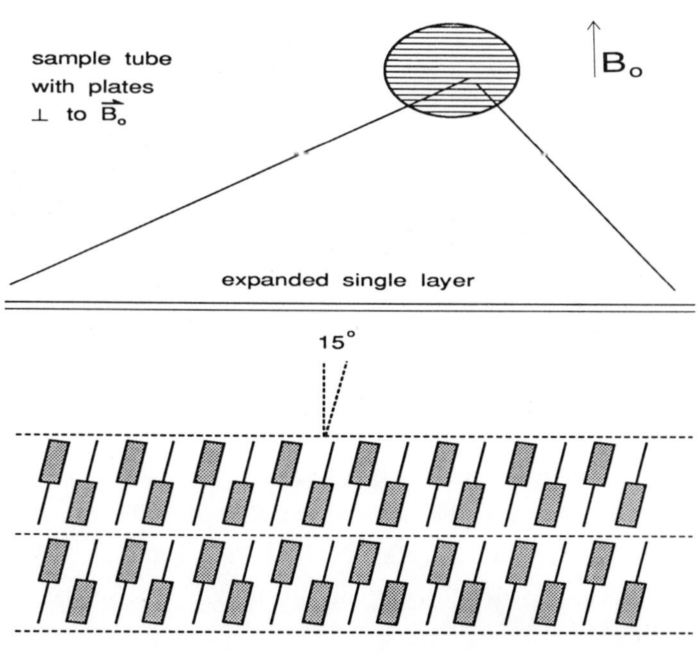

Figure 35. Schematic representation of the LC1 phase **oriented** between glass plates.

19. Conclusions on Semifluorinated Alkane

From the solid-state deuterium NMR lineshape data it can be stated that 1) the LC2 phase has restricted motion in the form of librations of the carbon chain; 2) in the LC1 phase the motion is averaged due to rotation around the long axis of the molecule; 3) the sample **does** align upon cooling in the presence of a magnetic field from the isotropic phase; 4) the lamella spontaneously align perpendicular to the magnetic field; 5) lamella orient between glass slides parallel to the glass surface; and 6) for LC1 lamella, the splitting value is consistent with the molecular axis being tilted at an angle of 18 degrees with respect to the lamellar plane.

20. References

1. Brandolini, A. J. in NMR Spectroscopy Techniques, p. 331, Dybowski, C. and Lichter, R. L. eds. Marcel Dekker, New York, 1987.
2. Schaefer, J. and Stejskal, E. O. *J. Am. Chem. Soc.*, **1976**, *98*, 1031.
3. Armeniades, C. D. and Baer, E. in Introduction to Polymer Science and Technology, 239, Kaufman, H. S., ed., John Wiley and Sons, New York, 1977.
4. Tonelli, A. E. *Macromolecules*, **1973**, *6*, 503.
5. Bailey, R. T.; North, A. M. and Pethrick, R. A. Molecular Motion in High Polymers, Clarendon Press, Oxford, England, 1980.
6. Spiess, H. W. *Colloid Polym. Sci.*, **1983** , *261*, 193.
7. Blum, F. D.; Durairaj, B. and Padmanabhan, A. S. *Macromolecules*, **1984**, *17*, 2837.
8. Jelinski, L. W.; Dumais, J. J.; Cholli, A. L.; Ellis, T. S.; Karasz, F. E. *Macromolecules*, **1985**, *18*, 1091.
9. Landry, C. J. and Henrichs, P. M. *Macromolecules*, **1989**, *22*, 2157.
10. Jelinski, L. W.; Dumais, J. J. and Engel, A. K. *ACS Symp. Ser.*, **1984**, *247*, 55.
11. Jelinski, L. W. in High Resolution NMR Spectroscopy of Synthetic Polymers in Bulk, chapter 10, Richard A. Komoroski, ed., VCH Publishers, New York, 1986.
12. Fyfe, C. A. Solid State NMR for Chemists, CFC Press, Ontario, Canada, 1984.
13. Li, F.; Yue, S. T. and Qiu, Z. W. *J. Macromol. Sci. - Chem.*, **1987**, *A24*, 167.
14. Abragam, A. Principles of Nuclear Magnetism, Oxford Science Publications New York, New York, 1961.
15. Pake, G.E. *J. Chem. Phys.*, **1948**, *16*, 327.
16. Hentschel, R. and Spiess, H. W. *J. Magn. Reson.*, **1979**, *35*, 157.
17. Burum, D.P.; Linder, M. and Ernst, R.R. *J. Magn. Reson.*, **1981**, *43*, 463.
18. Colletti, Ronald F., Ph.D. dissertation, University of Southern Mississippi, Hattiesburg, MS, 1990.
19. Spiess, H. W. *J. Chem. Phys.*, **1980**, *72*, 6755.

20. Schmidt, C.; Wefing, S.; Blümich, B. and Spiess, H. W. *Chem. Phys. Lett.*, **1986**, *130*, 84.
21. Schmidt, C.; Blümich, B. and Spiess, H. *J. Magn. Reson.*, **1988**, *79*, .
22. Mathias, L.J., and Colletti, R.F., "Facile Synthesis of α-Deuterated Acrylates and Activated Vinyls," *Macromolecules*, **1988**, *29*, 857-858.
23. Funchess, R. B. and Blum, F.D. *Polym. Prepr., Am. Chem. Soc. Div. Polym. Chem.*, **1990**, *31*, 544.
24. Lausch, M. and Spiess, H. W. *J. Magn. Reson.*, **1983**, *54*, 466.
25. Inoue, Y.; Nishioka, A. and Chujo, R. *J. Polym. Sci. Polym. Phys. Ed.*, **1973**, *11*, 2237.
26. Asakura, T. and Doi, Y. *Macromolecules*, **1983**, *16*, 786.
27. Torchia, D. and Szabo, A. *J. Magn. Reson.*, **1982**, *49*, 107.
28. Brill, R., *J. Prakt. Chem.*, **1942**, *161*, 49.
29. Arimoto, H.; Ishibashi, M. and Hirai, M. *J. Polym. Sci., Part A,* 1965, **3**, 317.
30. Gianchadani, J.; Spruiell, J. E. and Clark, E. S. *J. Appl. Polym. Sci.,* 1982, **27**, 3527.
31. Mathias, L. J.; Powell, D. G.; Autran, A.-P. and Porter, R. S. *Macromolecules,* 1990, **23**, 963.
32. Mathias, L. J.; Powell, D. G.; Porter, R. S. and Autran, J.-P. *Mater. Sci. Eng.*, 1990, **A126**, 253.
33. Weeding, T.L.; Veeman, W.S.; Gaur, H.A. and Huysmans, W.G.B. *Macromolecules*, 1988, **21**, 2028.
34. Hatfield, G.R.; Glans, J.H. and Hammond, W.B. *Macromolecules,* 1990, **23**, 1654.
35. Hirschinger, J.; Miura, H. and English, A.D., *Polym. Prepr., Am. Chem. Soc. Div. Polym. Chem.*, **1989**, *30(1)*, 312.
36. Schuttenberg, H. and Schulz, R.C., *Angew. Chem. Int. Ed. Engl.*, **1976**, *15(12)*, 777.
37. <u>Polymer Handbook</u>, Brandrup, J. and Immergut, E.H., Wiley Interscience, 1989.
38. Miura, H.; Hirschinger, J. and English, A.D. *Macromolecules*, **1990**, *23*, 2169.
39. Rabolt, J.F.; Russell, T.P. and Twieg, R.J. *Macromolecules*, **1984**, *17*, 2786.
40. Turberg, M.P. and Brady, J.E. *J. Am. Chem. Soc.*, **1988**, *110*, 7797.
41. Viney, C.; Russell, T.P.; Depero, L.E. and Twieg, R. *J. Mol. Cryst. Liq. Cryst.* **1989**, *168*, 63.
42. Höpken, J.; Pugh, C.; Richtering, W. and Möller, M. *Makromol. Chem.*, **1988**, *189*, 911.
43. Sozzani, P.; Behling, R.W.; Schilling, F.C.; Bruckner, S.; Helfand, E.; Bovey, F.A. and Jelinski, L.W. *Macromolecules* **1989**, *22*, 3318.
44. Rabolt, J.F.; Russell, T.P. and Twieg, R.J. *Macromolecules*, **1984**, *17*, 2786.
45. Samples were rotated by hand and visually set to the desired positions.

46. Samulski, E.T. *Polymer*, **1985**, *26*, 177.

47. Maret, G.; Schickfus, M.; Mayer, A. and Dransfeld, K. *Phys. Rev. Lett.*, **1975**, *35*, 397.

48. Luz, Z.; Poupko, R. and Samulski, E.T. *J. Chem. Phys.*, **1981**, *74*, 5825.

49. Maret, G. and Dransfeld, K. *Physica*, **1977**, *86-88B*, 1077.

50. Neugebauer, D.; Blaurock, A.E. and Worchester, D.L. *FEBS Lett.*, **1977**, *78*, 31.

51. Reeves, L.W. and Fujiwara, F.Y. *J. Am. Chem. Soc.*, **1976**, *98*, 6790.

52. Reeves, L.W. and Forrest, B.J. *Chem. Rev.*, **1981**, *81*, 1.

Chapter 3

APPLICATION OF NMR IMAGING TO POLYMERS: A TUTORIAL

JACK L. KOENIG

1. Basis of NMR Imaging

In conventional NMR spectrometers, the liquid or solid sample is placed in a homogeneous magnetic field, B_o, so that all chemically equivalent spins experience the same field and hence precess at the same Larmor angular frequency ω_o given by

$$\omega_o = \gamma B_o,$$

where γ is the magnetogyric ratio. Under these circumstances, the observed proton NMR spectrum of water consists of a single sharp line with an intrinsic width of about 0.1 Hz.

In contrast with conventional NMR spectrometry, NMR imaging involves placing a sample in a deliberately non-uniform magnetic field [1]. The homogeneous magnetic field is modified using a system of gradient magnetic coils which generate linear gradients on the order of a few G/cm (10^{-2} T/m). The purpose of the non-uniform field is to label or encode different regions of the sample linearly with different NMR frequencies. Since the magnetic field is varied in a known manner at specific positions within the sample, the frequency of the NMR signal indicates the spatial position of the resonating nuclei (Fig. 1). that is to say, in one dimension (1D) the position of the sample is related to a frequency by the relationship

---

JACK L. KOENIG - Department of Macromolecular Science, Case Western Reserve University, Cleveland, Ohio 44106.

Solid State NMR of Polymers, Edited by L. Mathias
Plenum Press, New York, 1991

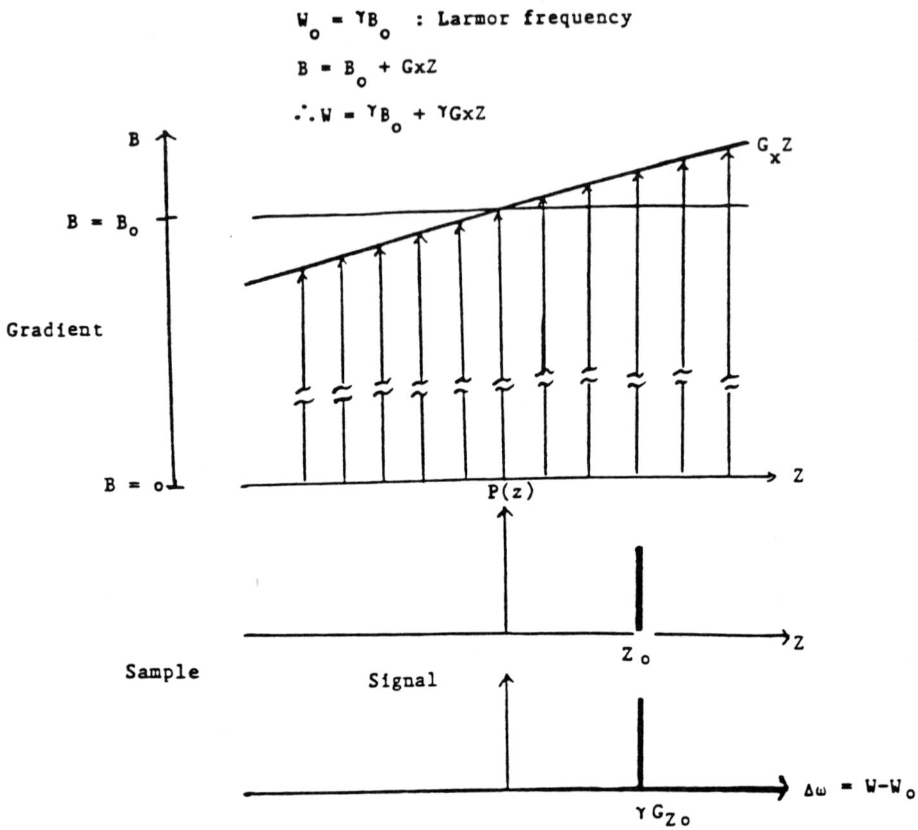

Figure 1. Diagram of the NMR imaging experiment.

$$\Delta\omega_z = \omega_z - \omega_o = \gamma\, G_z\, z,$$

where the magnetic field gradient $G_z = \partial B_z/\partial z$. A tailored rf pulse with a narrow frequency range is used to excite only those nuclei at a corresponding position in the z dimension. The amplitude of the NMR signal received from the z axis line is a measure of the number of resonant nuclei on that line, and so the NMR spectrum represents a graph of spin density versus distance (neglecting relaxation effects). The field gradient is in general described by a tensor with nine components, but for a large B_o we need only be concerned with the three components $G_\alpha = \partial B_\alpha/\partial\alpha$, where α = x, y, z.

In three dimensions, one operates with the sample immersed in a three-dimensional gradient field. The frequency spectrum (still obtained by Fourier transforming the FID) gives the number of resonating spins along a specific direction of the field gradient. In fact, each plane perpendicular to the direction of the field gradient has a different resonance frequency, and the signal intensity at that frequency will be proportional to the number of nuclei contained in that plane. In other words, the frequency spectrum is just a projection of the spin number density (neglecting relaxation effects for the moment) along the field direction. By electronically rotating the field gradient projections at different angles are taken, and from these projections an image is reconstructed using methods analogous to those utilized in x-ray tomography.

In the 2D Fourier imaging method (2DFT), a slice is defined in the XY plane by selective excitation [2]. A field gradient G_y is then applied along the Y-direction for a time t_y, after which a gradient G_x is applied along the X direction for a time t_x. The NMR signal is read as a function t_x to n data points. The process is repeated for a different time t_y, indeed for n different times t_y, yielding an array of n^2 data points. Let the proton density at any point (x,y) in the defined slice by $\rho(x,y)$. Then neglecting relaxation effects, the NMR signal from an element of area dxdy at (x,y) in the slice will be proportional to

$$\rho(x,y) \exp(i\gamma Bt)\, dxdy \tag{1}$$

where B is defined during the time t_x as

$$B = B_o + x\, G_x \tag{2}$$

and during the subsequent time t_y as

$$B = B_o + y\, G_y \tag{3}$$

After substituting (2) and (3) in (1) and demodulating the signal at an angular frequency $\omega_o = \gamma B_o$, the NMR signal from the element is proportional to

$$\rho(x,y) \exp[i\gamma(xG_xt_x + yG_yt_y)]\, dxdy.$$

For the whole slice the NMR signal is

$$S(t_x,t_y) = \iint \rho(x,y) \exp[i\gamma(xG_xt_x + yG_yt_y)]\, dxdy$$

on which a 2D Fourier transform can be performed giving an image of n x n pixels with values $\rho(x,y)$.

The spin-warp variation of 2D Fourier imaging successively increments G_y through n successive values [3].

Two-dimensional (2D) Fourier imaging can be extended to three dimensional imaging. One applies a gradient G_z along z for a time t_z, then G_y along y for a time t_y and finally G_x along x for a time t_x. The NMR FID is recorded to n data points during t_y. The procedure is repeated for n values of t_y and for each of these for n values of t_z. The NMR signal for the whole object is

$$S(t_x,t_y,t_z) = \iiint \rho(x,y,z) \exp[i\gamma(xG_xt_x + yG_yt_y + zG_zt_z)]\, dxdydz.$$

The 3D Fourier transform yields the 3D image $\rho(x,y,z)$.

2. Resolution in NMR Imaging

To obtain a spatially resolved image, the gradient-imposed spread in NMR frequencies, $(\Delta x)(\gamma G)$, must be greater than the natural spread in frequencies involved in the normal (absence of field gradient) linewidth $(\Delta\nu_{1/2})$, i.e.

$$(\Delta x)(\gamma G) > (\Delta\nu_{1/2}) > 1/T_2$$

where (Δx) is the spatial resolution and T_2 is the experimental linewidth. Current imaging hardware is based on biological tissues where T_2s > 30 msec so adequate resolution can be achieved using moderate gradient strengths: a γG of 0.1 kHz/mm.

However, this places a severe limitation on the type of samples which can be imaged with this type of hardware. In particular, we cannot currently image solids because the linewidths are so broad due to the very short T_2s that one cannot generate sufficiently large linear gradients. However, we can investigate the interactions of mobile liquids in solid polymers and derive spatial information about the polymer. We have achieved a spatial resolution of 70 μm and expect as further improvements we increase the signal-to-noise ratio.

3. Preliminary Examples of Images

NMR imaging is a method where the stimulated signal is spatially encoded using a linear magnetic gradient so that an image can be reconstructed showing the distribution of nuclei in the sample. Other than spatially encoding the signal, imaging works on the same principles as standard NMR. Interest in nuclear magnetic imaging applied to medical diagnosis has ballooned since it was discovered that diseased tissue has a different contrast in the magnetic image than normal tissue. The first body-scanning commercial MRI machine was demonstrated in 1981 and hundreds of MRI scanners are not in clinical use in hospitals and research laboratories. For medical purposes, the advantages of NMR imaging (called MRI by medical people in an effort to remove the frightening aspect of the term nuclear from the mind of the patient) are multiple. In particular, the images are not interfered with by bone as with x-ray methods. Consequently, MRI has found considerable use in the examination of the skull. Fig. 2 shows a clinical image of the head of a two-week old infant by the name of Christopher Stanley Koenig (the author's grandson). The lighter portions of the scan are the eyes and mouth. The contusion on the side of the image reflects a

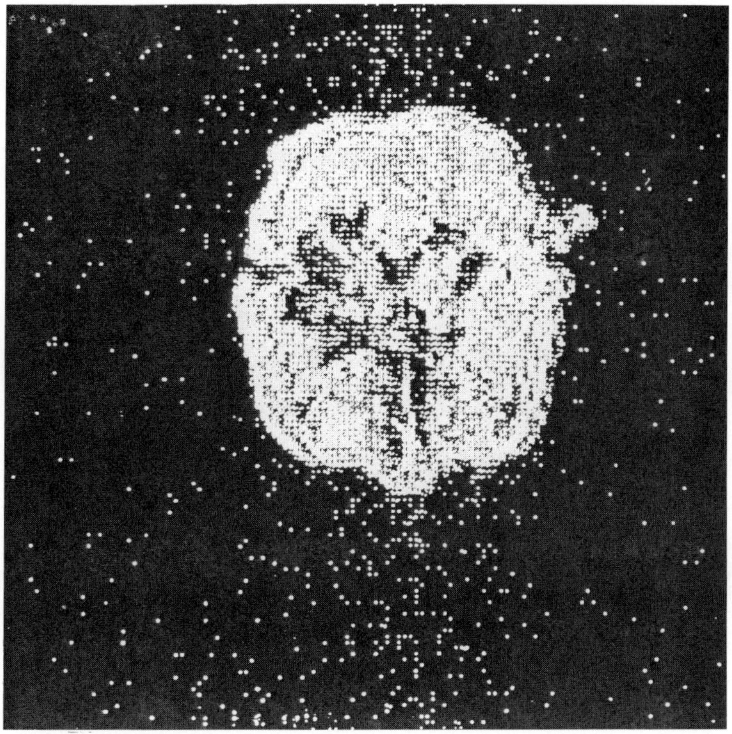

Figure 2. A clinical image of the head of a two-week old infant by the name of Christopher Stanley Koenig (the author's grandson).

"bump" on the baby's head. The MRI was obtained to ascertain whether any internal bleeding had occurred in the baby's head. Fortunately for the baby (and for the peace of mind of the parents and grandparents), no such bleeding was detected.

An NMR image of a cherry tomato is shown in Fig. 3a obtained on our Bruker MSL 300. This image demonstrates the general aspects of NMR imaging. The lighter regions correspond to higher concentrations of water. The outer rigid portion or rind of the tomato has lower water content while the inner softer portion is higher in water content. The dark portions are the seeds and the inner void of the tomato. To demonstrate the utility of NMR imaging, we performed the following experiment on the tomato. The tomato was aged at room temperature in the laboratory and dropped once (by accident). The image (Fig. 3b) was obtained again. The image of the tomato has changed reflecting the changes in the structure of the tomato. In particular the flat portion reflects the effect of the mechanical deformation due to the forces of gravity on the tomato. The tomato has obviously shrunk and become much more juicy reflecting the general observations one would visually make about a rotten tomato.

One final example, in Fig. 4, is the image of the eye of a cow taken on our Bruker MSL 300. The darker portion, indicative of less water, is the lens of the eye

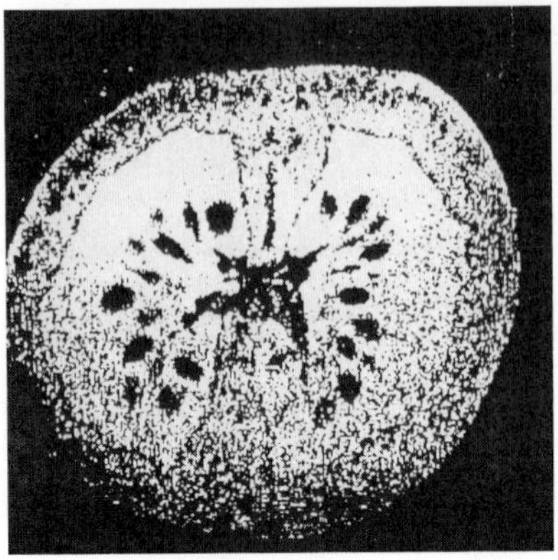

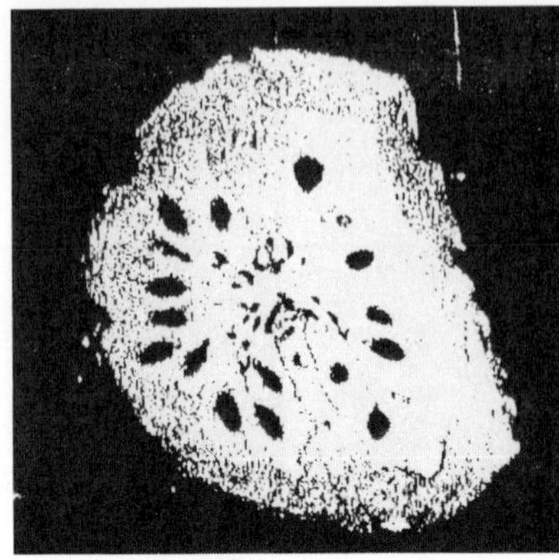

Figure 3. NMR image of a cherry tomato. A) Image of newly ripened tomato. B) Image of aged tomato which had been dropped once.

and the bright portions are the aqueous humor and vitreous humor of the eye. The optical nerve can be observed in the lower portion of the image. The image reflects the water content and its relative mobility.

4. Experimental Considerations for Image Formation

Spin-echo techniques have a unique position in NMR image applications [5]. The problem with NMR imaging is the long data collection time, mainly due to the long spin-lattice relaxation time T_1 and the large number of data points required to

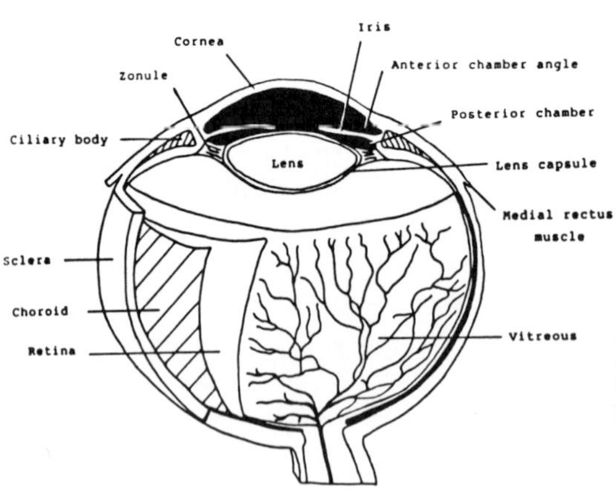

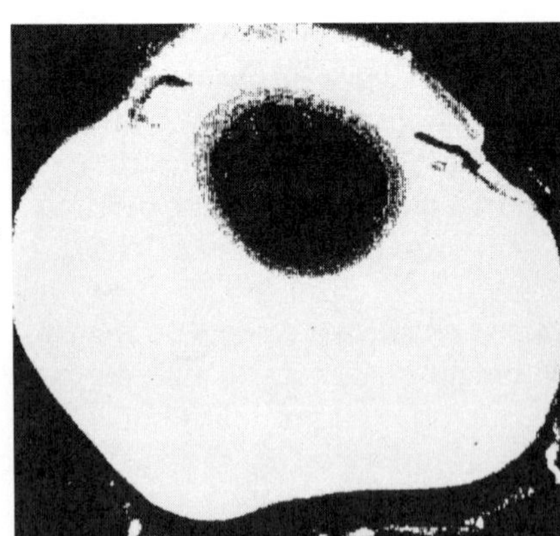

Figure 4. The image of the eye of a cow taken on our Bruker MSL 300.

reconstruct the image. Each measurement necessitates a time period of the order of T_1, which is about 0.5 s for aqueous systems, for the system to return to equilibrium magnetization.

The spin-echo image experiment (Fig. 5) starts with a creation of transverse magnetization with an application of a 90° rf pulse. A magnetic field gradient is switched on in the X direction. After some incremented time $t_{1/2}$ a 180° pulse is applied. The phase of the different isochromates is then inverted. Then the gradient is switched on with an opposite sign and is switched off after another $t_{1/2}$ period. At the end of the evolution period t_1 the different components of the magnetization are dispersed only due to their location. The dispersions due to either field inhomogeneity or chemical shift are refocussed.

When the T_1s and T_2s are different in the voxels of a heterogeneous sample, these difference can be exploited to develop contrast in the NMR images. Because differences in relaxation times determine image contrast, data on relaxation times are important in selection of the optimal rf pulse sequence for the imaging process of the selected sample. Relaxation times can be measured at any point of an image. The ability to accurately quantify relaxation rates is important for understanding and optimizing image contrast. Pure spin density T_1 and T_2 images can be computed from measurements using special pulse sequences. These fundamental images represent the inherent data in the system and can be recombined to reconstitute computed images for a given pulse sequence.

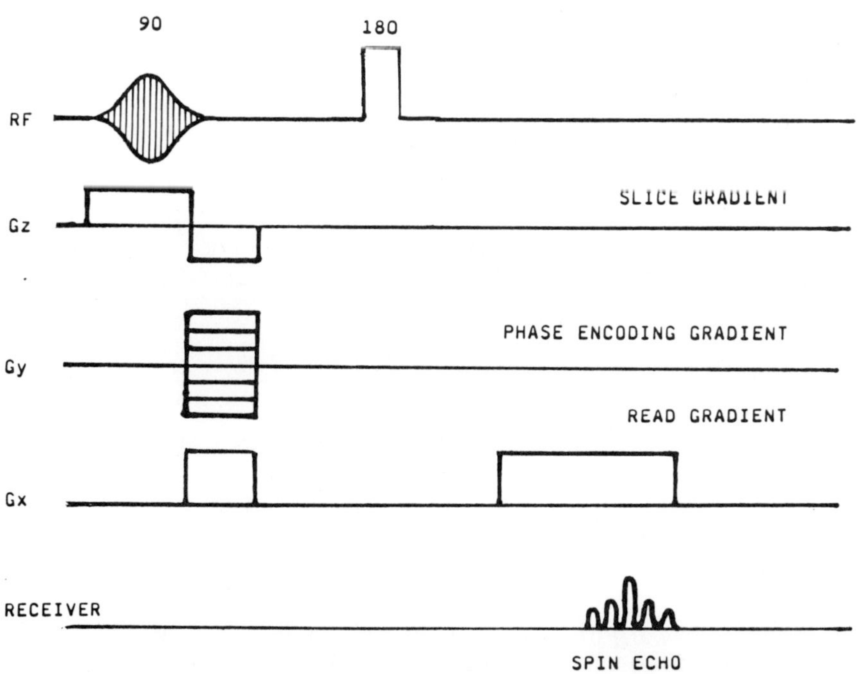

Figure 5. Pulse sequence for the spin-echo image experiment.

An NMR image is characterized by three factors: spatial resolution, object contrast level, and signal-to-noise. For any given resolution and contrast, S/N needs to reach only a level sufficient for confident detection. The signal from each volume element decreases as the resolution is enhanced, leading to an imaging time which is dependent on the sixth power of resolution. Resolution of an image is determined by the minimum volume that contains a sufficient number of nuclei to give a detectable signal. In practice, this is determined by the sample slice thickness, the number of encoding steps, and the number of data points required.

The signal-to-noise improves with higher magnetic field (according to a 7/4 power) and larger spin magnetization which is inversely proportional to the sample temperature.

5. Computer Processing of Images

Currently, most images are interpreted based on the art of visual science, namely examining pictures. For materials imaging this visual method is expensive, slow and often unreliable because of limited statistical sampling and subjective assessment by the analyst. The lack of the required precision and consistency by visual examinations suggests the need for digital analysis of the images and automation of the image analysis process.

For object identification, a variety of algorithms are required including: edge finding, line finding, noise removal, deblurring, classification, imaging matching and shape identification. Initially, the user examines the image and locates regions of interest, and then examines these regions at various levels of higher magnification.

First it is necessary to extract the features present in the image. This can be accomplished by using a very simple feature template. It is translated over the image and rotated about its own center. These templates are sometimes referred to as line-droppers because many of them are linear in their geometry. For example, a template consists simply of two "on" pixels separated by a fixed distance. Varying the distance yields a family of templates.

The simplest way to select objects is to use a threshold or simple boundary detection technique that will yield the outline of the object in a NMR image. The boundary detection algorithm traces the perimeter of the structure (Fig. 6). The resultant boundary is always a closed look composed of (x,y) coordinate pairs on the surface of the object. Boundary extraction is performed automatically on each slice until the entire sample is reduced to a collection of object loops with known spacing (the slice thickness).

Once the boundary information is collected, the loops on the adjacent levels must be connected to form an actual surface. In order to build a surface from a set of aligned boundaries, the loops must be interpolated by connecting adjacent loops

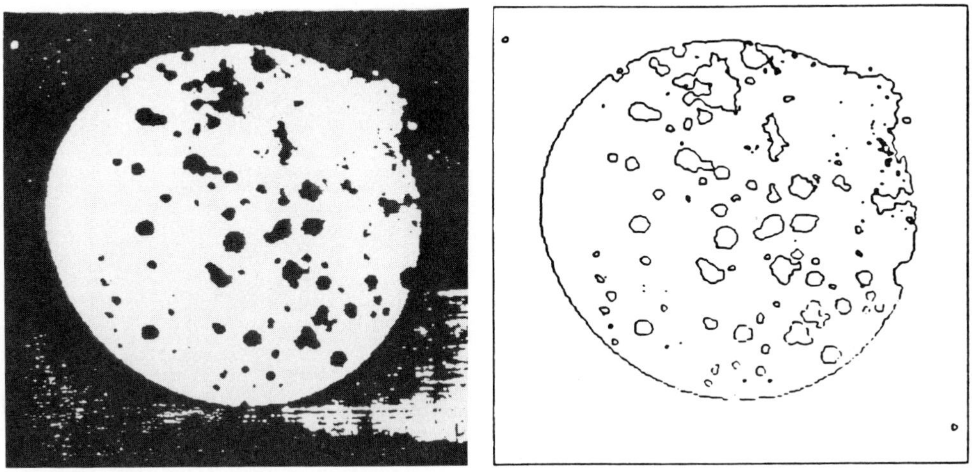

Figure 6. The boundary detection algorithm traces the perimeter of the holes in a polyurethane foam.

with planar pieces so that the resultant surface resembles a mosaic of adjoining polygons. These polygons are commonly called surface tiles or just tiles.

One can now proceed by combining the raw features to form derived features. The class of derived features is obtained by normalizing each of the raw feature measurements by the area of the corresponding image. The resulting features are geometrical probability functions that are translation- and rotation-invariant and relatively size-invariant (assuming the dimensions of the image are large compared to the dimensions of the template). Another class of derived features is made up of various ratios of the morphological measurements.

We send the images to a Microvax II for processing, enhancement and measurement. We have written a number of programs for the Microvax and Fig. 7 shows a listing of data conversion operations currently available to us.

6. Applications of NMR Imaging to Polymers

6.1. Introduction

The NMR imaging technique depends on the differences in the environment of the nuclei, and the environments are influenced by a wide variety of chemical and physical effects. The potential applications of the MRI technique to polymers are very large including:

Adsorption and diffusion of fluids in materials
Detection of internal defects and voids
Characterization of heterogeneous mixtures of different materials

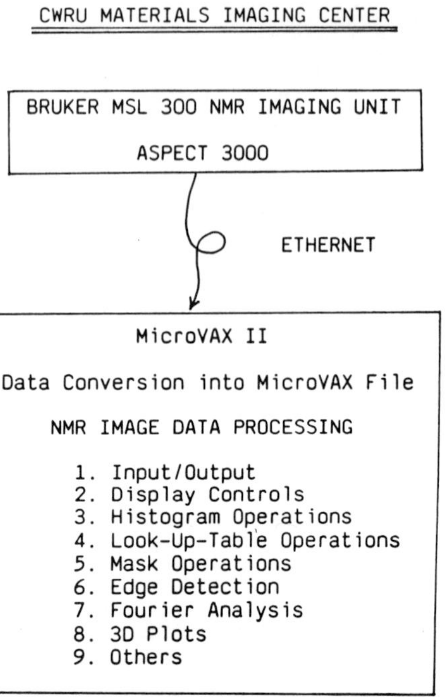

Figure 7. A listing of data conversion operations programmed on the Microvax currently available to us.

Study of molecular interactions between materials
Detection of internal gradients in materials
Determination of spatially distributed structural changes
Composition profiling from surfaces and other sources
Study variations in molecular mobility throughout the sample
Evaluation of homogeneity of mixing processes
Determination of internal flow processes.

6.2. Measurements of Internal Dimensions in Polymeric Systems

Since NMR imaging allows one to obtain the image of an internal slice of a polymeric system, internal dimensions can be measured if they are larger than the resolution of the technique (currently > 50 m at 300 MHz). An example is shown in Fig. 6a for a polyurethane foam filled with water. The light portions of the image represent water and the dark areas arise either from solid form or absence of material. With an edge detection algorithm, one can observe the pore outline as shown in Fig. 2b. Obviously, it is possible to observe the uniformity or lack of uniformity of the pore sizes in the foam. One can count the number of pores and measure their sizes from this computer-processed image. A histogram of the size of pores versus the number of pores having that size can be constructed. Such a histogram can be correlated with the chemical process variables which will lead to an understanding of the mechanism. Such knowledge can lead to better process control and an improved product.

6.3. Diffusion in Polymers

One of the applications of the NMRI technique is the ability to make spatially localized diffusion measurements for systems with unusual shapes, and with several diffusion species. A selected volume can be isolated which may be easily varied in size, shape, and location by varying the three gradient field strengths and/or the irradiation frequencies of the rf pulses, respectively.

Fickian and case II diffusion have been studied in our laboratory using NMR imaging and the results are reported in an accompanying paper [7]. Fickian diffusion of water in nylon has been reported [8] as well as deuterium oxide in cellulose [9].

6.4. Swelling of Polymers

The swelling of polymers with various liquids can be studied using NMR imaging [10]. Solvent absorption and swelling behavior have been used for the purpose of determining the crosslink density in elastomeric systems. The basis of the method is that the higher the crosslink density the less solvent is imbibed in the system and the smaller the degree of swelling. NMR imaging allows one to pursue this idea further by examining the homogeneity of the swelling process, that is, probing the homogeneity and spatial distribution of the crosslinks of the network system. Fig. 8 shows the benzene proton image using a spin-echo pulse sequence of a highly crosslinked sulfur-vulcanized rubber sample which has been swollen in

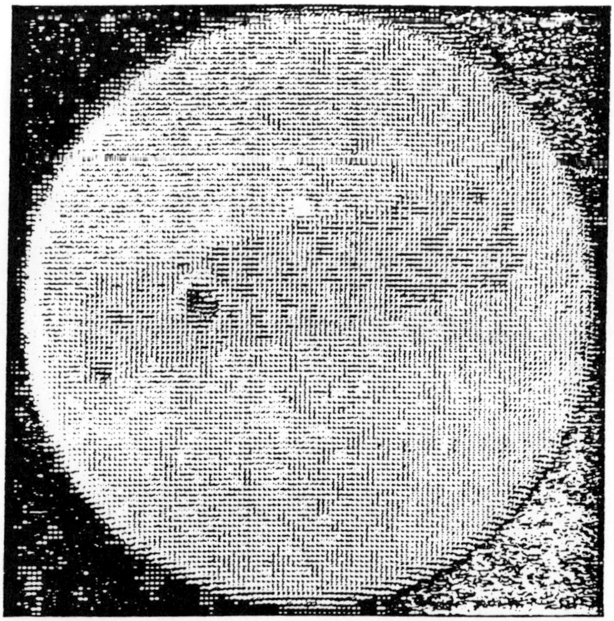

Figure 8. The benzene proton image using a spin-echo pulse sequence of a highly crosslinked sulfur-vulcanized rubber sample which has been swollen in benzene for 2 days. The black spot in the image is an air bubble artifact.

benzene for 2 days. The black spot in the image is an air bubble artifact. Fig. 9 shows a three-level magnified contour plot of a portion of the swollen rubber image. This contour plot indicates that there is a benzene background, an intermediate level of benzene indicating an intermediate level of crosslinking and regions of little benzene which is indicative of a high level of crosslinking. It is obvious that there is considerable inhomogeneity in the crosslinking for this rubber sample. Such inhomogeneities could arise from improper mixing, thermal gradients, or variations in the vulcanization chemistry.

6.5. Detection of Internal Imperfections in Manufactured Polymer Systems

NMR imaging is a means of detecting internal materials imperfections in fabricated articles [11]. Its applications in the field of polymeric materials are numerous and diverse. They include detection and imaging of subsurface defects including interfacial flaws and microcracks as well as detection and characterization of areas modified through the introduction of mobile foreign substances such as additives, degradation products and contaminants.

Due to the sophisticated structures of engineering articles of polymers and the complexity of the process and fabrication procedures, a rapid test is required to assure the quality and integrity of the manufactured article. Defective or damaged areas of

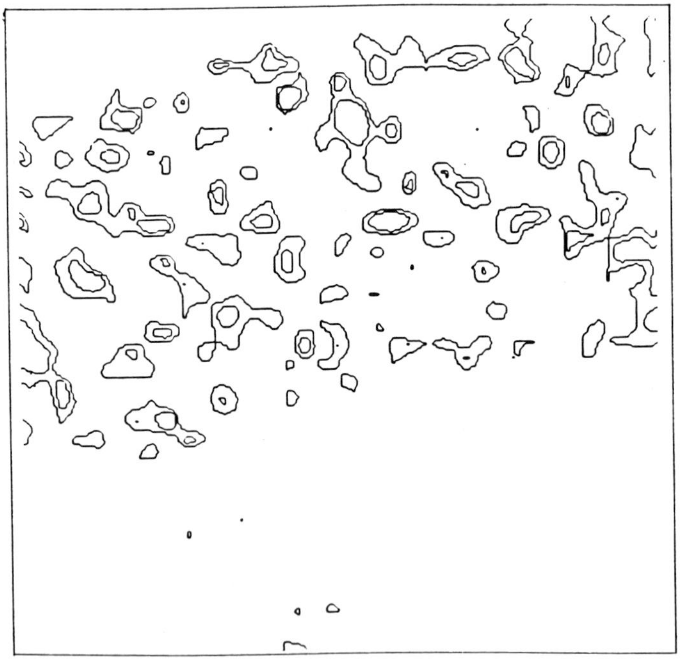

Figure 9. A three-level magnified contour plot of a portion of the swollen rubber image. This contour plot indicates that there is benzene background, an intermediate level of benzene indicating an intermediate level of crosslinking and regions of little benzene which is indicative of a high level of crosslinking.

the polymeric materials can be made to appear in the NMR image by sorbing of a liquid like water. Uniformity of the polymeric materials can be evaluated since improperly manufactured engineering articles would have different NMR images. A materials-acceptance criterion could therefore be written as a function of tolerances in the NMR images. Such an image-based materials-acceptance protocol would assure proper manufacture and performance of polymeric engineering components.

In Fig. 10, we show two images taken through the same rod but 0.5 cm apart. The rod was soaked in water and the light areas in the image represent void areas filled with water. The marker in the upper left hand portion of the image is 1 mm in diameter. Comparison of the size of the voids in the pultruded rod indicates some of the voids approach this magnitude in size. A computer comparison of the

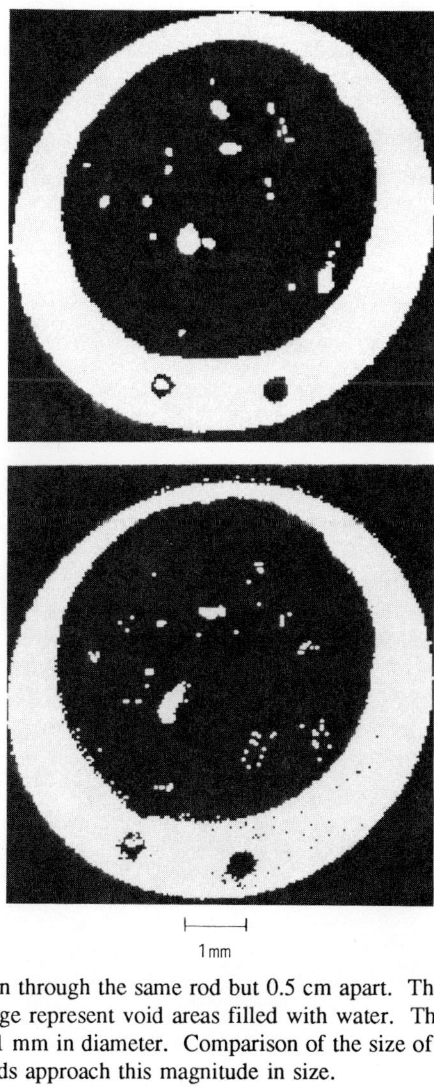

Figure 10. Two images taken through the same rod but 0.5 cm apart. The rod was soaked in water and the light areas in the image represent void areas filled with water. The marker in the upper left hand portion of the image is 1 mm in diameter. Comparison of the size of the voids in the pultruded rod indicates some of the voids approach this magnitude in size.

corresponding edge-enhanced images (Fig. 11) shows that some of the voids are tubular shaped, that is, a channel-like void region exists over the length dimension of 0.5 cm. Such tubular voids could result naturally from the nature of the pultrusion process through air entrapment or insufficient flow of the matrix material during the pull through of the fibers.

6.6. Detection of Bonding in Adhesive Systems

The adhesive in the bond can be detected in a water-based polyvinyl acetate-emulsion type of adhesive (e.g. Elmer's Glue, Borden Company). In Fig. 12, the image is shown for the adhesive (dried for 1.5 to 2.5 hrs) in a wood rod which contained a circular indentation which was filled with the adhesive. The image shows a lower level of brightness at the bonded interface and the circular indentation reflecting the presence of the water. As the curing proceeds, the second image shows a loss of the water at the bonded interface indicating curing of the adhesive during the drying/curing process. The final image shows a dark portion in the center of the circular indentation reflecting the presence of a large void due to the fact that the adhesive was initially only 60% solids; therefore, with drying, the space is no longer fully occupied and a large central void forms. From such studies, one should be able to detect the formation of voids during the curing process and determine the quality of the adhesive bonds.

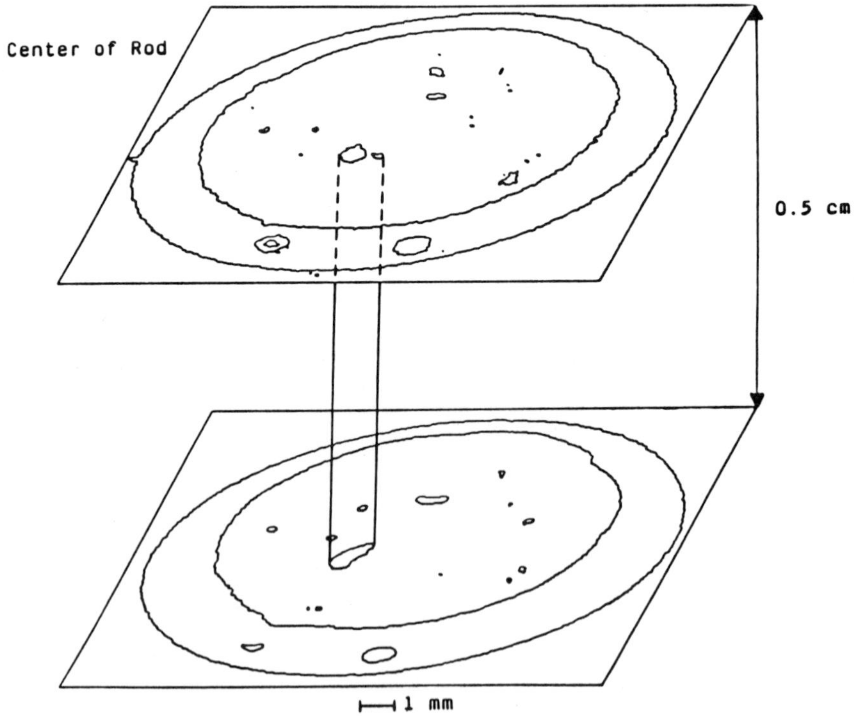

Figure 11. A computer enhanced image which shows that some of the voids are tubular shaped, that is, a channel-like void region exists over the length dimension of 0.5 cm.

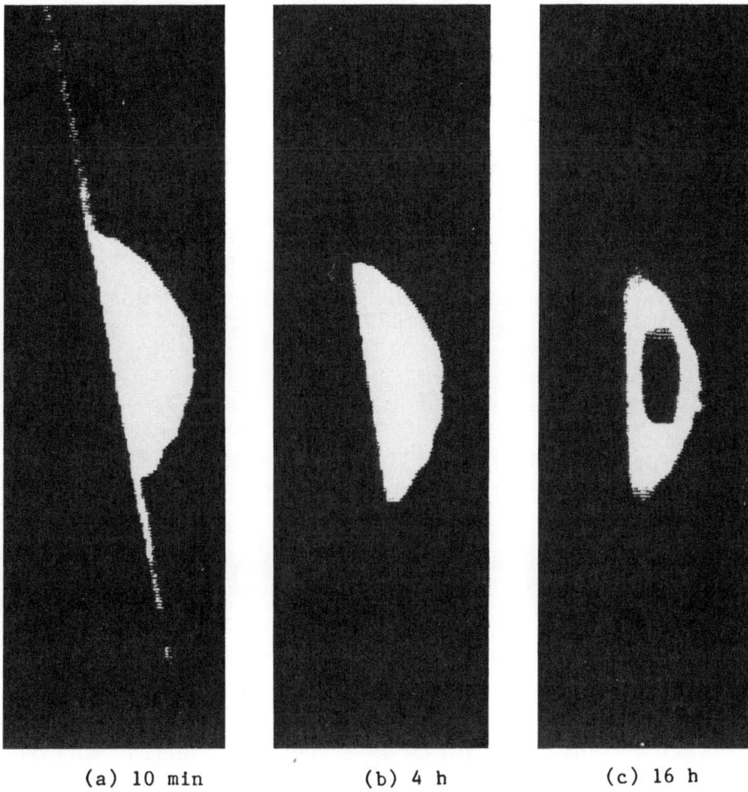

Figure 12.

7. Future Developments in NMR Imaging

As might be expected, much of the driving force for improvements in NMR imaging is coming from the medical field. In the medical field, one of the interests is to obtain rapid images such as images between heart beats. Consequently, one can logically expect that the technology for obtaining rapid images will develop and be transmitted to the more traditional NMR instruments such as found in polymer research laboratories. NMR imaging can then be used to study dynamic processes including flow [12].

Soon, we can expect to obtain simultaneous multislice images for three dimensional reconstruction and to decrease the data acquisition times. Multiecho images for determination of different levels of mobility will be a routine part of the imaging process.

One of the abiding interests in NMR imaging is the possibility of imaging different species in the same sample, that is, observing chemical shift imaging. The application of various spatially resolved NMR techniques for the observation of high resolution spectra has had limited success [13-18]. This is largely due to the mutually exclusive requirements of the highly homogeneous magnetic field which is necessary for the observation of chemical shift information, and the inhomogeneous field, usually applied as a linear magnetic field gradient, which is necessary to obtain spatially resolved data. Chemical-shift imaging techniques use pulsed magnetic field gradients which in the standard configuration of superconducting magnets generate sufficiently large eddy currents upon gradient removal to temporarily degrade the field homogeneity. This is one of the reasons why the implementation of high-resolution spectroscopy is difficult. The field homogeneity is recovered after a time delay of approximately 10 msec, so high resolution techniques utilize a spin echo. Consequently the experimental time is long. But progress is being made in this area and it is certain that chemical shift imaging will soon become a practical alternative [19-23].

For polymers, there is interest in the imaging of solids since one would be interested in the spatial distribution of chemical structure in solids. In a strongly protonated solid, where molecular motion is restricted, a typical value of the local dipolar field might be 5 G so that a gradient greater than 50 G/cm (0.5 T/m) would be needed to achieve a resolution of 1 mm. Presently, magnetic linear gradients of these magnitudes are not available. One approach is to use line-narrowing techniques to the solids prior to the image so the required gradients are smaller [24-27]. One method is to use magic angle spinning [28]. An alternate approach is to reduce the effective local field using a multiple-pulse line-narrowing sequence [29]. Recently, solids imaging using a surface coil has been reported [30]. All of these techniques require special skills in the experimental aspects of NMR imaging and will not find wide spread application.

The future of NMR imaging appears to be limited primarily by the cost of the equipment necessary for the experiment. With time and additional competition, the cost will decrease and NMR imaging will become as common to the polymer research laboratory as high resolution solution and solid-state NMR are today.

8. Summary

It is apparent from the few examples sited from the efforts in our laboratory that NMR imaging has a broad application to the study of polymers under a variety of circumstances. Coupled with solid-state and high resolution NMR, NMR imaging becomes an even more powerful tool. It is one of the few tools which allows a determination of the spatial distribution of structure, albeit presently limited to using liquids as the structure probe.

We transmit the images to a Microvax II for processing. Samples of external diameter of 60 mm can be imaged and sampling technique consists of placing the sample in the image probe. No other pretreatment is involved if the sample is the appropriate initial size.

The selection of the proper pulse sequences for a particular materials sample is critical to optimize image contrast and requires a number of preliminary tests. We are not generating proton density images as well as T_1 and T_2 modulated images of a variety of samples which reveal the structural distribution and the relative mobility of different regions of the sample. Many of the problems encountered in the polymer research laboratory are amenable to analysis using NMR imaging and this new technique should be considered.

9. General References

1. Smith, S. L., "Nuclear Magnetic Resonance Imaging," *Anal. Chem.*, *57*, 595A (1985).
2. Foster, M. A. and Hutchinson, J. M. S., "NMR Imaging: Method and Applications," *J. Biomed Eng.*, *7*, 171-82 (1985).
3. Andrew, E. R., "NMR Imaging," *Acc. Chem. Res.*, *16*, 114 (1983).
4. Manified, P. and Morris, P. G., *NMR Imaging in Biomedicine*, Academic Press, New York, 1982.
5. Hinshaw, W. S. and Lent, A. H., "An Introduction to NMR Imaging: From the Block Equation to the Imaging Equation," *Proc. IEEE*, *71*, 338 (1983).

10. References

1. (a) Mansfield, P. and Morris, P. G., *NMR Imaging in Biomedicine*, Academic Press, 1982.
   (b) Morris, P. G., *Nuclear Magnetic Resonance Imaging in Medicine and Biology*, Clarendon Press, Oxford, 1986.
2. Kumar, A.; Welti, D. and Ernst, R. R., *J. Mag. Res.*, *18*, 69 (1975).
3. Edelstein, W.; Hutchison, J. M. S.; Johnson, G. and Redpath, T., "Spin Warp NMR Imaging and Applications to Human Whole-Body Imaging," *Phys. Med. Biol.*, *25*, 751 (1980).
4. King, K. F. and Moron, P. R., "A Unified Description of NMR Imaging, Data-collection Strategies and Reconstruction," *Med. Phys.*, *11*, 1, 1984.
5. Frahm, J. and Hanicke, W., "Comparative Study of Pulse Sequences for Selective Excitation in NMR Imaging," *J. Magn. Res.*, *60*, 320 (1984).
6. Bates, R. H. T. and McDonnell, M. J., *Image Restoration and Reconstruction*, Clarendon Press, Oxford, 1986.
7. Weisenberger, L. A. and Koenig, J. L., "NMR and NMR Imaging Investigations of Case II Diffusion in Polymers," *Polymer Preprints*, *29*, 98 (1988).
8. Blackband, S. and Mansfield, P., "Diffusion in Liquid-solid Systems by NMR Imaging," *Solid State Phys.*, *19*, L49 (1986)

9. Pickup, S. B. and Blum, F. D., "Mutual Diffusion Coefficients of Deuterium Oxide in Cellulose Solutions Determined by Small Scale NMR Imaging," *Polymer Reprints*, *29*, 100 (1988).
10. Rothwell, W. P. and Gentempo, P. P., "Nonmedical Applications of NMR Imaging," *Bruker Reports*, *1*, 46 (1985).
11. Rothwell, W. P.; Holecek, D. R. and Kershaw, J. A., "NMR Imaging: Study of Fluid Absorption of Polymer Composites." *J. Polym. Sci. Polym. Lett. Ed.*, *22*, 241 (1984).
12. Lee, J. N.; Riederer, S. J.; Bobman, S. A.; Farzaneh, F. and Wang, H. Z., "Instrumentation for Rapid MR Image Synthesis," *Magn. Reson. in Med.*, *3*, 33, 1986.
13. Granot, J., "Selected Volume Spectroscopy (SVS) and Chemical Shift Imaging. A Comparison," *J. Mag. Res.*, *66*, 197 (1986).
14. Ordidge, J., Connely, A. and Lohman, J. A. B., "Image-Selected in vivo Spectroscopy (ISIS). A New Technique for Spatially Selective NMR Spectroscopy," *J. Magn. Res.*, *66*, 283 (1986),
15. Mansfield, P., "Spatial Mapping of the Chemical Shift in NMR," *Magn. Res. in Medicine*, *1*, 370-86 (1984).
16. Bottomley, P. A.; Foster, T. B. and Darrow, R. D., "Depth Resolved Surface Coil Spectroscopy (DRESS) for in vivo ^{1}H, ^{31}P and ^{13}C NMR," *J. Magn. Res.*, *59*, 338-42 (1984),
17. Maudsley, A. A.; Hilal, S. K.; Perman, W. H. and Simon, H. E., "Spatially Resolved High Resolution Spectroscopy by 'Four Dimensional' NMR," *J. Magn. Reson.*, *51*, 147 (1983).
18. Pykett, I. I. and Rosen, B. R., "NMR: In Vivo Proton Chemical Shift Imaging," *Radiology*, *149*, 197 (1983).
19. Volk, A.; Tiffon, B.; Mispelter, J. and Lhoste, J. -M., , "Chemical Shift-Specific Slice Selection. A New Method of Chemical Shift Imaging at High Magnetic Field," *J. Magn. Reson.*, *71*, 168, 1987.
20. Hall, L. D. and Rajanayagam, V., "Thin-Slice, Chemical-Shift Imaging of Oil and Water in Sandstone Rock at 80 MHz.." *J. Magn. Reson.*, *74*, 139, 1987.
21. Blackband, S.; Mansfield, P.; Barnes, J. R.; Clague, A. D. H. and Rice, S. A., *Soc. Pet. Eng. Form. Evaln.*, *1*, 31, 1986.
22. Hall, L. D.; Rajanayagam, V. and Hall, C., "Chemical-Shift Imaging of Water and n-Dodcecane in Sedimentary Rocks." *J. Magn. Reson.*, *68*, 185, 1986.
23. Hall, L. D. and Sukumar, S., "Three-Dimensional Fourier Transform NMR Imaging. High Resolution Chemical -Shift-Resolved Planar Imaging," *J. Magn. Reson.*, *56*, 314, 1984.
24. Chingas, G. C.; Miller, J. B. and Garroway, A. N., "NMR Images of Solids." *J. Magn. Res.*, *66*, 530 (1986).
25. Garroway, A. N.; Baum, J.; Munowitz, M. G. and Pines, A., "NMR Imaging in Solids by Multiple Quantum Resonance," *J. Magn. Res.*, *60*, 337 (1984).
26. Szeverenyi, N. M. and Maciel, G. E., "NMR Spin Imaging of

Magnetically Dilute Nuclei in the Solid State," *J. Magn. Res.*, *60*, 397 (1984).

27. McDonald, P. J.; Attard, J. J. and Taylor, D. G., "A New Approach to the NMR Imaging of Solids," *J. Magn. Res.*, *72*, 224 (1987).
28. Cory, D. G.; Reichwein, A. M.; van Os, J. W. M. and Veeman, W. S., "NMR Imaging of Rigid Solids," *Polymer Reprints*, *29*, 92 (1988).
29. Miller, J. B. and Garroway, A. N., "Line-Narrowing Techniques for NMR Imaging of Solids," *Polymer Reprints*, *29*, 90 (1988).
30. Miller, J. B. and Garroway, A. N., "NMR Imaging of Solids with a Surface Coil," *J. Magn. Res.*, *77*, 187 (1988).

Chapter 4

SOLID STATE ^{13}C NMR STUDIES OF THE STRUCTURES, CONFORMATIONS, AND DYNAMICS OF SEMI-CRYSTALLINE POLYMERS

A. E. TONELLI, M. A. GOMEZ, HAJIME TANAKA AND M. H. COZINE

1. Introduction

It has long been established [1] that the ^{13}C chemical shifts of polymers observed in high-resolution ^{13}C NMR spectra of their solutions are sensitive to their microstructures, i.e., stereoregularity, comonomer sequence, and defect structures. More recently [2,3] it has been demonstrated that the microstructural sensitivity of polymer ^{13}C chemical shifts has its origin in the local polymer chain conformation. Microstructural differences produce changes in the average local polymer chain conformation which in turn are manifested as different ^{13}C chemical shifts for the carbon atoms in the vicinity of each unique microstructure.

The γ-gauche effect [2], as illustrated in Fig. 1, successfully accounts for the microstructurally dependent ^{13}C chemical shifts exhibited by polymers in their high-resolution solution spectra. There are also several examples from the high-resolution spectra recorded on polymers in the solid state [4-10] that indicate the γ-gauche effect also importantly influences the ^{13}C chemical shifts of solid polymers.

Several authors [6,11,12] have reported that the amorphous carbons in semicrystalline polyethylene (PE) resonate 2-3 ppm upfield from the crystalline carbons. This observation is expected, because the crystalline carbons reside in the all-trans, planar zigzag conformation (no γ-gauche shielding), while the C-C bonds in

---

A. E. TONELLI, M. A. GOMEZ, HAJIME TANAKA, AND M. H. COZINE - AT&T Bell Laboratories, 600 Mountain Avenue, Murray Hill, NJ 07974.

Solid State NMR of Polymers, Edited by L. Mathias
Plenum Press, New York, 1991

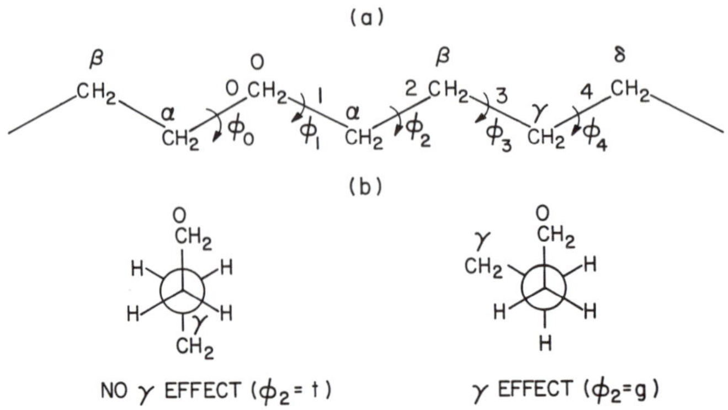

Figure 1. (a) Portion of a paraffinic hydrocarbon chain in the all-trans, planar zigzag conformation. (b) Newman projections along bond 2 in (a) illustrating the γ-gauche shielding effect.

the amorphous portions of PE possess some gauche character and therefore experience γ-gauche shielding (see Fig. 1).

Harris et al. [7] have observed the methylene carbon resonance in crystalline syndiotactic polypropylene (s-PP) to be a doublet split by 8.7 ppm, while for isotactic polypropylene (i-PP) the methylene carbon resonance is a singlet resonating midway between the s-PP methylene doublet. s-PP crystallizes [13] in the -TTGG- conformation where half the methylene carbons experience two γ-gauche effects, while the remaining half experience no γ-gauche interactions. i-PP crystallizes [14] in the -TGTG- conformation, where every methylene carbon experiences one γ-gauche shielding effect. As is observed, we would expect the methylene carbon resonances in crystalline s-PP to be split by 8-10 ppm and the methylene carbons of crystalline i-PP to resonate midway between them.

Poly(1-butene) crystallizes in three distinct helical conformations [15-22] characterized by 3/1, 11/3, and 4/1 monomer units/helical turn, and each backbone approximates the -TGTG- conformation. In passing from the 3/1 to the 4/1 crystal structure the backbone rotation angles open up from perfectly staggered positions (60° dihedral angles) to significantly nonstaggered values (85°). This results in a deshielding of ^{13}C resonances in the 4/1 form relative to the 3/1 form due to the reduction in the magnitude of γ-gauche interactions.

Aside from the three specific examples just discussed, several other reports [5,23,24] on the solid-state ^{13}C chemical shifts observed in crystalline polymers can and have been analyzed in terms of γ-gauche shielding effects. In addition, the motional characteristics of solid polymers can be studied by high resolution, solid state ^{13}C NMR. Through observation of spectral relaxation parameters, such as the spin-lattice relaxation time (T_1), the motions of polymer chains in one or more crystalline phases can be compared to the mobility of the amorphous chains or to each other.

The techniques of cross-polarization (CP), high power proton dipolar decoupling (DD), and rapid magic angle sample spinning (MAS) were applied at various temperatures to achieve high resolution ^{13}C NMR spectra from solid samples of several semi-crystalline polymers. Results were obtained for poly(diethyloxetane) (PDEO), poly(butylene terephthalate) (PBT), and isotactic polypropylene (i-PP). For each of these polymers, two or more crystalline polymorphs were studied to learn about the conformations, interchain packings, and dynamics of the crystalline chains in each polymorph.

2. Methods

^{13}C NMR spectra were recorded on a Varian XL-200 spectrometer operating at a static field of 4.7T. Variable-temperature MAS was achieved with a Doty Scientific Probe, which utilizes a double air bearing design. Aluminum oxide rotors with Kel-F end caps were routinely spun at 2-4 kHz. A 45 kHz rf field strength was used for decoupling, with a decoupling period of 200 ms. The optimal CP contact time of 2000 s was employed. No attempt was made to record the absolute chemical shifts of the resonances observed for PDEO and i-PP. Instead spectra were recorded consecutively on the same day without adjustment of the magnetic field and were then compared to obtain relative chemical shifts. Chemical shifts observed for PBT were compared to the resonance of poly(oxymethylene) (POM) which was placed in the rotor. Under CP conditions the POM resonance is expected [25] at 89.1 ppm downfield from TMS. Spin-lattice relaxation times, T_1, were measured for each carbon under the CP condition using the pulse sequence developed by Torchia [26] and without CP by the usual inversion-recovery method [27].

3. Results and Discussion

3.1. PDEO [28]

X-ray diffraction and calorimetric studies [29-31] of PDEO have demonstrated the existence of two distinct crystalline polymorphs produced by altering the conditions for crystal growth. Form I melts at 73°C and consists of PDEO chains in the all-trans T_4 conformation, while the T_2G_2 conformation is assumed by the PDEO chains in the form II crystals melting at 57°C.

```
              CH3-CH2
                  |
PDEO (-CH2-C-CH2-O-)
                  |
              CH3-CH2
```

The principal purpose of the present study is the correlation of high-resolution solid-state ^{13}C NMR spectra with the two different crystalline chain conformations adopted by PDEO in the form I and II crystals.

The PDEO employed in this work was obtained by cationic ring-opening polymerization of the corresponding monomer as described elsewhere [29]. Two fractions of the bulk polymer with number-average molecular weights (M_n) 8 x 10^5 and 5 x 10^4, as measured by osmometry, were studied.

The samples were melted and crystallized inside the NMR rotors in thermostatic baths set at predetermined temperatures for periods of time sufficient to assure complete crystallization as indicated by previous dilatometric studies [31]. PDEO in form I was obtained by crystallizing from the melt at 60°C for 2 days and in form II through crystallization at 0°C for 15 h. Crystallization at 35°C (for the low molecular weight sample) and at 20°C (for the high molecular weight sample) yielded comparable amounts of both crystalline forms in the same sample.

Small portions of the PDEO samples were removed before and after performing the NMR experiments, and X-ray diffraction and DSC measurements were conducted to verify the crystalline form obtained. Fig. 2 presents the CPMAS spectra

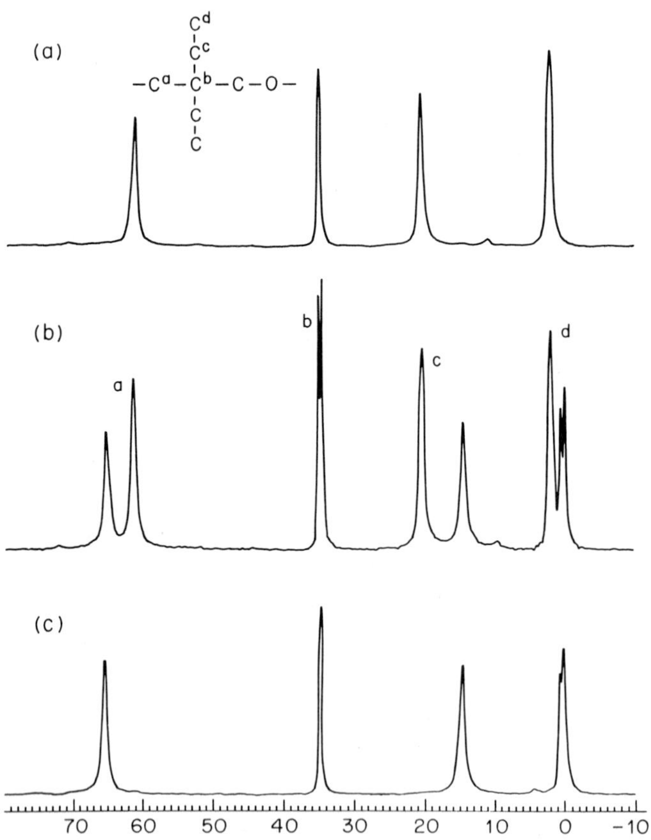

Figure 2. CPMAS spectra of PDEO (M_n=50000) crystallized at three different temperatures: (a) form I (60°C); (b) forms I and II (35°C); (c) form II (0°C). All three spectra were recorded at room temperature with no reference employed.

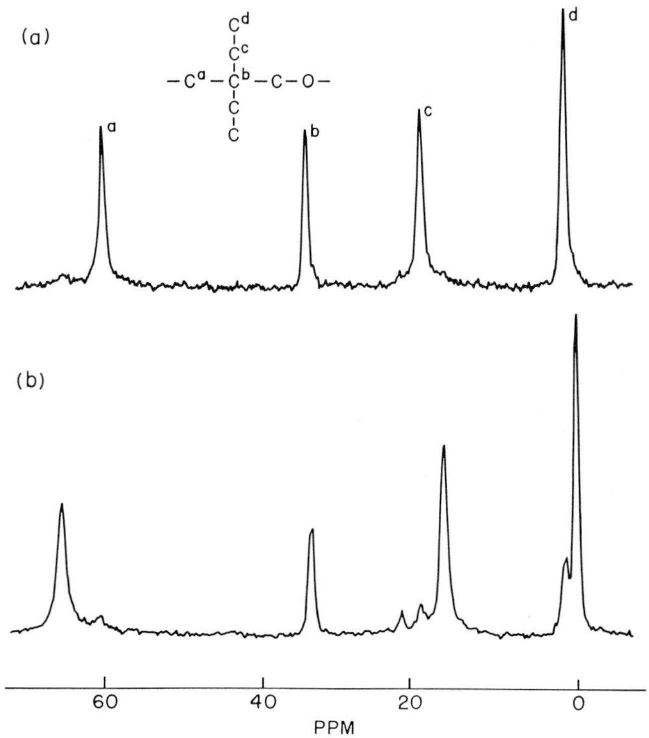

Figure 3. ^{13}C NMR spectra at 60°C of PDEO (M_n=800000) containing form I crystals obtained (a) with cross polarization (crystalline resonances) and (b) without cross polarization (amorphous resonances). No reference was employed.

Table I. ^{13}C NMR CHEMICAL SHIFTS (ppm)[t] FOR PDEO IN THE SOLID STATE

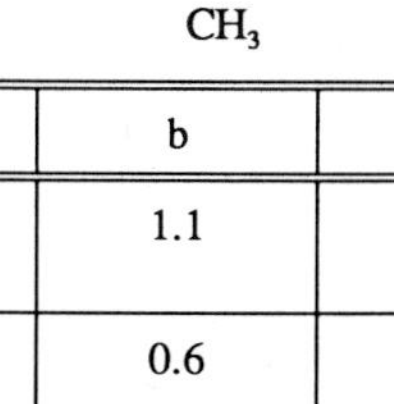

PDEO	a	b	c	d
form I (T_4, T_m= 73°C)	-5.1	1.1	3.1	1.7
form II (T_4G_2,T_m=57°C)	-1.2	0.6	-2.8	-0.3
amorphous	0.0	0.0	0.0	0.0

[t] ^{13}C NMR chemical shifts observed at 25°C in PDEO (M_n = 5 x 10^4) containing both form I and II crystals (see Fig. 2) and referenced to the resonances observed for the amorphous PDEO carbons of each type.

Table II. SPIN-LATTICE RELAXATION TIMES, T_1, FOR FORMS I AND II PDEO

	T_1,s			T_1,s	
carbon	**form I**	**form II**	**carbon**	**form I**	**form II**
>C<	62	15	CH_2 (side chain)	22	1
α-CH_2	40	14	CH_3	2	1

of PDEO (M_n = 50000) crystallized at three different temperatures from the melt to produce samples containing form I (Fig. 2a), form II (Fig. 2c), and form I and II (Fig. 2b) crystals. The spectra of PDEO (M_n = 800000) containing form I crystals, obtained with and without cross polarization, are compared in Fig. 3, where crystalline and amorphous resonances are seen to be easily discriminated. Table I presents the relative ^{13}C chemical shifts observed in forms I and II PDEO referenced to the corresponding resonances of amorphous PDEO.

Spin-lattice relaxation times, T_1, observed for each carbon in form I and II PDEO crystals are presented in Table II.

From the several examples discussed in the introduction the most pertinent among these is a solid-state ^{13}C NMR study of polyoxetane (PTO) and poly(3,3-dimethyloxetane) (PDMO) by Perez and VanderHart [24]. Both PTO and PDMO can be crystallized in at least two different crystalline forms with different chain conformations in each polymorph. Perez and VanderHart produced and reported spectra of the two crystalline modifications with T_2G_2 and T_3G conformations for each polymer.

While single resonances are observed for all the carbons in the T_2G_2 crystalline forms of both polymers, in the T_3G crystal forms two resonances are observed for the CH_2-O or α-methylene and methyl carbons (PDMO only).

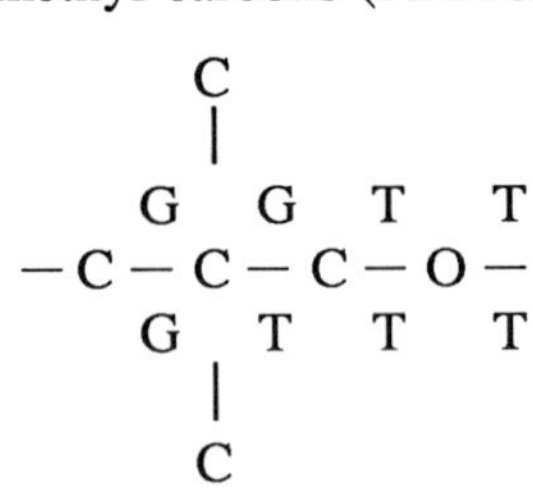

PDMO

In the T_3G conformation one α-methylene carbon experiences one γ-gauche interaction while the other α-methylene has no γ-gauche interactions. Both α-methylene carbons in the T_2G_2 crystalline forms experience a single γ-gauche interaction. On the basis of γ-gauche shielding effects we would expect both α-methylene carbons in the T_2G_2 forms to resonate near the most upfield α-methylene carbon in the T_3G forms, while the other T_3G α-methylene carbon should be downfield from these resonances by one γ-gauche interaction with oxygen. While the α-methylene carbons are observed to split by 2.2 (PTO) and 1.3 (PDMO) ppm in the T_3G crystal forms, the two α-methylene carbons in the T_2G_2 forms appear 1.8 (PTO) and 1.9 (PDMO) ppm further upfield from the most shielded T_3G α-methylene carbon resonance.

On the basis of a comparison [32] of the ^{13}C NMR spectra of *n*-alkanes and *n*-alkyl ethers, the shielding produced by an oxygen atom in a γ-gauche arrangement with the α-methylene carbons in PTO and PDMO would be expected to be at least -5 ppm. The fact that the α-methylene carbons are split by only 1.3-2.2 ppm in the T_3G crystal forms, coupled with the observation that both T_2G_2 α-methylenes resonate an additional 1.8-1.9 ppm upfield from the most shielded T_3G α-methylene carbon despite also being in a single γ-gauche arrangement, indicates that γ-gauche shielding interactions do not play the principal role in determining the solid-state ^{13}C NMR chemical shifts observed in PTO and PDMO.

This conclusion is further strengthened by the ^{13}C chemical shifts observed for the methyl carbons in PDMO which all resonate within a 1.1 ppm range in both crystal forms even though all methyls have one γ-gauche interaction with oxygen except one of the T_3G methyls which is γ-gauche to two oxygens. In addition, the β-methylene (PTO) and quaternary (PDMO) carbons, which have no γ-gauche interactions in either crystalline form, are separated by 1.0 (PTO) and 1.8 (PDMO) ppm between the T_2G_2 and T_3G forms. Clearly some source other than γ-gauche interactions, such as crystalline packing effects, must be sought to explain the ^{13}C chemical shifts observed by Perez and VanderHart [24] in crystalline PTO and PDMO samples.

Poly(3,3-diethyloxetane) (PDEO) crystallizes [29,30] into two polymorphs, I and II, with T_4 and T_2G_2 conformations, respectively. In Fig. 4 Newman projections about the four bonds attached to the quaternary carbon are presented. The two backbone bonds (1 and 2) are either both T or G in the T_4 and T_2G_2 conformations, respectively. In the form I, T_4 crystals the side-chain bonds are both T, while their conformations are not known in the form II, T_2G_2 crystals. The C-O bonds are T in both crystal forms.

Table III presents a summary of the number and kinds of γ-gauche interactions occurring in both PDEO crystal forms.

Based on these γ-gauche arrangements, we would expect the following behavior for the ^{13}C chemical shifts in PDEO forms I and II: (i) CH_3 should resonate at the same field in I and II, (ii) >C< should resonate at the same field in I and II, (iii) α-CH_2 II should resonate upfield from α-CH_2 I by one γ(O) or by γ(O)-γ(CH_3) depending on whether bonds 3,4=T or G in form II, and (iv) CH_2 (side chain) I should resonate upfield from CH_2 (side chain) II by one γ(O) or by γ(O)-γ(CH_3) depending on whether bonds 3,4=T or G in form II (see Fig. 4).

Table III. γ-GAUCHE INTERACTIONS IN FORMS I AND II PDEO
no. of γ-gauche interactions

carbon	form I (T_4)	form II (T_2G_2)
>C<	0	0
CH_3	2(CH_2)	2(CH_2)
α-CH_2	2(CH_3)	1(O) + if bonds 3,4 = T, 2(CH_3) if bonds 3,4 = G, 1(CH_3)
CH_2 (side chain)	2(O)	1(O) + if bonds 3,4 = T, 0(CH_3) if bonds 3,4 = G, 1(CH_3)

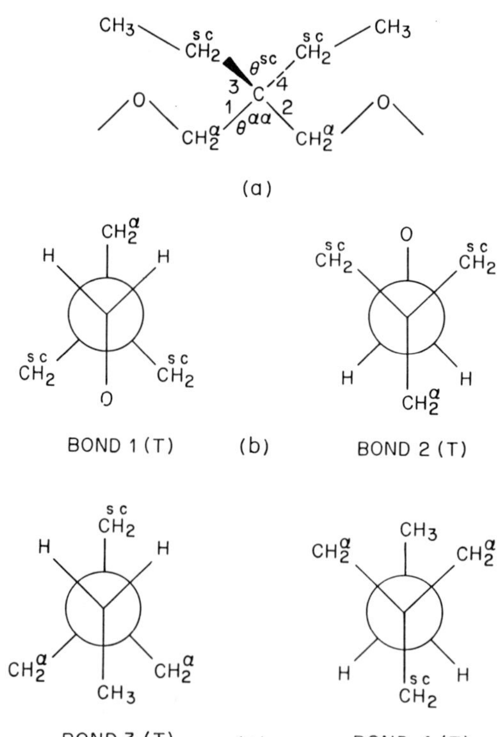

Figure 4. (a) PDEO chain structure. (b) Newman projections along the C-C backbone bonds [1,2] in PDEO. (c) Newman projections along the >C< to CH_2 side-chain bonds [3,4] in PDEO.

A comparison of the three CPMAS spectra of PDEO (M_n = 50000) in Fig. 2 makes it readily apparent that the ^{13}C chemical shifts expected for >C< and CH_3 are similar to those observed. However, instead of α-CH_2 II being upfield from α-CH_2 I it resonates 3.9 ppm downfield. Similarly, form II CH_2 (side chain) comes upfield from form I CH_2 (side chain) by 5.9 ppm instead of being downfield as expected. As observed for PDMO by Perez and VanderHart [24] the ^{13}C chemical shifts observed in forms I and II PDEO are not predominantly influenced by γ-gauche shielding effects.

It is natural to consider different packing of the PDEO chains in forms I and II as a possible source of the difference observed in their ^{13}C chemical shifts. The >C< and CH_3 carbons have identical γ-gauche interactions in both PDEO crystal forms, yet there are 0.5 (>C<) and 2.1 (CH_3) ppm chemical shift differences between the two polymorphs. Based on the ^{13}C chemical shift packing effects seen in *n*-alkane [33], PE [34], and i-PP [35] crystals, the differences observed between form I and II >C< and CH_3 chemical shifts do not seem atypical.

As another example, we may compare the ^{13}C chemical shifts of PTO in the T_2G_2 crystal form with those observed for its cyclic tetramer, c-$(TO)_4$, which also adopts [36] the T_2G_2 conformation in the crystalline state (see ref. 28). The α-CH_2 ^{13}C chemical shifts differ by 0.3-0.5 ppm and the β-CH_2 shifts by 1.4 ppm between crystalline PTO (T_2G_2) and c-$(TO)_4$. These are not unlike the packing effects observed on the >C< and the CH_3 carbons in PDEO. Thus it would appear that packing effects in PTO, PDMO, and PDEO can be as large as 2 ppm.

However, packing effects of this magnitude are not nearly sufficient to explain the observed differences in the ^{13}C chemical shifts of the α-CH_2 and CH_2 (side chain) carbons between forms I and II PDEO which are >4-6 ppm.

In addition to interchain packing effects, there must be some differences in intramolecular chain geometries between forms I and II PDEO. Possibly the valence angles differ significantly between the T_4 conformation in form I crystals and the T_2G_2 conformation in the form II crystals. Such valence angle differences would be expected [38] to produce large ^{13}C chemical shift effects; however, it is a bit more difficult to understand why the >C< and CH_3 carbons would not also be affected.

Let us suppose that the C^{α}-C-C^{α} backbone valence angle ($\theta^{\alpha\alpha}$) is sensitive to the rotational states of the C-C bonds (1 and 2 in Fig. 4) and adopts different values for the T and G rotational states. If $\theta^{\alpha\alpha}(G)$ > or < $\theta^{\alpha\alpha}(T)$ then we would expect [39] the valence angle C^{SC}-C-C^{SC} between ethyl side chains (θ^{SC}) to follow $\theta^{SC}(T)$ > or <$\theta^{SC}(G)$. The ^{13}C chemical shift of >C< should be independent of complimentary changes in $\theta^{\alpha\alpha}$ and θ^{SC}, because all four quaternary carbon substituents are methylene carbons. In addition, the methyl carbons are probably sufficiently removed from $\theta^{\alpha\alpha}$ and θ^{SC} to render their ^{13}C chemical shifts unaffected by changes in these valence angles.

Spin-lattice relaxation times, T_1, observed for each carbon in forms I and II PDEO crystals are compared in Table II. The most dramatic differences in T_1's are those observed for the side-chain methylene carbons, i.e., T_1 = 22 s (I) and 1 s (II). Apparently the side chains in the form II crystals possess considerable mobility, most probably as a result of rotations about side-chain bonds 3 and 4 (see Fig. 4). This implies that the ethyl side chains in form II crystals are conformationally disordered and are likely interconverting rapidly between T and G conformations.

In summary, we believe the large chemical shift differences observed for the methylene carbon resonances in the solid-state ^{13}C NMR spectra of forms I and II PDEO are likely the result of intramolecular geometrical differences between the chains residing in these two polymorphs. Valence angle distortion at the quaternary carbon, rather than conformationally sensitive γ-gauche effects or differences in interchain packing, seem the most likely source of the large chemical shift differences observed in the solid-state for the forms I and II methylene carbons in PDEO.

In addition, spin-lattice relaxation times observed for both crystalline forms of PDEO indicate that the side chains in the form II crystals are motionally labile and disordered. Rapid interconversion between T and G side-chain conformations seems a likely source of the short T_1's observed for the side-chain methylene carbons in form II PDEO.

3.2. PBT [41]

It was first observed [42-44] over a decade ago that the uniaxial extension of poly(butylene terephthalate) (PBT) fibers is accompanied by a crystal-crystal transition. In the relaxed or α-form crystals the molecular chain or fiber repeat is ~10% shorter than observed in the stretched or β-form crystals. The transition between the α and β forms of PBT produced by mechanical deformation (uniaxial stretching) is reversible [44-46] and only the α form is stable in the relaxed, unstretched state at ambient temperature.

X-ray structural studies have been reported [45,47-51] for both PBT crystal forms. Infrared and Raman spectroscopy [52] suggest a nearly trans-trans-trans sequence (ϕ_a, ϕ_b, ϕ_c) for the glycol residue in the extended structure (see Fig. 5). However, the crystal structures proposed by Yokouchi et al. [45] and Hall et al [48,59-51] depart significantly from the extended, all-trans glycol structure. All crystal structures proposed for the relaxed, contracted α form approximate a gauche-trans-gauche conformation for the glycol residue, although there are differences in detail among them.

As noted by Davidson et al. [53], the low scattering power of hydrogen atoms makes X-ray diffraction a technique less than suitable for defining the conformation of the glycol residues in PBT. Instead, these same authors applied broad-line ^{1}H NMR measurements to oriented PBT in both the relaxed α and strained β forms and determined the second moments of the proton line shapes as a function of specimen

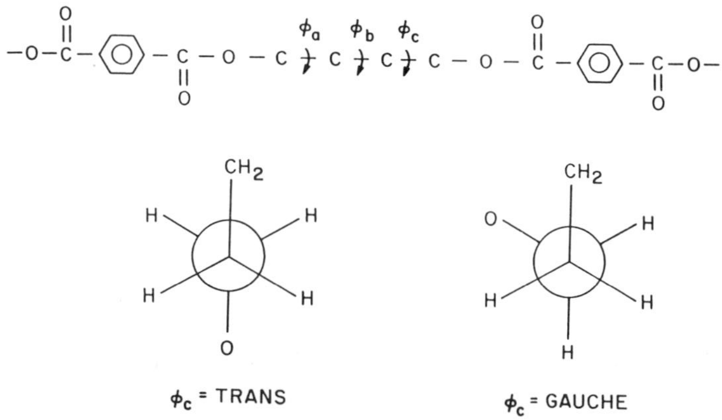

Figure 5. Schematic of PBT chain with torsions about C-C bonds indicated and Newman projections of the trans and gauche conformers about the terminal C-C bonds.

orientation. They found the ^{1}H NMR results to be consistent with a nearly fully extended (trans-trans-trans) conformation for the stretched β form and to agree quantitatively with the X-ray structure proposed by Hall and Pass [48]. However, their NMR results for the relaxed α form were not consistent with any of the proposed crystal structures and suggest instead that the conformation and orientation of the central methylene pairs in the glycol residues are not substantially altered in the strain-induced transformation from the α to β form.

More recently, Grenier-Loustalot and Bocelli [54] studied the structures of four PBT model compounds by X-ray diffraction and high-resolution ^{13}C NMR in the solid state. Single-crystal X-ray diffraction revealed that two of the PBT model compounds crystallized with trans-trans-trans glycol residues, one with a trans-trans-gauche glycol conformation, and the remaining compound had its glycol residue in the gauche-trans-gauche conformation.

In the high-resolution solid-state ^{13}C NMR spectra of the PBT model compounds they observed the central methylene carbons that are gauche to their ester oxygens to resonate 3.0-3.7 ppm upfield from those central methylene carbons adopting the trans arrangement (see Fig. 5). This is consistent with the often observed shielding of carbon nuclei whose γ-substituents are in a gauche arrangement, i.e., the γ-gauche effect [2,3,55]. The central methylene carbons in α form PBT were observed to resonate midway between the corresponding methylene resonances in the model compounds.

Grenier-Loustalot and Bocelli [54], due to the broadness of the central methylene resonance in PBT (>3 ppm), which is likely a consequence of contributions from carbon nuclei in both the crystalline and amorphous regions of their sample (see below), were unable to draw conclusions regarding the conformation of the glycol residue in α-PBT. A similar study was attempted by Havens and Koenig [56], but it too was plagued by broad resonances and, in addition, by an erroneous conformational assignment to one of their PBT model compounds as pointed out by Grenier-Loustalot and Bocelli [54].

Most recently, Perry et al. [57] employed high-resolution solid-state ^{13}C NMR techniques to study the crystalline conformations and dynamics of PBT chains in both the α and β crystalline forms. Their spectra also exhibited broad resonances, especially for the central methylene carbons (4-5 ppm). However, they concluded that as the amount of trans content in the glycol residue increases, the interior methylene resonance shifts to higher field, so the β form resonance moves upfield from the α form resonance. This is in direct opposition to the model compound study of Grenier-Loustalot and Bocelli [54] and to the expected order of chemical shifts based on the conformationally sensitive γ-gauche effect [2,3,55].

In an attempt to determine the conformations of PBT chains in their α and β form crystals, we have conducted variable-temperature, high resolution solid-state ^{13}C NMR studies. We find that above ca. 100°C the spectra are significantly better resolved, with resonance line widths not exceeding 1-2 ppm. This narrowing of resonances is apparently a consequence [58] of removing contributions made by the amorphous carbons, which no longer cross-polarize efficiently at temperatures well above the glass transition of PBT [59]. Comparison of the high-temperature, high-resolution, solid-state ^{13}C NMR spectra recorded for α- and β-PBT with those of the PBT model compounds reported by Grenier-Loustalot and Bocelli [54] permits us to draw several conclusions concerning the conformations of PBT chains in the relaxed α and stretched β crystals.

PBT in the form of pellets was obtained from Aldrich (19,094-2). The pellets were cryogenically ground to a fine powder, which was annealed at 150°C for 3 days to produce the α form [45,60]. Melt pressing of the pellets at 250°C produced thin films which were quenched into liquid nitrogen. Strips were cut from the film and placed in an Instron tensile testing machine. The strips were drawn to 300% elongation, held under tension, and annealed at 150°C for several hours to produce [45] the β-form sample.

X-ray diffraction photographs were recorded for both samples in the film form and confirmed that we did indeed produce both α- and β-PBT samples. The PBT samples were spun in aluminum oxide rotors with Kel-F [poly(chlorotrifluoroethylene)] end caps. PBT in the α form was placed in the rotor as a powder, while a strip of the β-form sample was wound under tension onto a spindle to form a spool, and the ends of the strip were glued to prevent relaxation to the α form. The spool wound with the β form strip was then inserted into the rotor and special end caps (Doty Scientific) were used to secure the ends of the spindle so that the β form PBT spool would rotate at the same speed as the rotor. A comparison of the CPMAS/DD spectra of α- and β-PBT recorded at 105°C is made in Fig. 6. Table IV contains the observed solid-state ^{13}C NMR chemical shifts, and ^{13}C spin-lattice relaxation times, T_1, measured in the solid state at 105°C, are given in Table V.

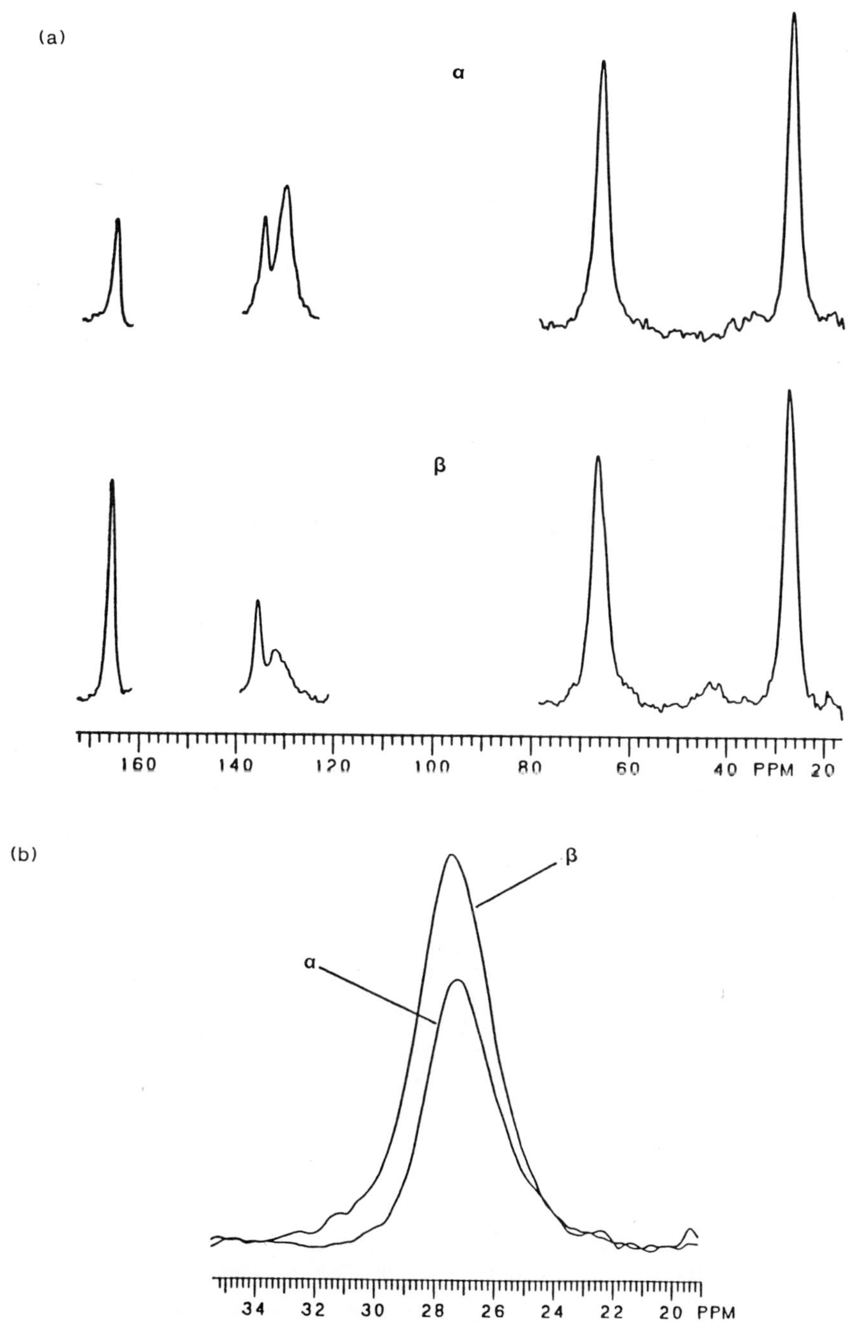

Figure 6. CPMAS/DD spectra measured at 105°C for α- and β-PBT. Full spectra (a) and expansion of the central methylene carbon regions (b).

Table IV. ^{13}C CHEMICAL SHIFTS OF α- AND β-PBT

	^{13}C, ppm vs TMS[a]		
carbon	α	β	**amorphous**[b]
CH_2	27.2	27.6	26.6
OCH_2	66.2	66.7	65.9
PAR	130.8	131.7	130.2
NPAR	135.2	135.2	134.9
C=O	165.6	165.5	166.0

[a]Measured at 105°C and referenced to the POM resonance at 89.1 ppm from TMS [55], although we have observed a 0.3 ppm downfield shift with respect to ambient temperature.

[b]Measured from spectrum obtained without CP, but with MAS/DD.

Table V. SPIN-LATTICE RELAXATION TIMES, T_1, FOR α- AND β-PBT

	T_1,[a]s		
carbon	α[b]	β[b]	**amorphous**[c]
CH_2	0.13	0.20	0.16
OCH_2	0.32	0.27	0.24
PAR	5.7	7.0	0.31
NPAR	18.9	12.7	---
C=O	18.0	15.0	---

[a]Measured at 105°C.
[b]Obtained under CP conditions with the Torchia [26] pulse sequence.
[c]Obtained without CP by the inversion-recovery method [27].

It is clear from the comparison of CPMAS spectra recorded at 105°C and presented in Fig. 6 and the corresponding chemical shifts listed in Table IV that, aside from the protonated aromatic carbons (PAR), the carbon nuclei in α- and β-PBT resonate at nearly identical frequencies. This observation is at variance with the results of Perry et al. [57], who found the CPMAS/DD spectra of α- and β-PBT at 20°C to be closely similar except for the resonances of the interior methylene carbons. The source of this disparity is revealed by comparing the spectra in Fig. 6 with those presented by Perry et al. [57]. The latter spectra are characterized by broad resonances, 2-3 times as broad as the resonances seen in Fig. 6, presumably a consequence of the different local environments, both conformational and packing, experienced by the carbon nuclei of the sluggish, amorphous PBT chains which result in a dispersion of chemical shifts.

Recording the CPMAS/DD spectra of PBT at elevated temperatures (105°C) results in enhanced resolution, because the amorphous carbons, which constituted 30-50% of the samples, are sufficiently above their glass transition (T_g = 50-55°C) [59] to be mobile enough not to cross-polarize efficiently [58]. Comparison of the spectra recorded with and without CP makes apparent that the crystalline and amorphous carbons resonate at similar frequencies, with the amorphous carbon resonances appearing upfield as expected. Thus at temperatures sufficiently close to T_g, where the amorphous PBT chains are relatively rigid and immobile, CPMAS/DD spectra would be expected [58] and in fact are observed to be significantly broadened by the overlap of crystalline and amorphous resonances.

The near coincidence of methylene carbon chemical shifts observed for α- and β-PBT strongly suggests that in both the relaxed (α) and strained (β) crystals the glycol residues of the PBT chains are adopting very similar conformations. Just what conformation is adopted by the glycol residues in crystalline PBT?

In Fig. 7 we present schematic structures of the four PBT model compounds studied by Grenier-Loustalot and Bocelli [54] and of PBT. The conformations of the glycol residues, as determined by X-ray diffraction [54], are indicated on each model compound structure as t (trans) or g (gauche). Also the chemical shifts (ppm versus TMS) of the central methylene carbons observed by CPMAS/DD ^{13}C NMR are listed above the central methylene carbon in each of the four PBT model compounds and in PBT itself. A comparison of the central methylene carbon chemical shifts clearly indicates that the glycol residues in both α- and β-PBT crystals are in the nearly extended trans-trans-trans conformation found by Grenier-Loustalot and Bocelli [54] for the PBT model compounds **3** and **4**.

If the glycol residues of both α- and β-PBT are nearly fully extended, then what conformational differences can account for the 10% increase in the fiber repeat of the β-form crystals that are formed upon extension of α-PBT? The ester bonds [C(=O)O] in PBT are likely trans planar, as they are in the four PBT model compounds. In each of the crystalline model compounds the CH_2-O bonds are also in the trans conformation. Because the chemical shifts of the central methylene carbons should be sensitive [2,3,55] to the conformations of these bonds (see Fig. 6 and 7), as well as to the conformations of the terminal CH_2-CH_2 bonds, the CH_2-O bonds in α- and β-PBT are also most likely trans.

If they were gauche in either or both polymorphs, then we would expect different chemical shifts for the central methylene carbons in α- and β-PBT, or chemical shifts reflecting the shielding [2] produced by γ-gauche carbonyl carbons. Instead, the chemical shifts of the central methylene carbons in both α- and β-PBT are very similar to the chemical shifts observed for the same carbons in model compounds **3** and **4**, where both the CH_2-O and terminal CH_2-CH_2 bonds are trans.

The only conformational degree of freedom that remains to distinguish the α and β forms of PBT is the relative orientation of the carbonyl groups in the

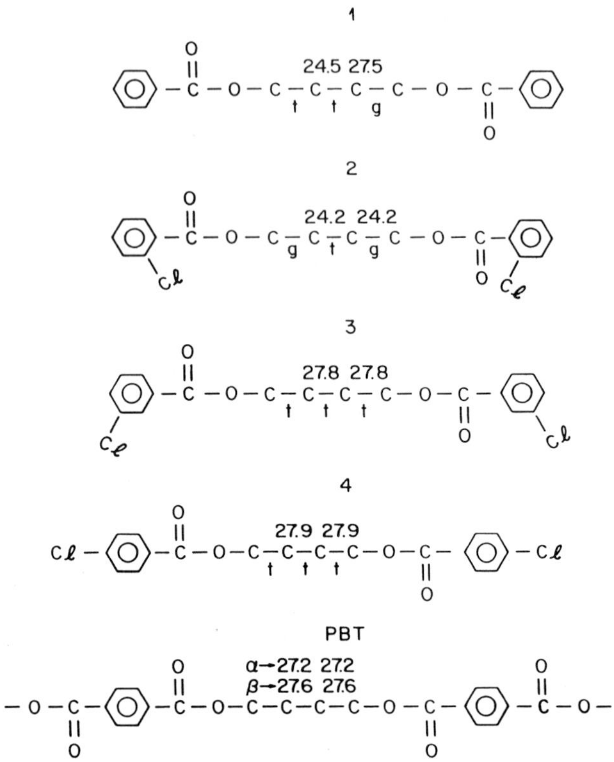

Figure 7. Schematic drawings of the four PBT model compounds studied by Grenier-Loustalot and Bocelli [54] using X-ray diffraction and high-resolution, solid-state ^{13}C NMR. The conformation of each glycol residues is indicated (t=trans, g=gauche) and the chemical shifts observed for the central methylene carbons are also listed. The structure of PBT is also presented and the chemical shifts observed in this work for the central methylene carbons in the α- and β-form crystals are indicated.

terephthaloyl residues, which are determined by the torsional angles about the sp^2-sp^2, carbonyl to aromatic C-C bonds (see Fig. 7). In terms of γ-gauche shielding effects [2,3,55], we would expect the chemical shifts of the protonated aromatic carbons (PAR) to reflect the conformations of these bonds.

In fact the 0.9 ppm chemical shift difference observed between the PAR's of α- and β-PBT is by far the largest difference between their high-resolution, solid-state ^{13}C NMR spectra (see Fig. 6). Thus we agree with the conclusion reached by Davidson et al. [58] via analysis of broad-line ^{1}H NMR data; changes in the conformation of the terephthaloyl residue, but not in the glycol residue, must accompany the solid-state transformation of PBT from the α to β form.

As we have deduced by the comparison of solid-state ^{13}C NMR chemical shifts observed for α- and β-PBT and several of their model compounds, all of the bonds in crystalline PBT are nearly trans except the bonds connecting the ester groups to the aromatic rings. The bonds in amorphous PBT chains would be expected to be a mixture of trans and gauche conformations, the latter producing upfield chemical shifts via the γ-gauche effect. It follows that the amorphous carbon nuclei should resonate upfield from the corresponding carbons in α- and β-PBT. The solid-state ^{13}C NMR chemical shifts presented in Table IV for the amorphous and crystalline carbons in PBT confirm this expectation.

Spin-lattice relaxation times, T_1, for both crystalline forms and amorphous PBT, as presented in Table V, serve to indicate the motional characteristics of solid PBT chains. The most striking observation is the near coincidence of the T_1's measured for the methylene carbons of the glycol residues in the α and β crystallites of PBT with those observed for the amorphous methylene carbons. By contrast, the T_1's measured for the crystalline PAR carbons are more than an order of magnitude longer than observed for the amorphous PAR carbons. It would appear that the methylene carbons of the glycol residue are undergoing significant motion independent of whether they are included in the α and β crystallites or not.

In agreement with Perry et al. [57], whose measurements were performed more than 80°C below ours (105°C), we do not find any significant differences between the spin-lattice relaxation times of the crystalline methylene carbons in α- and β-PBT. Similarly, the T_1's of the crystalline carbon nuclei belonging to the terephthaloyl residues in α- and β-PBT are also not markedly different. It does not appear that the α and β phases constrain the motions of their constituent PBT chains in any significantly different manner (also see Jelinski et al. [61] and Garbow and Schaefer [62] for further discussion of solid-state PBT motion).

High-resolution, solid-state ^{13}C NMR studies of α- and β-PBT have revealed several important features concerning the conformations and motions of PBT chains in both crystalline phases. We have found the glycol residues to be in the nearly extended (trans-trans-trans) conformations in both crystalline forms, while different orientations of the ester groups and phenyl rings probably account for the 10% difference in the fiber repeats of α- and β-PBT. In both crystals the methylene carbons are sampling rapid motions, which are significantly faster than the motions experienced by the carbons of the terephthaloyl residues.

3.3. i-PP [63]

Isotactic polypropylene (i-PP) is a stereoregular vinyl polymer that normally develops significant crystallinity below 200°C. The thermodynamically stable crystalline form, or α-form, consists of i-PP chains in the 3_1 helical conformation (...*TGTG*...) packed in a monoclinic unit cell [64,65]. It is believed that left- and right-handed helices are in close proximity. The metastable β-form crystals of i-PP contain hexagonically packed 3_1 helical chains, which are thought [65-66] to be arranged in groups of the same helical handedness (left or right) resulting in the distant packing of left- and right-handed chains.

The smectic form of i-PP is believed [64,68-73] to be only partially ordered compared to the α- and β-crystalline forms, though the i-PP chains in the smectic form remain in the 3_1 helical conformation [64,68]. Smectic i-PP is therefore primarily disordered in the intermolecular packing of its chains.

It was the purpose of the present investigation to learn more about the structures of the crystalline regions in the α-, β- and smectic forms of i-PP using high-resolution ^{13}C NMR as a structural probe.

Isotactic PP in the smectic form was made by cryogenic grinding of a Hercules Profax-6523 i-PP sample [74] as described by Lovinger et al. [60]. The α-form i-PP was obtained from the smectic sample by annealing for 1 h at 160°C. The β-form sample was made [75] by unidirectional crystallization at a growth rate of 10 m min^{-1} with a temperature gradient of 300°C cm^{-1}.

X-ray diffractograms of all three i-PP samples were recorded before and after the ^{13}C NMR experiments to ensure that the high-speed (3 KHz) magic angle spinning of the samples did not induce any solid-solid transitions.

Figure 8 presents the CPMAS/DD spectra of the three forms of i-PP recorded at ambient temperature. Table VI contains the observed solid-state ^{13}C NMR chemical shifts. ^{13}C spin-lattice relaxation times, T_1, measured in the solid state at ambient temperature, are given in Table VII. In addition, the T_1 values measured [76] for i-PP in solution are also presented in Table VII for comparison.

A comparison of the CPMAS ^{13}C NMR spectra in Fig. 8 and the relative chemical shifts presented in Table VI lead to several observations. First, both the methylene and methyl carbon resonances in α-form i-PP are split by ~ 1 ppm, as first reported by Bunn et al. [77]. The ratio of intensities of the downfield to the upfield component is ~ 2:1 for both carbon types. Bunn et al. [77] interpreted this splitting as due to the inequivalent sites, A and B, produced by pairing of helices of opposite handedness [64] (see Fig. 9a), which are also present in the ratio A:B=2:1. Note that the A sites correspond to a separation of 5.28Å between helical axes, while for the B sites the helices are 6.14Å apart.

In the spectrum of β-form i-PP (see Fig. 8b), on the other hand, each carbon exhibits a single resonance. The β-form methylene and methyl resonances are close to the upfield member of each pair of the same resonances in the α-form, which were attributed to the B sites (see Table VI). Fig. 9b indicates the interchain packing proposed [65] for β-form i-PP. Unlike the α-form packing [64], 3_1 i-PP helices of the same handedness are packed together in groups in the β-form crystals. The interhelical separation [68] is 6.36Å, very similar to the smallest interhelical separation involving the B sites of α-form crystals. Thus, the near-coincidence of the ^{13}C chemical shifts in β-form i-PP with those corresponding to the B sites in the α-form can be understood on the basis of similar interhelical separations in both packing modifications.

At this point it should be mentioned that Bunn et al. [77] have also reported the ^{13}C NMR CPMAS spectrum of β-form i-PP. They found, in opposition to our

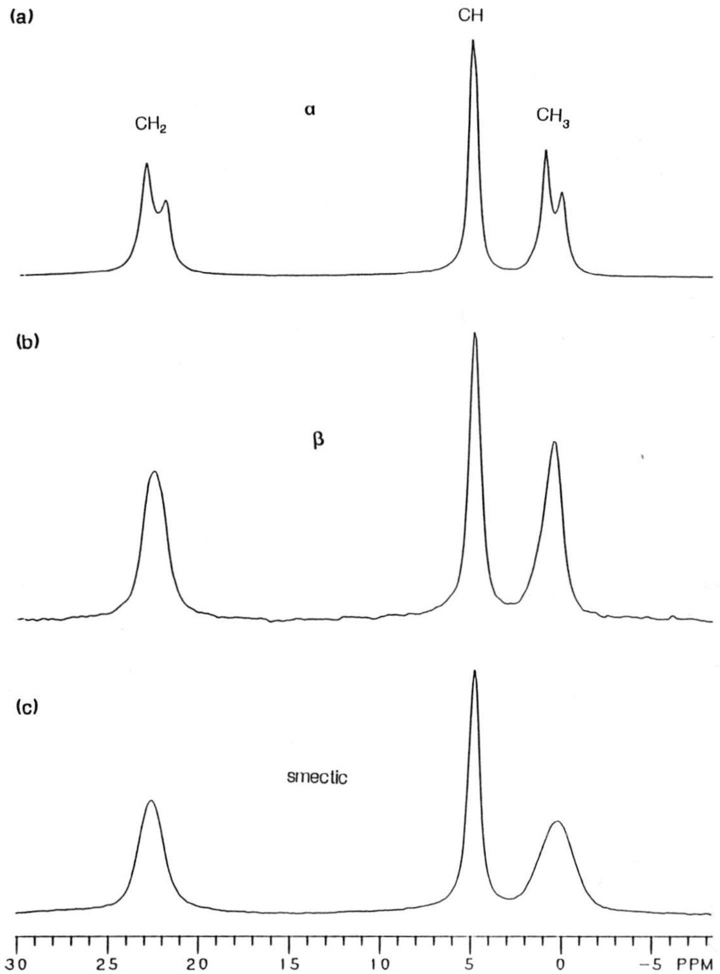

Figure 8. CPMAS spectra of i-PP in (a) α-form, (b) β-form and (c) smectic form. Spectra were recorded at ambient temperature with no reference employed.

Table VI. ^{13}C NMR CHEMICAL SHIFTS (ppm)[a] FOR i-PP IN THE SOLID STATE

i-PP	**CH_2**	**CH**	**CH_3**
α-form	0,1.07	0.20	0,0.88
β-form	0.37	0.02	0.27
smetic form	0.47	0	0.08

[a] ^{13}C NMR chemical shifts observed at ambient temperature and referenced to the most upfield resonance of each carbon type.

Table VII. ^{13}C T_1 RELAXATION TIMES FOR SOLID i-PP AT AMBIENT TEMPERATURE

	T_1 (s)		
i-PP	**CH**	**CH_2**	**CH_3**
α-form	37[a]	52[a]	0.32[a], 0.80[a], 0.48[a]
β-form	29[a]	34[a]	0.75[a], 0.44[b]
smetic form	22[a]	33[a]	0.75[a], 0.51[b]
solution[c]	0.40	0.20	0.75

[a]Measured with CP using the pulse method of Torchia [26]
[b]Measured by inversion-recovery [27] without CP
[c]Measured in solution at 46°C by Randall [76] using the inversion-recovery method [27]

results, that the chemical shifts of the A site methylene and methyl carbon resonances in α-form i-PP were closer than the B site resonances to those observed in their β-form i-PP. We were unable to obtain β-form i-PP following their preparation method, a finding confirmed by X-ray diffraction. The difficulties in obtaining pure β-form i-PP exclusively via thermal treatment is noted in the literature [75,78-81]. Instead special nucleating agents [80,81] or unidirectional crystallization [75] are employed. Consequently, Bunn et al. [75] may not have studied i-PP in the pure β-form.

^{13}C NMR chemical shifts recorded for the smectic form of i-PP are nearly coincident with those found for the β-form. This finding suggests that the local packing of 3_1 i-PP helices in the smectic form closely resembles that found in the β-form.

Having concluded that the local packing of chains is similar in the β- and smectic forms of i-PP, based on their observed ^{13}C chemical shifts, let us look at the spin-lattice relaxation time (T_1) behavior of the carbon nuclei in the three polymorphs of i-PP. The T_1 relaxation times of the crystalline carbons were obtained while cross-polarizing using the Torchia [26] pulse sequence. Smectic-form i-PP T_1 values (see Table VII) are very similar to those measured for the β-form crystals, while the α-form T_1 values appear unique. The T_1 values reported previously for α-form i-PP by Fleming et al. [82] are in good agreement with the values shown in Table VII.

The T_1 values obtained by the inversion-recovery method [27] are dominated by the relaxation of the amorphous carbons and are obtained without CP. As expected from the spectra obtained without CP (not shown [63]) only the T_1 of the methyl carbons are obtained by this method, and, within experimental error, they are the same for all three i-PP polymorphs. In addition, the T_1 values measured for the methyl carbons in the crystals and in solution are similar to the amorphous methyl spin-lattice relaxation times. Clearly the spin-lattice relaxation times of methyl carbons are dominated by their internal rotations and not by the segmental motions of the i-PP chains.

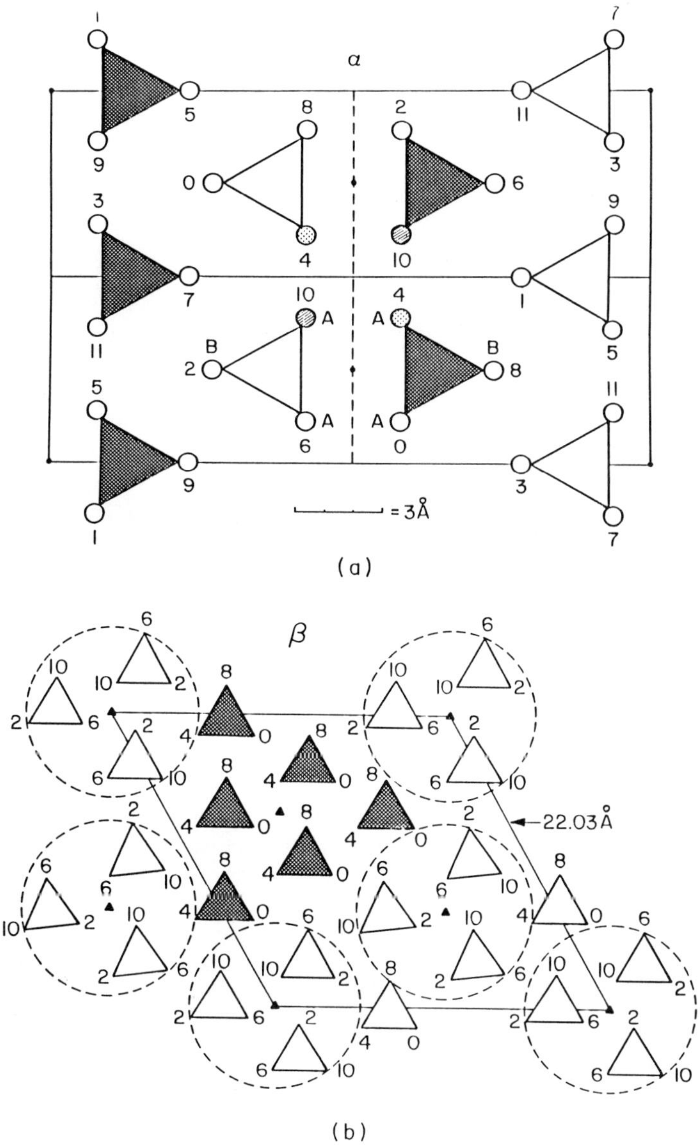

Figure 9. Crystal structures of (a) α-form [64] and (b) β-form [65] of i-PP. Full (RH) and open (LH) triangles indicate 3_1 helical i-PP chains of different handedness. A and B label the inequivalent sites discussed in the text, and are applicable to all three carbon types, because the CH-CH_2 bond is nearly parallel with the c-axis. Numerals at the triangle vertices indicate heights of methyl groups above a plane perpendicular to the c-axis in twelfths of c. The circles at the triangle vertices in (a) correspond to methyl carbons, and the cross-hatched and stippled pairs of circles correspond to the enmeshed A site methyls.

The two methyl resonances observed in the CPMAS spectrum of α-form i-PP (see Fig. 8a) relax at different rates with T_1=0.43 and 0.80 s for the downfield and upfield peaks, respectively. Having identified these resonances with the A and B packing sites in the α-form crystalline lattice (see Fig. 9a), it is worth mentioning that both the β- and smectic form crystalline methyls have T_1=0.75 s in agreement with the T_1 of the α-form, B site methyl carbons. This observation supports the conclusion, obtained previously from a comparison of ^{13}C chemical shifts, that the interhelical separation of chains is similar for the B sites in α-form crystals and in the β- and smectic-form i-PP crystals.

It is interesting that the α-form methyl carbons associated with the A and B packing sites exhibit T_1 relaxation times different by almost a factor of 2. The B site methyls are rotating twice as fast as the enmeshed methyl carbons at the A sites, because they are apparently on the fast side of the T_1 minimum [82].

The results of our solid-state ^{13}C NMR studies of the three polymorphs of i-PP are consistent with several known structural features of the α- and β-form crystals and permit us to infer the local chain packing structure in the smectic form. Both the ^{13}C chemical shifts and spin-lattice relaxation times observed for the smectic-form carbons indicate that the packing of their 3_1 helices is similar (at least on a very local scale) to the packing of i-PP chains in the β-form crystals.

This conclusion is consistent with the proposal of Gailey and Ralston [69] who suggested that smectic-form i-PP is composed of small (50-100Å) hexagonal or β-form crystals. Suggestions that the smectic form is composed of monoclinic, or α-form, microcrystals made by Bodor et al. [70], or a smectic form packing characterized locally by a core of α structure surrounded by chains in a pseudohexagonal arrangement made by Corradini et al. [71], are not consistent with the results of our ^{13}C NMR study. Also the suggestion made by Miller [72] and Zanetti et al. [73], that α-form paracrystallinity (distortions of the monoclinic lattice with loss of long-range order) characterizes the structure of smectic-form i-PP, is not supported by our results.

4. Conclusion

We believe that the information concerning the conformations, packings, and dynamics of these three different polymer chains in their various crystalline phases as obtained by the techniques of high resolution, solid state ^{13}C NMR has demonstrated the utility of this approach. CPMAS ^{13}C NMR investigations of solid polymers appear to be uniquely suited to the study of their structures, conformations, and dynamics.

5. References

1. Bovey, F. A., *High-Resolution NMR of Macromolecules*, Academic, New York, 1972.

2. Tonelli, A. E. and Schilling, F. C., *Acc. Chem. Res.*, *14*, 233, 1981.
3. Bovey, F. A., *Chain Structure and Conformation of Macromolecules*, Academic, New York, 1982.
4. VanderHart, D. L.; Earl, W. L. and Garroway, A. N., *J. Magn. Reson.*, *44*, 361, 1975.
5. Moller, M., *Adv. Polym. Sci.*, *66*, 59, 1985.
6. Earl, W. L. and VanderHart, D. L., *Macromolecules*, *12*, 762, 1979.
7. Bunn, A.; Cudby, M. E. A.; Harris, R. K.; Packer, K. J. and Say, B. J., *J. Chem. Soc., Chem. Commun.*, *15*, 1981.
8. Belfiore, L. A.; Schilling, F. C.; Tonelli, A. E.; Lovinger, A. J. and Bovey, F. A., *Macromolecules*, *17*, 2561, 1984.
9. Sacchi, M. C.; Locatelli, P.; Zetta, L. and Zambelli, A., *Macromolecules*, *17*, 483, 1984.
10. Terao, T.; Maeda, S. and Saika, A., *Macromolecules*, *16*, 1535, 1983.
11. Kitamaru, R.; Horii, F. and Murayama, M., *Polym. Bull.*, *7*, 583, 1982.
12. Ando, I.; Yamanobe, T.; Sorita, T.; Komoto, T.; Sato, H.; Deguchi, K.; and Imanari, M., *Macromolecules*, *17*, 1955, 1984.
13. Corradini, P.; Natta, G.; Ganis, P. and Temussi, P. A., *J. Polym. Sci., Part C*, *16*, 2477, 1967.
14. Turner-Jones, A.; Aizlewood, J. M. and Beckett, D. R., *Makromol. Chem.*, *75*, 134, 1964.
15. Natta, G.; Corradini, P. and Bassi, I. W., *Nuovo Cimento, Suppl.*, *15*, 52, 1960.
16. Miyashita, T.; Yokouchi, M.; Chatani, Y. and Tadokoro, H., Annual Meeting of the Society of Polymer Science, Japan, Tokyo, 1974, preprint p. 453. Quoted in: Tadokoro, H., *Structure of Crystalline Polymers*, Wiley-Interscience: New York, 1979; p. 405.
17. Turner-Jones, A., *J. Polym. Sci., Part B*, *1*, 455, 1963.
18. Petraccone, V.; Priozzi, B.; Frasci, A. and Corradini, P., *Eur. Polym. J.*, *12*, 323, 1976.
19. Zanetti, R.; Manaresi, P. and Buzzori, G. C., *Chim. Ind. (Milan)*, *43*, 735, 1961.
20. Danusso, F. and Gianotti, G., *Makromol. Chem.*, *61*, 139, 1963.
21. Geacintov, C.; Schottand, R. and Miles, R. B., *J. Polym. Sci., Polym. Lett. Ed.*, *1*, 587, 1963.
22. Miller, R. C. and Holland, V. F., *J. Polym. Sci., Polym. Lett. Ed.*, *2*, 519, 1964.
23. Schilling, F. C.; Bovey, F. A.; Tonelli, A. E.; Tseng, S. and Woodward, A. E., *Macromolecules*, *17*, 728, 1984.
24. Perez, E. and VanderHart, D. L., *Polymer*, *28*, 733, 1987.
25. Earl, W. L. and VanderHart, D. L., *J. Magn. Reson.*, *48*, 35, 1982.
26. Torchia, D. A., *J. Magn. Reson.*, *30*, 613, 1978.
27. Farrar, T. C. and Becker, E. D., *Pulse Fourier Transform NMR*, Academic Press, New York, 1971.
28. Gomez, M. A.; Cozine, M. H.; Schilling, F. C.; Tonelli, A. E.; Bello, A. and Fatou, J. G., *Macromolecules*, *20*, 1761, 1987.
29. Perez, E.; Gomez, M. A.; Bello, A. and Fatou, J. G., **Colloid Polym. Sci.**, *261*, 571, 1983.

30. Gomez, M. A.; Atkins, E. D. T.; Bello, A.; Fatou, J. G. and Upstill, C., *Polymer*, in press.
31. Gomez, M. A.; Fatou, J. G. and Bello, A., *Eur. Polym. J.*, *22*, 43, 1986.
32. Stothers, J. B., *Carbon-13 NMR Spectroscopy*, Academic, New York, 1972; Chapters 3 and 5.
33. VanderHart, D. L., *J. Magn. Reson.*, *44*, 117.
34. VanderHart, D. L. and Khoury, F., *Polymer*, *25*, 1589, 1984.
35. Bunn, A.; Cudby, M. E. A.; Harris, R. K.; Packer, K. J. and Say, B. J., *Polymer*, *23*, 694, 1982.
36. c-$(TO)_4$ actually adopts the $(T_2G_2T_2G_2)_2$ conformation in the crystal (37) due to its cyclic structure. However, both the α and β-methylene carbons in PTO (T_2G_2) and C-$(TO)_4$ experience identical numbers and types of -gauche interactions.
37. Groth, P., *Acta Chem. Scand.*, *25*, 725, 1971.
38. Reference 32, Chapter 4.
39. This is a consequence (40) of the conservation of total s and p character of the sp^3 molecular orbitals around >C<.
40. Huheey, J. E., *Inorganic Chemistry; Principles of Structure and Reactivity*, Harper and Row, New York, 1972, Chapter 4.
41. Gomez, M. A.; Cozine, M. H. and Tonelli, A. E., *Macromolecules*, *21*, 388, 1988, and *22*, 2015, 1989.
42. Boye, C. A., Jr. and Overton, J. R., *Bull. Am. Phys. Soc. Sci. 2*, *19*, 352, 1974.
43. Jakeways, R.; Ward, I. M.; Wilding, M. A.; Hall, I. H.; Desborough, I. J. and Pass, M. G., *J. Polym. Sci., Polym. Phys. Edn.*, *17*, 799, 1975.
44. Jakeways, R.; Smith, T.; Ward, I. M. and Wilding, M. A., *J. Polym. Sci., Polym. Lett. Edn.*, *14*, 41, 1976.
45. Yokouchi, M.; Sakakibara, Y.; Chatani, Y.; Tadokoro, H.; Tanaka, T. and Yoda, K., *Macromolecules*, *9*, 266, 1976.
46. Brereton, M. G.; Davies, C. R.; Jakeways, R.; Smith, T. and Ward, I. M., *Polymer*, *19*, 17, 1978.
47. Mencik, Z., *J. Polym. Sci., Polym. Phys. Ed.*, *13*, 2173.
48. Hall, I. H. and Pass, M. G., *Polymer*, *17*, 807, 1977.
49. Desborough, I. J. and Hall, I. H., *Polymer*, *18*, 825, 1977.
50. Stambaugh, B. D.; Koenig, J. L. and Lando, J. B., *J. Polym. Sci., Polym. Lett. Ed.*, *15*, 299, 1979; *J. Polym. Sci., Polym. Phys. Ed.*, *171*, 1053, 1979.
51. Hall, I. H., *ACS Symp. Ser.*, *141*, 335, 1980.
52. Ward, I. M. and Wilding, M. A., *Polymer*, *18*, 327, 1977.
53. Davidson, I. S.; Manuel, A. and Ward, I. M., *Polymer (British)*, *24*, 30, 1983.
54. Grenier-Loustalot, M. -F. and Bocelli, G., *Eur. Polym. J.*, *20*, 957, 1984.
55. Axelson, D. E., *High Resolution NMR Spectroscopy of Synthetic Polymers in Bulk*, Komoroski, R. E., Ed.; VCH, Deerfield Beach, FL, 1986, Chapter 6.
56. Havens, J. R. and Koenig, J. L., *Polym. Commun.*, *24*, 194, 1983.
57. Perry, B. C.; Koenig, J. L. and Lando, J. B., *Macromolecules*, *20*, 422, 1987.
58. Lyerla, J. R.; Komoroski, R. A. and Axelson, D. E., In Chapters 2-5 of ref. 55.
59. The T_g of PBT has been reported in the literature as varying from 22 to 80°C (see, for example: Lewis, O. G., *Physical Constants of Linear Homopolymers*,

Springer-Verlag, New York, 1968, pp 150-151. Lee, W. A. and Knight, G. J. in *Polymer Handbook*, Brandrup, J. and Immergut, E. H., Eds., Interscience, New York, 1966, p. III-79). DSC measurements performed on our α-PBT sample and a quenched amorphous PBT sample yielded T_g=50-55°C.
60. Lovinger, A. J.; Belfiore, L. A. and Bowmer, T. N., *J. Polym. Sci., Polym. Phys. Ed.*, *23*, 1449, 1985.
61. Jelinski, L. W.; Dumais, J. J.; Watnick, P. I.; Engel, A. K. and Sefcik, M. D., *Macromolecules*, *16*, 409, 1983.
62. Garbow, J. R. and Schaefer, J., *Macromolecules*, *20*, 819, 1987.
63. Gomez, M. A.; Tanaka, H. and Tonelli, A. E., *Polymer (British)*, *28*, 2227, 1987.
64. Natta, G. and Corradini, P., *Nuovo Cim. Suppl.*, *15*, 40, 1960.
65. Turner-Jones, A.; Aizlewood, J. M. and Beckett, D. R., *Makromol. Chem.*, *75*, 134, 1964.
66. Turner-Jones, A. and Cobbold, A. J., *Polymer Lett.*, *6*, 539, 1968.
67. Natta, G.; Peraldo, M. and Corradini, P., *Rend. Accad. Naz. Lincei*, *26*, 14, 1959.
68. Wyckoff, H., *J. Polym. Sci.*, *62*, 83, 1962.
69. Gailey, J. A. and Ralston, P. H., *SPE Trans*, *4*, 29, 1964.
70. Bodor, G.; Grell, M. and Kallo, A., *Faserforsch. Textiltech.*, *15*, 527, 1964.
71. Corradini, P.; Petraccone, V.; De Rosa, C. and Guerra, G., *Macromolecules*, *19*, 2689, 1986.
72. Miller, R. L., *Polymer*, *1*, 135, 1960.
73. Zannetti, R.; Celotti, G. and Armigliato, A., *Eur. Polym. J.*, *6*, 879, 1970.
74. ^{13}C NMR of a Profax solution indicated that this sample of i-PP contained 94% isotactic (mm) triads.
75. Lovinger, A. J.; Chua, J. O. and Gryte, C. C., *J. Polym. Sci., Polym. Phys. Edn.*, *15*, 641, 1977.
76. Randall, J. C., *J. Polym. Sci., Polym. Phys. Edn.*, *14*, 1693, 1976.
77. Bunn, A.; Cudby, M. E. A.; Harris, R. K.; Packer, K. J. and Say, B. J., *Polymer*, *23*, 694, 1982.
78. Varga, J. and Toth, F., *Makromol. Chem. Macromol. Symp.*, *5*, 213, 1986.
79. Shi, G.; Huang, B.; Cao, Y.; Hew, Z. and Han, Z., *Makromol. Chem.*, *187*, 643, 1986.
80. Leugering, H. J., *Makromol. Chem.*, *109*, 204, 1967.
81. Duswalt, A., *Am. Chem. Soc. Div. Org. Coat.*, *30*, 93, 1970.
82. Fleming, W. W.; Fyfe, C. A.; Kendrick, R. D.; Lyerla, J. R.; Vanni, H. and Yannoni, C. S., *Polymer Characterization by ESR and NMR*, Woodward, A.E. and Bovey, F.A., Eds., ACS Polym. Symp. Ser. 142, American Chemical Society, Washington, DC, 1980.

Chapter 5

CP/MAS ^{13}C NMR ANALYSIS OF POLY(ARYLENE METHYLENE)S

MICHAEL L. FROEHLICH AND MARTIN B. JONES

1. Introduction

Poly(arylene methylene)s [-(-ArCH$_2$-)-] (PAM) are readily synthesized by condensation of aromatic nuclei with formaldehyde (Eq.1) or by Friedel-Crafts alkylation polymerization (Eq.2,3). Depending on the conditions employed, either soluble or insoluble products may be obtained.

$$HArH + HCHO \xrightarrow[\text{solvent}]{H^+} -(ArCH_2-)- \quad (1)$$

$$HArCH_2X \xrightarrow[\text{solvent}]{\text{Lewis acid}} -(-ArCH_2-)- \quad (2)$$

$$HArH + RCl \xrightarrow[\text{solvent}]{\text{Lewis acid}} -(-ArCH_2-)- \quad (3)$$

(R=CH_2Cl, CH_3OCH_2, CH_3OCH_2CO)

MICHAEL L. FROEHLICH - Energy and Minerals Research Center and Department of Chemistry, University of North Dakota, Box 7185 University Station, Grand Forks, ND 58202; and MARTIN B. JONES - Department of Chemistry, Adams State College, Alamosa, CO 81102.

Solid State NMR of Polymers, Edited by L. Mathias
Plenum Press, New York, 1991

Nuclear magnetic resonance (NMR) spectroscopy has been widely employed for routine characterization (e.g., verification of aryl and methylene resonances) of soluble poly(arylene methylene)s. More extensive NMR studies of these materials have also been carried out. For example, Tong et al. conveniently followed the thermal conversion of dimethylene ether linkages of soluble *p*-cresol-formaldehyde condensates to methylene bridging groups by 300 MHz ^{1}H-NMR [1]. Hasan and Tsonis determined the extent of branching of polybenzyls prepared according to equation 2 above (Ar=Ph) under different experimental conditions via quantitative 200 MHz ^{13}C-NMR [2]. Characterization of insoluble poly(arylene methylene)s by ^{13}C cross-polarization and magic-angle sample spinning (CP/MAS) NMR has also been accomplished [3–5]. Fyfe [3] and Maciel et al. [4] have focused on PAM from phenol-formaldehyde resins, with particular emphasis on the structural changes which arise upon curing of the resins. Jones and coworkers [5] have employed CP/MAS for routine characterization of PAM from Friedel-Crafts reactions of dibenzofuran, N-methylcarbazole, and fluorene.

Herein we describe results of CP/MAS ^{13}C NMR investigations of PAM prepared from formaldehyde condensations and Friedel-Crafts alkylations of homocyclic aromatic nuclei.

2. Experimental

2.1. NMR Measurements

^{13}C CP/MAS NMR spectra were obtained by cross-polarization with flipback of the proton magnetization on a Chemagnetics M-100S spectrometer operating at 25.04 MHz. The Kel-F rotor was spun at 3-4 KHz with ice-cooled N_2 drive gas. The number of acquisitions ranged from 600-3000, CP contact times were 1-2.5 ms, and pulse delays were 3-6 s. For interrupted decoupling experiments, a delay of 50-60 s was introduced between the mix and acquisition periods of the cross-polarization pulse sequence. The T_{1H} values were determined with the inversion-recovery pulse sequence of Sullivan and Maciel [6], employing a three-parameter non-linear least-squares fitting program supplied by Chemagnetics.

2.2. General Procedure for Formaldehyde Condensation Polymerizations

The aromatic substrate (0.01 mole) and paraformaldehyde (0.01-0.03 mole equivalents) were dissolved at room temperature in glacial acetic acid (10-15 ml) in a three-necked round bottom flask equipped with condenser, mechanical stirrer, and thermometer. After thorough mixing, concentrated H_2SO_4 (2.3 ml) was added, and the reaction mixture was heated in an oil bath for 2 h at 60-80° C. The reaction was poured onto ice and allowed to warm to room temperature. The resulting suspension was neutralized with 5% $NaHCO_3$ and vacuum filtered. The filter cake was pulverized with a mortar and pestle and washed twice with distilled water. The solid material was separated into soluble and insoluble fractions by overnight Soxhlet extraction with appropriate solvents (acetone, 95% ethanol, and/or THF).

2.3. General Procedure for Friedel-Crafts Polymerizations

The aromatic substrate (0.01 mole) and chloromethylating reagent [CH_2Cl_2, CH_3OCH_2Cl, or CH_3OCH_2COCl] (0.01-0.02 mole) were dissolved in solvent [CH_2Cl_2,CS_2, or $CH_3CH_2NO_2$] (10-30 ml) in a three-necked round bottom flask equipped with condenser, magnetic stirrer, and thermometer. After thorough mixing, the solution was cooled to 5° C in an ice-water bath, and the Lewis acid [$AlCl_3$ or $SnCl_4$] (0.01-0.02 mole) was slowly added in portions. The mixture was gradually warmed to 40° C over 1/2 h and then stirred for 2-20 h. The contents of the reaction flask were poured onto a mixture of ice and 18% HCl. After warming to room temperature, the resulting suspension was vacuum filtered. The filter cake was pulverized with a mortar and pestle, triturated with boiling 18% HCl, and washed with boiling water until the washes were neutral. The solid material was separated into soluble and insoluble fractions by overnight Soxhlet extraction with appropriate solvents (acetone, 95% ethanol, and/or THF).

3. Results and Discussion

3.1. General

Table 1 gives data concerning the polymerization conditions and solubilities of the resulting poly(arylene methylene)s. Table 2 contains ^{13}C CP/MAS NMR data for these polymers. Representative ^{13}C CP/MAS NMR spectra for the polymers are shown in Figs. 1-3. For all samples, the following general spectral features were observed. Resonances assigned to aromatic carbons were clearly distinguishable from resonances assigned to methylene bridging carbons. The spectra of insoluble polymers, whether synthesized via formaldehyde condensation or Friedel-Crafts alkylation, were less highly resolved than those of soluble polymers. Formaldehyde condensation polymers had significantly longer T_{1H} values than did comparable Friedel-Crafts polymers prepared with Lewis acid catalysts (see Table 3). With $AlCl_3$ as the Lewis acid, the resulting polymers had shorter T_{1H} values than did materials from Friedel-Crafts reactions with $SnCl_4$ (see Table 3). Both observations may, at least in part, be explained by the unpaired electron content of the $AlCl_3$-prepared polymers (typically 1 X 10^{18} spins/g). ($AlCl_3$ can oxidize aromatic nuclei to radical cations, hence increasing radical content of the systems). Several workers have found that T_{1H} values for *p*-phenylene polymers and oligomers are markedly reduced in samples with increased radical content [7].

3.2. Naphthalene Polymers (Entries 1-5, Table 2)

For each sample, the resonances in the aromatic region are similar (see Fig.1). Comparison of the observed chemical shifts of the polymers synthesized in this study with estimated chemical shifts (calculated from appropriate model compounds and additivity parameters) for the aromatic carbon resonances of four different linear poly(naphthalenyl methylene)s (given in Table 4) does not permit unambiguous assignment of the regiochemistry of methylene bridging. However, the difference in

TABLE I. Polymerization Conditions and Solubility of PAM Samples

Entry	Substrate (mol)[a]	Alkylating Agent (mol)[a]	Acid Cat. (mol)[a]	Sol.[b]
1	Naph. (1)	CMME[c] (2)	$AlCl_3$ (3)	i
2	Naph. (1)	HCHO (1)	H_2SO_4 (4)	s
3	Naph. (1)	HCHO (2)	H_2SO_4 (4)	s
4	Naph. (1)	HCHO (2)	H_2SO_4 (4)	s*
5	Naph. (1)	HCHO (3)	H_2SO_4 (4)	s*
6	Biphenyl (1)	CMME (2)	$AlCl_3$ (2)	i
7	Biphenyl (1)	MAC (1)	$SnCl_4$ (2)	i
8	Biphenyl (1)	MAC (1.5)	$SnCl_4$ (1)	s*
9	Biphenyl (1)	CH_2Cl_2 (xs)	$AlCl_3$ (1)	s*
10	Biphenyl (1)	HCHO (2)	H_2SO_4 (4)	i
11	Biphenyl (1)	MAC (2)	$SnCl_4$ (2)	i
12	Biphenyl (1)	MAC (1)	$SnCl_4$ (1)	s*

[a] Numbers represent molar ratios
[b] i=insoluble; s=soluble in THF, acetone, $CHCl_3$; s*=soluble in THF only
[c] MAC=methoxyacetyl chloride; CMME=chloromethyl methyl ether

chemical shift of the methylene resonance for the Friedel-Crafts product (entry 1) relative to that of the soluble HCHO products (entries 2-5) suggests that the regiochemistry of the methylene linkage may be different for the two preparations (although the linewidth of entry 1 may cloud the interpretation). Previous researchers have employed ^{1}H-NMR data to suggest that the bridging methylene groups in naphthalene/formaldehyde resins are primarily located at α-positions [8]. Likewise, Al-Jumah and Fernandez reported that self-condensation of 1-(chloromethyl)naphthalene with $SnCl_4$ yielded a polymer with methylene linkages at α-positions (specifically the 1,5-positions) [9]. Very recent results from our laboratory reveal that the polymer from self-condensation of 2-(chloromethyl)-naphthalene with $SnCl_4$ is comprised of naphthalene rings linked at the 2 and 6 positions by methylene groups [10]. The ^{13}C solution NMR chemical shift for the methylene carbons in the 2,6-linked polymer is 3-4 ppm downfield from the methylene carbons in the α-linked polymer (40 ppm for the former, 36-37 ppm for the latter). Since the methylene resonance of entry 1 is downfield from those of entries 2-5, one may suggest that entry 1 is predominantly β-linked and that entries 2-5 are predominantly α-linked.

Comparison of the relative intensities of the peaks due to quaternary aromatic carbons (identified by use of interrupted decoupling experiments) with those due to protonated aromatic carbons provide qualitative assessments of crosslink density. For a linear polymer of naphthalene bridged by methylene groups, one would expect significantly larger peaks for the protonated carbons (ca. 125-128 ppm) than for the quaternary carbons (133-138 ppm). This is, in fact, observed for entries 2-5, which are soluble naphthalene/HCHO reaction products (e.g., see Fig. 1B and 1C). As the polymer becomes crosslinked, the number of quaternary aromatic carbons increases at the expense of the protonated aromatic carbons. For entry 1, an insoluble Friedel-Crafts product, the size of the quaternary peak is almost equal to that of the protonated peak, suggesting significant amounts of crosslinking (cf. Fig. 1A and 1B). Information about crosslinking was also offered by the aliphatic regions of the spectra. First, additional support for the crosslinking of entry 1 relative to entries 2-4 is given by the linewidth at half height for the methylene resonance. The insoluble Friedel-Crafts polymer has a $-CH_2-$ resonance that is almost twice as broad as that of the soluble HCHO products. Second, entry 5, a soluble extract from a polymerization carried out with a threefold molar excess of paraformaldehyde, has two distinct resonances in the $-CH_2-$ region (35.2 and 30.8 ppm). This is indicative of branches on the polymer chain. The lack of two resolved signals in the other soluble products (entries 2-4) may be explained by the stoichiometry of their preparation (equimolar and twofold excess of paraformaldehyde).

TABLE II. ^{13}C CP/MAS NMR Data for PAM Samples

Entry[a]	Chemical Shift, ppm (Integration)[b]	Linewidth of $-CH_2-$, Hz[c]
1	132.2 + 126.2 (10.5); 37.4 (1)	255
2	133.2 + 126.2 (10.4); 35.9 (1)	148
3	133.1 + 126.0 (10.1); 36.0 (1)	148
4	133.0 + 126.0 (8.3); 36.0 (1)	165
5	133.4 + 126.3 (7.2); 35.2 + 30.8 (1)	155, 98
6	140.2 + 128.7 (8.7); 41.5 (1)	220
7	139.7 + 128.5 (8.8); 40.8 (1)	185
8	140.6 + 135.7, 130.9 + 126.0 (12.4); 42.6 (1)	63
9	139.4 + 127.9 (9.1); 41.7 + 30.5 + 21.1 (1)	---
10	143.9 + 139.8, 126.2 + 121.0 (5.4); 43.0 + 36.9 (1)	235
11	141.2 + 137.4, 123.4 + 117.2 (6.1); 39.1 + 34.7 (1)	190
12	141.9 + 137.3, 123.8 + 118.5 (7.0); 41.1 + 33.8 (1)	83, 108

[a] Entry nos. correspond to those in Table 1
[b] Sidebands were included in the integration measurements
[c] Measured at half-height of the resonance. If two numbers are given, it means that the resolution is sufficient to distinguish two peaks below the half-height

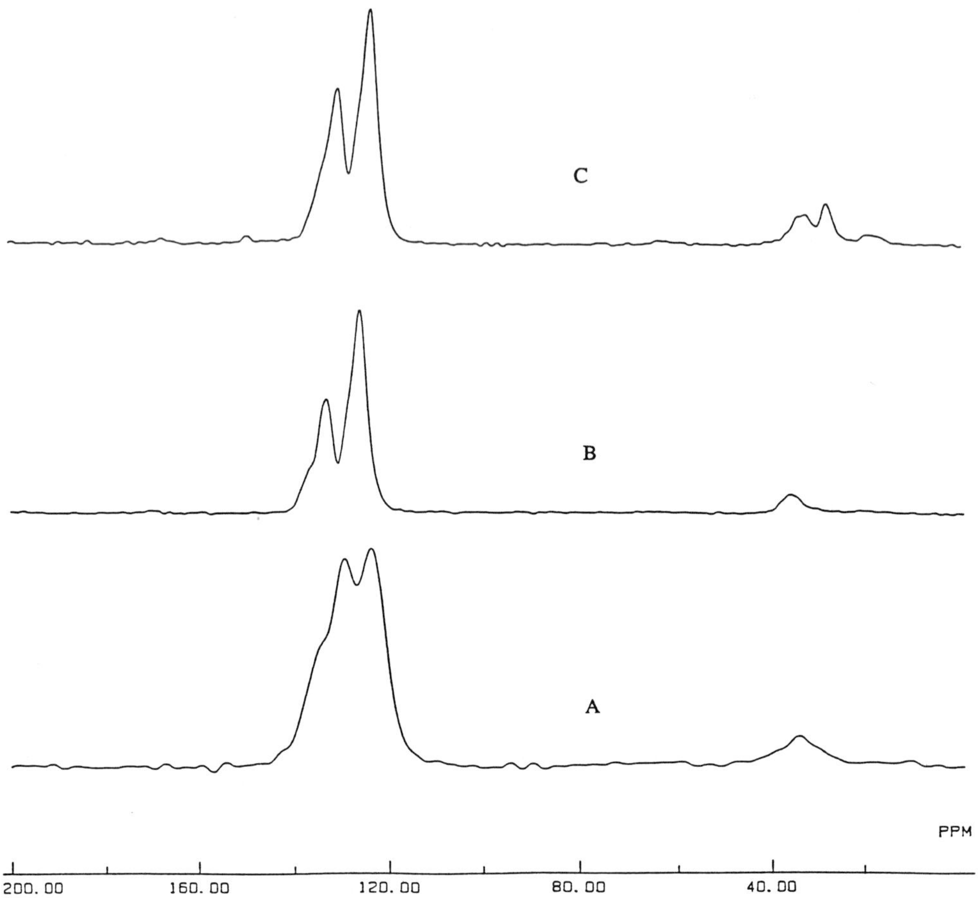

Figure 1. CP/MAS ^{13}C NMR spectra of poly(naphthalenyl methylene)s. (A) Entry 1, Tables 1 and 2; (B) Entry 2, Tables 1 and 2; (C) Entry 5, Tables 1 and 2.

3.3. Biphenyl Polymers (Entries 6-9, Table 2)

The regiochemistry of substitution would be expected to be 4,4' based on reactivity considerations. The soluble products have aromatic carbon resonances which agree well with those of the model compound 4,4'-dimethylbiphenyl (quaternary carbons at 136.5 and 138.5 ppm, protonated carbons at 126.9 and 129.6 ppm) [11] and with estimated chemical shifts for a linear poly(4,4'-biphenylyl methylene) (see Table 4). Biphenyl polymers with methylene groups bridging the 2 and 2' or 3 and 3' carbons would be expected to exhibit a resonance for C-1 at 141-142 ppm, based on the observed chemical shifts of 2,2'-and 3,3'-dimethylbiphenyl [11]. Thus, these polymers are most likely to be comprised of biphenyl nuclei bridged at the 4 and 4' positions with $-CH_2-$ groups. The resolution in the aromatic region for the spectra of the insoluble materials (entries 6 and 7) is insufficient to permit unequivocal assignment of the substitution pattern. However, the chemical shifts of the observed peaks are not in disagreement with 4,4'- as the principal substitution pattern.

As with the naphthalene products, the relative intensities of the peaks assigned to quaternary aromatic carbons (135-140 ppm) versus protonated aromatic carbons (126-130 ppm) of the biphenyl polymers is greater for the insoluble products than for the soluble products. These results are in agreement with increased branching and crosslinking present in the insoluble polymers.

3.4. Fluorene Polymers (Entries 10-12, Table 2)

^{13}C CP/MAS NMR of insoluble poly(fluorenyl methylene)s prepared via Friedel-Crafts reactions have been described in the literature [5b]. Both insoluble products (entries 10 and 11) and the soluble polymer (entry 12) examined for this study yield spectra in good agreement with a calculated spectrum of linear poly(2,7-fluorenyl methylene) (see Table 3). The increased intensity of the quaternary aromatic

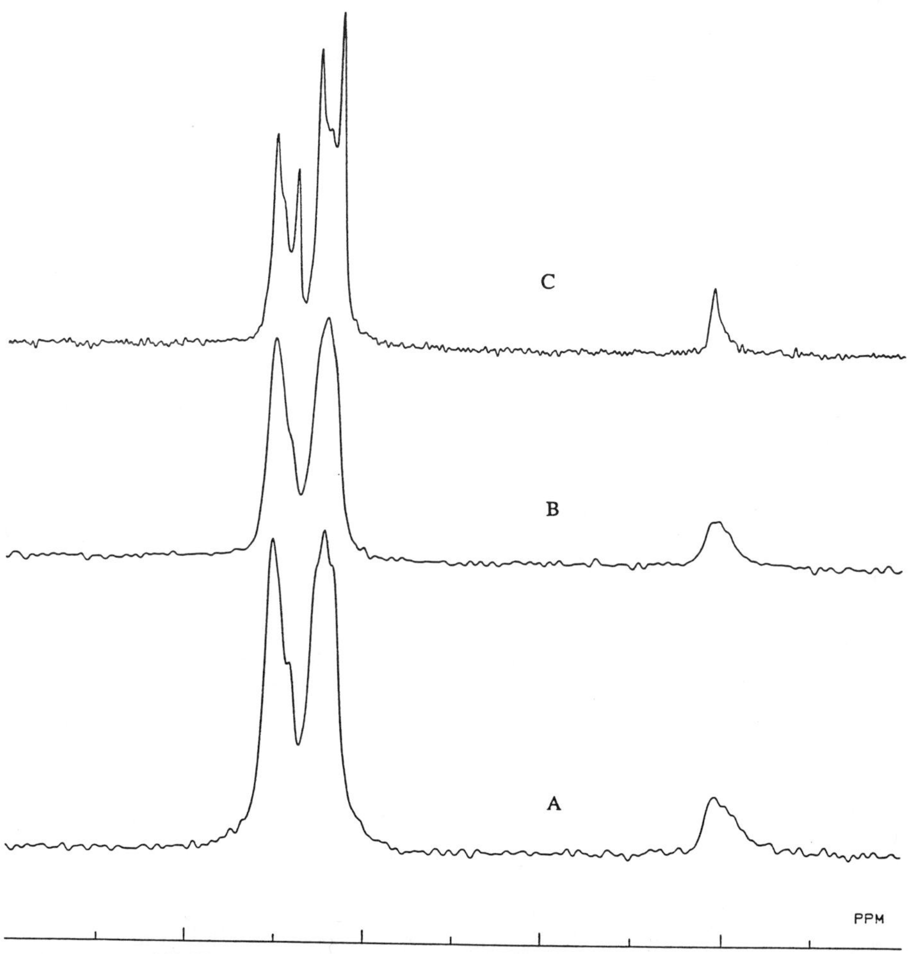

Figure 2. CP/MAS ^{13}C NMR spectra of poly(biphenylyl methylene)s. (A) Entry 6, Tables 1 and 2; (B) Entry 7, Tables 1 and 2; (C) Entry 8, Tables 1 and 2.

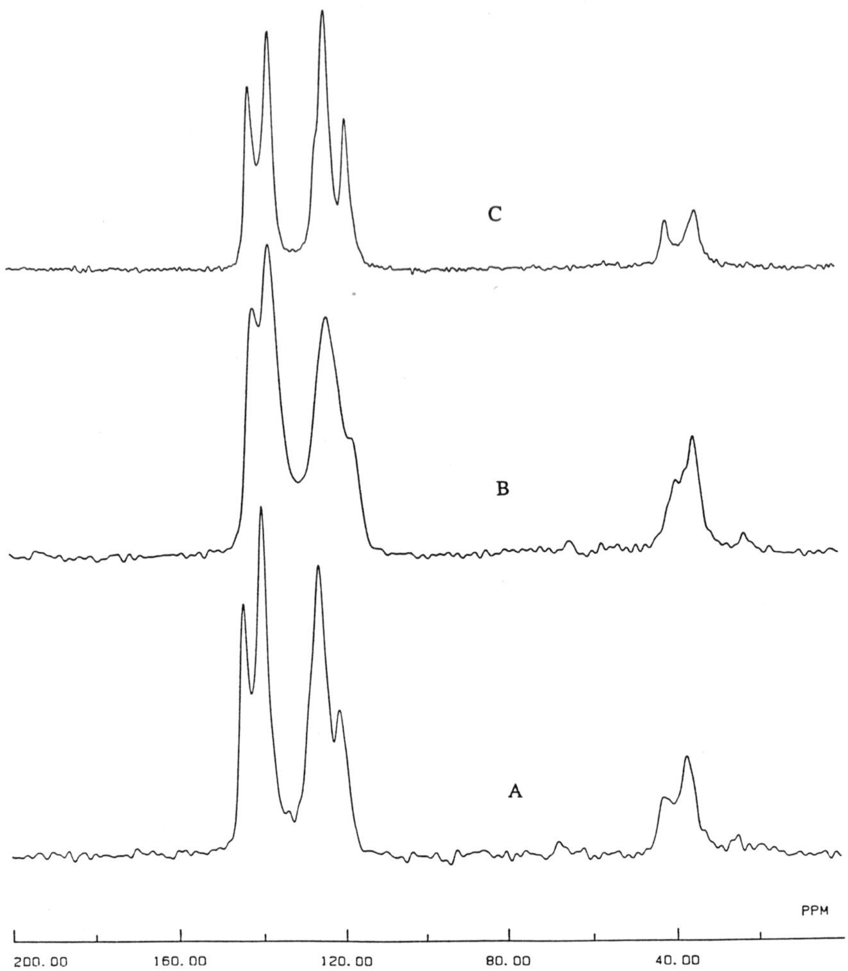

Figure 3. CP/MAS ^{13}C NMR spectra of poly(fluorenyl methylenes)s. (A) Entry 10, Tables 1 and 2; (B) Entry 11, Tables 1 and 2; (C) Entry 12, Tables 1 and 2.

TABLE III. T_{1H} Values for Selected Poly(arylene methylene)s

Entry[a]	**Sample**	**T_{1H}(s)**[b]
1	N/CMME/$AlCl_3$	1.60
4	N/HCHO/H_2SO_4	3.60
7	BP/MAC/$SnCl_4$	6.00
9	BP/CH_2Cl_2/$AlCl_3$	0.11
10	F/HCHO/H_2SO_4	5.42
11	F/MAC/$SnCl_4$	0.58

[a] Entry nos. correspond to those in Table I
[b] Standard deviation < 10%

TABLE IV. Estimated ^{13}C Chemical Shifts for Aromatic Carbons of Unbranched PAM[a]

	Chemical Shift, ppm from TMS											
	C-1	C-2	C-3	C-4	C-5	C-6	C-7	C-8	C-9	C-10	C-11	C-12
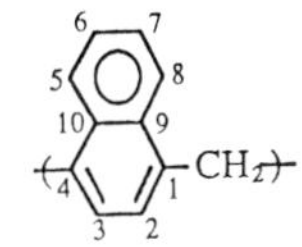	135.2	127.1	127.1	135.2	124.7	125.3	125.3	124.7	133.8	133.8		
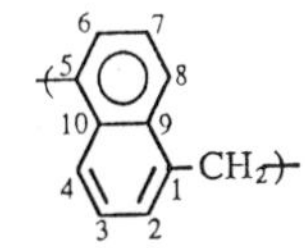	137.0	125.8	125.9	122.6	137.0	128.5	125.9	122.6	133.2	133.2		
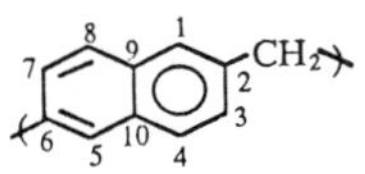	126.8	137.0	127.8	127.9	126.8	137.0	127.8	127.9	133.6	133.6		
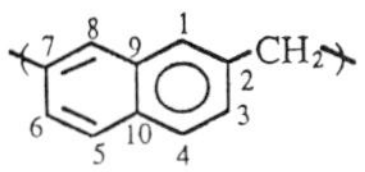	126.9	138.0	125.8	127.9	127.9	125.8	138.0	126.9	135.3	134.7		
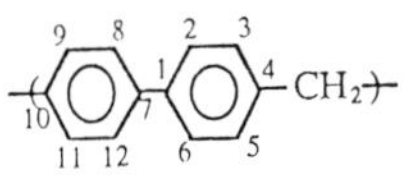	139.1	129.4	127.8	139.5	127.8	129.4	139.1	129.4	127.8	139.5	127.8	129.4
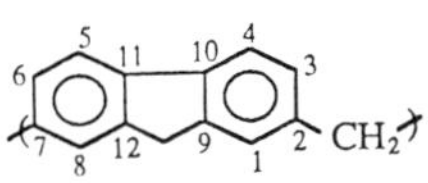	124.8	139.1	126.5	120.1	120.1	126.5	139.1	124.8	143.5	139.1	139.1	143.5

[a] Appropriate dimethyl aromatic used as basis; additivity parameters derived by comparison of toluene to diphenylmethane.

peaks (137–144 ppm) relative to the protonated aromatic peaks (117–126 ppm) for the insoluble polymers as compared to the soluble polymers is in agreement with crosslinked chains.

4. Summary

CP/MAS ^{13}C NMR has been shown to be useful for qualitatively assessing the extent of crosslinking and the regiochemistry of methylene substitution in poly(arylene methylene)s.

5. Acknowledgments

The authors gratefully acknowledge the United States Department of Energy which provided funds for the purchase of the Chemagnetics M-100S. We thank Steve Balaban, Julie Bouley, Julie Fettig, Kristen Johnson-Cousineau, and Natalie Thompson for synthesizing the samples.

6. References

1. Tong, S. N.; Park, K. Y. and Harwood, H. J., *J. Polym. Sci., Polym. Chem. Ed.*, *22*, 1097, 1984.
2. Hasan, M. U. and Tsonis, C. P., *J. Polym. Sci., Polym. Chem. Ed.*, *22*, 1349, 1984.
3. Fyfe, C. A., *Solid State NMR for Chemists*, CFC Press, Guelph, Ontario Canada, 1983, pp 307-311, and references therein.
4. (a) Maciel, G. E.; Chuang, I. S. and Gollob, L., *Macromolecules*, 43, *17*, 1081, 1984.
   (b) Hatfield, G. R. and Maciel, G. E., *Macromolecules*, *20*, 608, 1987.
5. (a) Rao, S. P.; Jones, M. B.; Baltisberger, R. J.; Burger, V. T. and Brown, C. E., *J. Polym. Sci. Polym. Lett. Ed.*, *21*, 551, 1983.
   (b) Nystuen, N. J. and Jones, M. B., *J. Polym. Sci., Polym. Chem. Ed.*, *23*, 1433, 1985.
   (c) Rolando, T. E. and Jones, M. B., *J. Polym. Sci., Polym. Lett. Ed.*, *24*, 233, 1986.
6. Sullivan, M. J. and Maciel, G. E., *Anal. Chem.*, *54*, 1615, 1982.
7. (a) Brown, C. E.; Jones, M. B.; Kovacic, P., *J. Polym. Sci., Polym. Lett. Ed.*, *18*, 653, 1980.
   (b) Murray, D. P.; Dechter, J. J. and Kispert, L. D., *J. Polym. Sci., Poly. Lett. Ed.*, *22*, 519, 1984.
   (c) Miller, J. B. and Dybowski, C., *Solid State Commun.*, *46*, 487, 1983.
8. Ouchi, T.; Otsuka, S. and Imoto, M., *Bull. Chem. Soc. Japan*, *48*, 918, 1975.
9. Al-Jumah, K. and Fernandez, J. E., *Macromolecules*, *20*, 1181, 1987.
10. Bouley, J. E. and Jones, M. B., *Pro. N. D. Acad. of Sci.*, *42*, 73, 1988.
11. Hasegawa, H.; Imanari, M. and Ishizu, K., *Bull. Chem. Soc. Japan*, *45*, 1153, 1972.

Chapter 6

RADIATION INDUCED MORPHOLOGICAL CHANGES IN A SEMI-CRYSTALLINE POLYMER: A ^{13}C NMR STUDY

A. L. CHOLLI, F. C. SCHILLING, AND A. E. TONELLI

1. Abstract

The structural and morphological changes occurring in poly(ethylene oxide) (PEO) when exposed to gamma-irradiation have been investigated by both solution and solid state ^{13}C NMR techniques. The most striking change in the solid state CP/MAS ^{13}C NMR spectrum is the growth of a resonance peak at 71.9 ppm as a function of radiation dose. This resonance peak is assigned to the crystalline carbons arising from the local rearrangements of polymer chains upon irradiation. The chemical shift separation of 1.4 ppm between the amorphous and crystalline resonances is attributed to differences in polymer chain conformations. Solution NMR data suggest the predominance of chain scission for samples irradiated in air, while vacuum-irradiated samples appear to be predominantly cross-linked.

2. Introduction

The effect of high-energy radiation on semi-crystalline polymers results in both chemical and physical changes. The most frequently observed chemical changes are the formation of intermolecular cross-links and main-chain scissions. These chemical changes have a direct impact on the physical properties of the irradiated polymers. Characterization of gamma-irradiated polymers is, therefore, essential to understand

---

A. L. CHOLLI - BOC Group, Inc., Murray Hill, NJ 07974; F. C. SCHILLING and A. E. TONELLI - AT&T Bell Laboratories, Murray Hill, NJ 07974.

Solid State NMR of Polymers, Edited by L. Mathias
Plenum Press, New York, 1991

their structure-property relationships. Because of the complex nature of the system, information regarding the structure of irradiated, solid polymers at the molecular level is rather limited.

In recent years, solid state ^{13}C NMR techniques have been used to study various aspects of semi-crystalline polymers. Using the combined techniques of cross-polarization (CP), high power decoupling, and magic angle sample spinning (MAS), it is possible to study the morphology, conformation, and molecular motions of semi-crystalline polymers in the solid state [1]. The basis behind the success of these applications is the ability of solid state NMR techniques to identify resonances for crystalline and amorphous carbons [2-6]. Until recently, these techniques have not been explored to study changes in the morphology of semi-crystalline polymers upon their gamma-irradiation [7].

Cholli et al. reported, based on the changes in the dipolar-dephased time decay constants (T_{DD}), that cross-linking in irradiated polyethylene (PE) takes place preferentially in the amorphous phase and occurs infrequently in the crystal interior [7]. It is believed that this kind of selectivity of the radiation attack in semi-crystalline polymers, originates from a preferential formation of radicals in the non-crystalline regions and possibly their subsequent migration to the crystal surface from the amorphous phase [8]. In this chapter, we focus on the effect of gamma-irradiation on the morphology of another semicrystalline polymer, poly(ethylene oxide) (PEO). A strong ^{13}C resonance, attributed to radiation induced morphological changes, has been identified and assigned to the crystalline carbons arising from the local rearrangements of polymer chains upon irradiation. Solution ^{13}C NMR data suggest the presence of chemical changes: polymer chain scission and oxidative reactions.

3. Experimental

3.1. Materials

The poly(ethylene oxide) (PEO) (molecular weight, 100,000) was obtained in the powder form from Aldrich Chemical Company. Two sets of samples were prepared: one set of samples was exposed to gamma-radiation in air at different dosages, and a second set was irradiated in vacuum. A ^{60}Co source was used for irradiation at room-temperature.

3.2. ^{13}C NMR Measurements

All ^{13}C NMR measurements were performed at room-temperature on a Bruker AC-270 MHz NMR spectrometer operating at 67.9 MHz.

3.3. Solution NMR

The spectral conditions were 90° pulse width, 5.5 μs; spectral width, 17857 Hz; number of data points, 32 K; recycle time, 10 s; and number of scans, 10,000.

To suppress heteronuclear NOE, the broad-band decoupler was turned on only during the data acquisition. Deuterated chloroform was used as the solvent.

3.4. Solid State NMR

A Doty Scientific probe and spinner assembly were used for solid state NMR measurements. Sample rotation at the magic angle (MAS), 54.7°, was 4500 Hz. Optimization of the Hartman-Hahn and magic angle settings were made using samples of adamantane and glycine, respectively. A typical contact time in the CP/MAS experiments was 2 ms. The proton 90° pulse width was 5 μs. All spectra were obtained using a spectral width of 20 KHz, a data size of 2 K, with an acquisition time of 51 ms. Unless otherwise noted, 10,000 scans were collected for each measurement. A delay of 200 μs was used in the dipolar-dephased experiment.

4. Results and Discussion

4.1. Narrow and Broad Resonances in Solid State ^{13}C NMR Spectrum of Poly(ethylene oxide) PEO

Figure 1a shows the room-temperature CP/MAS ^{13}C NMR spectrum for the solid PEO. In this spectrum, there are two components to the complex ^{13}C NMR line shape, a narrow resonance superimposed on a broad resonance. From such a spectrum alone, it is rather difficult to obtain any structural and morphological information about the polymer system. Additional experiments are required to assign the narrow and broad ^{13}C NMR peaks in the CP/MAS spectrum.

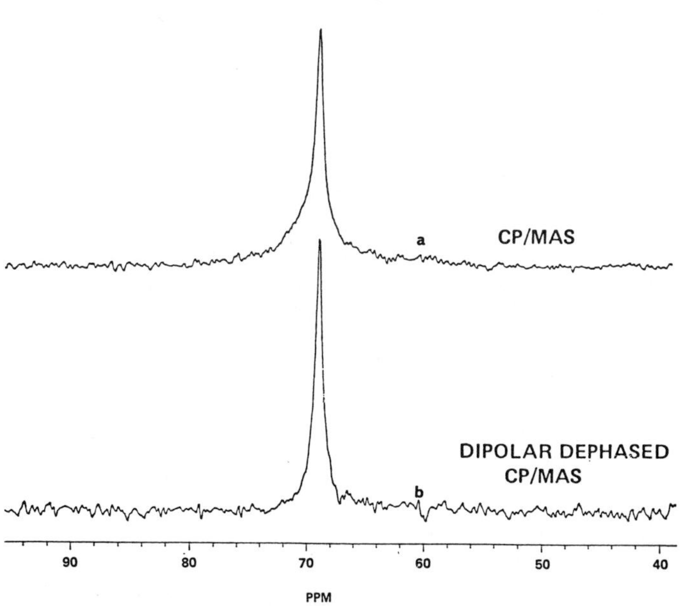

Figure 1. Solid state 67.9 MHz ^{13}C NMR spectra of PEO, a) CP/MAS experiment with a contact time of 2ms. b) Dipolar-dephased experiment with a delay of 200 μs.

Figure 1b shows the dipolar-dephased ^{13}C NMR spectrum. We observe only the narrow component of the spectrum in Fig. 1a. The dipolar-dephased experiment is the standard CP/MAS pulse sequence with a slight modification. A brief delay, in microseconds, without high-powered proton decoupling is inserted after the build-up of ^{13}C magnetization and before data acquisition with full proton decoupling [9]. This procedure has been used to identify resonances from different domains or phases of semicrystalline polymers [1,3,10]. Other applications of this pulse sequence include: i) the spectral editing of ^{13}C CP/MAS spectra to observe non-protonated carbon resonances [9], ii) two-dimensional, separated local field experiments by incrementing the interval delay [11], and iii) for measurement of T_2 for ^{13}C nuclei in plastic crystals [12].

The ability of the dipolar-dephased experiment to resolve resonances from different phases of a multi-phase polymer system depends on many factors. Among these are the strength of ^{1}H-^{13}C heteronuclear dipolar interactions and the presence of different molecular motions in the multi-phase polymer system. In addition, the difference between the isotropic chemical shifts of these domains is also an important factor in the study of multi-phase systems.

In a tightly coupled ^{1}H-^{13}C system, such as ^{13}C nuclei in the crystalline domain of a multi-phase system, the strength of the dipolar interaction is large compared to the interaction between nuclei in the non-crystalline region. The ^{13}C magnetization of these crystalline nuclei decays rapidly during the delay of the dipolar-dephased experiment [3]. The rapid loss of ^{13}C magnetization is mainly caused by the destructive interference of the collection of dipolar splittings [9]. Thus, the spectrum resulting from a suitable dipolar-dephasing delay corresponds to the mobile, non-crystalline component of the semicrystalline polymer. Hence, the narrow peak in Fig. 1b is assigned to the non-crystalline phase of PEO, and the broad peak in Fig. 1a to the crystalline region. This technique has also been used to identify amorphous resonance peaks in other semi-crystalline polymers, polyethylene (PE) [1,10] and poly(oxymethylene) (POM) [3].

An alternative approach to identifying the resonances of the crystalline and non-crystalline regions is to measure the spin-lattice relaxation times, T_1. To some extent, the carbon T_1's in solids depend on the physical state of the system. Molecular motions in the amorphous region result in reasonably short relaxation times, while the lack of molecular motions in the crystalline region lead to longer T_1's for the crystalline carbons. The reported T_1 value for the crystalline carbons of PEO is about 16 s, and 0.2 s for the amorphous carbons [13]. The T_1 value for crystalline PEO carbons is very small compared to those of other crystalline systems (in the range of several minutes to hours) [1,2,14]. This may indicate the presence of significant molecular motion in PEO crystallites. At room-temperature, the PEO sample is well above its glass-transition temperature, T_g (-85°C), and close to its melting temperature T_m (68°C) [15].

The difference in T_1 values for amorphous and crystalline phases of PEO was utilized in identifying resonances in Fig. 2. Both spectra a and b were obtained using the conventional ^{13}C $\pi/2$ repetitive pulsing technique, with the high power proton

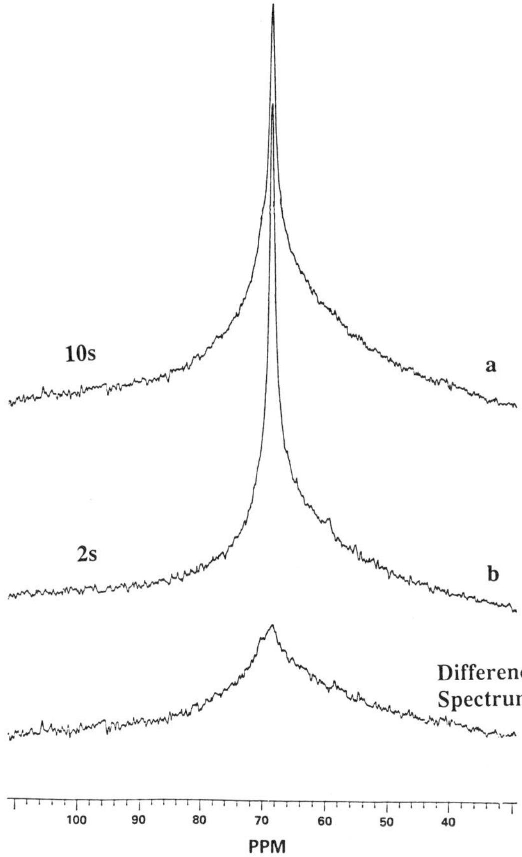

Figure 2. Magic angle spinning, dipolar decoupled 67.9 MHz ^{13}C NMR spectra with a recycle time of a) 10s and b) 2s. Spectrum c is the difference spectrum, a-b.

decoupling on only during the data acquisition. Recycle delay times were 10 s and 2 s for spectra a and b, respectively. Major contributions to the magnetization in spectrum b are from those carbons with short T_1s, i.e., amorphous PEO. In addition to magnetization from amorphous nuclei, in spectrum a, we observe build-up of ^{13}C magnetization from carbons with long T_1's, i.e., crystalline carbons. Increase in the magnetization due to carbons from the crystalline region is demonstrated from the difference spectrum, c in Fig. 2.

Alternatively, these two components can also be identified by varying the contact time of the CP/MAS experiment [13]. Dechter found that for short contact times, 100μs, the superimposed broad and narrow components are obtained; for long contact times, 800μs, the narrow component is obtained in the CP/MAS spectrum [13].

From the discussion above, it is clear that the broad peak arises from the crystalline domain carbons. By contrast, for most semi-crystalline polymers, the crystalline peak is narrow compared to the amorphous peak [2]. There are two main reasons for observing a broad, weak peak for the crystalline domains of PEO. First the efficiency of the CP/MAS experiment depends on abundant proton spins having a proton $T_{1\rho}$ of sufficient duration to allow complete polarization transfer to build-up ^{13}C magnetization. In the case of PEO, the proton $T_{1\rho}$ is near a deep minimum at room-temperature [16]. This results in the relaxation of the proton rotating-frame magnetization before appreciable carbon magnetization can be polarized. Therefore, ^{13}C peaks will be weak in the CP/MAS spectrum. Second, motional effects can also cause broadening of the peak. In PEO, the polymer chain motions in the crystalline region may be such that spin-decoupling is not efficient in narrowing resonances [14]. To overcome some of these broadening contributions to the ^{13}C resonance for crystalline domains, Fleming et al. have performed CP/MAS experiments at lower temperatures [17], where a substantial reduction of motion and greater homogeneity of the local environment lead to a narrowing of crystalline resonances.

Even though it is possible, by various solid state ^{13}C NMR experiments, to identify crystalline and amorphous resonances, it is still difficult to find the chemical shift separation between these peaks for PEO owing to the extreme breadth of the crystalline peak. The chemical shift difference between the two peaks can be used

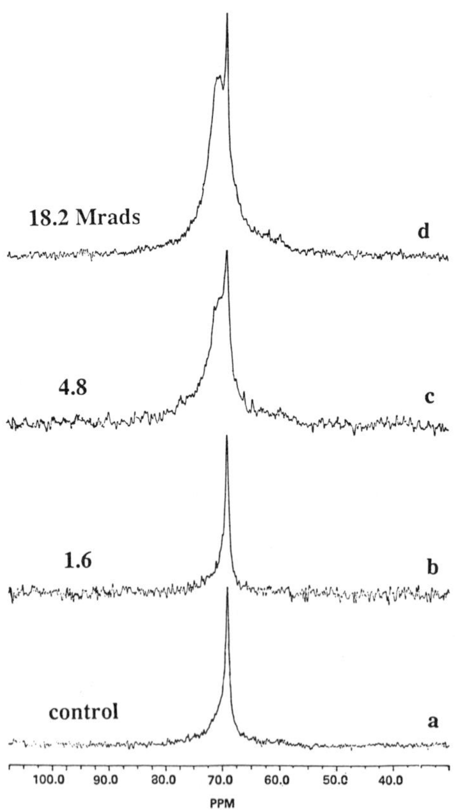

Figure 3. Solid state 67.9 MHz ^{13}C CP/MAS spectra for PEO samples irradiated in air: a) 18.2 Mrads, b) 4.8 Mrads, c) 1.6 Mrads, and d) control sample.

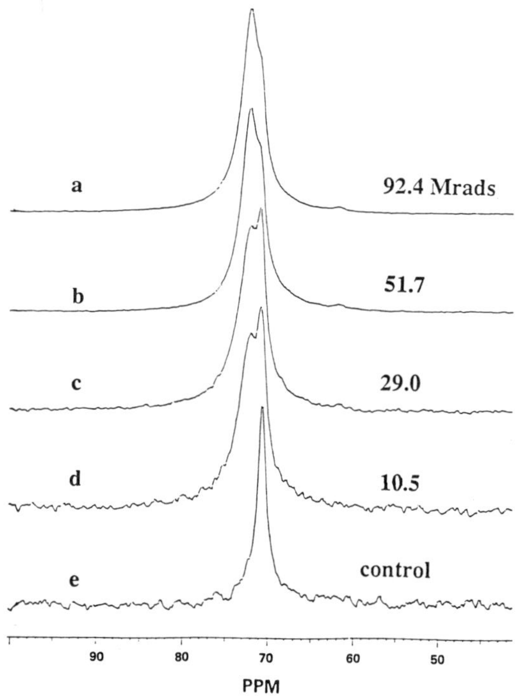

Figure 4. Solid state 67.9 MHz ^{13}C CP/MAS spectra for PEO samples irradiated in vacuum: a) 92.4 Mrads, b) 51.7 Mrads, c) 29.0 Mrads, d) 10.5 Mrads, and e) control.

to understand polymer chain conformations in both regions. A more detailed analysis will be discussed later concerning the sensitivity of chemical shifts to polymer chain conformation. First, we discuss the changes in CP/MAS spectra due to the effect of gamma-radiation on PEO.

4.2. Gamma-Irradiated PEO: Solid State ^{13}C NMR Spectra

Figure 3 shows the changes in the solid state ^{13}C CP/MAS spectrum of PEO at various levels of irradiation in air. These data show the growth of a broad down-field peak at 71.9 ppm. In the case of samples irradiated at higher doses, we observe more pronounced changes in the ^{13}C NMR data. To our knowledge, these are the first spectra to show a substantial change in the ^{13}C NMR spectra for gamma-irradiated semi-crystalline polymers. As we discuss later, these changes are directly related to modifications in their morphology. The chemical shift of the narrow component is the same as in the control sample (70.5 ppm). Similar observations were also made for the PEO samples irradiated in vacuum. In Fig. 4, we show ^{13}C CP/MAS spectra for the PEO sample irradiated in vacuum at 10.5, 29.0, and 92.4 Mrads. In the spectra for samples irradiated with more than 50 Mrads, the intensity of the down-field peak is more than the intensity of the up-field peak.

Using difference spectroscopy, it is possible to isolate the down-field peak from the up-field peak (Fig. 5). The chemical shift difference, 1.4 ppm, between the

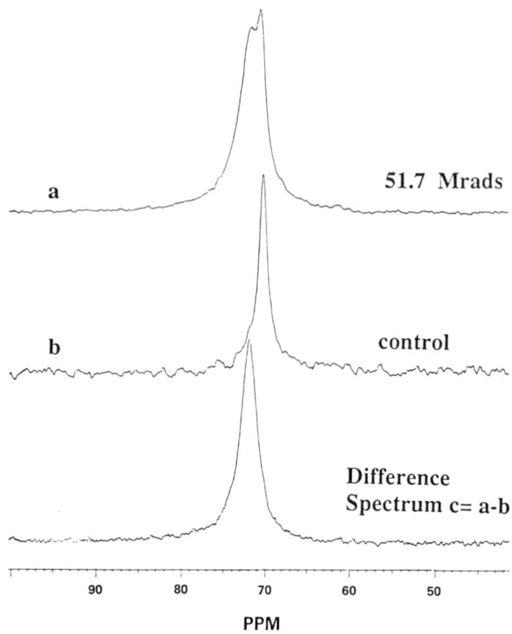

Figure 5. Solid state 67.9 MHz ^{13}C CP/MAS spectra of PEO samples: a) irradiated in vacuum at 51.7 Mrads, b) control sample, c) difference spectrum, a-b.

down-field and the up-field peaks can be easily determined from Figs. 3-5. It is interesting to observe that these gamma-irradiated PEO samples (irradiated in air) readily dissolve in chloroform, and imply that chain scission predominates over cross-linking. This permits measurement of high resolution ^{13}C NMR spectra for gamma-irradiated PEO samples. On the contrary, the PEO samples irradiated in vacuum were insoluble. This indicates the predominance of crosslinking.

4.3. High Resolution Solution ^{13}C NMR Spectra

High resolution, NOE suppressed ^{13}C NMR solution spectra for both control and 18.2 Mrad (in air) PEO samples are shown in Fig. 6. In this figure, both the up-field and down-field regions are separately plotted and also vertically expanded to show the presence of new peaks due to the effect of gamma-irradiation on the PEO polymer chains. In the spectrum for control PEO, apart from the strong main peak at 70.5 ppm, there are only two observable weak resonances, at 61.6 and 72.4 ppm. The intensity of these two peaks are enhanced in Fig. 6b for gamma-irradiated PEO. In addition to these two weak resonances, other weak resonances are also observed in both the up-field and down-field regions.

The quantitative results of ^{13}C NMR measurements on PEO both prior to and after 18.2 Mrad irradiation in air are presented in Table 1. These data were obtained from the spectra shown in Fig. 6. The results in Table 1 were obtained by dividing the intensity of a resonance peak by the total carbon intensity for the entire spectrum

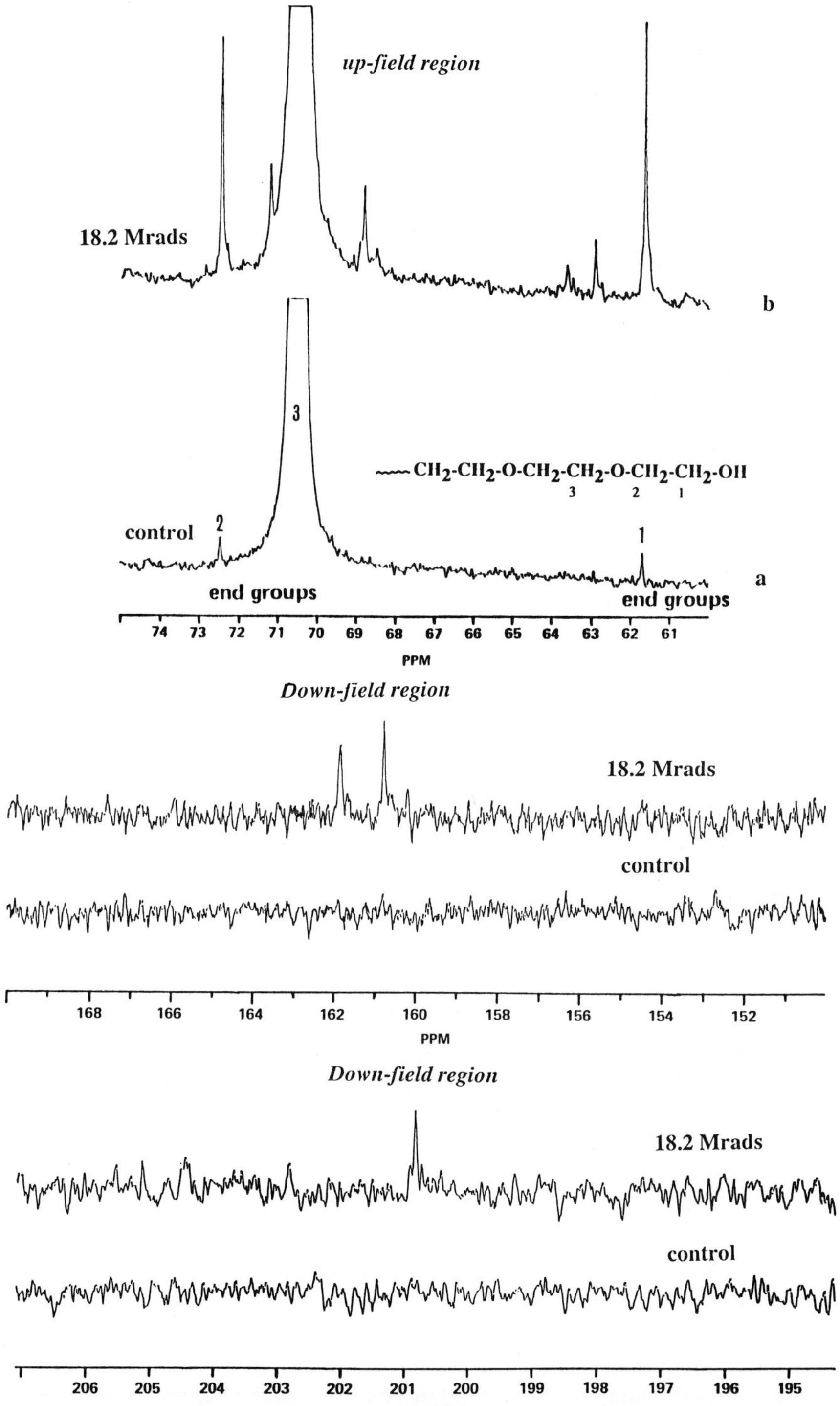

Figure 6. High resolution, solution ^{13}C NMR spectra showing up-field and down-field regions for irradiated PEO and control samples: a) control sample, b) 18.2 Mrads in air.

TABLE I. Concentrations of Structures Produced by Irradiation of PEO (18.2 Mrads in Air)

Resonance ppm	Structural assignments[x]	No. of Units per 10,000 Before Irradiation	Carbon Atoms After Irradiation
61.6	na	11.4	137.0
62.9	e2	---	27.4
63.5	a1, b2, g5	---	18.3
68.5	f2, g1	---	18.3
68.8	na	---	45.7
70.5	a3, b3, e3, f1	9982.9	9543.2
71.2	na	---	59.4
72.4	a2	5.7	100.5
73.5	d2, g3	---	y
160.9	z	---	18.3
161.8	z	---	18.3
200.8	b1, c1, d1	---	13.7

[x] see Fig. 7
y: very weak resonance
z: in the range of chemical shifts for anhydrides
na: not assigned

and multiplying by 10,000, yielding the number of structural units (corresponding to the resonance peak) per 10,000 carbon atoms.

Possible structural changes in the PEO polymer chain and corresponding chemical shifts calculated using a semi-empirical approach [18] are presented in Fig. 7. Chemical changes in PEO due to gamma-irradiation have been studied by many investigators and this has resulted in various reaction mechanisms and different conclusions [19]. For example, Pearson concluded, based on his work on low-molecular weight PEO, that gamma-irradiation causes cross-linking at low doses, after which chain degradation predominates [20]. On the contrary, Nitta and co-workers, based on their ESR work, suggested cross-linking in PEO, but no degradation [21]. Still others report that the radiation causes both cross-linking and main chain scissions, but their relative importance is mainly dependent on experimental conditions [19]. It appears, from the present work, that the main effect on PEO is chain scission, and also possibly oxidative degradation for samples irradiated in air.

Weak resonances at 61.6 and 72.4 ppm are assigned to the chain end group, ~CH_2-CH_2OH. It is evident from Table 1, that there is at least a ten-fold increase in the end group concentration for 18.2 Mrad PEO. The presence of resonances at 160.9, 161.8, and 200 ppm indicate the occurrence of oxidative reactions. Other kinds of possible end groups formed during gamma-irradiation and their corresponding calculated chemical shifts are shown in Fig. 7. Tentative assignments of the remaining resonances in the spectrum for gamma-irradiated PEO are also presented in Table 1.

4.4. Radiation Induced Morphological Changes

At this point it is instructive to examine closely the radiation effects on PEO, because small chemical changes may have pronounced effects on the morphology. These changes, both chemical and morphological, can be studied by NMR spectroscopy using solution and solid state ^{13}C NMR techniques. As discussed in the preceding section, irradiation in air causes polymer chain scission. This is evidenced, in the solution NMR spectrum, by an increase in the end group concentration. In addition, the solution spectra do not indicate the presence of cross-linking. These chemical changes are manifested differently in the solid state NMR spectra. The increased intensity of the down-field crystalline peak as a function of radiation dose indicates a change in the solid state morphology.

The major contributions to the intensity of the down-field peak in the ^{13}C NMR spectra for gamma-irradiated PEO may result from the local rearrangements of small polymer chains to form a modified crystalline order. Another possibility is the

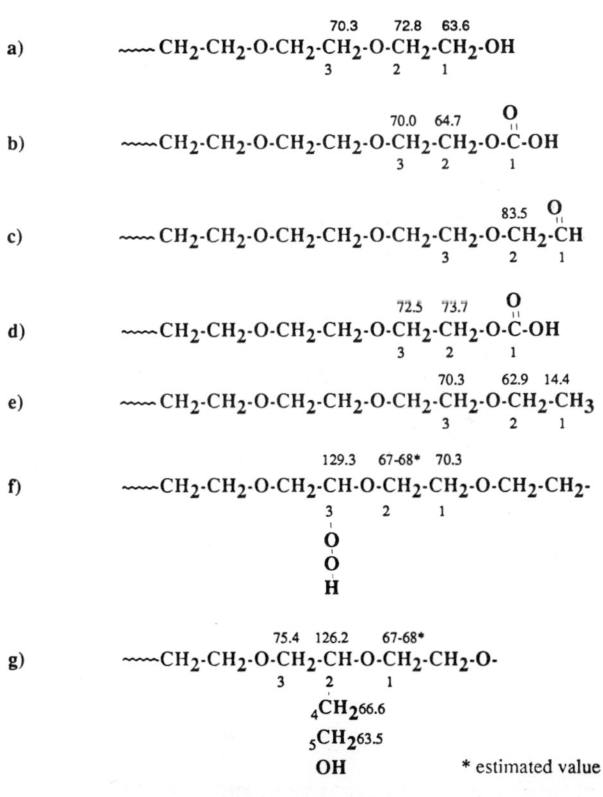

Figure 7. Possible structural entities for PEO irradiated in air and their calculated ^{13}C NMR chemical shifts. See ref. 18 for semi-empirical equations to calculate chemical shifts.

contribution from the distribution of chemical shifts due to various polymer chain lengths. However, the solution NMR data do not indicate such a distribution of chemical shifts for -(CH_2CH_2O)- units. Therefore, the appearance of the down-field peak is mainly associated with morphological changes.

In order to establish that the down-field peak is due to crystalline domain carbons, we used a rotational isomeric state (RIS) model [22-24] to calculate chemical shifts for crystalline and amorphous carbons. This method of solid state conformational analysis and the prediction of chemical shifts has been successfully employed for various polymers, including polyethylene [2], poly(oxymethylene) [3], 1,4-trans-polybutadiene [5], and poly (1-butene) [4].

According to Tadokoro and coworkers [25,26], the chain conformation of crystalline PEO is a 7_2 helix with a succession of nearly trans (C-C-O-C), trans (C-O-C-C), and gauche (O-C-C-O) conformations and is shown in Fig. 8.

As demonstrated by Newman projections in Fig. 8b, there are no shielding gamma-gauche effects for the methylene carbons of the crystalline region, whereas the methylene carbons of the amorphous region would be expected to be shifted up-field from the crystalline position by 2(.27) gamma-gauche arrangements for CH_2, i.e., 2(.27) (-5.0)=-2.7 ppm. The observed amorphous peak in the gamma-irradiated PEO is shifted -1.4 ppm from the crystalline peak. On the other hand, based on experimental data in Fig. 1, it is not possible to find the chemical shift difference between crystalline and amorphous peaks from the control sample, owing to the fact that the crystalline peak is very broad at room temperature and therefore it is difficult to measure its chemical shift. Our assignment of the down-field peak to the crystalline carbons is also consistent with the recent work by VanderHart and coworkers [27], on a very low molecular weight PEO sample.

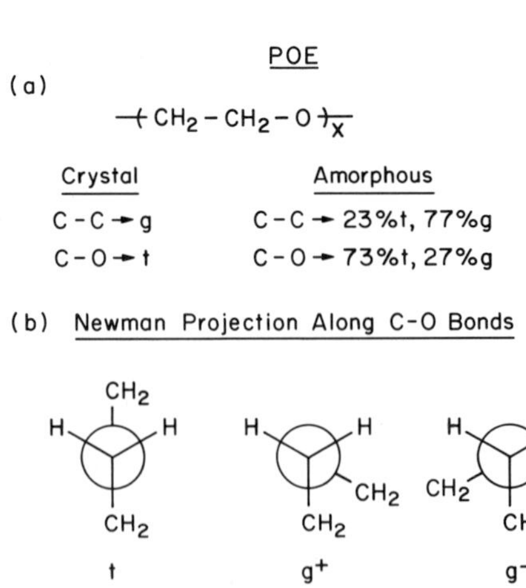

Figure 8. a) Portion of a PEO chain showing the chain conformation in the crystalline and amorphous regions. b) Newman projections illustrating the gamma-gauche effects on the CH_2 carbons.

It is of interest to note, from the present study that the crystalline peak in the CP/MAS spectrum narrows upon gamma-irradiation, though it is still broader than the amorphous peak. The time scale of those polymer chain motions which affect efficient cross-polarization and effective dipolar-decoupling may have also shifted for the irradiated samples compared to the PEO control. This explanation is speculative, and additional experimental results are required to show the change in proton $T_{1\rho}$'s is produced by gamma-irradiation.

In summary, changes in solid state ^{13}C NMR spectra are reflecting morphological alterations resulting from irradiation. The increased intensity of the down-field peak assigned to the crystalline carbons is attributed to the local rearrangements of polymer chains upon irradiation. In the case of PEO samples irradiated in air, a ten-fold increase in the end group concentration suggests the occurrence of chain scission, and the failure to observe resonances corresponding to cross-links indicates their rarity. After chain scission, newly formed, short-length polymers may rearrange themselves to form a regular order. The chemical shift difference between the crystalline (71.9 ppm) and amorphous (70.5 ppm) carbon resonance is consistent with the chemical shift difference expected between crystalline chains adopting a regular 7_2 helix, where all C-O bonds are trans, and the randomly-coiling amorphous chains, were both gauche and trans C-O bonds are rapidly interconverting.

5. References

1. Axelson, D.E., in *High Resolution NMR of Synthetic Polymers in Bulk*, R. A. Komorski Ed., VCH publishers, Deerfield Beach, FL, 1986.
2. Earl, W. L. and VanderHart, D. L., *Macromolecules*, *12*, 762, 1979.
3. Cholli, A. L.; Ritchey, W. M. and Koenig, J. L., *Spectrosc. Lett.*, *16*, 21, 1983.
4. Belfiore, L. A.; Schilling, F. C.; Tonelli, A. E., Lovinger, A. J. and Bovey, F. A., *Macromolecules*, *17*, 2561, 1984.
5. Schilling, F. C.; Bovey, F. A.; Tonelli, A. E.; Tseng, S. and Woodward, A. E., *Macromolecules*, *17*, 728, 1984.
6. Bunn, A.; Cudby, M. E. A.; Harris, R. K.; Packer K. J. and Say, B. J., *Chem. Commun.*, *15*, 1981.
7. Cholli, A. L.; Ritchey, W. M. and Koenig, J. L., *Appl. Spectrosc.*, *41*, 1418, 1987.
8. Lawton, E. J.; Balwit, J. S. and Powell, R. J., *J. Polym. Sci.*, *32*, 257, 1958.
9. Opella, S. J. and Frey, M. H., *J. Am. Chem. Soc.*, *101*, 5854, 1979.
10. Cholli, A. L.; Ritchey, W. M.; Koenig, J. L. and Veeman, W. S., *Spectrosc. Lett.*, *21*, 531, 1988.
11. Hester, R. K.; Ackerman, J. L.; Neff, B. L. and Waugh, J. S., *Phys. Rev. Lett.*, *36*, 1081, 1976.
12. Lippmaa, E. and Alla, M., *Chem. Phys. Lett.*, *37*, 260, 1976.

13. Dechter, J. E., *J. Polym. Sci., Polym. Lett.*, *23*, 261, 1985.
14. Lyerla, J. A., in *Methods in Experimental Physics*, Vol. *16A*, Academic Press, Inc., NY, 1980.
15. Boyer, R. F., in *Encyclopedia of Polymer Science and Technology*, Suppl. Vol. *II*, p 816, John Wiley and Sons, Inc., NY, 1977.
16. Connor, T. M. and Hartland, A. *J. Polym. Sci. A-2*, *7*, 1005, 1969.
17. Fleming, W. W.; Fyfe, C. A.; Hendrick, R. D.; Lyerla, J. R.; Vanni, H. and Yannoni, C. S., in *Polymer Characterization by ESR and NMR*, ACS Symposium Series, #142, Am. Chem. Soc., Washington, DC, 1980.
18. Pretsch, E.; Clerc, T.; Seibl, J. and Simon, W., *Table for Spectral Data for Structure Determination of Organic Compounds*, p c10, Springer-Verlag, NY 1983.
19. King, P. A., *Adv. Chem. Ser.*, *66*, 113, 1967, and references cited therein.
20. See reference 14 in 19 above.
21. See reference 12 in 19 above.
22. Mark, J. E. and Flory, P. J., *J. Am. Chem. Soc.*, *87*, 1415, 1965.
23. Mark, J. E. and Flory, P. J., *J. Am. Chem. Soc.*, *88*, 3702, 1966.
24. Flory, P. J., *Statistical Mechanics of Chain Molecules*, Interscience Publishers, NY, 1969.
25. Tadokoro, H.; Chatani, Y.; Kobayashi, M.; Yoshihara, T.; Murahashi, S. and Imada, K., *Rept. Progr. Polymer Phys. Japan*, *6*, 303, 1963.
26. Tadokoro, H.; Chatani, Y.; Yoshihara, T.; Tahara, S. and Murahashi, S., *Makromol. Chem.*, *74*, 109, 1964.
27. VanderHart, D. L., Private Communication.

Chapter 7

^{13}C NMR STUDIES OF POLY(VINYLIDENE FLUORIDE) BLENDS

R. A. GRINSTED AND J. L. KOENIG

1. Abstract

C-13 NMR was used to investigate the compatibility and subsequent separation of solution casted blends of poly(vinylidene fluoride)/poly(methyl methacrylate) and poly(vinylidene fluoride)/poly(vinyl acetate) with aging for a range of compositions. It was found that one phase existed for all compositions of the blends even after two months of aging at room temperature. There was an increase in the $T_{1\rho H}$ and T_{CH} values with aging indicating that a subtle separation between unlike chains in the blend was occuring though phase separation was not present.

2. Introduction

Nuclear magnetic resonance has been used to examine the molecular interactions and the dimensional level of homogeneity of a number of polymer blend systems, e.g. poly(phenylene oxide)/polystyrene [1], poly(vinyl methyl ether)/ polystyrene [2], poly(vinylidene fluoride)/poly(methyl methacrylate) [3-6] and poly(methyl methacrylate)/poly(styrene co-acrylonitrile) [7]. A number of parameters in NMR have been employed, e.g. spin-lattice relaxation times (T_1), spin-spin relaxation times (T_2), spin-lattice relaxation times in the rotating frame ($T_{1\rho}$) and cross relaxation times for varying blend compositions, thermal treatments and degrees of

---

R. A. GRINSTED AND J. L. KOENIG - Department of Macromolecular Science, Case Western Reserve University, Cleveland, Ohio 44106.

Solid State NMR of Polymers, Edited by L. Mathias
Plenum Press, New York, 1991

phase separation in order to better understand the nature of the interactions between polymers in a blend system. Understanding these interactions can help give some insight into the effect blending has on the improvement of the physical and chemical properties of a polymer.

Blends of poly(vinylidene fluoride) (PVF_2), which is known for its unique piezoelectric and pyroelectric properties, and poly(methyl methacrylate) (PMMA) have been shown to be compatible over a range of compositions even though PVF_2 is crystallizable. Since the blending of PVF_2/PMMA was first reported in 1971 [8], a variety of studies have been made using techniques such as differential scanning calorimetry [8-11], mechanical relaxation [12], Fourier transform infrared [13,14], X-ray [15], dilatometry [8] and dielectric relaxation [16] to characterize these blends. In particular, NMR methods which are sensitive to short range interactions have been utilized to study PVF_2/PMMA blends. Douglass and McBrierty [4] used transient Overhauser effects and F-19 resonance T_1, $T_{1\rho}$, and T_2 measurements over a large temperature range and determined that a large amount of the PVF_2 molecules were at nearest-neighbor distances in the amorphous regions with PMMA molecules, showing that there is a high degree of intimate mixing on the molecular level.

The first work with blends by solid state C-13 NMR was by Schaefer et al. using poly(phenylene oxide)/polystyrene blends [1] and polystyrene/polybutadiene block copolymer blends [17]. They showed that spin lattice relaxation in the rotating frame for protons attached to different carbons can be measured using high resolution C-13 NMR [18]. These proton spin lattice relaxation rotating frame parameters are very sensitive to the short range spatial proximity of interacting dipole moments of the protons and can be used in determining the homogeneity of the blend [19]. This method has been used by Lin and Ward [6,20] with high resolution C-13 NMR for PVF_2/PMMA and PVF_2/PVAc blends to observe intensity attenuations for melt extruded and dimethyl formamide and methyl ether ketone solution casted blends. Results indicated different degrees of intermolecular association in the blends. Tekely et al. [3] used cross polarization with variable contact times to examine local heterogeneities in compression molded PVF_2/PMMA blends under a variety of heat treatments to induce phase separation occuring from PVF_2 crystallization.

The purpose of this study is to investigate the effect of aging on the PVF_2/PMMA and PVF_2/PVAc blend systems using the method of cross polarization with variable contact times. Changes in the $T_{1\rho H}$ and T_{CH} values will be used to indicate the presence of phase separation in the blend.

3. Experimental

PVF_2 was obtained in powder form (Kynar 461) from Pennwalt Corp. with a $M_w = 534{,}000$. PMMA and PVAc beads were obtained from Polysciences Inc. with $M_w = 25{,}000$ and $M_w = 90{,}000$, respectively. Blends of PVF_2 with PMMA and PVAc having weight ratios from 20:80 to 80:20 were solution casted from one percent solutions of dimethyl formamide (DMF) on glass surfaces at 60° C under continuous vacuum for 24 hours. The blends were then quenched to room temperature and placed under vacuum an additional 24 hours at room temperature to remove residual solvent. This procedure proved effective since no solvent peaks were present in the NMR spectra. The films were then peeled off the glass surfaces and packed in 7mm aluminum oxide spinners with Kel-F caps. Aluminum oxide powder was added to aid in spinning and trihydroxy methyl silane ($CH_3Si(OH)_3$) was used as a chemical shift marker (-2.5 ppm).

Solid state C-13 NMR experiments using cross polarization, dipolar decoupling, and magic angle spinning were run on the blends. A C-13 frequency of 75.47 MHz on a Bruker MSL300 spectrometer was used for the PVF_2/PMMA blends and a C-13 frequency of 37.725 MHz on a Nicolet NTC150 spectrometer was used for the PVF_2/PVAc blends. Experiments at the magic angle were performed at a spinning speed of 3.6 kHz and 3.0 kHz for the PVF_2/PMMA and PVF_2/PVAc blends respectively. The PVF_2/PMMA blends analyzed on the MSL300 required 600 scans to obtain a good signal to noise ratio while the PVF_2/PVAc blends studied on the NTC150 needed 2000-2400 scans.

The magnetic field was shimmed with an adamantane sample spinning at 2.0 kHz. The FID was maximized for area and length at a spectral size of 8K and a time domain of 4K. The 90° pulse was then set by obtaining the difference between the null points for the 360° and 720° pulses and dividing by 4. The match was set by maximizing the signal for adamantane. The magic angle was checked with a potassium bromide (KBr) sample at a Br-79 resonance of 75.195 MHz and the sidebands were maximized. The blend sample was then placed in the probe, tuned and a variable contact time cross polarization experiment was set up. The pulse sequence is shown in Fig. 1.

Cross polarization experiments with variable contact times were used to make T_{CH} (time constant for transfer of magnetization) and $T_{1\rho H}$ (spin lattice relaxation time in the rotating frame) measurements. Contact times from 10 microseconds to 10 milliseconds were used and pulse widths for 90° pulses were 3.2 microseconds and 3.55-3.8 microseconds for the PVF_2/PMMA and PVF_2/PVAc blends respectively. Delays between scans (dead times) were 2 and 3 seconds respectively for the PVF_2/PVAc and PVF_2/PMMA blends. The T_{CH} and $T_{1\rho H}$ values were determined by

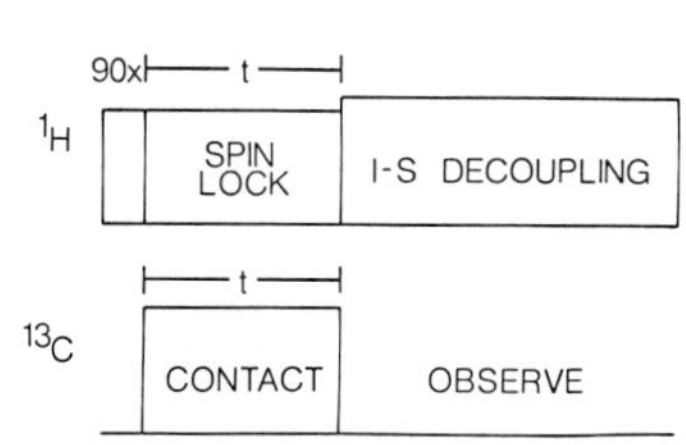

Figure 1. Pulse sequence for the variable contact time cross polarization experiment.

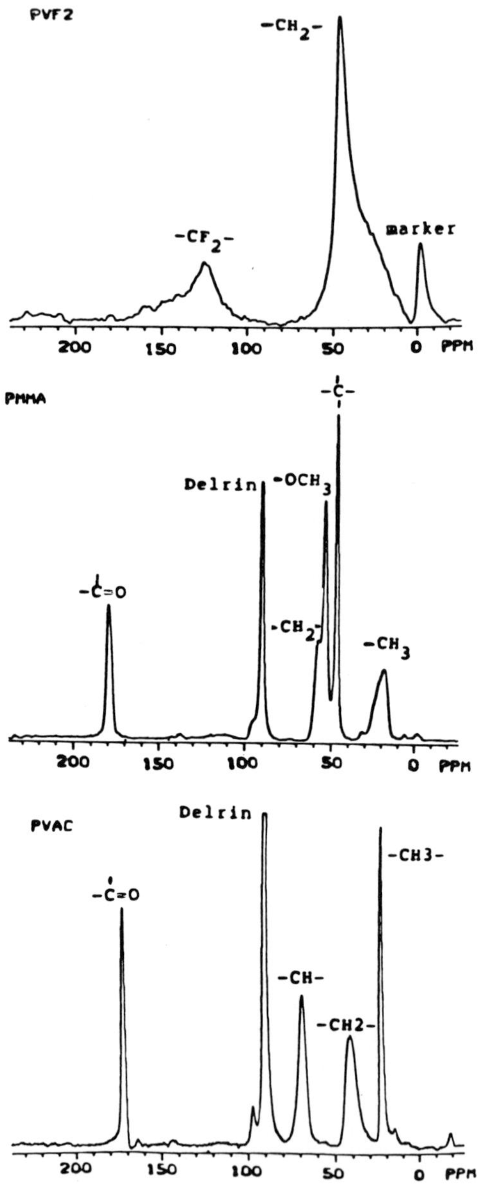

Figure 2. The C-13 NMR spectra of PVF_2, PMMA and PVAc with their respective resonance assignments.

computer analysis. The C-13 spectra of PVF_2, PMMA and PVAc are shown in Fig. 2 with their resonance assignments.

4. Results and Discussion

The change in polarization with cross polarization contact time for a 20:80 PVF_2/PMMA blend is shown in Fig. 3. The proton $T_{1\rho}$'s were calculated from the slope of the peak intensities of the blend spectra versus cross polarization contact time. By plotting ln (I/I_{max}) versus contact time (Fig. 3) and using the longer contact times, the $T_{1\rho H}$'s can be found by taking the negative of the inverse of the slope (-1/slope). The time constant for transfer of magnetization from proton to carbon or cross polarization constant, T_{CH} can also be found from Fig. 3 and is related to the inverse of the slope using the shorter contact times (usually less than 100 microseconds) [18]. A steeper slope is an indication of faster transfer of magnetization (shorter T_{CH}) or faster relaxation by spin diffusion (shorter $T_{1\rho H}$) depending on the part of the ln (I/I_{max}) versus contact time curve being used. A rigid system usually has a smaller TCH than a non-rigid system and the system would normally relax faster, so the $T_{1\rho H}$ would be smaller also. Phase separation or inhomogeneities in the system can be detected using the $T_{1\rho H}$ part of the ln (I/Imax) versus contact time curve. For a phase separated system the curve will be bent, having two slopes, each related to the $T_{1\rho H}$ for its respective phase. If the curve was linear, this would be an indication of efficient spin diffusion and one phase would prevail.

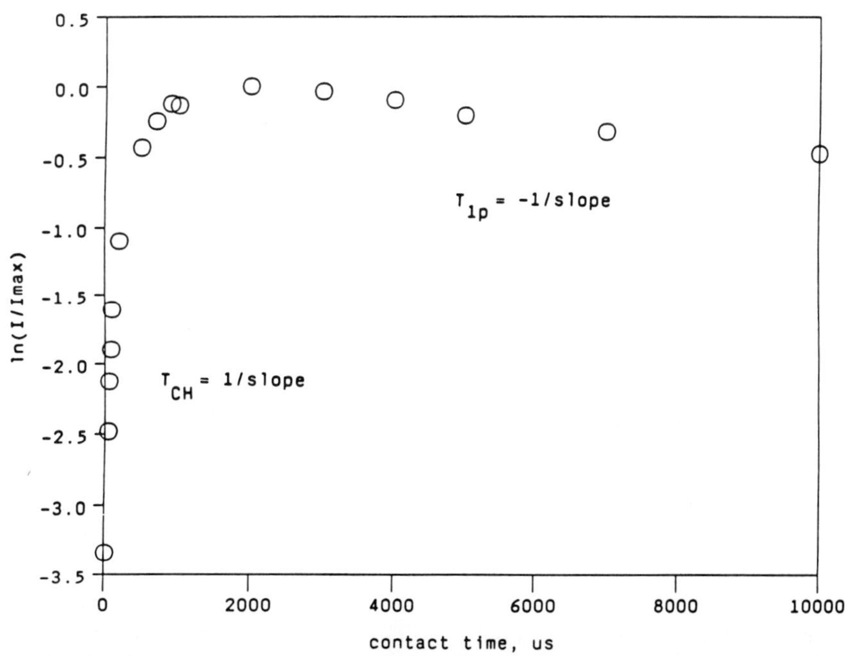

Figure 3. Natural log of the peak intensities of the carbon resonances for a 20:80 PVF_2/PMMA blend scaled to the maximum intensity versus cross polarization contact time. The $T_{1\rho}$ and T_{CH} values for the blend can be determined from the inverse slopes of the plots.

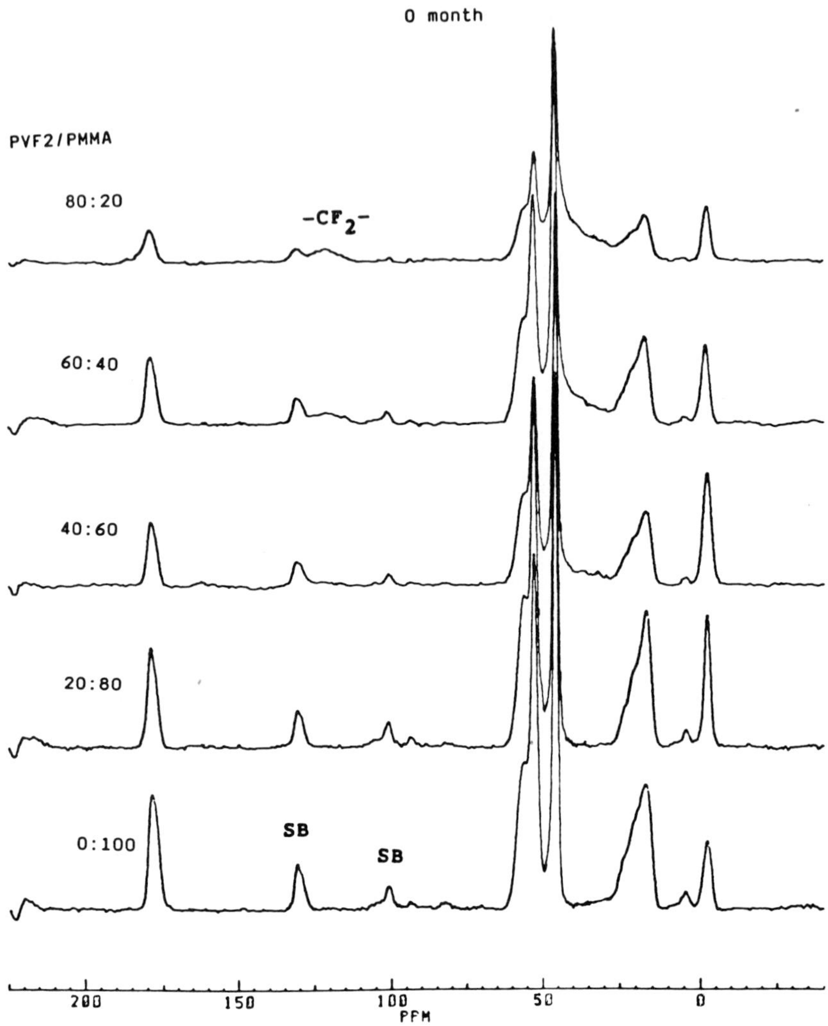

Figure 4. Spectra for 0:100, 20:80, 40:60, 60:40 and 80:20 PVF_2/PMMA blends with no aging and a CP contact time of 1 ms.

4.1. PVF_2/PMMA Blend System

PVF_2 with its broad chemical shift in the methylene region (45 ppm), resulting from head to head and tail to tail defects, appears to dominate the lineshapes of the quaternary (46 ppm) and methyl (16 ppm) carbons of PMMA in the blend as PVF_2 content increases as shown in Fig. 4 for the initial blends at a contact time of 1 ms. This is seen also for the aged blends (Fig. 5). The carbonyl carbon of PMMA increases in relative intensity in the blends as PMMA content increases and for higher PVF_2 contents the CF_2 carbon resonance (121 ppm) begins to show up.

The effect of blend composition on T_{CH} of the carbon resonances of PMMA is shown in Fig. 6. The T_{CH} is weighted against PVF_2 concentration in the blend and

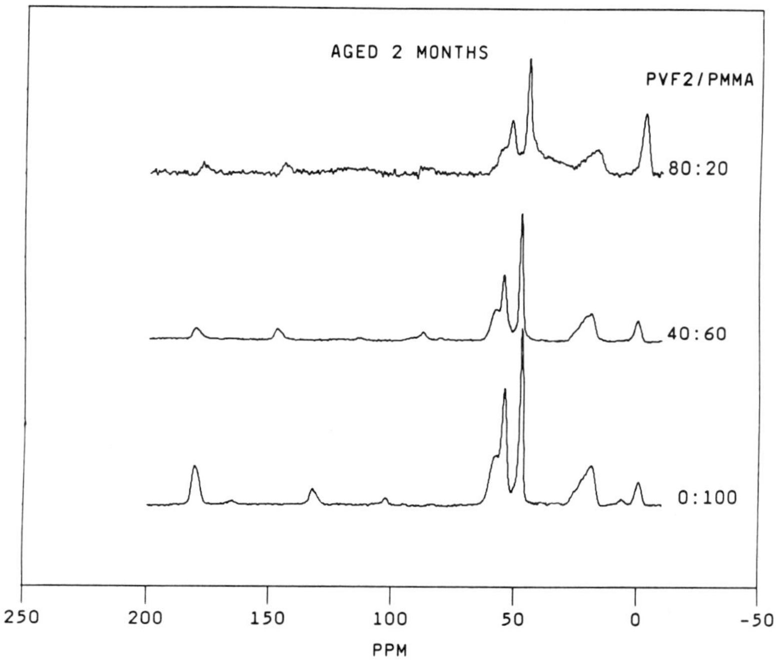

Figure 5. Spectra for 0:100, 40:60 and 80:20 PVF_2/PMMA blends with 2 months aging and a CP contact time of 1 ms.

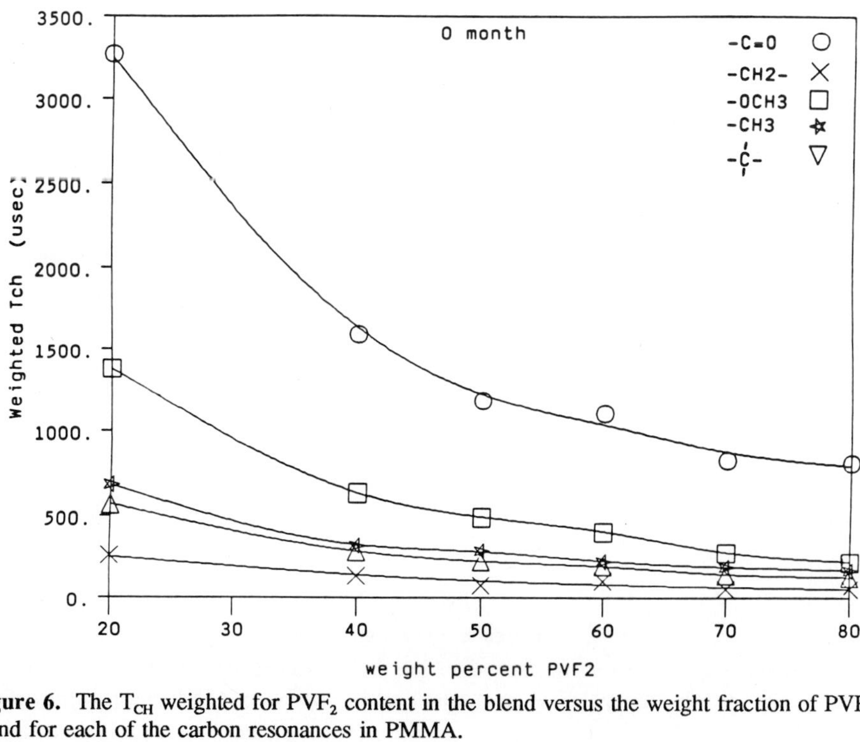

Figure 6. The T_{CH} weighted for PVF_2 content in the blend versus the weight fraction of PVF_2 in the blend for each of the carbon resonances in PMMA.

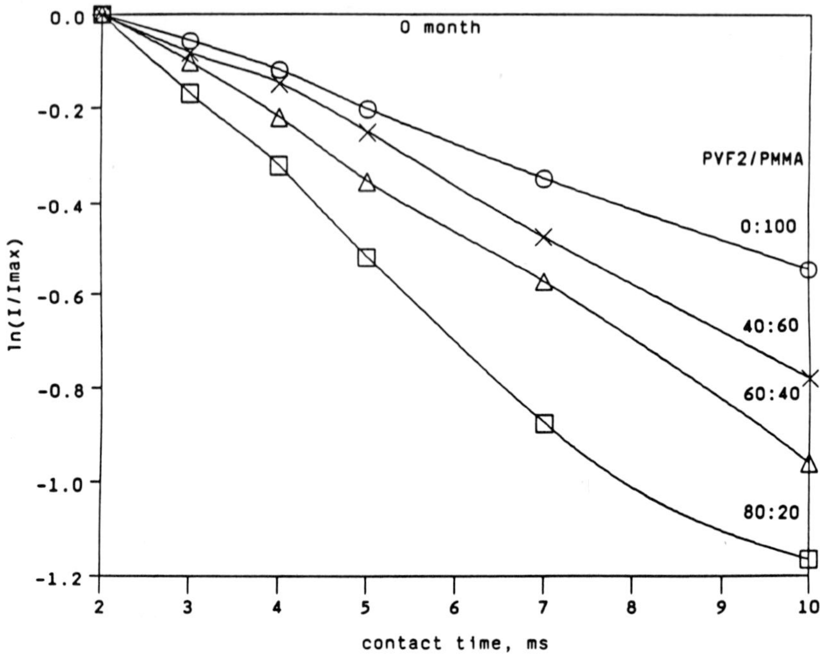

Figure 7. Natural log of the peak intensities of the carbon resonances of PMMA scaled to the maximum intensity versus cross polarization contact time for various compositions of the blend with no aging.

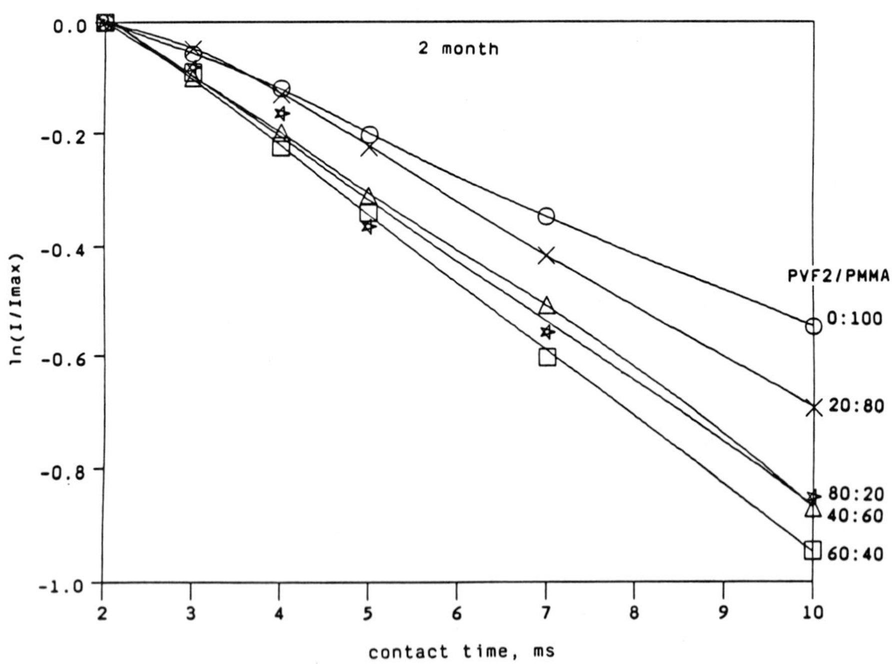

Figure 8. Natural log of the peak intensities of the carbon resonances of PMMA scaled to the maximum intensity versus cross polarization contact time for various compositions of the blend with two months aging.

is a decreasing function of PVF_2 content. The blend becomes more rigid as PVF_2 concentration is increased and rigid materials usually cross polarize more efficiently than mobile materials. This effect on the T_{CH} with PVF_2 concentration indicates that intimate mixing between PVF_2 and PMMA is present. A similar effect also occurs for the blends that have been aged 2 months.

In Figs. 7 and 8, the changes in $T_{1\rho H}$ values with composition of the blend are shown. The increasing slope with increasing PVF_2 content shows that $T_{1\rho H}$ decreases with an increase in weight percent PVF_2 in the blend. This effect on the $T_{1\rho H}$ value shows that intimate mixing exists between PVF_2 and PMMA in the blend. The straight line, within 10 percent error for each composition in Figs. 7 and 8, indicates that only one $T_{1\rho H}$ is present for each composition of the blend. This implies that PMMA does not form domains or occlusions in the amorphous phase of PVF_2 [3], so no detectable phase separation is apparent. The main cause of the decreasing $T_{1\rho}$ values in the blends as PVF_2 composition is increased is either that the protons of PMMA have direct dipole interactions with PVF_2 or that there is sufficient PVF_2 as nearest-neighbors to alter the molecular motion of PMMA [3]. In either case, this agrees with results by Douglass and McBrierty [4] from cross-relaxation measurements which state that a large number of PVF_2 molecules see PMMA molecules at nearest neighbor distances [3].

Intimate mixing is also present in the blends that were aged 2 months as well as the initially made blends (Fig. 8). The aged blends show a linear behavior within 10 percent error for the ln (I/I_{max}) versus contact time curve and also show no detectable phase separation. This demonstrates that mixing is still intimate enough to have PMMA relaxed efficiently by PVF_2 or that the molecular motions are still altered by blending [3].

If the ln (I/I_{max}) versus contact time curves for a 20:80 PVF_2/PMMA blend (Fig. 9) at no aging and 2 months aging are compared, it is found that there is an increase in the $T_{1\rho H}$ values with aging. This relationship was evident with all compositions tested and indicates that even though one phase was present, PMMA is relaxed less efficiently by PVF_2 or the molecular motions of PMMA are not altered as greatly as the aging process progresses. This may indicate that a subtle separation between PVF_2 and PMMA is occurring with time.

4.2. PVF_2/PVAc Blend System

PVF_2 does not seem to dominate the lineshape of the region around 45 ppm in the PVF_2/PVAc blends (Figs. 10 and 11) as it did with the PVF_2/PMMA blends. There is a broadening in this region though for the higher PVF_2 content blends. There also seems to be an attenuating effect on the relative intensities of the carbonyl and methyl carbons to the other carbons of PVAc with increasing PVF_2 content.

In Figs. 12 and 13, the effect of composition on the $T_{1\rho H}$ values for the PVF_2/PVAc blends is shown. The increasing slope with increasing PVF_2 content is evident, as was for the PVF_2/PMMA blends, that $T_{1\rho H}$ decreases with an increasing

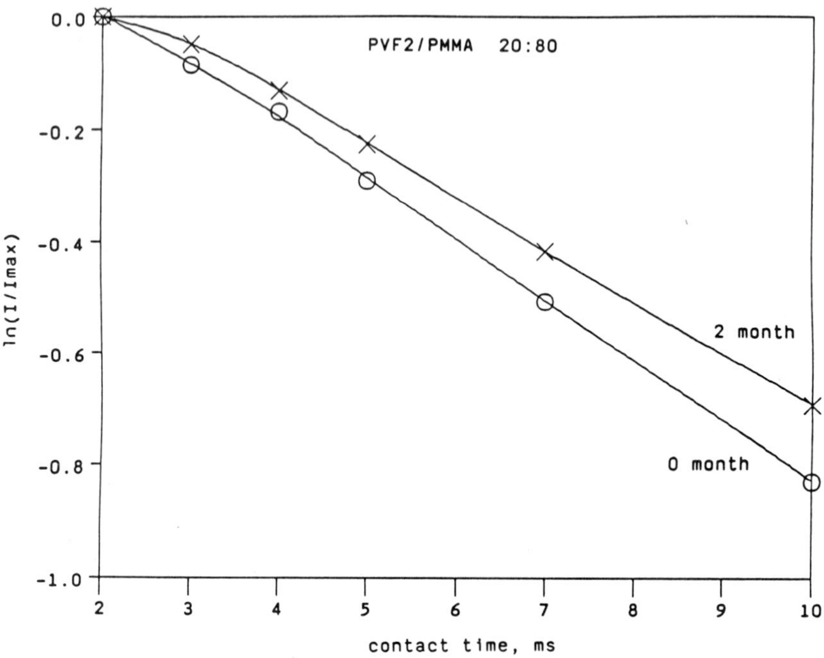

Figure 9. Natural log of peak intensities of carbon resonances of PMMA scaled to maximum intensity versus CP contact time for 20:80 PVF_2/PMMA blend at zero and two months aging.

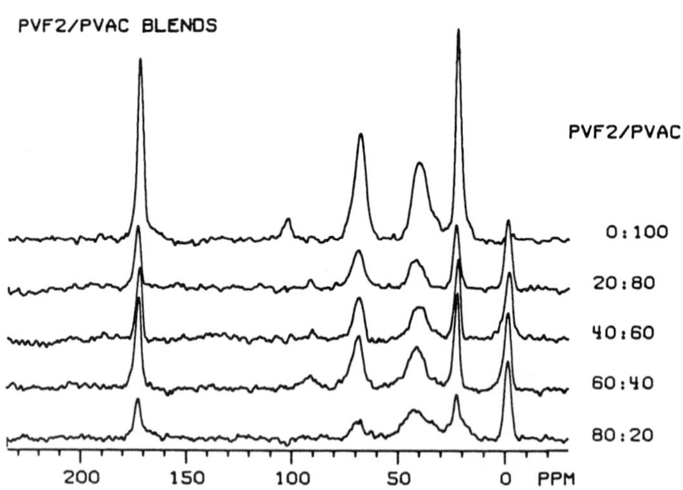

Figure 10. Spectra for 0:100, 20:80, 40:60, 60:40 and 80:20 PVF_2/PVAc blends with no aging and a CP contact time of 2 ms.

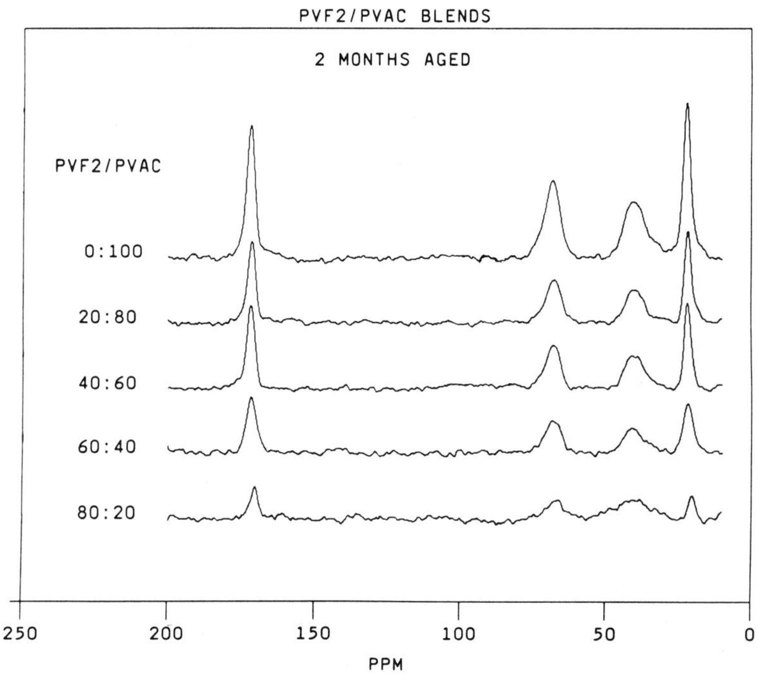

Figure 11. Spectra for 0:100, 20:80, 40:60, 60:40 and 80:20 PVF_2/PVAc blends with 2 months aging and a CP contact time of 2 ms.

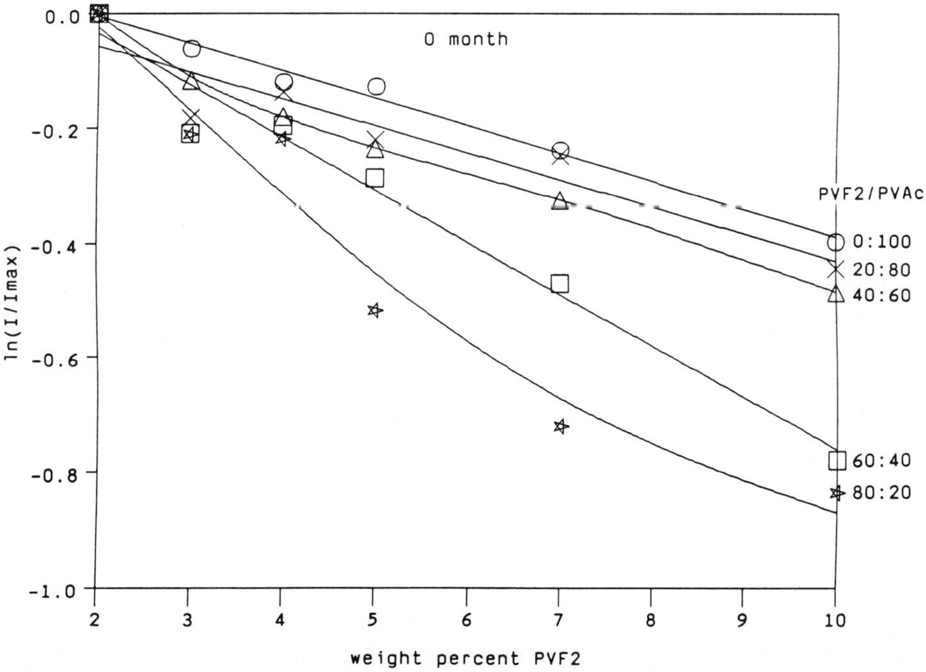

Figure 12. Natural log of the peak intensities of the carbon resonances of PVAc scaled to the maximum intensity versus cross polarization contact time for various compositions of the blend with no aging.

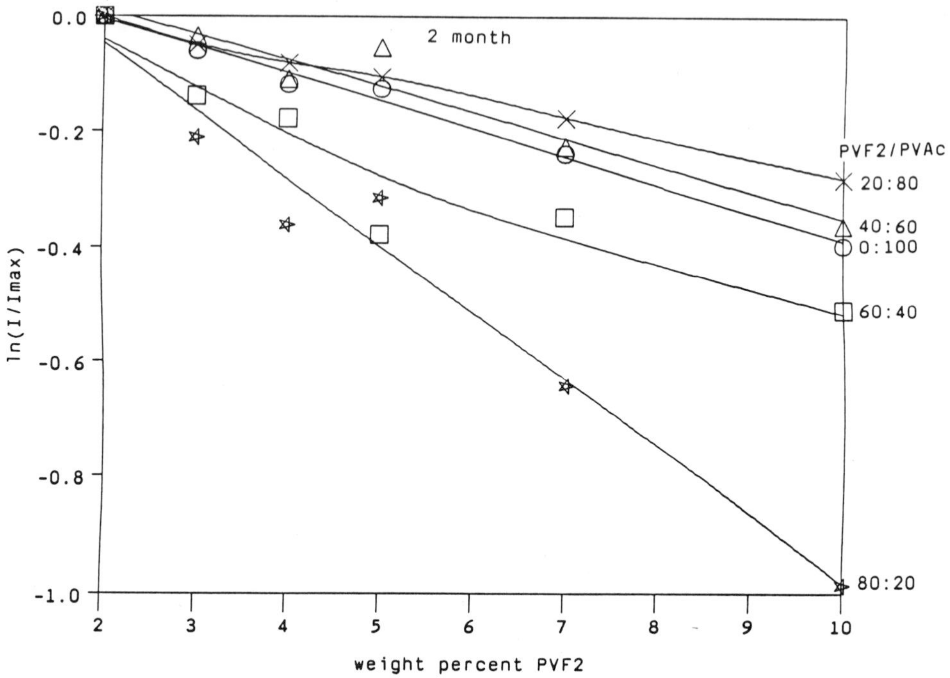

Figure 13. Natural log of the peak intensities of the carbon resonances of PVAc scaled to the maximum intensity versus cross polarization contact time for various compositions of the blend with 2 months aging.

weight percent PVF_2 in the blend. This is an indication that there is intimate mixing between the PVF_2 and PVAc chains in the blend and may be caused by the protons of PVAc having direct dipole interactions with PVF_2 or that there is sufficient PVF_2 as nearest-neighbors to alter the molecular motion of PVAc.

At each composition in Figs. 12 and 13, a straight line is found, within 10 percent experimental error, suggesting that only one $T_{1\rho H}$ present. This implies, as with the PVF_2/PMMA blends, that PVAc has not formed occlusions or domains in the amorphous phase of PVF_2 and no apparent phase separation has occurred.

Intimate mixing is also present in the blends that were aged two months, as shown in Fig. 13. The $\ln(I/I_{max})$ versus contact time plot has a linear behavior within 10 percent error, showing that there is no apparent phase separation after aging. This indicates either that mixing is still intimate enough to have PVAc relaxed efficiently or that its molecular motions are still affected by blending with PVF_2.

If the $\ln(I/I_{max})$ versus contact time curves for a 60:40 PVF_2/PVAc blend (Fig. 14) at no aging and 2 months aging are compared, it is found as with the PVF_2/PMMA blends, that there is an increase in the $T_{1\rho H}$ values with aging. This occurred for all compositions tested. Therefore, even though there is still a linear relationship between $\ln(I/I_{max})$ and contact time, either the molecular motions of PVAc are not affected as greatly as they were initially or they are not relaxed as efficiently by PVF_2. This may indicate that a subtle separation is occuring between PVF_2 and PVAc in the blend.

5. Conclusions

It has been shown that PVF_2/PMMA and PVF_2/PVAc blends were intimately mixed and compatible over all compositions tested and that one phase apppears to exist. It has also been demonstrated that one phase and intimate mixing are still present in the aged blends though there is a lengthening of the $T_{1\rho}$ values toward the PMMA and PVAc values with aging. This would indicate that, over the compositions tested, polymers in the blends existed in the amorphous phase and little or no crystalline PVF_2 domains were detected. This may be a consequence of the method of solution casting chosen or thermal treatment used to expel the solvent.

6. Acknowledgement

The authors would like to thank the Edison Polymer Innovation Corporation for their support and funding of this research.

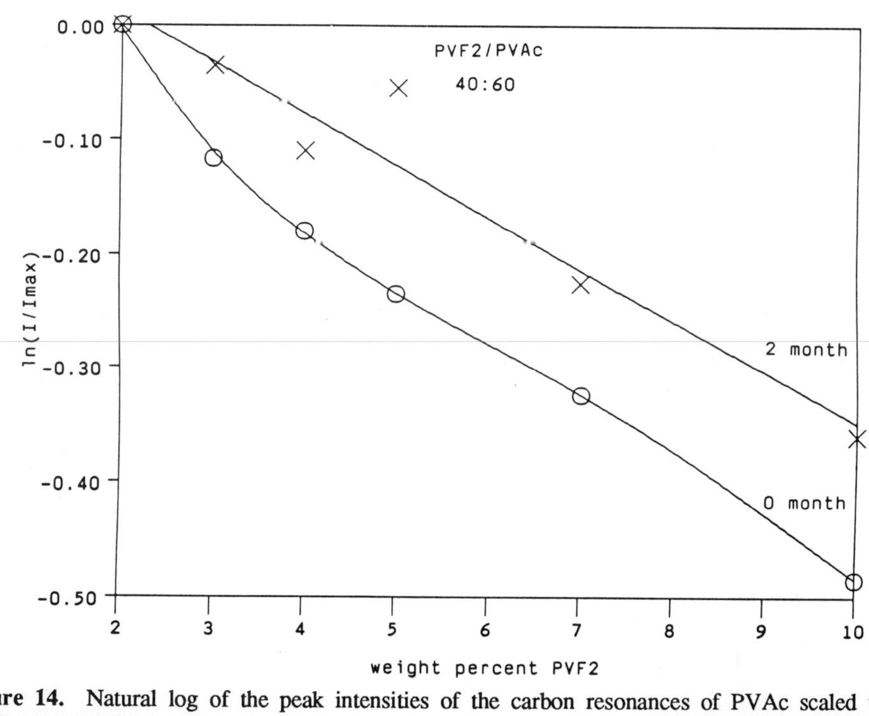

Figure 14. Natural log of the peak intensities of the carbon resonances of PVAc scaled to the maximum intensity versus cross polarization contact time for a 60:40 PVF_2/PMMA blend at zero and two months aging.

7. References

1. Stejskal, E. O.; Schaefer, J.; Sefcik, M. D. and McKay, R. A., *Macromolecules*, *14*, 275, 1981.
2. Nishi, T.; Wang, T. T. and Kwei, T. K., *Macromolecules*, *8*, 227, 1975.
3. Tekely, P.; Lauprêtre, F. and Monnerie, L., *Polymer*, *26*, 1081, 1985.
4. Douglass, D. C. and McBrierty, V. J., *Macromolecules*, *11*, 766, 1978.
5. Bovey, F. A.; Schilling, F. C.; Kwei, T. K. and Frisch, H. L., *Macromolecules*, *10*, 559, 1977.
6. Ward, T. C. and Lin, T. S., *Adv. Chem. Ser.*, *206*, 59, 1984.
7. McBrierty, V. J.; Douglass, D. C. and Kwei, T. K., *Macromolecules*, *11*, 1265, 1978.
8. Noland, J. S.; Hsu, N. N. C.; Saxxon, A. and Schmidt, J. M., *Adv. Chem. Ser.*, *15*, 99, 1971.
9. Nishi, T. and Wang, T. T., *Macromolecules*, *8*, 909, 1975.
10. Aubin, M. and Prud'homme, R. E., *J. Polym. Sci., Polym. Phys. Edn.*, *19*, 1245, 1981.
11. Roerdink, E. and Challa, G., *Polymer*, *19*, 173, 1978.
12. Hourston, O. J. and Hughes, I. D., *Polymer*, *18*, 1175, 1977.
13. Coleman, M. M.; Zarian, J.; Varnell, D. F. and Painter, P. C., *J. Polym. Sci., Polym. Lett. Edn.*, *15*, 745, 1977.
14. Leonard, C.; Halary, J. L.; Monnerie, L.; Broussoux, D.; Servet, B. and Micheron, F., *Polym. Commun.*, *24*, 110, 1983.
15. Wendorff, J. H., *J. Polym. Sci., Polym. Lett. Edn.*, *18*, 439, 1980.
16. Rashmi, G. K. Narula and Pillai, P. K. C., *J. Macromol. Sci. Phys.*, *B26(2)*, 185, 1987.
17. Schaefer, J.; Sefcik, M. D.; Stejskal, E. O. and McKay, R. A., *Macromolecules*, *14*, 188, 1981.
18. Schaefer, J.; Stejskal, E. O. and Buchdahl, R., *Macromolecules*, *10*, 384, 1977.
19. Schaefer, J.; Stejskal, E. O.; Sefcik, M. D. and McKay, R. A., *Phil. Trans. Roy. Soc. Lond. A*, *299*, 593, 1981.
20. Lin, T. S. and Ward, T. C., American Chemical Society Meeting, 1983, Wash. D.C.

Chapter 8

SOLID STATE NMR DETECTION OF MOLECULAR-LEVEL MIXING PHENOMENA IN STRONGLY INTERACTING POLYMER BLENDS AND PHASE-SEPARATED COPOLYMERS

LAURENCE A. BELFIORE, THOMAS J. LUTZ AND CHIHMIN CHENG

1. Abstract

This research contribution addresses mixing phenomena in polymer blends that exhibit strong intermolecular association. Molecular-level observations of specific interactions between dissimilar blend components have been obtained from high-resolution solid state proton and carbon-13 NMR experiments at ambient temperature. Results illustrate mixing effects on the isotropic chemical shifts of the critical component in a completely or partially phase-mixed blend. Perturbations in the NMR spectra result from conformational changes, hydrogen bonding, molecular complexation, or altered packing geometries that occur concomitantly with the mixing process. More convincing evidence that two components of a strongly interacting blend reside in a near-neighbor environment is obtained from the measurement of proton spin diffusion between dissimilar species. Proton spin diffusion is measured directly via the high-resolution CRAMPS experiment (Combined Rotation and Multiple Pulse Spectroscopy) in a molecular complex of poly(ethylene oxide) and resorcinol, and indirectly via the carbon-13 spin system in a modified version of the Goldman-Shen experiment for microphase-separated copolymers. One of the primary objectives of this research endeavor is to bridge the gap between macroscopic and molecular-level probes of phase behavior and intermolecular association in mixtures that form molecular complexes. Hence, in addition to the molecular-level solid state

---

LAURENCE A. BELFIORE and CHIHMIN CHENG - Department of Chemical Engineering, Colorado State University, Fort Collins, Colorado 80523. THOMAS J. LUTZ - IBM Faculty Loan Program, Boulder, CO 80503.

Solid State NMR of Polymers, Edited by L. Mathias
Plenum Press, New York, 1991

NMR results, implications from a macroscopic phase of the polymer-blend research are discussed. In summary, we have measured the thermal and mechanical properties of strongly interacting systems, whose interaction sites have been characterized via solid state NMR. The success of this proposed structure-property relationship scheme, which bridges microscopic fundamentals (via NMR) with practical macroscopic (DSC and stress-strain) measurements, depends on our ability to understand material properties at a level where continuum hypotheses are no longer valid.

2. Introduction

Mixing phenomena in polymer blends is an area of practical importance because the solid state phase behavior of a blend governs its physical properties [1,2]. In this contribution, macroscopic mechanical and thermal properties of strongly interacting polymer blends are measured and correlated with microscopic aspects of the mixing process. At the site-specific level of probing interactions between dissimilar species in a blend, one can gain appreciation for the kinds of functional groups that potentially give rise to exothermic energetics and favorable mixing [3-6]. High-resolution solid state NMR spectroscopy is the analytical technique of choice used herein to fingerprint the components of a strongly interacting blend and monitor changes in conformation, packing geometry, hydrogen bonding, or other electron transfer reactions which accompany the formation of a molecular complex.

Thermal analysis via differential scanning calorimetry is extremely useful for generating phase diagrams and interpreting solid state NMR spectra of blends in a two-phase region [7,8]. When phase separation occurs, it is important that one of the components (the critical component) must be present in both phases and that the carbon-13 NMR spectrum of this critical component must contain at least one signal that is sensitive to, or can detect the presence of, the other component in the blend. These conditions are satisfied for the mixture of poly(ethylene oxide), PEO, and resorcinol in Fig. 1 (spectrum B) which contains 67 mole % resorcinol. The phenolic carbon resonance of the critical component (resorcinol) is sensitive to local environment, and is capable of detecting the presence of the polymer in a molecular complex via a 3 ppm chemical shift perturbation relative to the phenolic ^{13}C resonance of resorcinol in its undiluted crystalline state [5,8]. Other spectral changes are also observed as a consequence of molecular-level mixing. The PEO-resorcinol system is unusual because it exhibits bi-eutectic phase behavior with eutectic compositions in the vicinity of 10 and 50 mole % resorcinol [9]. The molecular complex, stabilized by optimal formation of intermolecular hydrogen bonds, exhibits a local melting point maximum on the liquidus phase boundary for a mixture composition that is almost midway between the two eutectic compositions [9].

A second example of strong intermolecular association is found in a strategically selected polymer-ionomer blend via transition metal complexation [10]. In this case, the interaction between poly(4-vinylpyridine) and a zinc-neutralized copolymer of ethylene and methacrylic acid is localized in the amorphous phase [6].

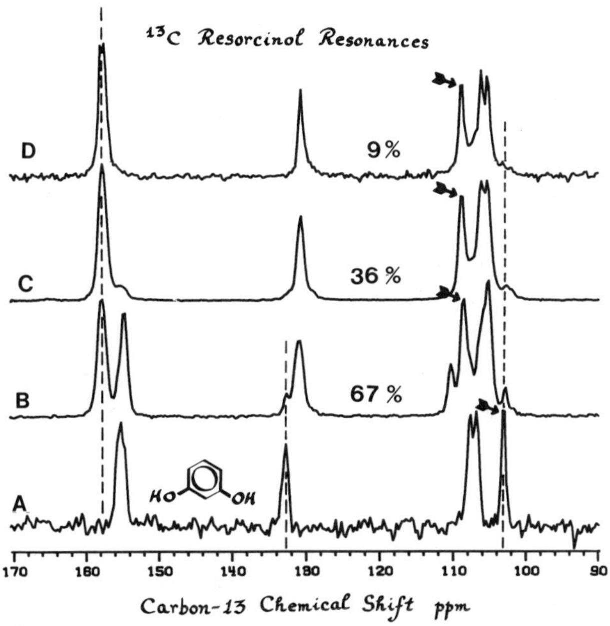

Figure 1. High-resolution carbon-13 solid state NMR spectra of methanol-cast blends of poly(ethylene oxide) and resorcinol in the aromatic chemical shift region via cross-polarization. a) undiluted crystalline resorcinol, b) 67 mole% resorcinol, c) 36 mole % resorcinol, d) 9 mole % resorcinol. The dashed line on the left side of the figure represents the chemical shift of the phenolic carbons of resorcinol in the molecular complex with PEO. The arrows on the right side of the figure indicate the aromatic resonance of resorcinol that is ortho to both hydroxyl groups.

Mechanical fracture stress measurements are sensitive, at the microscopic level, to the pyridine-nitrogen/zinc-counterion molar ratio in these blends. Uniaxial tensile deformation experiments suggest that the fracture response is optimized when the nitrogen/zinc molar ratio is between 2:1 and 3:1. This synergistic effect on mechanical properties is consistent with the carbon-13 NMR detection of strong intermolecular complexation between macromolecular components at the site-specific level [6].

Molecular proximity between dissimilar components on the order of a few angstroms is synonymous with microscopic mixing phenomena and the formation of molecular complexes [6]. Spin diffusion between dipolar-coupled protons in dissimilar blend components represents a sophisticated diagnostic probe of spatial proximity [6,11]. In this contribution, we apply two well-known techniques to monitor communication between dipolar-coupled protons and address the question of molecular proximity. The first example exploits the high-resolution nature of solid state proton NMR spectra via the CRAMPS experiment [6,11,12]. The strategically chosen blend is a molecular complex of poly(ethylene oxide) and resorcinol whose proton spectrum is well-resolved [5,6]. The second example illustrates a technique that is reserved for the measurement of proton spin diffusion in microphase-separated blends and copolymers when the ^{1}H CRAMPS experiment does not provide useful information about molecular proximity [13,14]. In this case, commercial block and random copolymers that contain both rigid and mobile domains are investigated. Both types of spin-diffusion experiments rely on through-space spin-spin communication between protons to provide *qualitative* information about dipolar distances. When these distances are on the order of 2-5 angstroms, as they are in the PEO-resorcinol molecular complex, time-dependent proton spin diffusion between dissimilar molecules (nearest neighbors) occurs (and most likely equilibrates) on a time scale of 10^{-4} seconds [6]. In contrast, transient *interdomain* communication between dipolar-coupled protons is at least one-to-two orders of magnitude slower (1-10 ms), as illustrated herein for industrially important microphase-separated copolymers of styrene and butadiene (KRATON®), and ethylene with methacrylic acid (SURLYN®) [14]. The critical parameter in these proton-spin-diffusion experiments is the mixing time during which spin exchange or magnetization transfer takes place in the presence of homonuclear dipolar couplings. Due to the complexity of the blends and copolymers investigated, however, a detailed analysis of dipolar distances between dissimilar molecules or copolymer microdomains is not attempted.

3. Experimental Methods

3.1. Polymer Blend Preparation

Physical mixtures of poly(ethylene oxide), PEO, and resorcinol were prepared from methanol solutions at 35°C. Upon evaporation of the solvent in a fume hood, the residue (PEO and resorcinol) was dried under vacuum at 70°C for 15-20 hrs. Blend composition was controlled by accurately weighing the starting materials before dissolution in methanol. Thermogravimetric analysis of the solid polymer blend residues yielded mixture compositions that were within 3% of the weights of the starting materials. An alternate route to produce blends of PEO and resorcinol relies on the driving force for these components to precipitate from aqueous solution [15]. Since each pure component is soluble in water, the formation of a milky precipitate (when aqueous solutions of the pure components are mixed) suggests that the overall mixture composition lies within a two-phase region of the ternary aqueous phase diagram. It should be mentioned that the precipitation method produces strongly interacting blends whose compositions are difficult to control because the starting materials are partitioned between the precipitate and the supernatant solution in unpredictable proportions (unless one has detailed information about the ternary phase diagram of PEO, resorcinol, and water). In light of this difficulty, we have chosen the methanol-solution route to prepare blends of PEO and resorcinol without the formation of a precipitate. Blends of these strongly interacting components have also been cast from tetrahydrofuran, or a mixture of acetone and water without precipitate formation [16].

Polymer-ionomer blends of a zinc-neutralized copolymer of ethylene and methacrylic acid with poly(vinylpyridine) were prepared by a direct molten mixing procedure. Films approximately 0.20 mm (0.008 inch) in thickness were compression molded at 225°C using a Carver Laboratory Press.

The sources and physical characteristics of all materials investigated in this study are listed in Table I.

3.2. Nuclear Magnetic Resonance Spectroscopy

Proton-enhanced dipolar-decoupled carbon-13 NMR spectra of strategically selected polymer blends and copolymers in the solid state were obtained under high-resolution conditions on a modified Nicolet NT-150 spectrometer at the NSF-supported Regional NMR Center, Colorado State University, Fort Collins, Colorado. The carbon frequency was 37.735 MHz and magic-angle spinning was performed at

TABLE I

Material	**Source**	**Characterization**
Resorcinol	Kodak	Recrystallized white powder, melting range 109-111°C (via hot-stage microscopy)
Poly(ethylene oxide)	Scientific Polymer Products	low MW ≈ 3,400; flakes high MW ≈ 900,000; powder
Poly(2-vinylpyridine)	Scientific Polymer Products	MW ≈ 40,000; powder
Poly(4-vinylpyridine)	Scientific Polymer Products	MW ≈ 50,000; powder
Random Copolymer of Ethylene and Methacrylic Acid (NUCREL®)	DuPont	85 wt. % ethylene, 15 wt. % methacrylic acid, unneutralized, Melt Index = 60 decigram/min @ 190 °C; pellets
Zinc Ionomer (SURLYN® 1706)	DuPont	85 wt. % ethylene, 15 wt. % methacrylic acid, 60% of COOH groups neutralized with zinc; Melt Index = 0.7 decigram/min @ 190 °C; pellets
SBS Tri-block Copolymer (KRATON® D-1101)	Shell	31 wt. % Polystyrene, Polystyrene end-block MW ≈ 17,500, Polybutadiene mid-block MW ≈ 78,000; granules
Hydrogenated SBS Tri-block Copolymer (KRATON® G-1651)	Shell	32 wt. % Polystyrene, Polystyrene end-block MW ≈ 29,000; Ethylene/Butylene mid-block MW ≈ 122,000; granules

3600 Hz. The spectrometer incorporates a home-built cross-polarization/magic-angle spinning (CP/MAS) unit, including the probe. The spinner system is a modified version of Wind's [17] with a sample volume of 0.3 cm^3. A proton 90° pulse width of 5 μs was employed, corresponding to an r.f. field strength of 50 kHz. The r.f. field was maintained at 50 kHz during cross-polarization and subsequent high-power ^1H decoupling. The ^{13}C free induction decay (FID) was accumulated via 2K time-domain data points using quadrature detection. Prior to Fourier transformation, the signal-averaged FID was zero-filled to 4K. The spectral width encompassed a ±10 kHz frequency range, and 5 Hz line broadening was employed. Following Stejskal and Schaefer [18], spin-temperature alternation in the rotating frame was used to suppress the build-up of artifacts which may occur in proton-enhanced spectra. The sample temperature was maintained at 15 ± 2°C by passing the spinner air through a copper cooling coil immersed in an ice bath. Carbon-13 chemical shifts were referenced externally to the methyl resonance of hexamethylbenzene, 17.355 ppm deshielded from tetramethylsilane (TMS) [19].

High-resolution solid state proton NMR spectra of poly(ethylene oxide), resorcinol, and their 2:1 molecular complex were obtained via the CRAMPS experiment (Combined Rotation and Multiple Pulse Spectroscopy) at 187 MHz in collaboration with Dr. C. E. Bronnimann of Colorado State University's Regional NMR Center [20]. Multiple pulse decoupling of the ^1H- ^1H dipolar interaction was accomplished via the BR-24 pulse train [21], employing a 1.2 μs proton 90-degree pulse width and a 108 μs cycle time. Two-dimensional proton-spin-diffusion spectroscopy via the ^1H CRAMPS experiment [6,11] allows one to monitor dipolar communication between chemically inequivalent protons under high-resolution conditions. In this experiment, multiple pulse decoupling (BR-24) was employed during the chemical shift evolution period (t_1) and the data acquisition period (t_2). The data in Figs. 8 and 9 reveal two-dimensional contour representations of the NMR spectra from the ^1H spin-diffusion experiment after Fourier transformation with respect to the time-domain parameters t_1 and t_2 mentioned above. Proton spin diffusion was operative during a mixing period of either 40 μs (Fig. 9) or 100 μs (Fig. 8) in the presence of ^1H- ^1H dipolar *couplings*, which provide the mechanism for through-space spin-spin communication. The mixing period occurs after the chemical shift evolution (preparation) period (t_1) and before data acquisition (t_2). More specific details concerning the high-resolution ^1H CRAMPS experiment [20], and the two-dimensional proton spin diffusion experiment [11] can be found in recent literature publications.

An in-depth description of the modified Goldman-Shen experiment [14] for measurement of proton spin diffusion between two domains of dissimilar mobility (in commercial phase-separated copolymers, for example) is provided in the Results and Discussion section.

3.3. Thermal Analysis

Differential scanning calorimetry (DSC) and thermogravimetric analysis of PEO-resorcinol mixtures were recorded on a Perkin-Elmer Thermal Analyzer in collaboration with Tom Geist at IBM Corporation, Boulder, Colorado. Thermograms were recorded at a heating rate of 20 degrees C/min. As mentioned above, thermogravimetric analysis of the blends was employed to verify mixture composition based on weights of the starting materials. The DSC data, representing macroscopic characteristics of high-temperature phase behavior, are needed to interpret molecular-level mixing results from high-resolution solid state NMR. The *bimodal* melting behavior of the blend of PEO and resorcinol in Fig. 2, indicative of both eutectic and liquidus melting for a mixture whose composition is slightly different from the eutectic composition, is correlated with the NMR detection of two distinct ^{13}C resonance signals for the phenolic carbons of resorcinol. This molecular-level observation of mixing effects via solid state NMR is attributed to hydrogen bonding and/or conformational differences between the small molecule in separate phases of the blend.

3.4. Mechanical Properties

Stress-strain properties of polymer-ionomer blends (transition-metal complexes) were obtained on an Instron Model 8501 servohydraulic mechanical testing system at a uniaxial tensile deformation rate of 0.5 inch/min. The initial specimen length was 0.4 inch and the temperature was uncontrolled at ambient conditions.

4. Results and Discussion

4.1. High-Resolution Carbon-13 NMR Spectra of Strongly Interacting Blends--Mixing Effects and Macroscopic Correlations

The overall objective of the research presented in this section is to bridge the gap between macroscopic and molecular-level probes of intermolecular association in strongly interacting blends. Hence, in addition to the solid state NMR results

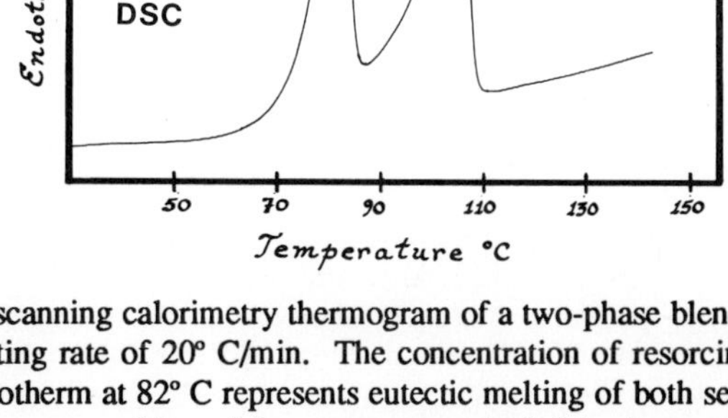

Figure 2. Differential scanning calorimetry thermogram of a two-phase blend of poly(ethylene oxide) and resorcinol at a heating rate of 20° C/min. The concentration of resorcinol in the binary mixture is 67 mole %. The endotherm at 82° C represents eutectic melting of both semicrystalline phases, and the peak at 105° C represents melting of the excess resorcinol-rich phase on the liquidus boundary.

described herein, we have implemented a macroscopic phase of this polymer-blend research to measure the mechanical properties of strongly interacting macromolecular systems, whose interaction sites have been characterized via solid state NMR. Thermal analysis via Differential Scanning Calorimetry allows one to characterize the phase behavior of semicrystalline complexes, particularly from the standpoint of first-order thermodynamic phase transitions that differ from the pure-component melting phenomena. Knowledge of the phase diagram for a strongly interacting binary mixture, generated via Differential Scanning Calorimetry, is required to understand and interpret the carbon-13 NMR spectra of blends in a two-phase region [7,8].

Examples of microscopic mixing effects on the ambient-temperature carbon-13 solid state NMR spectra of strategically selected polymer blends are illustrated in Figs. 1, 3, 4 and 5. The blends whose NMR spectra appear in Fig. 1 contain poly(ethylene oxide) and resorcinol (meta-dihydroxybenzene), which form a molecular complex when the molar ratio of polymer repeat unit to small molecule is 2:1 [9]. Figs. 3, 4 and 5 illustrate mixing phenomena in blends of poly(vinylpyridine) with zinc-neutralized ionic copolymers of ethylene and methacrylic acid. This latter system represents a second class of strongly interacting materials--a consequence of transition metal (zinc) complexation with the pyridine nitrogen lone electron pair [10].

Some of the aromatic ^{13}C NMR signals of resorcinol in Fig. 1 are chemically shifted in the blends with PEO (spectra B, C, D) relative to undiluted resorcinol (spectrum A). Detectable and interpretable spectral changes are observed for the phenolic carbon signal (≈ 3 ppm perturbation indicated by the dashed line on the left side of Fig. 1, 155-160 ppm) and the aromatic carbon that is ortho to both hydroxyl

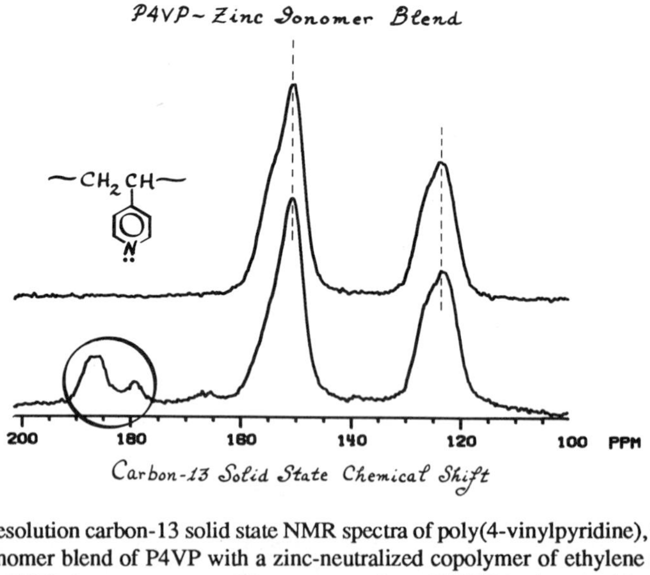

Figure 3. High-resolution carbon-13 solid state NMR spectra of poly(4-vinylpyridine), upper spectrum, and a polymer-ionomer blend of P4VP with a zinc-neutralized copolymer of ethylene and methacrylic acid (SURLYN® 1706), lower spectrum. The concentration of P4VP in the transition-metal complex is 29 weight %. The circle on the left side of the lower spectrum highlights the carboxyl carbon resonance of the ionic copolymer in the complex.

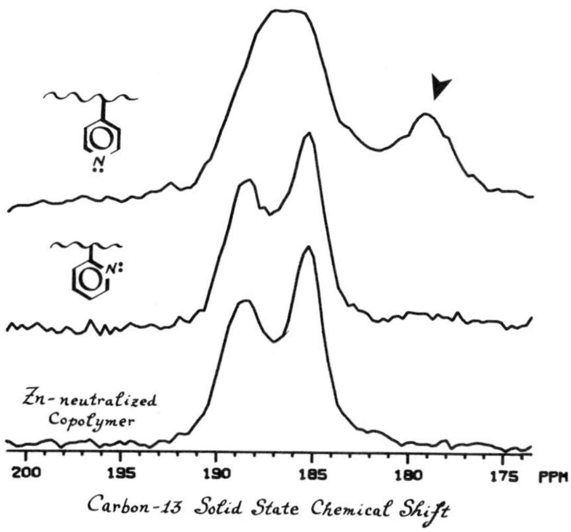

Figure 4. High-resolution carbon-13 solid state NMR spectra of polymer-ionomer blends in the carboxyl carbon chemical shift region via cross-polarization. **Lower spectrum;** undiluted zinc-neutralized copolymer of ethylene and methacrylic acid (SURLYN® 1706). Weight fraction of methacrylic acid in the copolymer is 0.15, and 60% of the carboxylic acid groups are neutralized with zinc. **Middle spectrum;** blend of the zinc-neutralized copolymer (SURLYN® 1706) with poly(2-vinylpyridine). The concentration of P2VP in the blend is 33 weight %. **Upper spectrum;** blend of the zinc-neutralized copolymer (SURLYN® 1706) with poly(4-vinylpyridine). The concentration of P4VP in the transition-metal complex is 29 weight %. The resonance indicated by the arrow is attributed to the formation of a complex between the zinc counterion and the nitrogen lone electron pair.

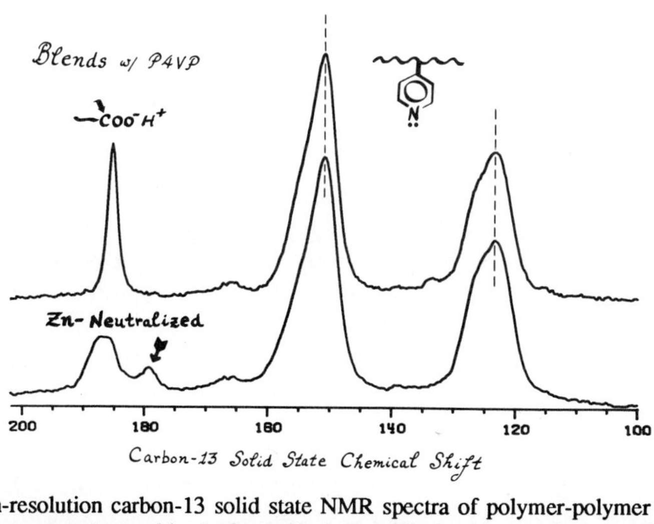

Figure 5. High-resolution carbon-13 solid state NMR spectra of polymer-polymer blends via cross polarization. **Upper spectrum;** blend of poly(4-vinylpyridine) with a random copolymer of ethylene and methacrylic acid (NUCREL®). Weight fraction of methacrylic acid in the copolymer is 0.15, and the weight fraction of P4VP in the blend is 0.24. **Lower spectrum;** blend of P4VP with the zinc-neutralized copolymer of ethylene and methacrylic acid (SURLYN® 1706). The concentration of P4VP in the transition-metal complex is 29 weight %. The carboxyl carbon resonance indicated by the arrow is attributed to the formation of a strongly interacting complex.

groups (≈ 6 ppm perturbation indicated by the dashed line and the arrows on the right side of Fig. 1, 100-110 ppm). Based on the proposed structure of the molecular complex between PEO and resorcinol [9], these three carbon sites in resorcinol are in close proximity to the macromolecular chain--a consequence of hydrogen bonding of the hydroxyls to the ether oxygen of PEO. Of particular interest is the presence of two phenolic carbon resonances for resorcinol (spectrum 1B) in the blend with PEO containing 67 mole % resorcinol. The NMR data suggest the presence of two distinguishable molecular environments (possibly due to morphological inequivalences in the conformation of resorcinol), with the signal to the right of the dashed line (≈ 155 ppm, on the left side of Fig. 1) representing an environment (or conformation) of resorcinol in the blend that is essentially "NMR-indistinguishable" from that in undiluted resorcinol. To support the claim that NMR is capable of detecting two semicrystalline phases of the PEO-resorcinol mixture, specific heat vs. temperature data (from Differential Scanning Calorimetry) are illustrated in Fig. 2 for the 67 mole % (resorcinol) blend. The thermogram indicates two distinct first-order melting processes at 82°C (eutectic) and 105°C (liquidus) characterized by phase-transition temperatures that differ from those of undiluted PEO (T_m = 59°C, for a nominal molecular weight of 3400) and undiluted resorcinol (T_m = 114°C) [8]. In correlating the results from NMR and DSC for the 67 mole % (resorcinol) blend, the phenolic carbon resonance at 158 ppm (indicated by the dashed line on the left side of Fig.1) is assigned to resorcinol in one of the semicrystalline phases that melts at the eutectic temperature (82°C), whereas the ^{13}C resonance to the right of the dashed line (which is essentially "NMR-indistinguishable" from undiluted resorcinol at 155 ppm) is representative of the higher-melting resorcinol-rich phase (T_m = 105°C), whose melting temperature on the liquidus phase boundary is closer to that of pure resorcinol. The NMR and DSC data are consistent with the approximate phase diagram of this strongly interacting binary mixture [8,9].

The second phase of our overall scheme to bridge macroscopic and molecular-level probes of mixing phenomena and intermolecular association focuses on ^{13}C solid state NMR and the stress-strain properties of blends which have the potential to interact via transition-metal complexation [6,10]. The NMR spectra of these blends are illustrated in Figs. 3, 4 and 5. The components of these strategically chosen blends are poly(vinylpyridine), with nitrogen either in the 2-or 4-position on the aromatic ring, and a random copolymer of ethylene and methacrylic acid (15 wt. % acid) in which the carboxyl groups in the methacrylic acid segments are partially (60%) neutralized with zinc. The aromatic and carboxyl carbon chemical shift regions of the solid state NMR spectrum are illustrated in Fig. 3 for polymer-ionomer blends containing poly(4-vinylpyridine), P4VP. The carboxyl carbon resonance envelope highlighted in Fig. 3 (lower spectrum) is unique to the ionomeric component. Hence, chemical shift overlap between dissimilar components is nonexistent in the 100-to-200 ppm region for these blends. The carbon-13 chemical shift region of primary interest in Figs. 4 and 5 focuses on the *C*OOH and *C*OOZn resonances in the methacrylic acid segments of the zinc-neutralized copolymer. These two ^{13}C signals have chemical shifts of 185 and 189 ppm, respectively, in the undiluted copolymer as illustrated in Fig. 4 (lower spectrum). These assignments for the zinc-neutralized (*C*OOZn, 189 ppm) and unneutralized (*C*OOH, 185 ppm) carboxyl carbon resonances have been documented in a previous study [22]. Solid state NMR suggests that zinc complexation with the pyridine nitrogen lone electron pair is operative in the blend with poly(4-vinylpyridine) as evidenced by the new carboxyl carbon signal in Fig. 4 (upper spectrum), centered between 179 and 180 ppm. This ^{13}C resonance is absent when the ionic copolymer is blended with poly(2-vinylpyridine), as illustrated in Fig. 4 (middle spectrum), suggesting that steric hindrance prevents the formation of the complex. Furthermore, this new carboxyl carbon signal (lower spectrum in Fig. 5), assigned to the transition-metal complex formed between the zinc cation and the pyridine nitrogen, is absent also when poly(4-vinylpyridine) is blended with the unneutralized random copolymer of ethylene and methacrylic acid that is void of zinc (see the upper spectrum in Fig. 5). In this respect, it should be mentioned that Fourier transform infrared studies of hydrogen bonding in polymer-polymer blends can detect intermolecular association between the unneutralized carboxylic acid group in poly(ethylene-co-methacrylic acid) and the pyridine group in poly(2-vinylpyridine) [3].

If solid state NMR can detect unequivocally the formation of transition-metal-induced complexes in strategically selected blends [10], then the ability to correlate this molecular-level identification of intermolecular association with macroscopic results is a high-priority goal of this research effort. Unidirectional (stress-strain) tensile testing has revealed mechanical fracture stresses (lbs/in^2) of 4.3 x 10^3 for the ionomeric (Zn-neutralized random copolymer of ethylene and methacrylic acid) blend with poly(4-vinylpyridine), (upper spectrum in Fig. 4), and 2.7 x 10^3 for the blend with poly(2-vinylpyridine), (middle spectrum in Fig. 4). These mechanical results support the NMR detection of a complex in the former blend, since strong intermolecular complexation between macromolecular components at the microscopic level is expected to produce a blend with a greater resistance to mechanical fracture at the macroscopic level. However, the fracture stress of each ionomeric blend is less than that of the undiluted ionic copolymer (4.6 x 10^3 psi), indicative of blends that are not, for the most part, intimately mixed. The NMR spectrum in Fig. 4 (upper spectrum) hints also at such a phase-separated morphology because only a small fraction of the *C*OOH/*C*OOZn resonance envelope of the ionic copolymer is chemically shifted to the 179-180 ppm region in the blend with poly(4-vinylpyridine), due to complexation between zinc and nitrogen.

A more complete concentration-dependent study of the ambient-temperature mechanical properties of blends containing the partially neutralized zinc ionomer (SURLYN® 1706) with poly(4-vinylpyridine) is illustrated in Fig. 6. For comparison, the mechanical data at 90 mole % P4VP are representative of the blend whose ^{13}C NMR data are shown in the upper spectrum of Fig. 4 (the mole % of SURLYN® 1706 is based on the moles of Zn^{++} cations that neutralize the carboxylic acid group). In this case, the available zinc cations in the ionic copolymer are overwhelmed by the presence of an excess of pyridine nitrogens. Consequently, the inferior mechanical properties of the blend reflect this imbalance. Attempts to balance the molar ratio of zinc cations to pyridine nitrogens have produced a slight increase in the engineering fracture stress and a dramatic increase in the elastic modulus, with an acceptable

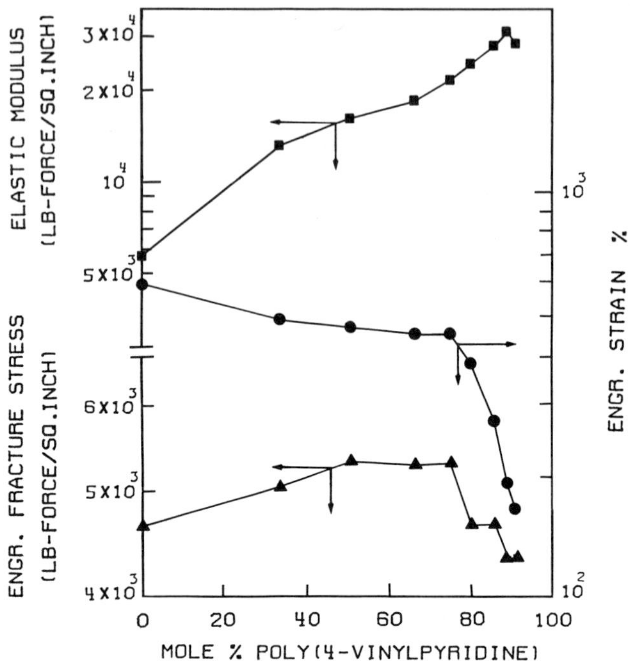

Figure 6. Mechanical properties of polymer-ionomer blends containing poly(4-vinylpyridine) and SURLYN® 1706. All stress-strain experiments were performed at a uniaxial tensile deformation rate of 0.5 inch/min. The molar concentration of nitrogen in P4VP is calculated with respect to the moles of Zn^{++} in SURLYN® 1706 (i.e. 67 mole % N corresponds to a 2:1 molar ratio of N to Zn^{++}). (•) ultimate strain at failure, (■) modulus of elasticity, (▲) fracture stress.

decrease (from 600% to 450%) in ultimate engineering strain (see Fig. 6). Carbon-13 NMR studies on a blend of SURLYN® 1706 and poly(4-vinylpyridine) having a better balance of interacting functional groups (75 mole % P4VP) have revealed that the carboxyl carbon resonance envelope of the ionomeric component is very similar to that in the upper spectrum of Fig. 4 at 90 mole % P4VP [23].

4.2 Molecular Proximity and Spin Diffusion via ^{1}H CRAMPS: Combined Rotation and Multiple Pulse Spectroscopy

Two applications of high-resolution ^{1}H solid state NMR are presented herein to support the concept of a molecular complex between poly(ethylene oxide) and resorcinol when the components are present in a 2:1 molar ratio. Both applications exploit the high-resolution nature of ^{1}H CRAMPS [20], and the simplicity of the solid state proton spectra for the PEO-resorcinol blend [5,6]. The first study provides a proton-analog of the ^{13}C NMR detection of microscopic mixing effects in strongly interacting systems. The one-dimensional ^{1}H CRAMPS spectrum of the molecular complex of PEO and resorcinol is presented in Fig. 7. In support of the isotropic ^{13}C chemical shift perturbations observed for mixtures of these components (see Fig. 1), the aromatic proton resonance envelope of resorcinol is less shielded (by approximately 1 ppm) in the molecular complex (lower spectrum in Fig. 7) relative to the undiluted crystalline state of the small molecule (upper spectrum in Fig. 7). Surprisingly, the hydroxyl proton resonance is not affected to the same degree -- a result which is difficult to rationalize considering that the electronic distribution about the hydroxyl proton should be strongly distorted due to hydrogen-bonding interactions [24,25]. Nevertheless, when proton signals from dissimilar blend components are resolved, the ^{1}H CRAMPS experiment provides a second independent source of high-resolution isotropic chemical shift information to correlate with molecular-level mixing effects on ^{13}C NMR and the macroscopic thermal measurements discussed in the previous section for this strongly interacting complex [6,8].

Of primary importance, in this section, is a discussion of intermolecular proton spin diffusion (via CRAMPS) between dissimilar species in a mixture. These experiments provide substantial evidence that dissimilar molecules (or chain segments) reside in a near-neighbor environment when such a morphology is thermodynamically or kinetically favored [6,11]. Chemical shift resolution of proton signals in the two blend components is required for (i) the generation of a "magnetization gradient" which drives the proton-spin-exchange process via proton dipolar couplings, and (ii)

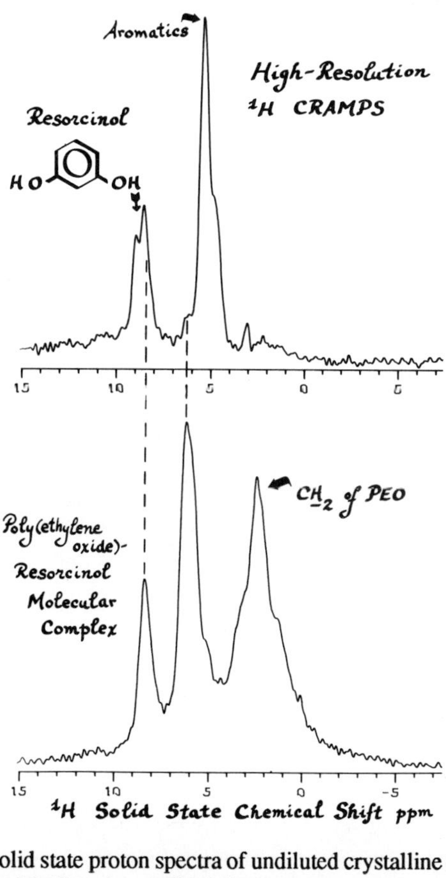

Figure 7. High-resolution solid state proton spectra of undiluted crystalline resorcinol (upper spectrum) and the molecular complex of PEO and resorcinol (lower spectrum) via the CRAMPS technique. The concentration of resorcinol in the molecular complex is 33 mole %.

the detection of "spin exchange" in a two-dimensional contour representation of the data. In this respect, the components of the molecular complex between PEO and resorcinol have been strategically selected for analysis via the two-dimensional ^1H spin diffusion experiment because (i) the ^1H solid state spectrum of each component is rather simple [5], (ii) spectral overlap between dissimilar blend components is essentially non-existent [6] or, at most, minimal, and (iii) molecular-level mixing of dissimilar molecules or chain segments is favorable based on chemical structure, thermodynamic driving forces [26-28] and the presence of interacting functional groups [5,29]. In reference to requirement (iii) mentioned above, the ^1H CRAMPS experiment is capable of monitoring proton spin diffusion in intimately mixed as well as phase-separated blends [11]. However, the duration of the spin-diffusion mixing period necessary to detect "off-diagonal spin-exchange contours" (generated by proton dipolar couplings) depends on the solid state morphology of the blends [6,11]. Once again, the molecular complex of PEO and resorcinol [9] provides an illustrative example of two-dimensional proton-spin-diffusion spectroscopy in Figs. 8 and 9. The high-resolution ^1H spectrum of the solid complex is quite simple as illustrated in Fig. 7. Representative one-dimensional projection spectra in Figs. 8 and 9 mimic the high-resolution nature of the blend spectrum in Fig. 7. This allows one to distinguish contours of the methylene protons of PEO from the aromatic and hydroxyl proton contours of resorcinol. After 100 μs of proton dipolar interaction, both *inter*molecular and *intra*molecular spin diffusion can be observed in Fig. 8 via the off-diagonal contours indicated by the arrows. This illustrates that proton magnetization transport between PEO and resorcinol is operative on the 100-μs time scale, suggesting molecular-level mixing of the two dissimilar components. In contrast, the data in Fig. 9 reveal that *inter*molecular spin diffusion between PEO and resorcinol is absent when homonuclear ^1H dipole-dipole interactions (couplings) are exerted for 40 μs. However, *intra*molecular (aromatic-hydroxyl) couplings within resorcinol are strong enough to be observed at this shorter spin-diffusion mixing time (40 μs).

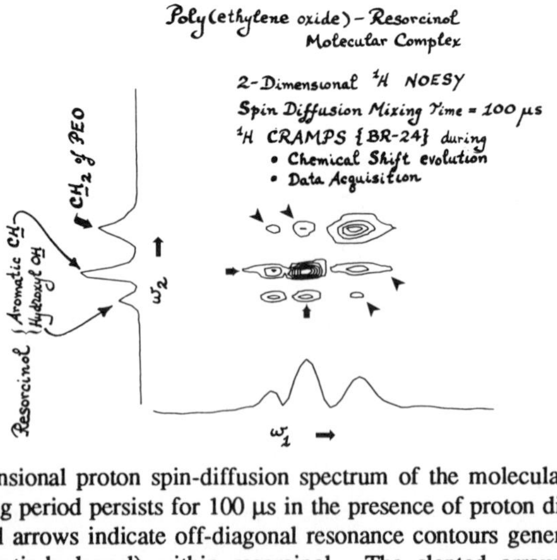

Figure 8. Two-dimensional proton spin-diffusion spectrum of the molecular complex of PEO and resorcinol. The mixing period persists for 100 μs in the presence of proton dipolar interactions. The horizontal and vertical arrows indicate off-diagonal resonance contours generated via proton dipolar communication (aromatic-hydroxyl) within resorcinol. The slanted arrows indicate off-diagonal resonance contours generated via intermolecular dipolar couplings between PEO and resorcinol.

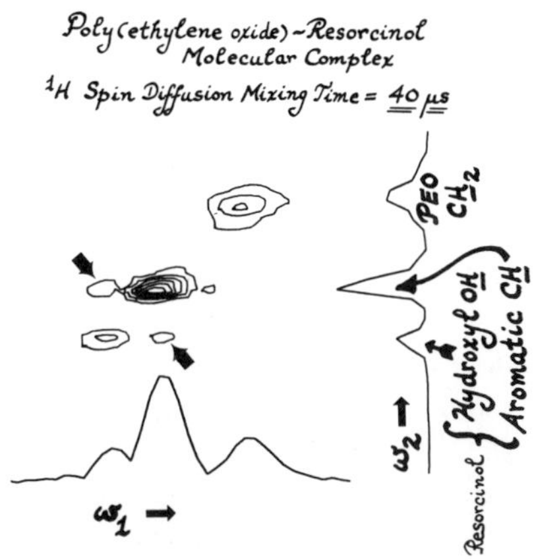

Figure 9. Two-dimensional proton spin-diffusion spectrum of the molecular complex of PEO and resorcinol. In this case, dipolar couplings are exerted for 40 μs during the mixing period. Only the aromatic-hydroxyl off-diagonal proton resonance contours, generated via dipolar communication within resorcinol, can be observed at this shorter spin-diffusion mixing time.

In summary, the time scale of the spin-diffusion process is of paramount importance in an analysis of molecular proximity--differentiating between intimate (microscopic) mixing at the segment level and phase separation with the formation of microdomains. The observation of off-diagonal spin-exchange contours after 100 μs of dipolar interaction in the molecular complex formed between PEO and resorcinol (Fig. 8) supports the concept of intimate mixing of dissimilar molecules at the 2-5 angstroms level.

In the following section, we present evidence (see Figs. 11 and 12) that *interdomain* proton spin diffusion is operative between (i) disordered (thin) crystalline lamellae and the amorphous matrix of a semicrystalline ionic copolymer (SURLYN® 1706), and (ii) rigid spherical occlusions and the mobile matrix in a microphase-separated triblock copolymer of styrene and butadiene (KRATON®), on the 10-20 ms time-scale.

4.3. Interdomain Communication via Magnetic (^{1}H) Spin Diffusion: The Modified Goldman-Shen Experiment

This technique is reserved as morphological probe of microphase-separated solid state blends and copolymers that do not allow a study of proton spin diffusion via the ^{1}H CRAMPS experiment. In proton-spin-diffusion studies from which domain sizes and molecular proximity may in principle be determined, it is attractive to observe proton resonances directly [6,11] via ^{1}H NMR. This was demonstrated in the previous subsection. However, solid state ^{1}H spectra of polymers are often insufficiently resolved to fully separate signals from different domains. In this respect, we have modified the Goldman-Shen experiment [13,14], as described below, such that ^{1}H spin-diffusion is monitored indirectly via carbon-13 sites directly bound to the protons of interest. The success of this morphological probe relies on the ability to resolve ^{13}C resonances in each phase. Of critical importance is the establishment of a magnetization, or spin-temperature, gradient, necessary to drive the spin-diffusion process. This is readily facilitated by a substantial difference in mobility between the two domains of interest, and the consequent difference between ^{1}H spin-spin relaxation rates in the two domains. It should be mentioned that the Goldman-Shen experiment is not useful to study spin diffusion and molecular proximity in microphase-separated systems when transverse spin-spin relaxation rates are not sufficiently different in the two domains of interest [30].

The modified Goldman-Shen experiment is illustrated schematically in Fig. 10. This experiment is a "second-generation" analog of the technique proposed by Goldman and Shen [13] almost two decades past. Initially, an ^{1}H 90°-pulse produces proton polarization in the transverse plane. After the time delay designated by T_1 in Fig. 10, proton magnetization in the rigid domains is severely attenuated by spin-spin relaxation processes. Proton polarization which survives the T_1 time delay is returned to the ± z-axis by the second 90°-pulse. At the beginning of the T_2-time interval, a gradient in proton magnetization between the rigid and mobile domains has been prepared. However, the modified Goldman-Shen experiment does not sample proton magnetization directly and, hence, cannot detect spatial characteristics of the spin-temperature gradients in the proton spin manifold. If proton dipolar interactions are operative across domain boundaries, which depends strongly on the strength of dipolar couplings in the interfacial regions, then the initial magnetization gradient will decrease with time, analogous to phenomenological transient diffusion-like processes [31], and proton polarization will be redistributed between the rigid and mobile domains until spin-temperature equilibration is achieved. By incorporating ^{1}H-^{13}C cross polarization [32,33] in the pulse sequence together with magic-angle sample

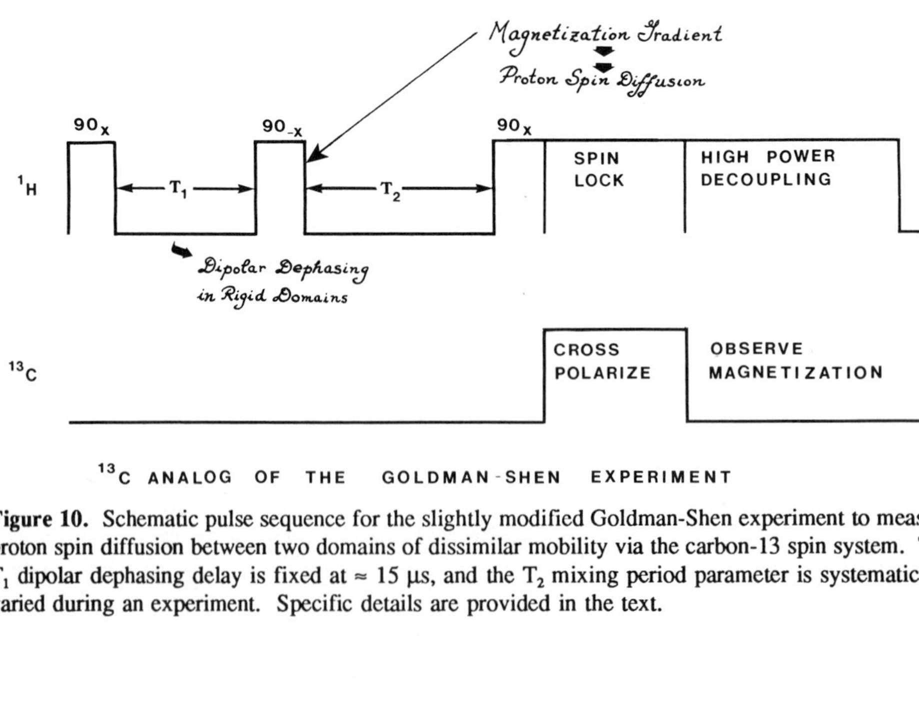

Figure 10. Schematic pulse sequence for the slightly modified Goldman-Shen experiment to measure proton spin diffusion between two domains of dissimilar mobility via the carbon-13 spin system. The T_1 dipolar dephasing delay is fixed at ≈ 15 μs, and the T_2 mixing period parameter is systematically varied during an experiment. Specific details are provided in the text.

spinning and high-power ^{1}H-^{13}C dipolar decoupling (during data acquisition), one can map the effect of proton spin diffusion between two microdomains of dissimilar mobility onto the carbon-13 nuclear spin manifold. If the thermal mixing time during cross polarization is relatively short (on the order of 50-100 μs), then one can effectively thwart long-range ^{1}H-^{13}C dipolar communication and obtain carbon intensities that are proportional to the magnetization of directly attached protons. Results for random ionic copolymers of ethylene and methacrylic acid, and tri-block copolymers of styrene and butadiene (or completely hydrogenated butadiene) [14], indicate that a dipolar dephasing delay of approximately 10-20 μs is sufficient to nullify proton polarization in the rigid domains. Proton polarization in the mobile domains is reduced to a much lesser degree during the dephasing interval because ^{1}H homonuclear dipolar couplings in the mobile regions are partially averaged by molecular motion. Thus, a magnetization or spin-temperature gradient exists between the rigid and mobile regions of the sample at the beginning of the mixing period, T_2. Proton magnetization is subsequently redistributed between the two domains by spin diffusion during the mixing period. Several experiments are performed in which the T_2 parameter is systematically varied. While the proton flip-flop terms in the dipolar Hamiltonian produce a redistribution in ^{1}H polarization between the rigid and mobile proton reservoirs during the mixing period [13], both reservoirs also approach spin-temperature equilibration with the lattice at a rate which is presumably much slower than the spin-diffusion rate. The effect of spin-lattice relaxation is nulled by storing proton magnetization alternately along the ± z-axis at the beginning of the mixing period and subtracting alternate signals in the time domain before Fourier transformation. It should be mentioned that the proposed experiment fails to monitor an undistorted proton-spin-diffusion phenomenon if the spin-diffusion rate is comparable to the spin lattice relaxation rate.

Proton spin diffusion data from the modified Goldman-Shen experiment are illustrated in Fig. 11 for the zinc-neutralized (60%) copolymer of ethylene and methacrylic acid (15 wt. % acid). This ionic copolymer forms a transition-metal complex with poly(4-vinylpyridine), as discussed in a previous section of this contribution. Carbon-13 magnetization from the ethylenic CH_2 segments in the crystalline domains is generated via proton-proton dipolar coupling (across interfacial boundaries) to the mobile amorphous CH_2 segments, and subsequent ^{1}H-^{13}C cross polarization *within* the crystallites. The proton dipolar dephasing period was 15 μs, the mixing time was varied from 0.1 μs to 100 ms with emphasis on the short millisecond time scale, the ^{1}H-^{13}C contact time was 75 μs, and proton magnetization was stored alternately along the ± z-axis at the start of the mixing period to suppress spin-lattice relaxation processes. Redistribution of proton polarization between the two domains occurs on the 10-to-20 millisecond time scale. This claim is supported by the transient behavior of the proton-spin-diffusion process illustrated in Fig. 11, based on integral carbon-13 intensities of the crystalline ethylenic CH_2 signal obtained via deconvolution of the NMR spectra. Such a curve-fitting scheme was employed because of the overlap between the crystalline and amorphous CH_2 resonances in the ionic copolymer [22]. The data in Fig. 11 reveal that the proton-spin-diffusion process is governed primarily by one time constant of 3.3 ms.

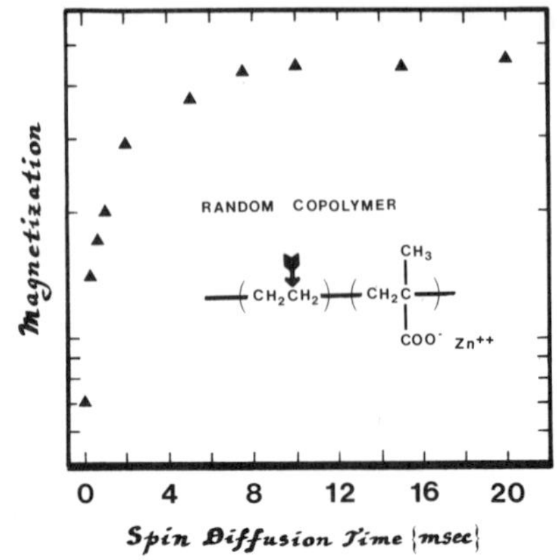

Figure 11. Proton spin-diffusion data from the slightly modified Goldman-Shen experiment. The material is a random copolymer of ethylene and methacrylic acid in which the carboxyl groups are partially neutralized with zinc counterions (SURLYN® 1706). The data points represent carbon-13 magnetization of the crystalline CH_2 segments, generated via proton dipolar communication (across interfacial boundaries) with the amorphous ethylene segments. The horizontal time axis corresponds to the mixing period (designated by T_2) in Fig. 10.

Modified Goldman-Shen results are illustrated in Fig. 12 for phase- separated commercial triblock copolymers of styrene and butadiene (SBS), and styrene with completely hydrogenated butadiene (S-EB-S). In these cases, protonated aromatic carbon-13 magnetization, unique to the rigid styrene endblocks, is generated via proton-proton dipolar coupling (across interfacial boundaries) to the mobile butadiene or ethylene/butylene midblock, and subsequent ^{1}H- ^{13}C cross polarization *within* the rigid polystyrene occlusions. Redistribution of proton polarization between the mobile midblock and the rigid endblocks occurs on the 10-to-20 millisecond time scale, analogous to the proton-spin-diffusion results in Fig. 11 for the ionic copolymer of ethylene and methacrylic acid. In the commercial triblock copolymers, the time-dependent interdomain communication process between the mobile matrix and the rigid styrene occlusions has been modeled phenomenologically via coupled unsteady state radical (Fickian) diffusion equations appropriate to a concentrically spherical (core-shell) morphology. Details of the mathematical simulation and parameter estimation can be found in reference 14.

Internal proton-spin-temperature equilibration between (i) the crystalline and amorphous regions of the ionic copolymer (SURLYN® 1706) and (ii) the rigid and mobile blocks of the KRATON® thermoplastic elastomers, on the order of 10-to-20 milliseconds, might be considered fast relative to a macroscopically phase-separated blend [11] or copolymer, but slow relative to a strongly interacting mixture [6]. This is expected for (i) a polyethylene-like material whose spherulitic superstructure is disrupted by the presence of the atactic comonomer (methacrylic acid), resulting in a fringed micellar pattern of thin disoriented ethylenic crystallites [34], and (ii) spherical polystyrene microdomains having an average diameter of approximately 300 angstroms dispersed in a rubbery matrix [35]. For mixtures that are amenable to a study of intimate contact between dissimilar molecules via a two-dimensional application of the ^1H CRAMPS experiment, proton spin diffusion within the molecular complex of poly(ethylene oxide) and resorcinol occurs on the 100-*micro*second time scale, as discussed in the previous subsection, whereas Caravatti et. al., [11] report the absence of such communication on the 100-*milli*second time scale for an *immiscible* blend of polystyrene and poly(vinyl methyl ether), cast from chloroform.

5. Conclusions

The most important aspects of the work described herein are summarized generically in the following paragraph.

When polymer blend components are tailored such that *strong* intermolecular association between dissimilar chain segments is favorable, then perturbations in the isotropic resonance envelopes of nuclear sites that participate in the interaction will, most likely, occur. If blend composition dictates that two solid state phases coexist at ambient temperature, then high-resolution carbon-13 NMR experiments should be capable of detecting the critical component in both phases of the binary mixture, unless very dilute concentrations of the critical component present a signal-to-noise problem in one of the phases. Thermodynamic phase diagrams, generated via Differential Scanning Calorimetry, are extremely useful to interpret the carbon-13 NMR spectra of polymer blends in a two-phase region. Under favorable conditions related to chemical shift resolution, the direct observation of spin diffusion via ^1H CRAMPS between dipolar-coupled protons in dissimilar molecules or chain segments on a time scale of 10^{-4} seconds or less provides convincing evidence for molecular-

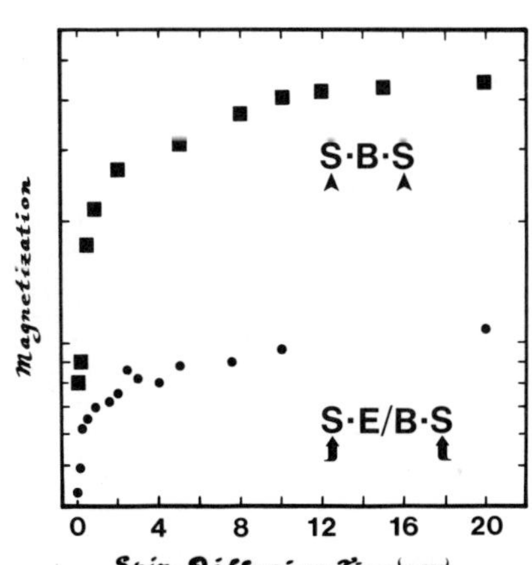

Figure 12. Proton spin-diffusion data from the slightly modified Goldman-Shen experiment. The materials are commercial triblock copolymers of styrene and butadiene (KRATON® D-1101, upper curve), and styrene with completely hydrogenated butadiene (KRATON® G-1651, lower curve). The data points represent carbon-13 magnetization of the styrenic ring, generated via proton dipolar communication (across interfacial boundaries) with the mobile butadiene (SBS) or ethylene/butylene (S-EB-S) matrix. The horizontal time axis corresponds to the mixing period (designated by T_2) in Fig. 10.

level mixing in strongly interacting blends. It has been demonstrated that *interdomain* proton spin diffusion is operative on the 1-10 ms time scale in microphase-separated copolymers. In this respect, the time scale during which proton spin diffusion occurs in the presence of homonuclear dipole-dipole couplings is a critically important parameter in any analysis of molecular proximity.

6. Acknowledgements

Financial support during the course of this research from the Colorado Advanced Materials Institute, located at the Colorado School of Mines in Golden, is gratefully acknowledged. The solid state NMR experiments were performed at the Regional NMR Center, Colorado State University, funded by NSF under Grant #CHE-8746548. Dr. C. E. Bronnimann, of CSU's Regional NMR Center is acknowledged for obtaining the high-resolution proton NMR data via the CRAMPS experiment. Tom Geist, at the IBM Corporation in Boulder, Colorado is acknowledged for obtaining the thermal data discussed herein via Differential Scanning Calorimetry and Thermogravimetric Analysis. The NUCREL® and SURLYN® samples were kindly supplied by Drs. Ruskin Longworth and George Hoh of E. I. DuPont de Nemours and Company in Wilmington, Delaware. The KRATON® samples were obtained courtesy of Drs. Richard Gelles and John Greak of Shell Development Company in Houston, Texas. Finally Ms. Melissa Taylor is gratefully acknowledged for her secretarial assistance.

7. References

1. "Polymer Blends and Composites in Multiphase Systems," Han, C. D., Ed., *Adv. Chem. Ser.*, *206*, 1984.
2. "Multicomponent Polymer Materials," Sperling, L. H. and Paul, D. R., Eds., *Adv. Chem. Ser.*, *211*, 1986.
3. Lee, J. Y.; Painter, P. C. and Coleman, M. M., *Macromolecules*, *21*, 954, 1988.
4. Lichkus, A. M.; Painter, P. C. and Coleman, M. M., *Macromolecules*, *21*, 2636, 1988.
5. Belfiore, L. A., *Polymer Preprints*, *28(2)*, 114, 1987.
6. Belfiore, L. A., *Polymer Preprints*, *29(1)*, 17, 1988.
7. Patwardhan, A. A. and Belfiore, L. A., *Polym. Engr. Sci.*, *28(14)*, 916, 1988.
8. Belfiore, L. A., *Polymer Preprints*, *29(1)*, 430, 1988.
9. Myasnikova, R. M.; Titova, E. F. and Obolonkova, E. S., *Polymer*, *21*, 403, 1980.
10. Peiffer, D. G.; Duvdevani, I.; Agarwal, P. K. and Lundberg, R. D., *J. Polym. Sci., Polym. Lett. Ed.*, *24*, 581, 1986.
11. Caravatti, P.; Neuenschwander, P. and Ernst, R. R., *Macromolecules*, *18*, 119, 1985.

12. Caravatti, P.; Neuenschwander, P. and Ernst, R. R., *Macromolecules*, *19*, 1889, 1986.
13. Goldman, M. and Shen, L., *Physical Review*, *144(1)*, 321, 1966.
14. Belfiore, L. A. and Patwardhan, A. A., *ACS Proceedings, Div. Polym. Mater. Sci. Engr.*, *54*, 638, 1986.
15. Lin, P.; Clash, C.; Pearce, E. M.; Kwei, T. K. and Aponte, M. A., *J. Polym. Sci., Polym. Phys. Ed.*, *26*, 603, 1988.
16. Cheng, C., M.S. Thesis, Colorado State University, 1989.
17. Wind, R. A.; Anthonio, F. E.; Duijvestijn, M. J.; Smidt, J.; Trommel, J. and DeVette, G. M. C., *J. Magn. Reson.*, *52*, 424, 1983.
18. Stejskal, E. O. and Schaefer, J., *J. Magn. Reson.*, *18*, 560, 1975.
19. Earl, W. L. and VanderHart, D. L., *J. Magn. Reson.*, *48*, 35, 1982.
20. Bronnimann, C. E.; Hawkins, B. L.; Zhang, M. and Maciel, G. E., "^{1}H CRAMPS as an Analytical NMR Technique for Solids," *Analytical Chem.*, *60*, 1743, 1988.
21. Burum, D. P. and Rhim, W. K., *J. Chem. Phys.*, *71*, 944, 1979.
22. Belfiore, L. A. and Shah, R. J., *ACS Proceedings, Div. Polym. Mater. Sci. Engr.*, *54*, 490, 1986.
23. Belfiore, L. A., unpublished results.
24. Imashiro, F.; Maeda, S.; Takegoshi, K.; Terao, T. and Saika, A., *Chem. Phys. Lett.*, *92(6)*, 642, 1982; *99(2)*, 189, 1983.
25. Takegoshi, K.; Naito, A. and McDowell, C. A., *J. Magn. Reson.*, *65*, 34, 1985.
26. Stejskal, E. O.; Schaefer, J.; Sefcik, M. D. and McKay, R. A., *Macromolecules*, *14*, 275, 1981.
27. Gobbi, G. C.; Silvestri, R.; Russell, T. P.; Lyerla, J. R.; Fleming, W. W. and Nishi, T., *J. Polym. Sci., Polym. Lett. Ed.*, *25*, 61, 1987.
28. Belfiore, L. A.; Patwardhan, A. A. and Lenz, T. G., *Industr. Engr. Chem. Res.*, *27*, 284, 1988.
29. Belfiore, L. A., *Polymer*, *27*, 80, 1986.
30. Havens, J. R. and VanderHart, D. L., *Macromolecules*, *18*, 1663, 1985.
31. Carslaw, H. S. and Jaegar, J. C., "Conduction of Heat in Solids," 2nd Ed., Oxford University Press, New York, 1984.
32. Hartmann, S. R. and Hahn, E. L., *Physical Review*, *128*, 2042, 1962.
33. Pines, A.; Gibby, M. G. and Waugh, J. S., *J. Chem. Phys.*, *59*, 569, 1973.
34. Longworth, R. in "Ionic Polymers," Holliday, L., ed., Wiley, New York, 1975, Ch 2.
35. Helfand, E. and Wasserman, Z. R., *Macromolecules*, *11*, 960, 1978.

Chapter 9

CP-MAS NMR SPECTRA OF POLY(N-VINYLCARBAZOLE) AND OF ITS CHARGE TRANSFER COMPLEXES: PRELIMINARY INVESTIGATION

ALMERIA NATANSOHN

1. Introduction

Poly(N-vinylcarbazole) (PNVC) presents photoconductive properties in the UV region [1]. Translation of these properties into the visible region can be achieved by preparing a charge transfer complex with an electron acceptor substance. These complexes are photoconductive at the wavelength of their absorbance. One of them, with 2,4,7-trinitro-9-fluorenone (TNF) as electron acceptor, was extremely successful as the photosensitive element in photocopying machines [2,3]. Since this discovery, there has been a constant interest in polymer charge transfer complexes, mainly in correlating the polymer chemical structure with the charge transfer complexation and photoconductivity. Pearson, Turner and Ledwith reviewed the field in 1979 [4] and subsequent papers appeared on different photoconductive carbazole-containing (co)polymers [5-10].

The geometry of these charge transfer complexes appears to play a very important role in determining their photoconductive properties. Analysis of the dependence of the photocurrent on the film composition, and correlations with the electronic charge transfer absorption band, suggest the presence of a combination of complexed and uncomplexed components [4]. Also, the photoconductive properties are strongly dependent on the polymer microstructure. The mode of attachment of the active groups into the polymer and the relative orientation of these groups imposed

ALMERIA NATANSOHN - Department of Chemistry, Queen's University, Kingston, Ontario, K7L 3N6, Canada

Solid State NMR of Polymers, Edited by L. Mathias
Plenum Press, New York, 1991

by the polymer backbone is extremely important in determining the carrier mobility [11]. It is clear that in-depth analysis of the degree of interaction in these complexes combined with an understanding of the polydispersity of the charge transfer interaction (when polymer complexes are involved) are vital in the study of their photoconductive properties.

Charge transfer complexes of carbazoles and carbazole-containing polymers were extensively analyzed using electronic spectroscopy. The charge transfer band in such spectra is usually asymmetric due to two possible transitions that contribute to it. Molecular orbital calculations on the complex of PNVC with chloranil (CHL) demonstrate that there are two possible parallel plane alignments for the carbazole and CHL ring that contribute to the two charge transfer bands in the electronic spectra [12]. Another method of analysis of charge transfer complexes in carbazole-containing copolymers is by measuring chemical shifts in the NMR spectra of these compounds. In a series of copolymers containing substituted carbazole as donor and 3,5-dinitrobenzoic acid derivatives as acceptor [13] it was demonstrated that the chemical shift of the aromatic protons belonging to the acceptor structural unit can be correlated with the degree of charge transfer interaction in solution. Carbon chemical shifts are much more sensitive to charge transfer complexation [14], but the information available from both electronic and NMR spectra is complicated by the presence of the solvent.

One of the possible methods to investigate the charge transfer complex geometry at the molecular level in solid state is cross-polarization magic-angle-spinning NMR spectroscopy. To my knowledge, however, there is only one such study on small molecular complexes in the literature [15]. That study reveals the multiple potentials of this approach. First, chemical shift differences are observed between complexed and uncomplexed molecules, and they are assigned to the degree of electron transfer in the ground state. Second, when unprotonated substances are used, intermolecular cross-polarization transfer can be obtained, indicating molecular association. And finally, that study only suggested measurement of relaxation times and correlation with molecular or co-operative movements. This last point has started to be extensively applied to many polymer blend systems [16]. Carbon-13 chemical shift changes are also known for solid clathrates [16a].

This paper is first in a series that will investigate charge transfer complexation in different polymer systems by means of CP-MAS NMR spectroscopy. For the preliminary investigation the first two polymer charge transfer complexes chosen were PNVC-TNF and PNVC-CHL. PNVC-TNF is the best photoconductive polymer charge transfer complex, while PNVC-CHL was the subject of many studies concerning its spatial arrangement. Furthermore, CHL has no protons, an ideal case for intramolecular cross-polarization studies.

2. Experimental

PNVC (Aldrich) was purified by reprecipitation from different systems (THF-water; chloroform-acetone), but its CP-MAS spectra were not different from the initial material. TNF (Polysciences) and CHL (Aldrich) were used as received. N-ethylcarbazole (NEC) (Aldrich) was recrystallized four times from ethanol. Small molecular complexes (NEC-TNF and NEC-CHL) were obtained by a few procedures. First, hot, saturated solutions of the components were mixed and the resulted complex was filtrated and dried. Second, acetone solutions of the 1:1 NEC-TNF complex would yield crystals by standing for a few days. Third, plain slow evaporation of THF from a 1:1 complex solution would yield a powder that was considered to be the solid complex. Polymer complexes were obtained by slow evaporation of the common solvent, usually as a film. Both polymer films (PNVC-TNF and PNVC-CHL) were analyzed by DSC to identify possible uncomplexed acceptor molecules present in the film. PNVC-CHL had a stable glass transition at ca. 160° C (PNVC alone had a Tg of ca. 220° C). PNVC-TNF presented a multiple melting in the first run, and then a stable melting-glass transition peak at about 170° C. The disappearance of the first melting peak (just below 160° C) was therefore considered to be due to "total" complexation of "free" TNF molecules still present in the system before heating.

DSC scans were performed on a Mettler TA-3000 instrument. CP-MAS NMR spectra were recorded on a Bruker CXP-200 spectrometer using a Doty probe at room temperature. Standard pulse sequences were used for measuring proton $T_{1\rho}$ (variable contact time) and carbon T_1 (with cross-polarization). Chemical shifts were measured from an internal reference - a small piece of crosslinked polyethylene that has a sharp signal at 32.85 ppm [17].

3. Results and Discussion

3.1. PNVC-TNF System

CP-MAS NMR spectra of carbazole-containing polydiynes are reported in the literature [18,19] and the assignments of the aromatic signals are straightforward. Figure 1 presents the carbon spectra of PNVC in $CDCl_3$ solution [1a] and in solid state [1b]. Carbon signals assignments were made in another paper [20] using J-modulated spin echo and two-dimensional techniques. The solid state carbon spectrum is less resolved than the solution spectrum, and the assignments can be made by direct comparison and are identical to those already reported. Another proof for the suggested assignment is given during $T_{1\rho}H$ measurements, where in the initial stage there is a difference in the buildup of the signals for the protonated and unprotonated carbons. Figure 2 illustrates this for the signals centered at 142 ppm (purely unprotonated: 8a and 9a carbons signal) and 110 ppm (purely protonated: 1 and 8 carbons signal). In the rest of the spectrum there is no separation between these kinds of carbon. Not enough data points were measured in this range of contact times in order to determine T_{CH} [21], but it is obvious that the buildup of the magnetization is more rapid for protonated carbons.

The spectrum of TNF is much more complicated, because each of the 13 carbons present gives a separate signal. Assignments based on J-modulated spin-echo

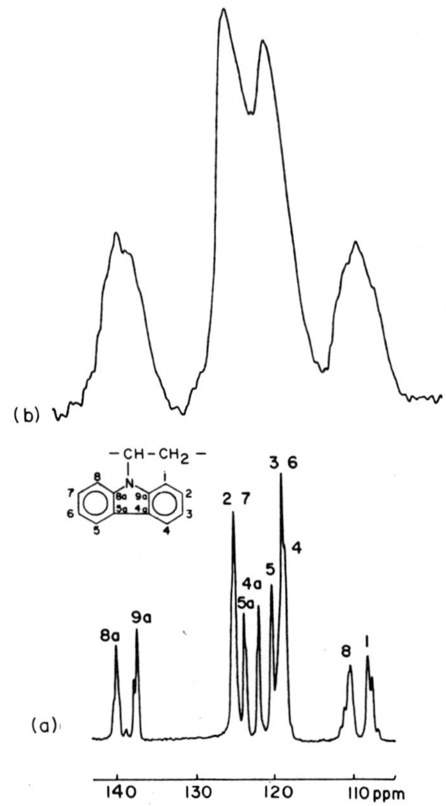

Figure 1. Carbon NMR spectra of PNVC (aromatic region): a. in $CDCl_3$ solution; b. obtained with CP-MAS in solid state with a 3.8 μs pulse, 3 ms contact time, 10 s delay, 200 scans and ca. 4 kHz spinning rate.

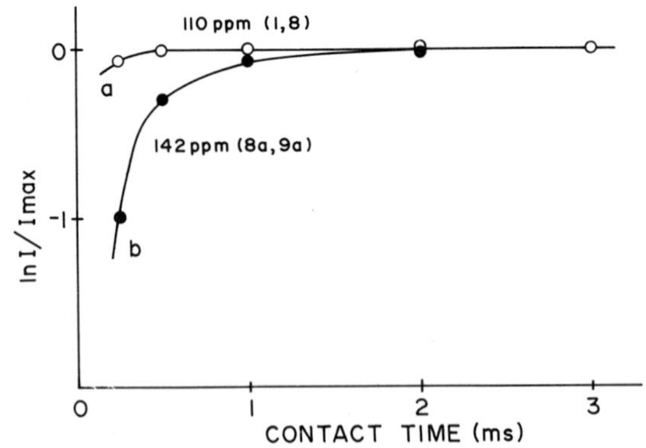

Figure 2. Buildup of the magnetization in solid state for two signals of the PNVC spectrum: a. carbons 1 and 8 (110 ppm); b. carbons 8a and 9a (142 ppm).

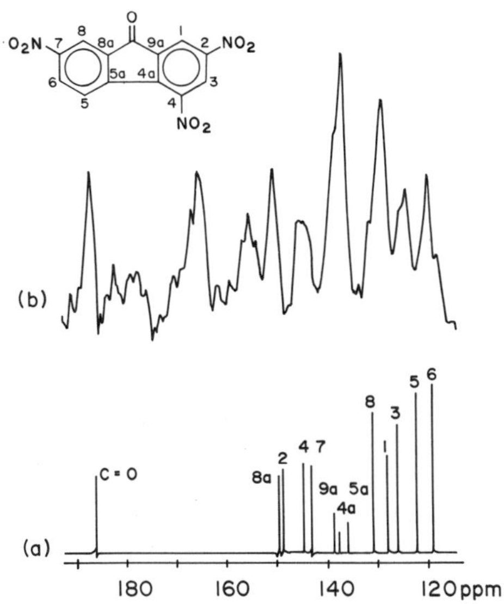

Figure 3. Carbon NMR spectra of TNF: a. in DMSO-d_6 solution; b. obtained with CP-MAS in solid state with a 3.8 μs pulse, 20 ms contact time 15 s delay, 1500 scans and ca. 4 kHz spinning rate.

and two-dimensional spectra were presented in a separate paper [22] and are written in Fig. 3a.

The assignments for the solid state spectrum are assumed to correspond to the solution assignments, even if the chemical shifts are quite different in this case for a few carbons. They are presented in Fig. 3b. The differences in the chemical shifts are supposed to appear due to solid-state organization of the molecules that would favor the appearance of deshielding effects. There is no independent proof for this assumption at this stage, and direct assignment in solid state is extremely complicated from the experimental point of view. However, a very simple method to prove which carbons are protonated and which are not is the one presented before for PNVC. Fig. 4 illustrates the magnetization buildup for the two types of carbons.

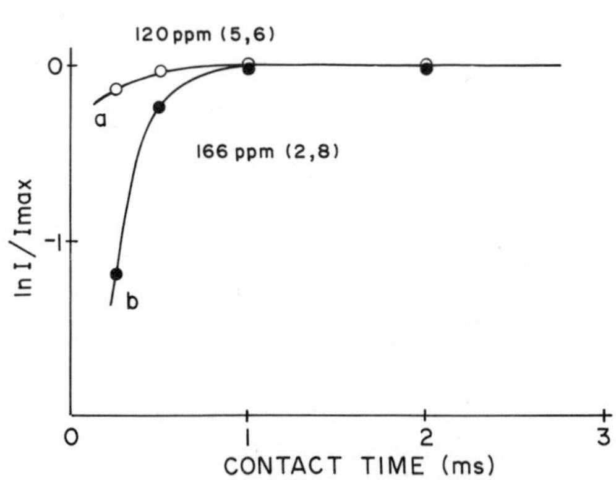

Figure 4. Buildup of the magnetization in solid state for two signals of the TNF spectrum: a. carbons 5 and 6 (120 ppm); b. carbons 2 and 8a (166 ppm).

PNVC has a rather poor spectrum in solid state, the resonances are broad probably due to the polydispersity in configuration and conformation. For more precise assignments, NEC can be used as the model compound for the structural unit of PNVC. However, as it can be seen in Fig. 5, the solid state spectrum of NEC has an unexpected complexity. There are only 6 lines in the solution ($CDCl_3$) spectrum of NEC (Fig. 5a). Each one splits into many more in solid state (Fig. 5b). For example, carbons 1 and 8 have at least 5 lines in the spectrum 5b, while carbons 8a and 9a give a broad signal, similar with the polymer spectrum. As a whole, the NEC spectrum is better resolved than that of PNVC, but its assignment is by no means an easy task. Attempts are being made to grow single crystals of NEC and to study their X-ray diffraction pattern, because this complex NMR spectrum suggests different crystalline arrangements of individual molecules.

Changes in chemical shifts are expected when complexation occurs. Yannoni et al. [15] observed a small downfield shift of the aromatic carbons belonging to the electron donor substance. No chemical shift was reported for the electron acceptor, but this was because the acceptor was usually protonless and did not have any CP spectrum in uncomplexed state. For proton spectra in solution, the acceptor seemed the most sensitive to complexation [13]. Carbon spectra of complexes in solution exhibited large upfield chemical shifts for both the donor and the acceptor [14] suggesting that the shift might be due to the π-π orbital overlapping of the two aromatic compounds that would generate aromatic shielding effects in both and not to an increase-decrease of electronic density caused by partial electron transfer from the donor to the acceptor in the ground state.

The NMR spectrum of the 1:1 NEC-TNF complex is presented in Fig. 6 together with the spectra of NEC and TNF, for comparison. The spectrum of the complex contains many lines and the only certain assignment is that of the carbonyl

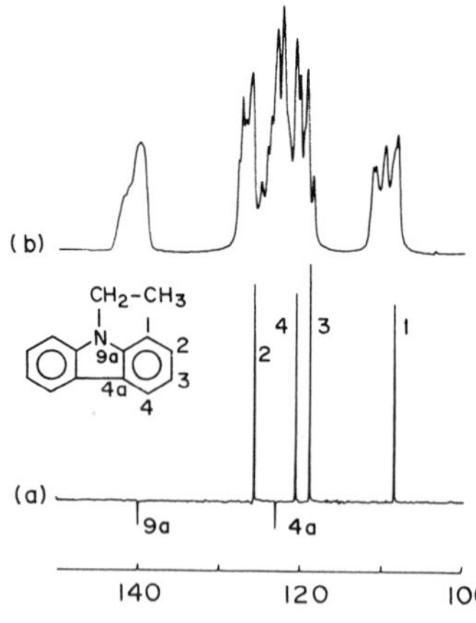

Figure 5. Carbon NMR spectra of NEC (aromatic region): a. in $CDCl_3$ solution; b. obtained with CP-MAS in solid state with a 3.8 μs pulse, 3 ms contact time, 10 s delay, 100 scans and ca. 4 kHz spinning rate.

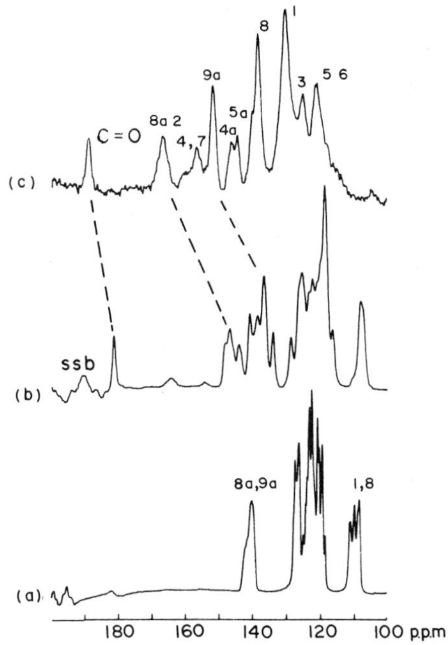

Figure 6. Solid state carbon spectra (aromatic region) of: a. NEC; b. 1:1 complex NEC-TNF obtained by slow evaporation of THF; c. TNF. Spectra collected in the same conditions as Figure 5b, except for c: same as Figure 3.

carbon, which is shifted upfield by 5 ppm. The next signal of a comparable intensity appears at 149 ppm, 17 ppm upfield from the signal of carbons 8a and 2 in the uncomplexed TNF. It is reasonable to suppose that this is the same signal, probably resolved now in its two components (8a and 2). Two other signals that can be tentatively assigned in the spectrum of the complex, based only on their relative intensity, are the 9a carbon (shifted upfield by 15 ppm) and the 1 carbon (upfield 10 ppm). It is impossible to say if some of the NEC signals are shifted in the complex, as compared with the spectrum of pure NEC. The only apparent change seems to be a partial collapse of the lines corresponding to carbons 1 and 8 into an unresolved, narrower signal (110 ppm).

Figure 7 presents the NMR spectra of two PNVC-TNF complexes: one obtained by slow evaporation of THF (b) and the other obtained after the film has been subjected to 180° C for ca. 20 minutes (c). Spectra of the PNVC (a) and TNF (d) are also included. The spectra of these polymer complexes seem to contain less information than the spectra of NEC-TNF. Signals appearing from complexed TNF can be identified between 135 and 155 ppm, but the resolution is poor enough to allow any distinction between these and the 8a-9a signal of PNVC. The carbonyl signal in (b) is very small and is not shifted from the TNF spectrum. In the spectrum of the heated sample the carbonyl signal is masked by a rather intense spinning side band. The only certain shift is that of 2 and 8a carbons signal, upfield by ca. 13 ppm, less than in the small molecular complex. In this case, the two PNVC signals that are

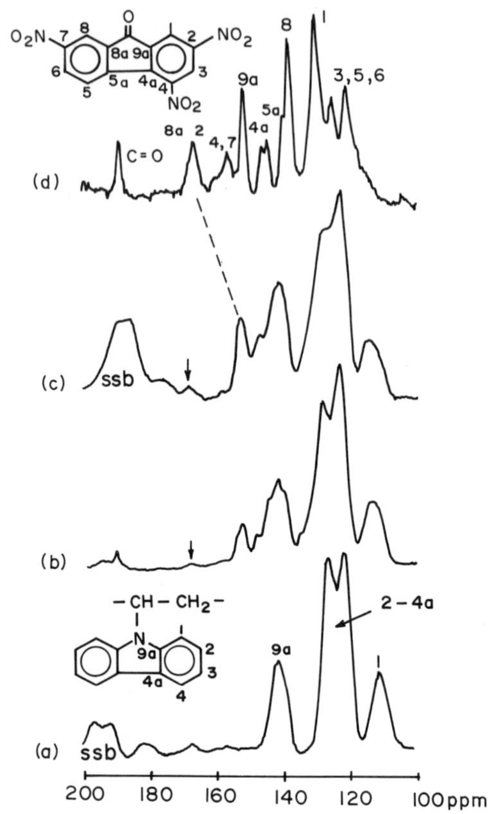

Figure 7. Solid state carbon spectra (aromatic region) of: a. PNVC; b. 1:1 complex PNVC-TNF obtained by slow evaporation of THF; c. same complex heated at 180° C; d. TNF; a and b: same conditions as in Figure 1; c: same as Figure 3.

still recognizable (110 ppm and 122 ppm) seem to be slightly shifted downfield, but as the signal of carbons 1 and 8 suggests, this could be a consequence of the apparent broadening of the signal. Another observation is that even if the two spectra in Fig. 7b and c belong to the same polymer film, there seems to be "more" TNF in the heated complex (i.e. the intensities of the complexed TNF peaks are higher). The buildup of the magnetization for three of the signals of PNVC-TNF complex is presented in Fig. 8. It is interesting to note that buildup of the signals belonging to TNF is much slower than in pure state and that they do not reach a maximum even at 4 ms contact. Also, the intensities of these signals are much higher in the complex, indicating intermolecular cross-polarization, i.e., the carbons of TNF are in contact with the protons of PNVC and are receiving magnetization from them.

There are many papers in the literature analysing different aspects of molecular dynamics in solid state by measuring different relaxation times and their temperature dependence. In this paper only $T_{1\rho}$ for protons and T_1 for carbons are measured and compared for complexed and pure materials. $T_{1\rho}H$ values are very simple and direct indication of intimate mixing, as pioneering work by Schaefer et al. demonstrated

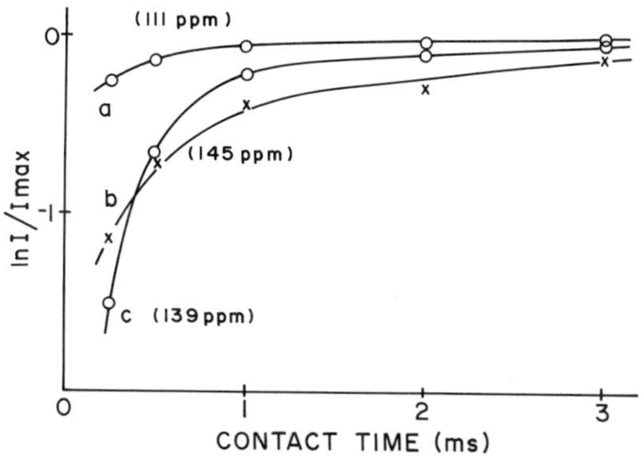

Figure 8. Buildup of the magnetization in solid state for three signals of the PNVC-TNF complex spectrum: a. carbons 1 and 8 of PNVC (111 ppm); b. a signal of complexed TNF (145 ppm); c. a combination signal that probably has a TNF and a PNVC component (139 ppm).

[23]. T_1C can be used to obtain information about the sample crystallinity, and it was extensively used in polyethylene by a large number of authors [24]. Table I summarizes the values obtained for the two measured relaxation times. It is interesting to note that TNF presents two values for $T_{1\rho}H$, which is rather unusual for a proton-rich compound. Having only 5 protons in a large molecule, TNF can probably be considered in a special case of "dilute" spin systems in which not all the carbons have direct access to the proton reservoir. The complex NEC-TNF has a single $T_{1\rho}H$ value for all carbon signals, indicating intimate mixing and cooperative motion of the two components of the complex. Different $T_{1\rho}Hs$ are found for the unheated polymer complex, which indicates imperfect mixing. By heating, the complex becomes more homogenous, yielding a single $T_{1\rho}H$ value for all carbons. It seems rather peculiar that the value is higher than the individual values of the two components, but at this stage it is difficult to decide if this finding is significant. The T_1C values are not expected to be the same for the same molecule and, as Table I shows, sometimes there are two values, sometimes there is a whole range of T_1C values (TNF). All the complexes seem to present intermediate values between those of the components.

TABLE I. Relaxation Time Values

Compound	$T_{1\rho}H$ (ms)	T_1C (s)
NEC	600	250
1:1 NEC-TNF	500	300 and larger
TNF	27 and 60	500-800
1:1 PNVC-TNF obtained by slow evaporation	19, 45 and 100	43 and 70
1:1 PNVC-TNF heated	200	2.5, 12 and 100
PNVC	180	35 and 55

3.2. PNVC-CHL System

Solid state spectra of a 1:1 PNVC:CHL complex are identical to those of PNVC. Also, all measured relaxation times do not change. No signals due to CHL appear in the spectra, even after thermal treatment. This suggests no complexation at the molecular level. Model compound studies confirm the absence of complex invariably leads to a colorless or off-white mixture of separately crystallized NEC and CHL. The solution mixture is dark blue, but no complex could be obtained in solid state. Therefore, it is normal to obtain a solid state spectrum of NEC, having the same relaxation times as pure NEC, when analysing the solid "complex".

4. Conclusions

CP-MAS NMR spectroscopy is a powerful method that could provide important information on polymer charge transfer complexes. There are differences in chemical shifts between the complexed and uncomplexed states and more research is needed to elucidate the dependence of these shifts on the degree of complexation. Intermolecular cross-polarization exists in systems where one component is proton-rich and the other has only a few protons (PNVC-TNF). This phenomenon should also be present where protonless substances are involved, but it still has to be demonstrated for polymers. Proton $T_{1\rho}$ is related to intermolecular interaction and can be used as a direct measurement of the cooperative motion of the two components of the complex. Other relaxation times will certainly be relevant to the polymer complexes analysis.

5. Acknowledgements

The author is grateful to Queen's University for funding this research through the Advisory Research Committee and the Principal's Development Fund.

6. References

1. Pearson, J. M. and Stolka, M., "Poly(N-vinylcarbazole)", Gordon 1981.
2. Shaffert, R. M., *IBM J. Res. Dev.*, *15*, 75, 1971.
3. Shattuck, M. D. and Vahtra, U., U. S. Pat. 3,484,237 (1969) (to IBM).
4. Pearson, J. M.; Turner, S. R. and Ledwith A., in "Molecular Associations", Vol. 2, R. Foster Ed., Academic, NY, 1979.
5. Natansohn, A. and Flaisher, H., *J. Polym. Sci., Polym. Lett. Ed.*, *22*, 579, 1984.
6. Tazuke, S.; Inoue, T.; Saito, S.; Hirota, S. and Kokado, H., *Polym. Photochem.*, *6*, 221, 1985.
7. Oshima, R.; Uryu, T. and Seno, M., *Macromolecules*, *18*, 1043, 1985.

8. Subramaniam, P.; Sasakawa, T.; Ikeda, T; Taxuke, S. and Srinivasan, M., *J. Polym. Sci., Part A: Polym. Chem.*, *25*, 1463, 1987.
9. Horhold, H. H.; Rathe, H.; Helbig, M. and Opfermann, J., *Makromol. Chem.*, *188*, 2083, 1987.
10. Tsutsumi, N.; Yamamoto, M. and Nishijima, Y., *J. Polym. Sci., Part B: Polym. Phys.*, *25*, 2139, 1987.
11. Pearson, J. M. and Stolka, M., "Poly(N-vinylcarbazole)", Gordon and Breach Science Publishers, NY, 1981.
12. Landman, U.; Ledwith, A.; Marsh, D. G. and Williams, D. J., *Macromolecules*, *9*, 833, 1976.
13. Natansohn, A., *Polymer Bulletin*, *9*, 67, 1983.
14. Natansohn, A., *J. Polym. Sci., Polym. Chem. Ed.*, *22*, 3161, 1984.
15. Blann, W. G.; Fyfe, C. A.; Lyerla, J. R. and Yannoni, C. S., *J. Am. Chem. Soc.*, *103*, 4030, 1981.
16. For a very recent paper see: Chu, C. W.; Dickinson, L. C. and Chien, J. C. W., *Polymer Bulletin*, *19*, 265, 1988, and references therein.
16a. Lee, F.; Gabe, E.; Tse, J. S. and Ripmeester, J. A., *J. Am. Chem. Soc.*, *110*, 6014, 1988; Komoto, T.; Ando, I.; Nakamoto, Y. and Ishida, S., *J. Chem. Soc., Chem. Commun.*, 135, 1988.
17. Hunter, B., personal communication.
18. Eckert, H.; Yesinowski, J. P.; Sandman, D. J. and Valazquez, C. S., *J. Am. Chem. Soc.*, *109*, 761, 1987.
19. Yesinowski, J. P.; Eckert, H.; Sandman, D. J. and Valazquez, C. S., *ACS Symp. Ser.*, *337*, 230, 1987.
20. Natansohn, A., *Polymer Preprints*, *29(1)*, 162, 1988; and *J. Polym. Sci., Part A. Chem.*, *27*, 1989.
21. Axelson, D. E. and Russell, K. E., *Progr. Polym. Sci.*, *11*, 221, 1985.
22. Natansohn, A., *Polymer Bulletin*, *21*, 217, 1989.
23. Stejskal, E. O.; Schaefer, J.; Sefcik, M. D. and McKay, R. A., *Macromolecules*, *14*, 275, 1981.
24. For a review see: Axelson, D. E., in "High-Resolution NMR Spectroscopy of Synthetic Polymers in Bulk", Komoroski Ed., VCH Publishers, Inc., 1986, p 157.

Chapter 10

^{13}C NMR OF CROSSLINKED POLY(METHACRYLIC ANHYDRIDE)

DIANA D. GRANGER

1. Abstract

Solution and solid state ^{13}C nuclear magnetic resonance (^{13}C NMR) spectra for poly(methacrylic anhydride) (PMAH) polymers with various tacticities and thermal histories were correlated with infrared spectra and gel permeation chromatographs to detect anhydride crosslinks. Under most baking conditions, thin films of isotactic poly(*tert*-butyl methacrylate) (PtBMA) formed spherulitic crystals upon conversion to isotactic PMAH (i-PMAH), whereas PMAH prepared from 83% syndiotactic PtBMA or by the free-radical cyclopolymerization of methacrylic anhydride remained amorphous. Amorphous polymers gave broad solid state ^{13}C NMR peaks, but crystalline i-PMAH yielded multiple carbonyl and α-methyl peaks. The i-PMAH spectra were compared to those for amorphous PMAH and a model compound to assign chemical shifts to PMAH crosslink α-methyl and carbonyl carbons.

2. Introduction

Poly(methacrylic anhydride), or PMAH, may be prepared by the methods diagrammed in Fig. 1. The thermal decomposition of poly(methacrylic acid) (PMAA) or certain of its esters is shown in Fig. 1a; Fig. 1b illustrates steps in the free-radical cyclopolymerization of the PMAH monomer, methacrylic anhydride (MAH). Products

---

DIANA D. GRANGER - Hughes Research Laboratories, 3011 Malibu Canyon Road, Malibu, California 90265.

Solid State NMR of Polymers, Edited by L. Mathias
Plenum Press, New York, 1991

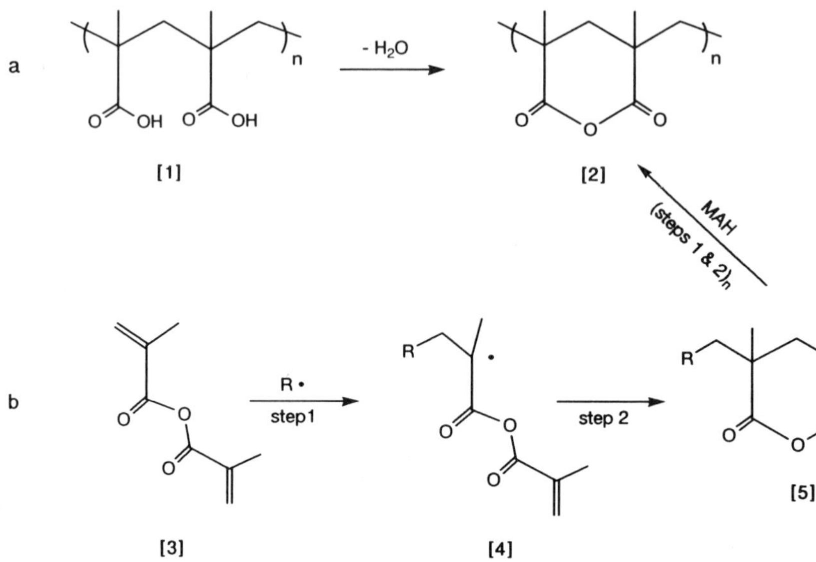

Figure 1. The synthesis of PMAH: thermal degradation of PMAA (a) and cyclopolymerization of MAH (b).

of both methods were studied in this research. The polymeric precursor used for the thermal degradation method was the *tert*-butyl ester of PMAA (PtBMA). To differentiate the products of the two synthesis routes, the PMAH derived from MAH cyclopolymerization will be termed "cy-PMAH" throughout this chapter.

PMAH properties vary with tacticity, molecular weight, and structure irregularities. Anhydride crosslinks joining polymer chains (Fig. 2a) are common structure irregularities which appreciably change polymer physical properties. The greater the crosslink density, the less soluble the polymer; yet, the 100% anhydride character of the polymer is preserved. When hydrolyzed (Fig. 2b), crosslinks become single PMAA units on each chain involved.

Cy-PMAH crosslinks form during the cyclopolymerization process by the free-radical initiation and chain propagation of a pendent vinyl group (PVG). PVGs are produced by the propagation of a new monomer unit before the cyclization step of the

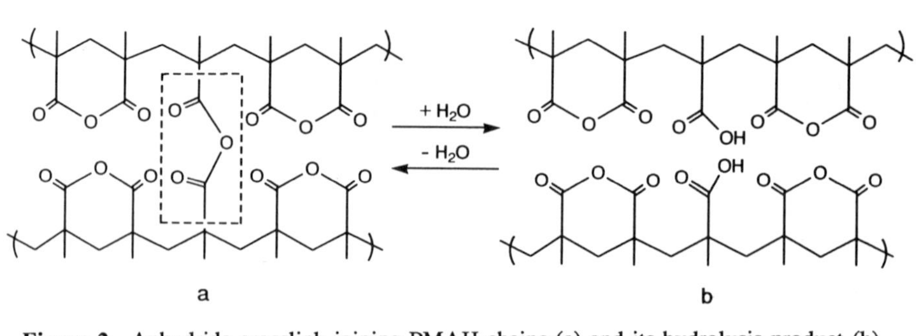

Figure 2. Anhydride crosslink joining PMAH chains (a) and its hydrolysis product (b).

previous addition (step 2 of Fig. 1b). Tiers and Bovey [1] prepared cy-PMAH that was insoluble and swollen in all solvents and concluded that cy-PMAH is inherently highly crosslinked. Hwa and Miller [2] showed that cy-PMAH crosslinking was dependent on monomer purity and polymerization conditions, and that soluble polymer could be reproducibly synthesized.

Whether PMAH derived from PMAA or its esters contains crosslinks has been controversial. Its insolubility in common solvents led Grant and Grassie to conclude that PMAH prepared from PMAA at 192.5° C must contain some crosslinks [3]; however, Miura and Hideyama reported that PMAH prepared from PtBMA at 210° C was soluble in such solvents as *N,N*-dimethylformamide (DMF), *N,N*-dimethylacetamide (DMAc), and dimethylsulfoxide (DMSO) [4]. Miller and Granger [5] observed that PMAH derived from PtBMA at 230° C or higher was swollen and insoluble in DMAc, whereas PMAH prepared at 210° C dissolved readily in the solvent even though its infrared (IR) spectra showed it was completely converted to anhydride. The diminished solubility of the PMAH that was baked at higher temperatures suggests an increased crosslink density.

PMAH crosslinks are formed during PtBMA thermal conversion by the reaction of PMAA units on different polymer chains (Fig. 2b). Statistically, the random and independent elimination of alkene from single ester units on the PtBMA chain (Fig. 3a) would generate about 13.5 mol% of product PMAA units that are odd-number residuals after the other acid units paired to form intramolecular cyclic anhydrides [6]. The intermolecular reaction of these lone acid groups would form anhydride crosslinks; however, Grant and Grassie [7] described an autocatalytic PtBMA thermal degradation mechanism by which such crosslinking seems less likely to occur. According to this mechanism, a newly-formed acid group catalyzes the de-esterification of its neighbor in the transition state shown in Fig. 3b. As the two acid groups complete the conversion to anhydride, some of the steric hindrance is relieved for the next adjacent ester. This next unit more easily de-esterifies and, in turn, catalyzes the formation of its own dehydration reactant. The reaction is propagated down the chain in a zipper-like fashion, diminishing the probability of lone acid units remaining to form intermolecular crosslinks.

PMAH derived from PtBMA at a temperature higher than the minimum required to convert it to anhydride shows decreased solubility. This suggests a

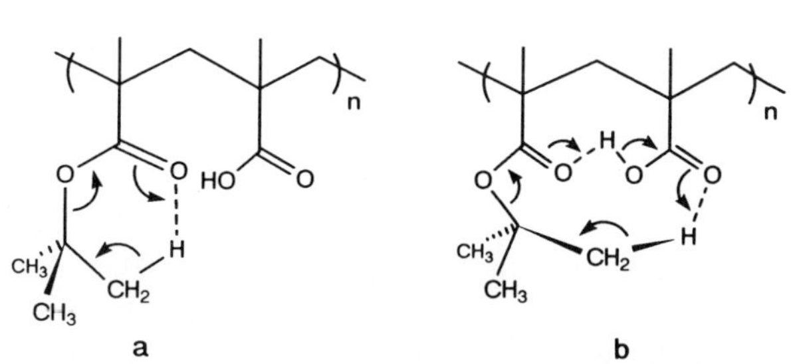

Figure 3. Mechanisms for the de-esterification of PtBMA units: random (a) and autocatalytic (b).

mechanistic change at higher temperature. Grant and Grassie described the autocatalytic mechanism for PtBMA at 200° C; however, random single-unit de-esterification may be the kinetically favored mechanism at a higher conversion temperature. Random de-esterification should create more residual lone acid groups as crosslink sites, resulting in higher crosslink density for polymers converted at the higher temperature.

PMAH crosslinks are difficult to identify by IR spectroscopy. Open chain anhydride carbonyls absorb IR doublets in the same frequency range as do ring anhydrides; however, the relative intensities are reversed [3]. The IR spectra of PMAA heated to form PMAH show predominately cyclic anhydride structure [3], as do those of PMAH derived from heated copolymers of MAA with tBMA [8]; however, it is improbable that the small (ca. 1 mol%) carbonyl concentration participating in intermolecular crosslinks would be discernible in such spectra.

Crosslinking renders most polymers insoluble and requires that NMR spectroscopy be obtained in the solid state. Fyfe and McKinnon [9] published the solid state ^{13}C NMR spectrum for PMAH prepared from PMAA at 250° C. The single carbonyl resonance peak provided no crosslinking evidence; however, the crosslink carbonyl chemical shift may have been obscured by the broad carbonyl peak of the 22.6 MHz solid state spectra.

This research was to answer four questions: (1) whether crosslinks are formed when PtBMA is heated to form PMAH, (2) whether the crosslink density is increased by a higher conversion temperature, (3) whether the polymer tacticity affects crosslink density, and (4) whether crosslinks can be detected using ^{13}C NMR.

3. Experimental

3.1. Monomers

The *tert*-butyl methacrylate (tBMA) monomer used to prepare isotactic PtBMA (i-PtBMA) was prepared by the Heyboer and Staverman method [10], starting with methacryloyl chloride from Aldrich Chemical Co. The product was purified to >99% by vacuum distillation. The monomer used for syndiotactic PtBMA (s-PtBMA) synthesis was purchased from Monomer-Polymer Laboratories and purified to >98.5% by washing with base followed by vacuum distillation. Methacrylic anhydride (MAH) was purchased as 94% MAH from Polysciences, Inc., and used as received.

3.2. Polymers

The polymerization conditions are summarized in Table I. The i-PtBMA polymers were prepared by anionic polymerization initiated by n-butyllithium in toluene at -72 to -60° C [11]. The syndiotactic triads of s-PtBMA were enriched by photopolymerization at low temperature [12]. The reaction flask was placed in an ice

Table I. The Polymerization Conditions for PtBMA and cy-PMAH

Batch	Solvent[s]	Initiator[i]	Temp. °C	$[M]_o$ M	$[I]_o$ Mx10^3	Time hr	% Conv.
i-PtBMA#1	tol	nBuLi	-56±7	0.39	100.	0.35	74
i-PtBMA#2	tol	nBuLi	-63±5	0.512	5.1	1.80	78
i-PtBMA#3	tol	nBuLi	-68±2	0.758	13.9	1.82	22
s-PtBMA	tol	AIBN	10±5	3.5	9.3	34	44
c-PtBMA#1	MEK	AIBN	80	2.0	8.0	5.0	70
c-PtBMA#2	tol	AIBN	80	3.3	4.98	18	92
cy-PMAH#1	bz	BPO	80	0.639	4.94	0.67	66
cy-PMAH#2	bz	BPO	80	0.195	1.65	6.0	83
cy-PMAH#3	clf	BPO	61	0.611	26	6.0	89

[s]Solvents: tol = toluene; MEK = methyl ethyl ketone; bz = benzene; clf = chloroform
[i]Initiators: AIBN = 2,2′-azo-bis(isobutyrylnitrile); nBuLi = n-butyllithium;
BPO = benzoyl peroxide.

bath and the initiator was decomposed by exposing the reaction mixture to the 2200-4000 Å ultraviolet light generated by a 200-watt Canrad-Hanovia immersion lamp encased in a quartz sleeve. C-PtBMA was prepared by conventional solution polymerization techniques. Cy-PMAH was prepared by the cyclopolymerization of MAH monomer in benzene or chloroform with benzoyl peroxide as the initiator [13]. Cy-PMAH#PS was purchased from Polysciences, Inc.

Isofine PM-15 Electron Beam Resist was obtained as a 15% (w/v) solution in methyl isobutyl ketone from Micro-Image Technology, Ltd. of Ridding Derby, England. The solvent was removed by evaporation under a dry nitrogen stream followed by vacuum desiccation at room temperature to constant weight.

3.3. Sample Preparation

i) Spin-coating: PtBMA thin films were prepared by spin-coating 10% w/v solutions in 2-ethoxyethyl acetate onto 1" silicon wafers with a Headway Model EC101 photoresist spinner. The spin-casting solvent was removed by baking at 60-80° C in the vacuum desiccator for 30-60 minutes. The cy-PMAH thin films were prepared by spin-casting from DMF under N_2 positive pressure.

ii) Baking: PtBMA was converted to PMAH by baking it either in bulk or as a thin film on silicon in a Fisher Isotemp Model 281 vacuum oven for the specified times and temperatures. The oven temperature was set and controlled by an Omega RTD digital proportional temperature controller with ±1° C resolution. The bake times were controlled by an I^2R digital timer model TD99A; unless specified otherwise, all bake times were followed by slow cooling to room temperature in an unopened oven. Bake temperatures under 100° C were obtained in a GCA/Precision Scientific heated volume desiccator evacuated to about 5 torr.

3.4. Characterization

i) Molecular weight: The weight-average (M_w) and number-average (M_n) molecular weights of the soluble PtBMA and cy-PMAH polymers were measured by gel permeation chromatography (GPC) using a Waters Associates Model ALC-202/401 liquid chromatograph and differential refractometer with Waters Associates Ultrastyragel Model 10681 columns against four monodisperse polystyrene standards with M_w's of 1,840,000, 598,800, 111,000, and 20,400.

ii) Tacticity: The relative proportions of isotactic (i), heterotactic (h), and syndiotactic (s) triads for PtBMA polymers were determined by the integration of ^{13}C NMR configurational triads for the carbonyl and alkoxy quaternary carbons at 176-178 and 80-81 ppm [14].

3.5. Spectroscopy

i) IR: Spectra were recorded with a Phillips Pye Unicam SP3-300A-SW infrared spectrophotometer in the "Spectroset" mode with a 7-minute scan time. Thin film solid state spectra of PtBMA and their PMAH thermal products were recorded on silicon wafer substrates. The wafers were polished on both sides into a wedge shape to minimize standing wave effects. The sample reference was air.

ii) Solution ^{13}C NMR: Spectra were obtained at 20.12 MHz using an IBM NR/80 NMR free induction decay spectrometer with proton decoupling and processor-controlled Fourier transformation of the data. A 30° pulse width (8.0 μs) and a 2.0 s relaxation delay seconds allowed for the slower relaxation times encountered with similar polymers [15]. The 6024 Hz spectral width free induction decay was acquired for 2.72 seconds for the specified number of scans, typically 1000 for PtBMA and 8000 or more for PMAH. The polymers were 5-15% w/v concentration in 99.9+% deuterated solvents purchased from Wilmad. The solvents were chloroform-d ($CDCl_3$), *N,N*-dimethylsulfoxide-d6 (DMSO-d6), *N,N*-dimethylformamide-d7 (DMF-d7), or mixtures of these solvents. The chemical shifts for the polymers were referenced to those of the solvents: the 77.0 ppm for $CDCl_3$, 39.5 for DMSO-d6, and 162.7 for DMF-d7 [16].

iii) Solid State NMR: Solid state ^{13}C cross-polarized/magic-angle spinning (CP/MAS) NMR spectra were obtained at the Southern California Regional NMR Facility, located at the California Institute of Technology in Pasadena and supported by NSF Grant CHE84-40137. The spectra were obtained at room temperature on an in-house manufactured double-resonance high-field spectrometer controlled by a Nicolet 1280 data system and a 293B pulse programmer and operating at a ^{13}C frequency of 50.36 MHz [17]. Samples were measured in a CP/MAS probe from

Table II. The Molecular Weight and Tacticity of the PtBMA Polymers.

Polymer	M_w x10^{-3}	M_n x10^{-3}	MWD	^{13}C NMR triads			%S
				i	h	s	
i-PtBMA#1	33.6	14.3	2.34	1.00	0.00	0.00	0
i-PtBMA#2	513	306	1.68	1.00	0.00	0.00	0
i-PtBMA#3	74.1	27.6	2.68	1.00	0.00	0.00	0
s-PtBMA	203	89.6	2.26	0.01	0.32	0.67	83
c-PtBMA#1	50.8	22.8	2.22	0.01	0.48	0.51	75
c-PtBMA#2	280	---	---	0.03	0.39	0.58	78

Doty Scientific, Inc. The 7 mm O.D. sapphire rotors were spun at the maximum speed for each sample (3.2-3.8 kHz). The transmitters, Amplifier Research Models 200L (for ^{13}C) and 500L (for ^{1}H), were equipped and operated with blanking pulses. ^{1}H 90° pulse lengths of 5 μs were used, corresponding to a radio frequency field strength of 1.2 mT. Cross-polarization contact times were 5 ms and the Hartmann-Hahn matching condition was set with use of an adamantane standard, the methylene peak of which also provided a secondary chemical shift reference at 38.56 ppm from Me_4Si. High-power decoupling was applied during the acquisition period of 20 ms. The cross-polarization pulse program incorporated spin temperature inversion to suppress artifacts and proton flip of the ^{1}H magnetization into the z direction by an additional 90° pulse at the end of the acquisition period. Line broadening was typically 2-5 Hz, and resolution enhancement was achieved using a negative exponential multiplier factor of -10 to -30 Hz.

4. Results and Discussion

4.1. Polymer Characterization

PtBMA polymer weight-average (M_w) and number-average (M_n) molecular weights, molecular weight distribution (MWD, defined as M_w/M_n), and tacticity are listed in Table II. The 100% isotactic character of each i-PtBMA batch was demonstrated by resonance peak singularity in the 174-175 ppm carbonyl region of its ^{13}C NMR spectrum in $CDCl_3$. The s-PtBMA polymer was found [14] to have 83% syndiotactic dyads. For comparison, two PtBMA polymers (c-PtBMA), prepared by conventional free-radical polymerization, were found to be 75% and 77% syndiotactic, respectively. Both c-PtBMA polymers had heterotactic triad fractions higher than s-PtBMA.

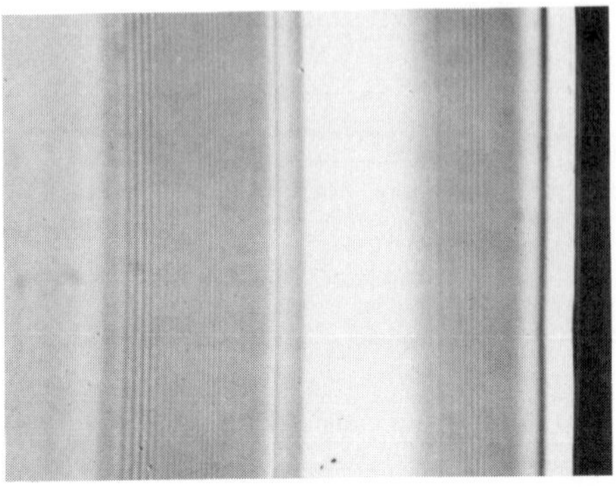

+0

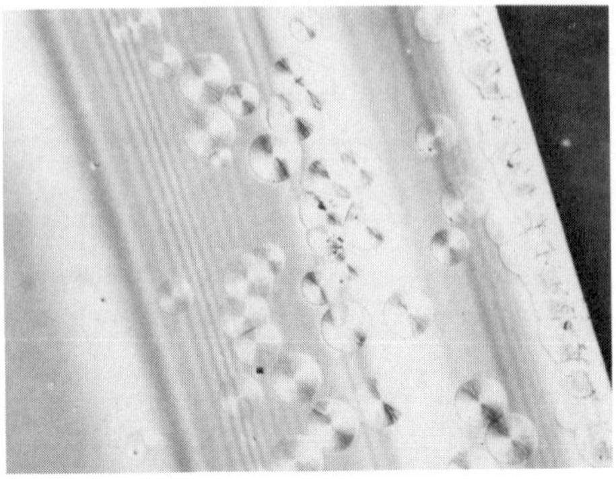

+3

+1

+4

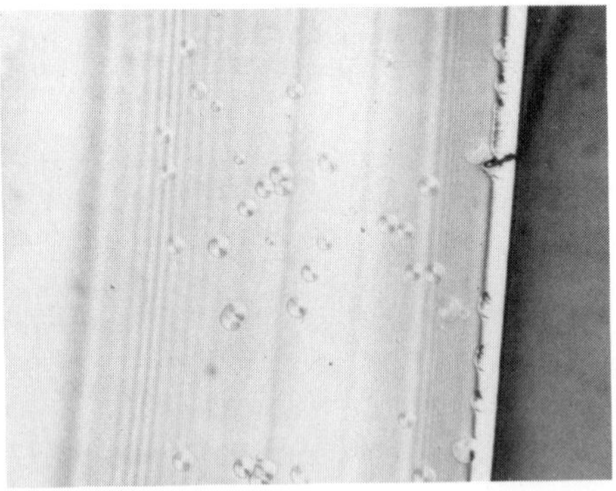

+2

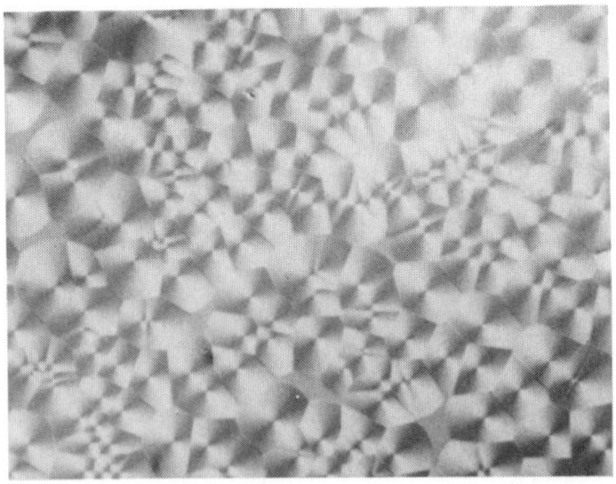

+5

Figure 4. Spherulite crystals in i-PMAH#2 thin films on silicon. Cross-polarized micrographs were magnified 83X on 4X5" film. Sample numbers and qualitative ratings of crystal formation are 921-8 (+0); 921-5 (+1); 921-6 (+2); 921-12 (+3); 889 (+4); 921-4 (+5).

4.2. Crystallinity in i-PMAH

The spherulitic crystals shown in Fig. 4 formed in i-PMAH thin films prepared from i-PtBMA. The crystalline areas dissolved more slowly and were more resistant to concentrated ammonium hydroxide hydrolysis than the amorphous areas. Thin films of s-PMAH, c-PMAH, and cy-PMAH remained amorphous in all situations studied.

Thin i-PtBMA#2 films were baked from one to fifteen hours at 180 to 230° C in either air or vacuum. After baking, the samples were either removed from the oven and cooled quickly, or they were cooled to room temperature slowly in the unopened oven. The crystallization extent from these treatments, listed in Table III, was rated qualitatively according to the examples shown in Fig. 4. Factors which affected crystallization were, in order of their importance: high temperature, a long heating time, a vacuum environment, and a slow cooling rate.

4.3. Crosslink Hydrolysis in Cy-PMAH

Anhydride crosslinks form during the cyclopolymerization process by the free-radical chain propagation of PVGs (similar to Structure 4 in Fig. 1) that remain when

Table III. The Formation of i-PMAH Spherulites Under Various Baking Conditions

Sample	Temp °C	Time hr	Environment		Cooling Rate		Rating
			Vac	Air	Slow	Fast	
889	230	4.0	X		X		+4
921-0	70	0.50	X			X	+0
921-1	180	15.0		X			+0
921-2	200	1.0		X		X	+0
921-3	200	2.0		X		X	+0
921-4	230	4.0	X		X		+5
921-5	230	1.0		X		X	+1
921-6	230	4.0		X		X	+2
921-7	230	4.0		X	X		+2
921-8	201	4.0	X		X		+0
921-9	220	4.0	X		X		+2
921-10	210	4.0	X		X		+0
921-11	215	4.0	X		X		+2
921-12	230	4.0		X			+3
921-13	200	4.0				X	+0

Rating: +5 = maximum crystallization; +0 = no crystals.

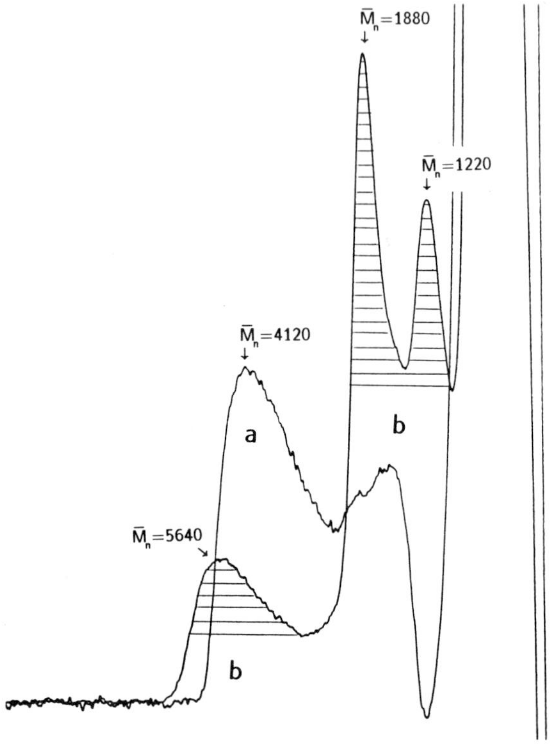

Figure 5. GPC curves for cy-PMAH#2 taken after dissolution: after 2 weeks (a); after 6 months (b).

MAH units fail to cyclicize [18]. PMAH becomes soluble only after these crosslinks are hydrolyzed [18,19].

Cy-PMAH dissolution behavior demonstrated that crosslinks were broken as the polymers dissolved. For example, a cy-PMAH#2 DMF solution was spin-cast onto a silicon wafer and vacuum dried at 70° C. Although the solid polymer required seven days to dissolve at room temperature, a second coat could not be added because the coating solvent redissolved the first film within the two-minute spin time. Intermixing occurred even after baking the first coating for four hours at 230° C - a baking condition which rendered PMAH derived from c-PtBMA insoluble in DMAc [19]. Apparently the solid cy-PMAH contained crosslinks that hydrolyzed during dissolution. Once hydrolyzed, the polymer remained soluble after drying, allowing the first film to readily dissolve in the spin coating solvent while adding the second coat. Baking did not recreate a sufficient crosslink density to render the thin film insoluble.

The cy-PMAH#3 polymer required more than seven weeks to dissolve in tetrahyrofuran (THF) at room temperature. Once dissolved, its M_w was close to that of cy-PMAH#2, a batch of the same age that dissolved in less than a day. The decreased dissolution rate probably resulted from a higher crosslink density in the solid cy-PMAH#3 that required more time to hydrolyze before the polymer was soluble.

Crosslink hydrolysis was further evidenced by molecular weight changes. After six months' storage in THF solution, cy-PMAH#2 produced two low-MW GPC peaks (Fig. 5b) in addition to the original single peak (Fig. 5a). These new peaks in the fast-dissolving cy-PMAH#2 batch cannot be explained except as crosslink hydrolysis products. The M_n associated with the three peaks in the six-month-old solution are integer multiples of PMAH tetramer MW. The newer peaks were two and three tetramers, respectively, and the original peak was nine.

As the cy-PMAH polymers aged, their dissolution rates increased and they acquired an odor similar to methacrylic acid. Polymers which required elevated temperatures or extended time periods to dissolve in DMF or THF readily dissolved when retested six months later. The aging process could involve MAA production by the hydrolysis of PVGs that failed to crosslink by chain propagation during cyclopolymerization. The PVG hydrolysis products are PMAA polymer chain units and MAA monomer in the polymer matrix. It is unclear whether the increased dissolution rate is associated with the presence of MAA, but MAA inclusion was confirmed in the ^{13}C NMR spectra taken of aged cy-PMAH polymers. The increase in cy-PMAH dissolution rate with aging coupled with the appearance of new, low-MW peaks in the GPC is evidence that the polymers contained short segments joined by anhydride crosslinks that hydrolyzed during dissolution.

4.4. Solution ^{13}C NMR of cy-PMAH

Butler and Matsumoto acquired the cy-PMAH ^{13}C NMR spectrum in DMSO-d6 and assigned chemical shifts to cyclic anhydride carbonyls (171.9 ppm), backbone quaternary carbons (42.7 ppm), and methyl carbons (24.9 and 22.9 ppm) [20]; however, solution ^{13}C NMR spectra revealed that the cy-PMAH polymers prepared for this research had hydrolyzed to become P(MAH-MAA) copolymers. As in the cy-PMAH#1 spectrum shown in Fig. 6, the each cy-PMAH carbonyl region contained peaks from PMAA units (179 ppm), PMAH units (172 ppm), and MAA monomer (168 ppm). Narrow peaks assigned to MAA monomer appeared in the vinyl (138 and 125 ppm) and methyl (18 ppm) regions. The monomer is believed to be a PVG hydrolysis product.

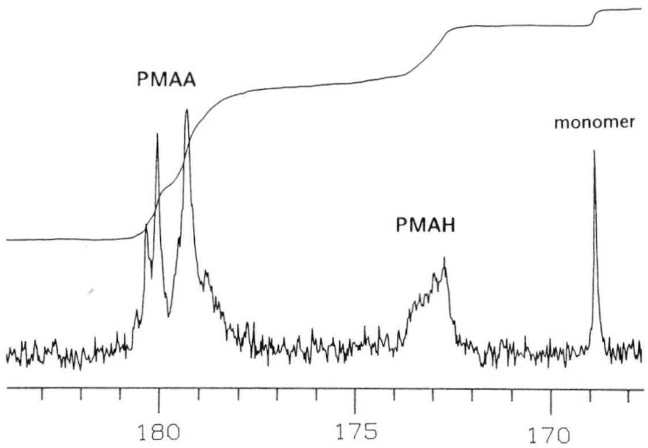

Figure 6. The ^{13}C NMR carbonyl region for cy-PMAH#1 in DMF-d7.

The cy-PMAH#1 carbonyl peak areas were in the ratio 8.9:3.1:1.0. Each carbonyl represents one monomeric unit in the polymer (one PMAH unit is then two monomeric units); therefore, the cy-PMAH#1 P(MAH-MAA) copolymer is 74 mol% PMAA. The resolution of multiple acid carbonyl peaks is believed to reflect copolymer compositional triads.

4.5. Solid State ^{13}C NMR of cy-PMAH

The solid state ^{13}C NMR spectra are shown in Fig. 7. As in the solution spectra, the cy-PMAH carbonyl region contains multiple peaks, showing that the polymers had partially hydrolyzed in the solid state. Superimposed in Fig. 8, the solution and solid state cy-PMAH chemical shifts essentially coincide except for the downfield shift of the PMAA carbonyl peak from 179 to 185 ppm, which is also downfield from the 181 ppm position reported [9] for solid, atactic PMAA homopolymer.

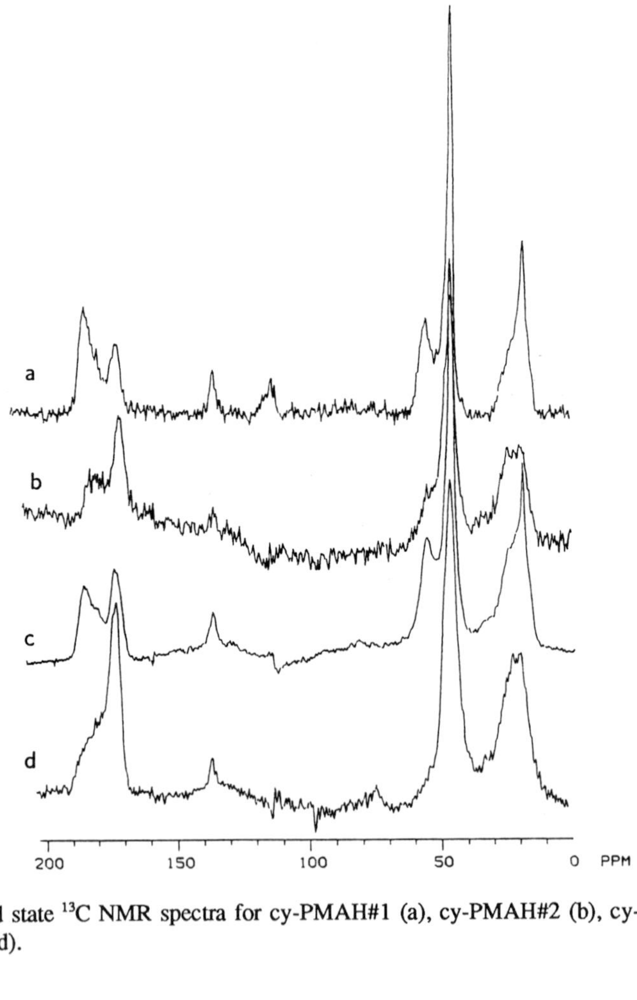

Figure 7. Solid state ^{13}C NMR spectra for cy-PMAH#1 (a), cy-PMAH#2 (b), cy-PMAH#3 (c), and cy-PMAH#PS (d).

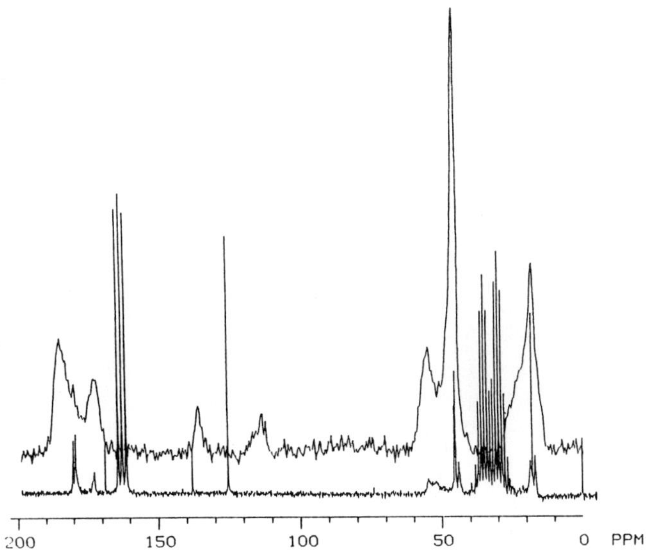

Figure 8. Solid state ^{13}C NMR spectrum for cy-PMAH#1 superimposed over its solution spectrum.

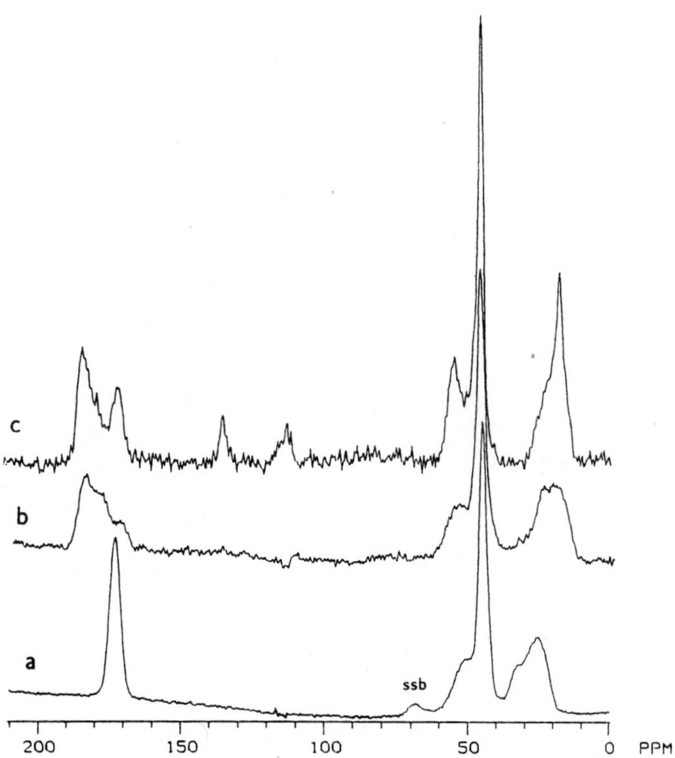

Figure 9. Solid state ^{13}C NMR spectra for cy-PMAH#1 after baking at 5 torr: 220° C (a); 80° C (b); no bake (c).

Solid state ^{13}C NMR spectra for cy-PMAH#1 after treatments to eliminate two probable components are compared with that of the untreated polymer in Fig. 9. Treatments included two steps, the first of which removed MAA monomer by vacuum baking at 80° C. The monomer's vinyl peaks and sharp α-methyl peak at 18 ppm were absent in the new spectrum (Fig. 9b). The 185 ppm carbonyl peak remained, but the 173 ppm peak was diminished, indicating that the MAA monomer carbonyl peak had been shifted downfield from its position at 168 ppm to superimpose the anhydride carbonyl peak position in the spectrum for the untreated polymer (Fig. 9c).

After MAA monomer removal, cy-PMAH#1 PMAA units were converted to anhydride by vacuum baking the polymer at 220° C for four hours. The solid state ^{13}C NMR polymer spectrum (Fig. 9a) after this treatment showed a singular carbonyl peak at 172 ppm; the spectrum also showed methylene, quaternary, and α-methyl carbon peaks at 50, 44, and 25 ppm, respectively. The α-methyl peak has a shoulder at 31 ppm. The absence of the 185 ppm peak confirmed its assignment to the PMAA carbonyl. The downfield shifts for the PMAA and MAA acid carbonyls in the "cy-PMAH" P(MAH-MAA) copolymers is probably related to hydrogen bonding, which also caused an acrylic acid carbonyl peak downfield shift [21].

4.6. Crosslinking in Baked PtBMA

The PMAH dissolution rate in DMF-d7 for solution NMR sample preparation was affected by the baking temperature and tacticity of the PtBMA precursors. S-PtBMA baked at 200° C dissolved completely only after six months' storage at room temperature. When baked at 250° C, another sample of the same polymer required ten months to dissolve, leaving gel particles in the solution for several months before they also were dissolved. Eighteen to twenty-two months were required to dissolve i-PtBMA#2 baked at 200° C, and quite a bit of gel remained. An i-PMAH sample derived from baking i-PtBMA#2 at 250° C appeared unchanged the first ten months, and merely softened somewhat during the second year.

Although the cyclic structure of its backbone is relatively rigid, PMAH readily dissolves in polar solvents if it is not crosslinked [2]. The identical IR spectra (Fig. 10) for PMAH derived by baking s-PtBMA in vacuum at 200, 220, and 250° C show that anhydride conversion is complete at these temperatures; therefore, the polymers should be soluble in DMF. Miller and coworkers [19] attributed the diminished solubility of PMAH thin films derived from PtBMA at 230° rather than at 200° C to an increased crosslink density generated by the more vigorous baking conditions and to a slightly reduced free volume. True, Ouano [22] showed that baking poly(methyl methacrylate) (PMMA) above its glass transition temperature (T_g) lowered its dissolution rate by increasing the film density, but this reasoning would not apply to PMAH. Because its baking temperature is well above its T_g of 158° C [8], the maximum PMAH film density is achieved upon conversion; therefore, the lower dissolution rate is caused by crosslinking.

Ouano [22] found that isotactic PMMA dissolved in amyl acetate much more quickly than syndiotactic PMMA. The difference was attributed to the greater

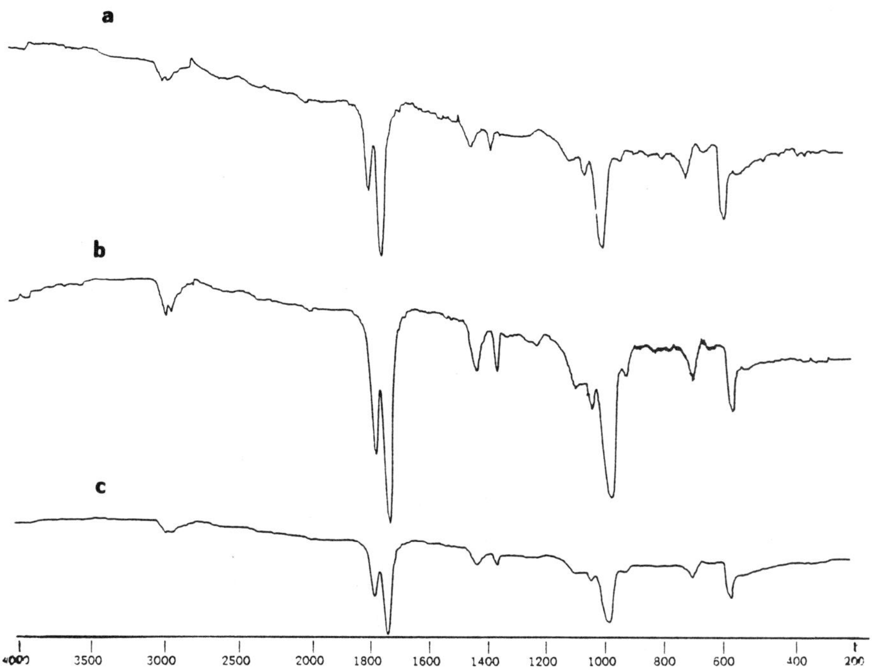

Figure 10. IR spectra for s-PtBMA thin films converted to PMAH at 200° (a), 220° (b), and 250° C (c).

segmental mobility in the isotactic polymer, as evidenced by its much lower T_g. The reason for the decreased dissolution rate of i-PMAH is less certain. An appreciable decrease in segmental mobility is unlikely because of the cyclic structure of the anhydride rings; rather, the more reasonable cause is the cohesive energy of its crystalline structure [23] and increased packing density. The degree of i-PMAH crystallinity increases with increased baking temperature; therefore, the decreased solubility of i-PMAH baked at the higher temperature may be attributed to increased crystallinity regardless of crosslink density changes.

4.7. ^{13}C NMR Spectra for Baked PtBMA

The solution ^{13}C NMR spectrum for s-PMAH derived by vacuum baking s-PtBMA for 4 hours at 200° C overlays its solid state spectrum in Fig. 11. The two PMAA and PMAH carbonyl peaks at 179.0 and 173.7 ppm, respectively, are in a 1:1.7 ratio, and show that s-PMAH had become a P(MAH-MAA) copolymer with 35-40 mol% PMAA units during the 6 months required to dissolve it in DMF-d7. In contrast, its solid state ^{13}C NMR spectrum contained a single carbonyl peak at 171.8 ppm and resembles that (Fig. 9a) for baked cy-PMAH. The PMAA unit carbonyl peaks identified in its solution spectrum are absent in the solid state spectrum; therefore, the slow dissolution of s-PMAH is related to its hydrolysis in solvent.

The solid state ^{13}C NMR spectra (Fig. 12) for i-PMAH derived by baking i-PtBMA in vacuum four hours at 250° C or 12 hours at 200° C were similar to that

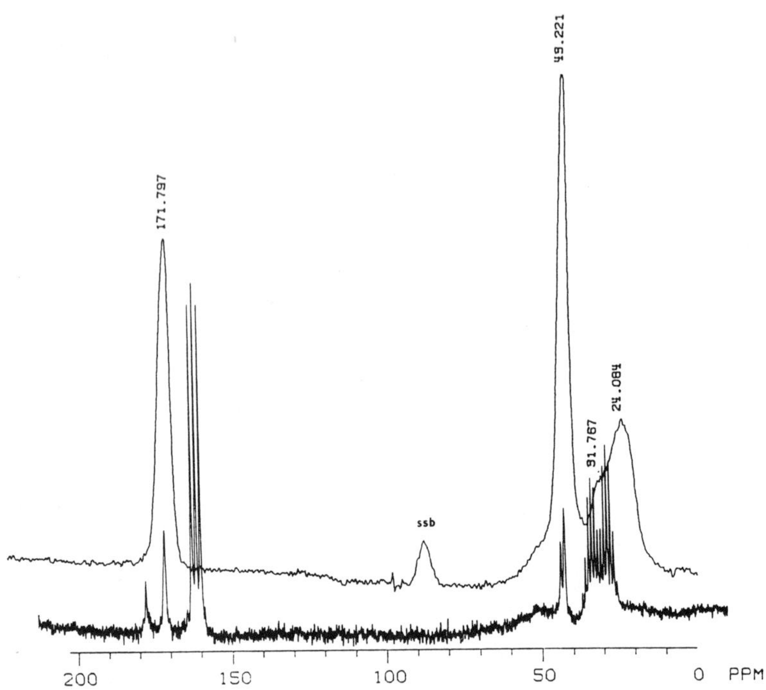

Figure 11. ^{13}C NMR spectrum for s-PMAH in DMF-d7 superimposed on its solid state spectrum.

of s-PMAH except for splitting observed in the four main resonance peaks. The twin peaks at 49.3 and 50.6 ppm in the i-PMAH#2 spectrum (Fig. 12a) represent methylene groups outside and inside the anhydride ring. The quaternary carbon at 44.0 ppm has a 42.5 ppm shoulder, and the carbonyl resonance at 172.8 ppm also has a shoulder at 174.8 ppm. The α-methyl carbon resonance resolves three peaks centered at 21.7, 27.7, and 32.9 ppm, of which only the upfield peak matches the chemical shift reported [9] for PMAH derived from conventional PMAA at 250° C.

Solid state ^{13}C NMR spectroscopy resolves molecular conformation in crystalline materials [24]. Derived from 100% isotactic PtBMA, i-PMAH must also be 100% isotactic; therefore, the multiple carbonyl and methyl resonance peaks are not configurational triads, but represent stereoregular polymer conformations in the crystalline areas that place some methyl groups closer to adjacent ring carbonyls. The multiple carbonyl peaks reflect the carbonyl complements to these methyl groups.

4.8. Crosslink Chemical Shifts

There is a shoulder on the α-methyl peak at 32 ppm in the solid state ^{13}C NMR spectrum for s-PMAH (Fig. 11). A similar shoulder is discernible in Fyfe and McKinnon's [9] solid state spectrum for PMAH derived from PMAA at 250° C, and one also appeared at about 31 ppm in the solid state spectrum for baked cy-PMAH#1 (Fig. 9a). Resolution enhancement of the carbonyl and α-methyl regions in the s-PMAH and baked cy-PMAH solid state spectra resolved no additional features. The overlaid spectra shown in Fig. 13 demonstrate that the shoulder is absent for soluble cy-PMAH in solution. The solid state ^{13}C NMR α-methyl region for four crosslinked PMAH polymers is shown in Fig. 14. A crosslinked model compound was sought to determine whether its spectrum would also contain a peak near 31 ppm.

The Isofine PM-15 Resist (PM-15) from Micro-Image Technologies is a 1:1 mixture of two PMMA copolymers. One copolymer contains 10 mol% PMAA units, and the other contains 10 mol% poly(methacryloyl chloride) (PMACl) units. Heating the polymer mix causes the PMAA and PMACl units to react to form anhydride crosslinks, producing a "crosslinked PMMA" homopolymer [25]. Elemental analysis of PM-15 baked in bulk at 180° C for 30 minutes revealed that only 0.024 wt% chlorine remained from the PMACl units. The theoretical chlorine content in the unbaked copolymer mixture is 1.8 wt%; therefore, the reaction was 99% complete. The baked PM-15 solid state ^{13}C NMR spectrum is shown in Fig. 15. The additional peaks were very small owing to the low crosslink density (<10 mol%), but a shoulder at 171.4 ppm on the PMMA carbonyl peak and a small peak downfield from the

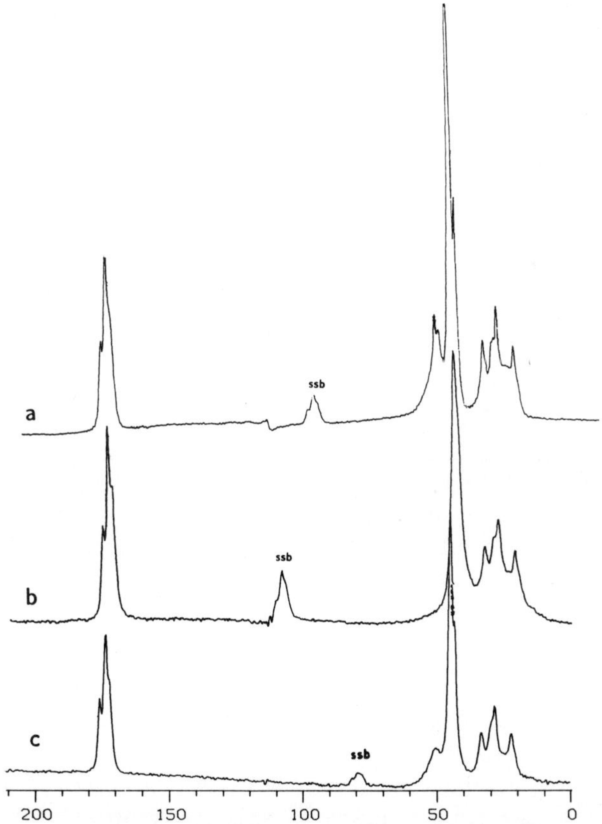

Figure 12. Solid state ^{13}C NMR spectra for i-PMAH: i-PtBMA#2 (a) and i-PtBMA#3 (b) baked 4 hours at 250° C; i-PtBMA#3 baked 12 hours at 200° C (c).

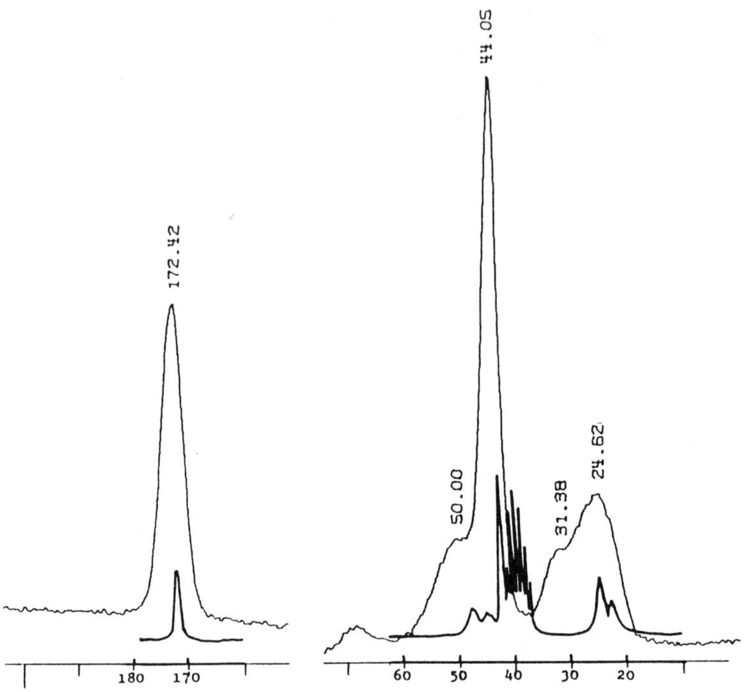

Figure 13. Solid state ^{13}C NMR carbonyl and α-methyl regions for baked cy-PMAH overlaid by the solution spectrum for soluble cy-PMAH in DMSO-d6 from Butler and Matsumoto [20] (spectrum reproduced by permission of *Journal of Polymer Science*, John Wiley & Sons, New York).

PMMA α-methyl peak at 30.8 ppm were confirmed and assigned to anhydride crosslink carbonyl and α-methyl carbons, respectively.

The expanded resolution-enhanced α-methyl carbon and carbonyl peaks for i-PMAH#3 are shown in Fig. 16. The resonance peak at 32.2 ppm is assigned to the crosslink α-methyl carbon. The carbonyl resonance resolves into three peaks centered at 170.7, 172.1, and 174.3 ppm in a peak height ratio roughly corresponding to the α-methyl peak areas. The upfield peak is positioned near the PM-15 crosslink chemical shift and is assigned to the crosslink carbonyl carbon.

5. Conclusions

The initial insolubility of s-PMAH in DMF-d7 is characteristic of crosslinked polymers; that it eventually dissolves after a long period of time is consistent with the slow hydrolysis of the crosslinks in the solvent. The appearance of new low-MW peaks in the cy-PMAH GPC after six months in THF, and of acid units in the carbonyl region of the s-PMAH solution ^{13}C NMR spectrum after six months in DMF, confirm this conclusion. The sample baked at the higher temperature took longer to dissolve because it had a greater crosslink density, consistent with the hypothesis that

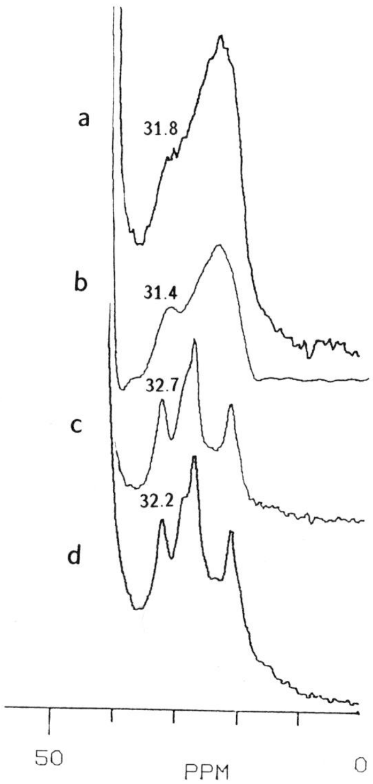

Figure 14. Solid state ^{13}C NMR α-methyl region for PMAH: s-PMAH baked 4 hours at 250° C (a); cy-PMAH#1 baked 4 hours at 220° C (b); i-PMAH#3 baked 4 hours at 200° (c) and 250° C (d).

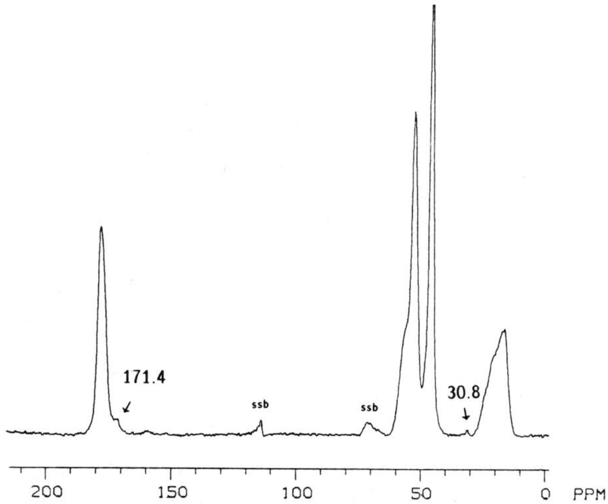

Figure 15. Solid state ^{13}C NMR spectrum for PM-15 crosslinked PMMA showing chemical shifts for crosslink carbonyl and α-methyl carbons.

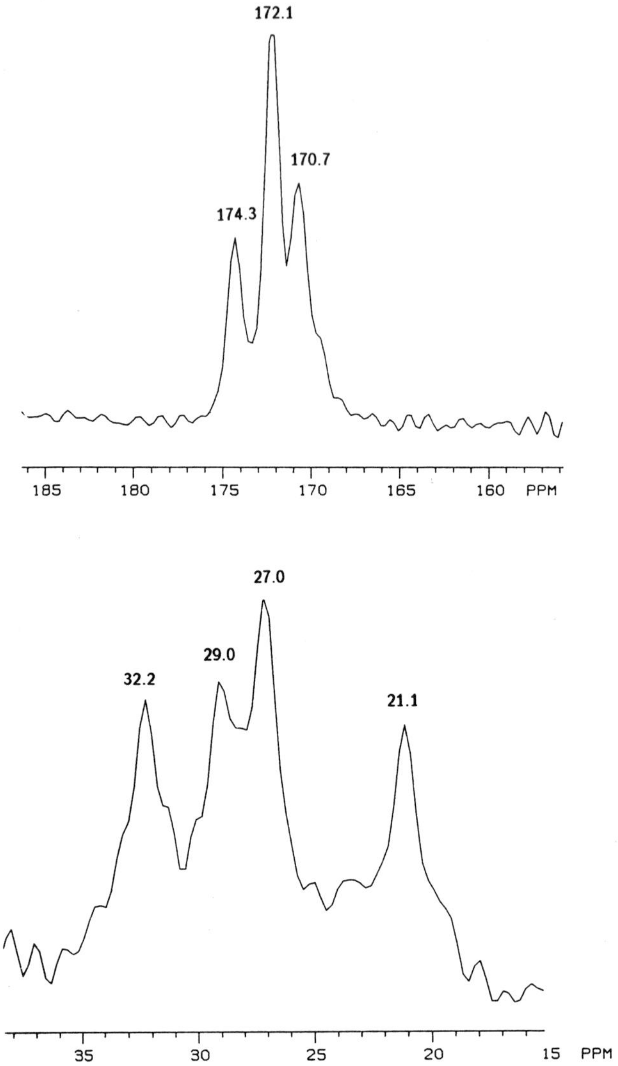

Figure 16. Resolution-enhanced carbonyl (a) and α-methyl (b) i-PMAH solid state ^{13}C NMR peaks.

there would be a change from the autocatalytic mechanism to random de-esterification at higher temperatures. A similar decrease in solubility in i-PMAH with increased baking temperature may be associated with its crystallinity, which also increases with baking temperature.

The ^{13}C NMR peak which appears at 31 ppm only in solid state spectra for crosslinked PMAH is assigned to the α-methyl group of anhydride crosslinks. The chemical shifts for PMAH crosslink carbonyl and α-methyl carbons at 171 and 31 ppm, respectively, were confirmed using a crosslinked PMMA model compound.

6. Acknowledgments

I am very grateful to Hughes Aircraft Company for a generous Staff Master's Fellowship, to Dr. Leroy J. Miller of Hughes Research Laboratories for invaluable technical guidance, and to my Thesis Committee at the California State University at Northridge (CSUN): Drs. Edward Rosenberg, Lydia Savedoff, and Francis L. Harris. Mr. Stuart Dimock assisted in obtaining some of the solution NMR spectra at CSUN. Mr. Scott Ross operated the solid state ^{13}C NMR spectrometer at the California Institute of Technology Regional NMR Center in Pasadena, and Drs. Hellmut Eckert and James Yesinowski helped plot the spectra.

7. References

1. Tiers, G. V. D.; Bovey, F. A., *J. Polym. Sci.*, *47*, 479, 1960.
2. Hwa, J. C. H.; Miller, L., *J. Polym. Sci.*, *55*, 197, 1961.
3. Grant, D. H.; Grassie, N., *Polymer*, *1*, 125, 1960.
4. Miura, A.; Hideyama, S., *J. Vac. Sci. Technol.*, *16(6)*, 1996. 1979.
5. Miller, L. J.; Granger, D. D., Hughes Research Laboratories, Malibu, CA (unpublished results).
6. Elias, H. G., *Macromolecules*, Vol. 2, Plenum: New York, pp 815-818, 1977.
7. Grant, D. H.; Grassie, N., *Polymer*, *1*, 445, 1960.
8. Lai, J. H., *Macromolecules*, *17(5)*, 1010, 1984.
9. Fyfe, C. A.; McKinnon, M. S., *Macromolecules*, *19(7)*, 1909, 1986.
10. Heyboer, J.; Staverman, A. J., *Recueil*, *69*, 787. 1950.
11. Azimov, Z. A.; Korotkov, A. A.; Mitsengendler, S. P., *Izv. Akad. Nauk SSSR, Ser. Khim.*, *1*, 55 (Russ.), 1964; *Bull. Acad. Sci., USSR Div. Chem. Sci.*, *46* (Eng. transl.), 1964.
12. Fox, T. G.; Garrett, B. S.; Goode, W. E.; Gratch, S.; Kincaid, J. F.; Spell, A.; Stroupe, J. D., *J. Am. Chem. Soc.*, *80*, 1768, 1958. Fox, T. G.; Schnecko, H. W., *Polymer*, *3*, 575, 1963.
13. Hwa, J. C. H.; Fleming, W. A.; Miller, L., *J. Polym. Sci.*, *2(5)*, 2385. A 1964.
14. Suzuki, T.; Yamada, O.; Watanabe, Y.; Takegami, Y., *Polym. J.*, *14(5)*, 343, 1982.
15. Hatada, K.; Kitayama, T.; Okamoto, Y.; Ohta, K.; Umemura, Y.; Yuki, H., *Makromol. Chem.*, *179*, 485, 1978.
16. "Deuterated Solvents - Handy Reference Data", Isotope Div., Merck & Co.
17. Eckert, H.; Yesinowski, J. P.; Sandman, D. J.; Valazquez, C. S., *J. Am. Chem. Soc.*, *109*, 761, 1987.
18. Gray, Jr., T. F.; Butler, G. B., *J. Macromol. Sci., Chem.*, *A9(1)*, 45, 1975.
19. Miller, L. J.; Brault, R. G.; Granger, D. D.; Jensen, J. E.; van Ast, C. A.; Lewis, M. L., *J. Vac. Sci. Technol.*, *B7(1)*, 68, 1989.

20. Butler, G. B.; Matsumoto, A., *J. Polym. Sci., Polym. Lett. Ed.*, *19*, 167, 1981.
21. Brouwer, H.; Stothers, J. B., *Can. J. Chem.*, *50*, 601, 1972.
22. Ouano, A. C., *Prepr. Pap.- Am. Chem. Soc., Div. Ctgs. and Plstcs.*, *48*, 42, 1983.
23. Billmeyer, F.W., *Textbook of Polymer Science*, 2nd ed., Wiley-Interscience, New York, 1971, p 24.
24. Veregin, R. P.; Fyfe, C. A.; Marchessault, R. H., *Macromolecules*, *19*, 2379, 1986.
25. Ryan, C. F.; Fleischer, P. C. Jr., *J. Phys. Chem.*, *69(10)*, 3384, 1965.

Chapter 11

APPLICATION OF SOLID STATE ^{13}C NMR SPECTROSCOPY TO SULFUR VULCANIZED NATURAL RUBBER

JACK L. KOENIG AND MLADEN ANDREIS

1. Introduction

High resolution ^{13}C Nuclear Magnetic Resonance (NMR) spectroscopy can be used to characterize sulfur vulcanized natural rubber systems in the solid state [1-7]. Because solid state ^{13}C NMR is a very powerful technique for structure elucidation, it will provide further understanding of the relationships between the crosslink structures and physical properties [8-15]. Samples prepared under a variety of conditions can be analyzed for evidence of crosslink formation with the aid of appropriate pulse sequences in the NMR experiments. All the NMR experiments utilize magic angle spinning and, either Bloch decays with high power decoupling or one-pulse low power decoupling experiments which yield good high resolution spectra. Various types of crosslinks can be distinguished, and it is possible to quantify the amounts of monosulfidic, disulfidic and polysulfidic crosslinks that are found in the different systems. In addition, molecular changes associated with crosslink formation, such as cis-trans isomerism can be observed.

It has been observed in some instances that the modulus and the crosslink density go through a maximum and continue to decrease with additional cure times (a result of loss of network structure by non-oxidative thermal aging). This process, which can occur in addition to the maturing reactions, is known as reversion [16-18]. Reversion occurs when the desulfurization reactions are faster than the crosslinking reactions. The researchers at the Malayasian Rubber Producers Research Association

---

JACK L. KOENIG AND MLADEN ANDREIS - Department of Macromolecular Science, Case Western Reserve University, Cleveland, Ohio 44106.

Solid State NMR of Polymers, Edited by L. Mathias
Plenum Press, New York, 1991

(MRPRA) have divided these sulfur maturing reactions into two categories; see Fig. 1. The left side is the desulfurization of polysulfides to di- and monosulfidic crosslinks. This pathway has been shown to be affected by the Zn-accelerator complex (found in accerelated sulfur vulcanizations mixes). The other routes are characterized as thermal decomposition, where the crosslinks and the sulfuration species decompose into conjugated species, cyclic sulfides, shorter sulfur crosslinks and main chain modifications [17]. Chen et al., [19] have shown that for certain accelerated sulfur cured systems, the reversion process is accompanied by the formation of a trans butadiene-like structure, in addition to formation of trienes and cyclic sulfides. In this paper, we have studied by solid state ^{13}C NMR a series of sulfur-cured natural rubbers (accelerated and unaccelerated) in search of the structural changes that occur during curing and in the desulfurization reactions.

2. Experimental

Samples of natural rubber (Rubber Research Institute of Malaya) with 10% of sulfur (NR10-0) were cured in a temperature controlled hydraulic press at 150° C and

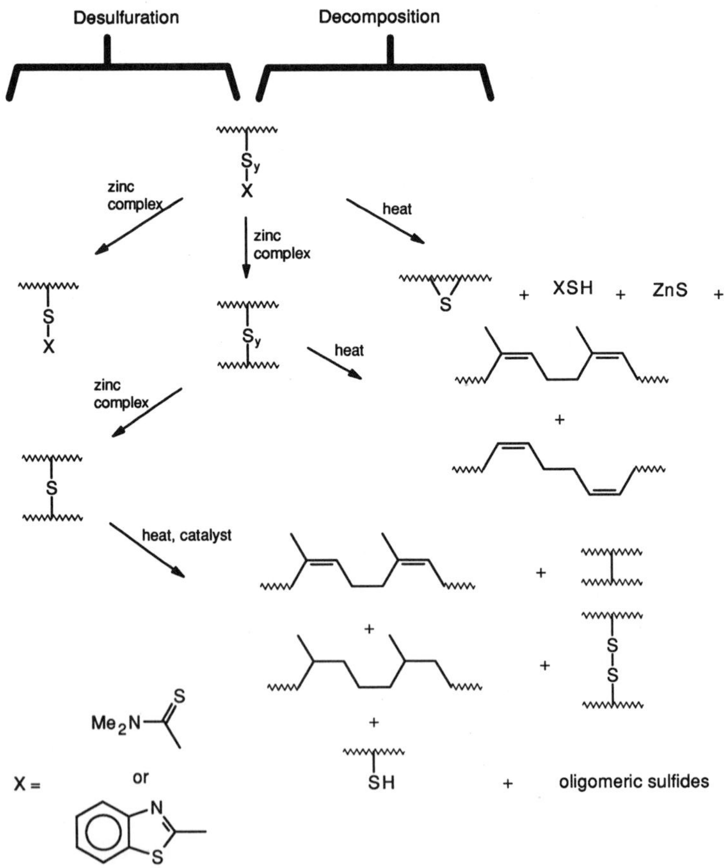

Figure 1. The crosslinking and crosslink decomposition scheme for natural rubber vulcanizations.

13.8 MPa (2000 psi). The curing times were 15, 22.5, 30, 60, 90 and 120 minutes, respectively (NR10-15, NR10-22, NR10-30, NR10-60, NR10-90 and NR10-120). Additional series of samples containing 5%, 3% and 1% of sulfur, respectively, (NR5-0, NR3-0 and NR1-0) are also cured for curing times up to 90 minutes. The ^{13}C NMR measurements were performed on a modified Nicolet Technology NT-150 Fourier Transform spectrometer operating at 37.7 MHz for ^{13}C. The rubber vulcanizates were rolled and fit into the spinners. Magic angle spinning (MAS) (with a frequency of approximately 3 kHz) and dipolar decoupling (with radio-frequency field strength $\gamma B_1/2\pi$ = 60-70 kHz) were used simultaneously to obtain the spectra.

Gated high power decoupling (GHPD) spectra were recorded by applying a 9O° pulse (pulse widths are typically 3.7 μs) and a 5 sec delay time between pulses. The spin-lattice relaxation times (nT_1) were measured using a standard inversion recovery pulse sequence (180°-τ-90°-) modified with high power dipolar decoupling. Fourteen different τ values in the range from 10 s to 4 s were chosen. An artificial line broadening of 10 Hz is used to smooth the spectra before the Fourier transform. The temperature was controlled by a Doty Co. variable temperature unit with an accuracy of 1 degree.

The curve fitting procedure employs a standard Levenberg-Margdt algorithm. For simulation of the negative peaks in the inversion recovery spectra, the peak position and widths are fixed and the signal intensities are optimized. All programs are written in FORTRAN 77.

3. Results

The ^{13}C NMR spectra of natural rubber cured with sulfur for 15 to 120 min consist of two clearly separated spectral regions: the low field (olefinic) and the overlapped high field (aliphatic) region. A large number of new resonances, as compared to spectra of the original rubber, reflect changes in the chemical structure introduced by the ^{13}C NMR vulcanization process. Figure 2 shows the GHPD ^{13}C spectrum of a highly crosslinked vulcanized rubber (NR10-090) recorded at 293 K.

3.1. Alphatic spectral region

The high field aliphatic spectral region of vulcanized samples exhibits a number of overlapping aliphatic resonances, arising mainly from the different types of methine, methylene and methyl groups. All the observed chemical shifts (δ) of the structural units in a sulfur vulcanized rubber are summarized in Table I. The aliphatic resonances show relatively small differences in chemical shifts which, with the greater solid state line widths (particularly in the spectra of highly crosslinked samples at lower temperatures), cause overlap of absorption "wings" and a significant contribution of the neighboring high intensity signals to the observed intensity of the weak absorptions. In addition, weak resonances surrounded by strong neighboring signals are obscured, and their contributions to the observed spectra are not detectable.

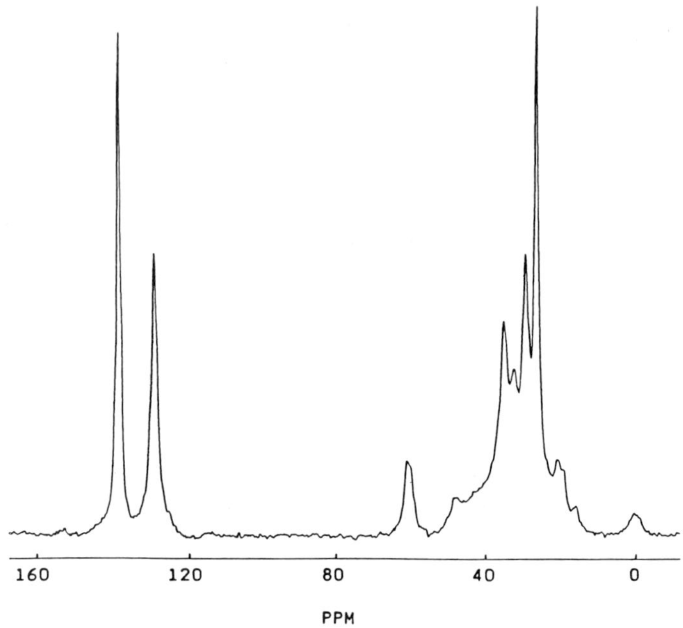

Figure 2. The ^{13}C NMR GHPD spectra of the sample NR10-090 at 293 K.

Table I. ^{13}C CHEMICAL SHIFTS (δ) OF STRUCTURAL UNITS IN A SULFUR VULCANIZED RUBBER

δ/ppm	C-atom	Structural Unit
134.8	C-α	cis-1,4-polyisoprene
134.4[a]	C-α	trans-1,4-polyisoprene
125.3	C-β	cis-1,4-polyisoprene
123.9[a]	C-β	trans-1,4-polyisoprene
57.9	CH(γ)	Structure III
57.9	C(α)	Structure VI
56.8	CH(γ)	Structure I
56.8	CH(δ)	Structure IV
56.8	CH(β)	Structure V
44.6	CH(δ)	Structure II
44.6	CH_2(β)	Structure VI
44.6	CH_2(γ)	Structure VI
42.9	CH_2(ε)	Structure VIII
40.1	C-γ	trans-1,4-polyisoprene
40.1	CH_2(γ)	Structure II

Table I. *(Continued)*

δ/ppm	C-atom	Structural Unit
38.8	CH(α)	Structure V
36.4	CH_2(δ)	Structure III
34.2	CH_2(δ)	Structure I
32.5	C-γ	cis-1,4-polyisoprene
32.5	CH_2(δ)	Structure IV
~31.0	CH_2(δ)	Structure VI
~30.5	CH_2(δ)	Structure V
30.0	C-δ	trans-1,4-polyisoprene
~30.0	CH_3(ε)	Structure VII
~30.0	CH_2(γ)	Structure VIII
26.7	C-δ	cis-1,4-polyisoprene
25.2[b]	CH_2	cyclic sulfide structures
23.6	C-ε	cis-1,4-polyisoprene
~23.5	CH_3(ε)	Structure II
~21.8[b]	CH_3(ε)	cyclic sulfide structures
~20.0	CH_3(ε)	Structure I
18.0[b]	CH_3-	cyclic sulfide structures
16.1	C-ε	trans-1,4-polyisoprene
16.1	CH_3(ε)	Structure IV
16.1	CH_3(ε)	Structure V
13.1	CH_3(ε)	Structure III

[a] The small amount of trans-olefinic asbsorptions not resolved from the corresponding cis-olefinic resonance.
[b] See Text.

In order to elucidate the absorption intensities, a computer decomposition is applied to the high field spectral region. Determination of the peak intensities in the inversion recovery spectra by computer fitting of the resonances is important in the selective analysis of spectra that consist of overlapping methyl and methylene/methine signals. The inverted magnetization of methyl groups recovers more slowly compared to the methylene or methine magnetizations, due to their relatively longer T_1 values. The T_1 relaxation time for $-CH_3$ groups is approximately five times longer than for $-CH_2-$ groups. Consequently, a complete recovery of the $-CH_2-$ magnetization is observed at the inversion recovery delay time (t) longer than 500 ms, while the $-CH_3$ magnetization has a zero value at $\tau \sim 200$ ms. This difference in recoveries of the magnetizations can be used to suppress the strong methyl absorption and detect weak signals which are normally obscured.

Figure 3 shows a part of the inversion recovery ^{13}C NMR spectra of sample NRl0-120 at 338 K for varied delay times.

The simulation of inversion-recovery spectra is done by the iterative fitting through a series of 14 different t values until the "peaks" in the difference betweeen the simulated and experimental results are minimized for all the spectra. The fitting procedure is first carried out for the spectra recorded at 353 K, in which the enhanced mobility increases the resolution. The signals at particular chemical shifts in simulated spectra are then applied to the simulation of the spectra recorded at lower

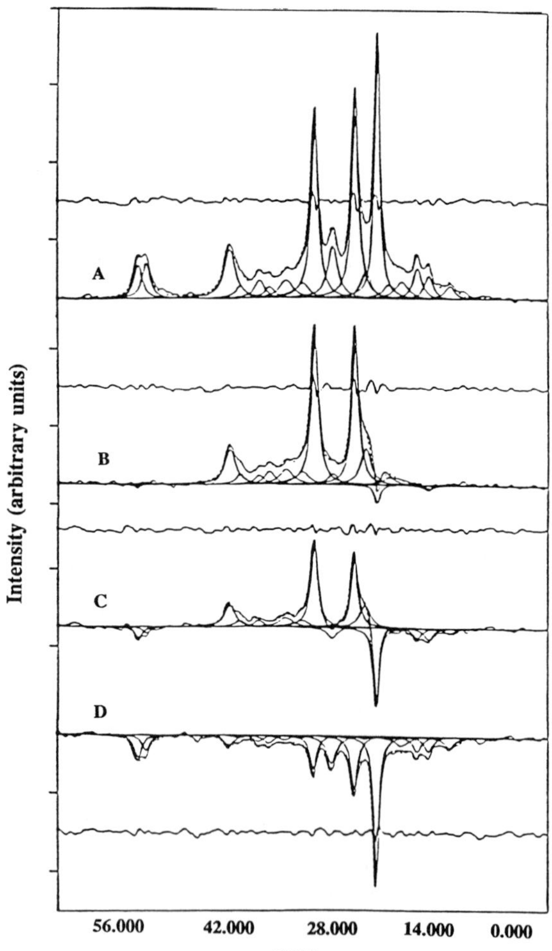

Figure 3. The ^{13}C NMR inversion recovery spectra of the sample NR10-120 at 338 K. The τ values for the sequence 180-τ-90° are 4 s (A), 0.2 s (B), 0.1 s (C) and 0.03 s (D). This figure shows curves fit to the data and the resulting difference spectra.

temperatures, in which the decreased resolution results in heavily overlapped resonance areas. The application of the same chemical shifts for simulation of the spectra recorded at different temperatures, associated with the negligible differences between the simulated and experimental results for all the t values, confirms the existence of the proposed resonances. Furthermore, the observed presence of at least two different T_1's for some resonances indicates that they still consist of several overlapped signals. Figure 4 shows the aliphatic part of inversion recovery spectra for the sample NRl0-090 at 295 K simulated with 18 resonances.

Difference spectra between the experimental and calculated spectra are shown in Fig. 4, indicating that the results of fitting the aliphatic region with 18 absorptions are in good agreement with the experimental data. However, the intensities of the difference spectra between the experimental and simulated inversion recovery spectra also indicate the existence of a few additional signals. For example, there is another resonance at appproximatly 25.5 ppm, which has the relaxation time typical of the $-CH_2-$ group. However, a simulation of the experimental spectra with the assumption of the existence of two or more obscured signals, having very close chemical shifts and unknown line widths is inexact. In some spectra, particularly of rubbers cured with smaller amounts of sulfur, a weak high field resonance can be observed at about 14.0 ppm.

The detection of the proposed structures and the quantitative analysis of absorption areas suggest the existence of the following reaction types in the

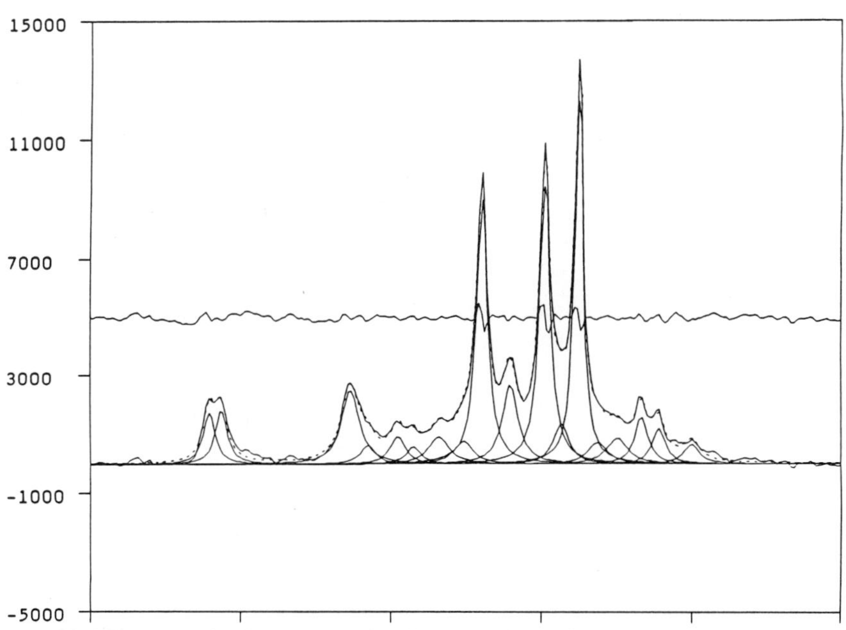

Figure 4. Experimental (-------) and calculated (———) aliphatic part of the ^{13}C NMR GHPD spectra of samples NR10-090 at 293 K.

vulcanization process: (1) formation of sulfur crosslinks at the methylene (C-γ and C-δ) and methyl (C-ε) carbons, (2) formation of sulfur crosslinks at the quarternary C-α and methine C-β carbon with the saturation of double bond, (3) crosslinking with the sulfur at the C-α carbon associated with breaking of the polyisoprene main chain, (4) formation of five-membered heterocyclic rings (unsaturated and saturated), (5) cis-trans isomerization, (6) formation of sulfur crosslinks at the methylene C-γ carbon of the trans-isomer, and (7) chain scission. It was observed that chain scission is slight at relatively short curing times and becomes significant at longer curing times. For example, the following molecular weights of natural rubber main chains (M_n = 2.4 x 10^5) cured with 10% of sulfur at 140° C for 2 h, 4 h and 24 h are found to be 2.0 x 10^5, 1.6 x 10^5 and 0.65 x 10^5, respectively [28].

The observed structural units of vulcanizates are defined by NMR only for the carbons removed three chemical bonds from sulfur atoms, for which the shielding effect can be detected in the solid state. A more accurate determination of the length of sulfidic crosslinks is also limited due to the insensitivity of substituent-effect values on the number of sulfur atoms. It is not possible to distinguish disulfidic crosslinks from tri- or polysulfidic crosslinks, and the mono- and di-/poly-sulfidic bridges have the same β- and γ-chemical shift substituent effects. Thus, the distribution of structural units in the network and lengths of sulfidic bridges for the structures shown in Fig. 5 can not be determined from the spectra.

3.2. Relaxation Measurements

The application of ^{13}C NMR relaxation techniques to crosslinked materials in the solid state has been limited until the development of modern solid state high resolution and multiple pulse techniques, which enable the detection of ^{13}C nuclei and their relaxation parameters. ^{13}C NMR relaxation data for linear polyisoprene at particular temperatures were reported in the solid state [9,10,11]. A study of molecular motion in linear and filled vulcanized natural rubber at room temperature by the multiple pulse ^1H NMR technique has also been carried out [12].

Variable temperature ^{13}C T_1 relaxation measurements yield information concerning the local motions for individual carbons, allowing the interpretation of data in terms of the main chain and side group motions in the bulk polymer. Although the ^{13}C T_1 data at particular temperatures have been reported for a large number of polymer systems, only a few studies have been made on the temperature dependence of spin-lattice relaxation times in the solid-state.

Here the effect of crosslinking on the molecular dynamics in natural rubber is investigated by the complementary application of variable temperature solid state ^{13}C NMR relaxation measurements and the computer simulation of overlapped spectral region.

Although the T_1 relaxation times reflect high frequency motions, the observed changes in line shapes also indicate the existence of low frequency molecular motions in the wider kHz region, comparable to the magic angle spinning frequency (10^0 kHz)

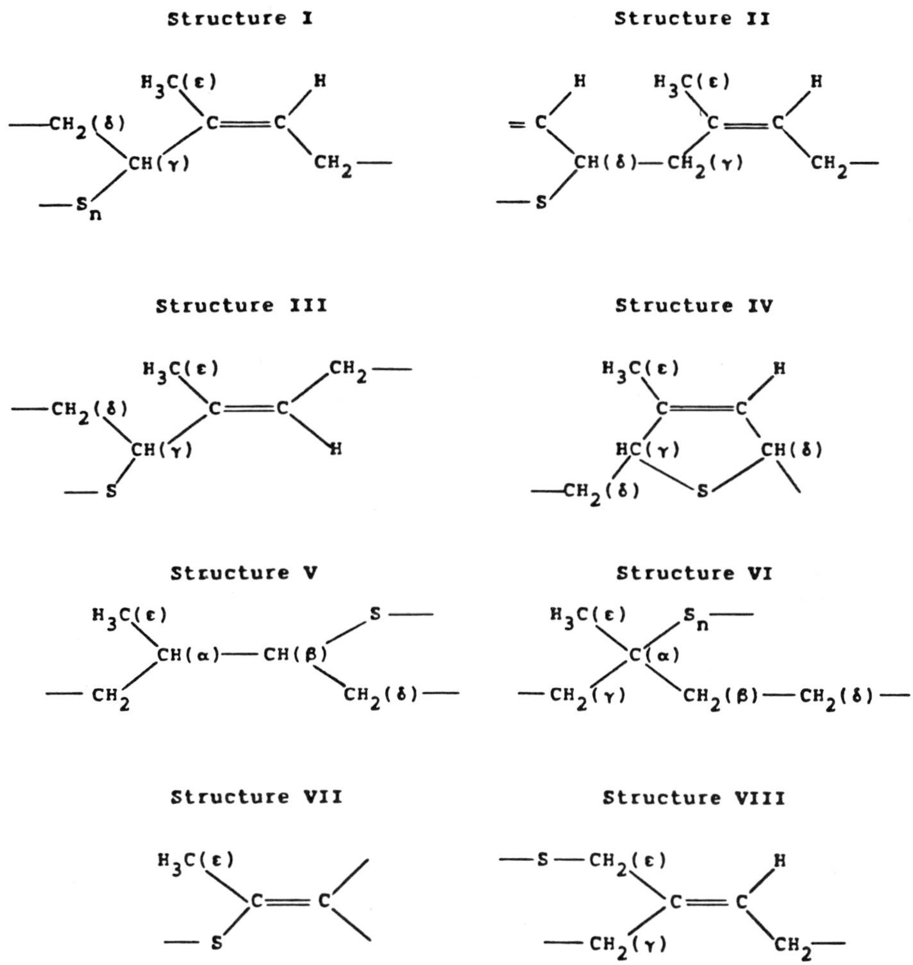

Figure 5. Proposed structural units for sulfur vulcanized natural rubber.

and/or the high power dipolar decoupling frequency (10^1 kHz). The changes are also observed in gated high power decoupled spectra, where the possible distortions caused by the inversion recovery pulse sequence are excluded. However, a general trend of increase in line widths (decrease of T_2 relaxation time) with increasing curing time is observed. The multiple pulse ^1H NMR experiment also shows that narrowed lines in the filled, vulcanized rubber are broader than in the natural rubber, suggesting the existence of residual dipolar ionteractions associated with moderately rapid segmental motions [25].

Due to the low natural abundance of ^{13}C nuclei, the spin-diffusion process between them is minimal and different carbons exhibit different relaxation times. Thus, it is possible to separate T_1 times for each carbon, and observe the respective main chain and side group relaxations. The main relaxation process for carbons is the dipolar relaxation by neighboring protons. Since the dipolar interaction is proportional to the inter nuclear distance by r_{CH}-6, the number of directly bonded protons, n, is the major contribution to the relaxation times nT_1. From the signal intensities of the components in a computer simulated spectra, the ^{13}C spin-lattice relaxation times are determined. The temperature dependence of the nT_1 for the uncrosslinked natural rubber (NR10-0) is shown in Fig. 6.

The nT_1 curves for all polyisoprene carbons have similar temperature dependences with the T_1 minimum at about 303 K. The observed minima for the ^{13}C relaxation at 37.7 MHz are in agreement with the reported ^{1}H T_1 minima for Hevea rubber at the resonance frequencies of 20 MHz (286 K) and 28 MHz (294 K) [7]. For comparison, the ^{13}C nT_1 values for natural rubber, observed at different frequencies at approximately same temperature, are listed in Table II.

The nT_1 curves for backbone carbons are parallel, while the methyl group exhibits a slightly broader V-shaped curve on the high temperature side, possibly

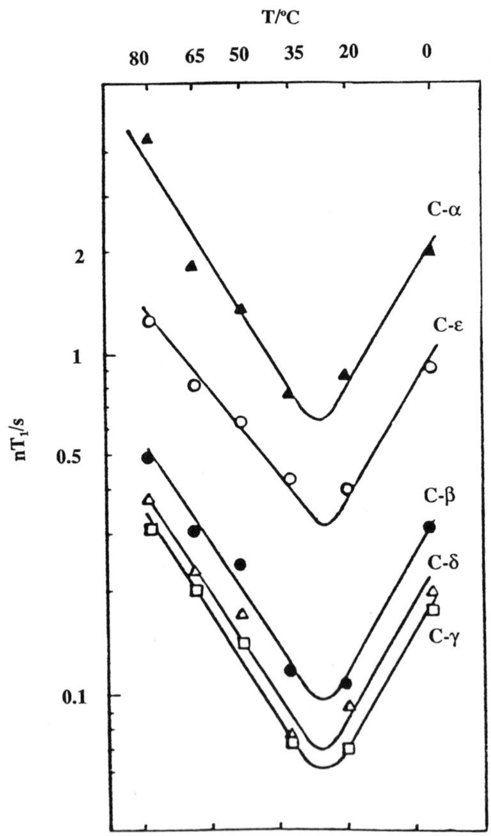

Figure 6. The temperature dependence of ^{13}C nT_1 values for sample NR10-0: C-α (▲), C-β (•), C-γ (□), C-δ (△) and C-ε (○).

Table II. ^{13}C nT_1 RELAXATION TIME (msec) FOR CIS-POLYISOPRENE

Carbon	22.6 MHz[a] T = 40 °C	37.7 MHz T = 35 °C	67.9 MHz[b] T = 40 °C
α	700	758	940
β	95	117	166
γ	50	73	114
δ	55	75	110
ε	350	427	656

[a] Data from J. Schaefer, Macromolecules, **5**, 427 (1972).
[b] Data from R. A. Komoroski, J. Maxfield and L. Mandelkern, Macromolecules, **10**, 545 (1977).

indicating a distribution of correlation times (Fig. 6). Three protons directly bonded to the Cγ atom do not increase the NT_1 value of the CH_3- group by a factor of three, as expected, because of the rotational freedom of the methyl group, which reduces the effectiveness of the dipolar coupling. The longest T_1 observed for the non-protonated C-ε carbon indicates the presence of dipolar relaxation by the protons on neighboring protonated carbons.

The temperature dependence of the nT_1 relaxation times for the main chain C-β and the side group C-ε carbons in polyisoprene networks are shown in Fig. 7.

Samples cured for times shorter than 30 minutes show an increase in the T_{1min} to longer values and slight shifts in positions of the T_{1min} of the C-β relaxation to higher temperatures. A broadening of the V-curves, i.e. an increase in the T_1 values at the minimum and a decrease of the T_1 values away from the minimum can be attributed to a distribution of τ_c in a heterogeneous network. All the other main chain carbons exhibit similar behavior as the degree of crosslinking increases, indicating that the effect of network formation is the same for all the backbone atoms. The vulcanizates cured for 60, 90 and 120 min., respectively, show significant shifts in the T_{1min} to higher temperatures as well as V-curve broadening. This is a consequence of greater motional restrictions induced by the crosslinking as compared to the samples cured for 15, 22 and 30 minutes. The changes in the relaxation curves suggest that the main chain motions for the lightly crosslinked samples are not significantly affected, although a wider distribution of t_c in the heterogeneous network can exist. Curing times longer than 30 minutes result in more restricted backbone motions (NR10-60, NR10-90 and NR10-120).

The ^{13}C T_1 results show that the broadenings of the v-curves are accompanied with the shifts if T_{1min} to higher temperatures for all the samples. At the same time V-curves for backbone carbons of samples cured to 30 min. do not show any significant shift of the minima to higher temperatures. The data indicate that the motional restrictions imposed by the vulcanization are more pronounced for the

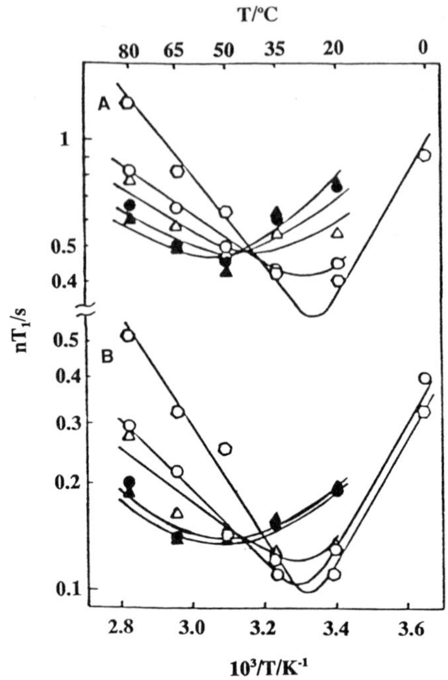

Figure 7. The temperature dependence of ^{13}C nT_1 values for the: (A) C-ε and (B) C-β carbon: NR10-0 (◯), NR10-15 (o), NR10-22 (▵), NR10-60 (•), and NR10-120 (▴).

methyl group than for the main chain carbons, although the methyl group is not directly involved in the crosslinked structures. It can be assumed that the relatively short crosslinks (mono- and di-sulfidic) at the neighboring carbons impose spatial restrictions on the methyl group reorientation.

A detailed analysis of the effect of crosslinking on the model parameters requires a large number of data points over a wide temperature range, and cannot be satisfactorily made with these rubber samples because of the relatively narrow temperature range in which data can be obtained. Temperature-induced crosslinking at temperatures higher than 80° C is observed during the 12h experiment and by the measurements at lower temperatures. At temperatures below 10° C the broad lines in overlapped spectral region do not allow an accurate spectral decomposition, particularly for highly crosslinked samples.

4. Conclusions

Motional restrictions introduced by crosslinks are investigated from the temperature dependence of nT_1 relaxation times for individual carbons. The v-curves for all polyisoprene signals and for the detectable signals arising from the network units exhibit a similar general trend with increased curing time: curve broadening, shift of the minima to higher temperatures and increase of the T_{1min} values. All the backbone carbons show quantitatively similar effect of vulcanization on the spin-lattice relaxation. At shorter curing times motional restrictions for the methyl side group are more pronounced as compared to the main chain carbons.

5. References

1. Patterson, D. W.; Koenig, J. L. and Shelton, J. R., *Rubber Chem. Technol.*, *56*, 971, 1983.
2. Zaper, A. M. and Koenig, J. L., *Rubber Chem. Technol.*, *60*, 252, 1987.
3. Zaper, A. M. and Koenig, J. L., *Rubber Chem. Technol.*, *60*, 278, 1987.
4. Patterson, D. W. and Koenig, J. L., *Appl. Spectrosc.*, *41*, 441, 1987.
5. Komoroski, R. A., *Rubber Chem. Technol.*, *56*, 959, 1983.
6. Komoroski, R. A.; Shockcor, J. P.; Gregg, E. C. and Savoca, J. L., *Rubber Chem. Technol.*, *59*, 328, 1986.
7. Clough, R. S. and Koenig, J. L., to be published.
8. Duch, W. M. and Grant, D. M., *Macromolecules*, *3*, 165, 1970.
9. Tanaka, Y.; Sato, H. and Seimiya, T., *Polym. J.*, *7*, 264, 1975.
10. Tanaka, Y. and Sato, H., *Polymer*, *17*, 113, 1976.
11. Tanaka, Y. and Sato, H., *Polymer*, *17*, 413, 1976.
12. Tanaka, Y.; Sato, H.; Ogura, A. and Nagoya, I., *J. Polym. Sci., Polym. Chem. Ed.*, *14*, 73, 1976.
13. Werstler, D. D., *Rubber Chem. Technol.*, *58*, 1191, 1980.
14. Schaefer, J.; Chin, S. H. and Weissman, S. I., *Macromolecules*, *5*, 798, 1972.
15. Schaefer, J. and Stejskal, E. O., *J. Am. Chem. Soc.*, *98*, 1031, 1976.
16. Scheele, W., *Rubber Chem. Technol.*, *34*, 1306, 1961.
17. Bateman, L.; Moore, C. G.; Porter, M. and Saville, B., "Chemistry of Vulcanization" in *The Chemistry and Physics of Rubber-like Substances*, Bateman, L., Ed., MacLaren and Sons Ltd., London, 1963, pp 449-561.
18. Porter, M., "Vulcanization of Rubber" in *Organic Chemistry of Sulfur*, Oae, S., Ed., Plenum Press, New York, 1977, pp 71-118.
19. Chen, C. M.; Koenig, J. L.; Shelton, J. R. and Collins, E. A., *Rubber Chem. Technol.*, *54*, 734, 1981.

Chapter 12

A SOLID-STATE ^{13}C NMR STUDY OF POLYSILOXANE SIDE-CHAIN LIQUID CRYSTALLINE POLYMERS CONTAINING 2-[4-(2S)-METHYL-BUTOXY)-PHENYL]-5-(ω-ALKYL)-1,3,2-DIOXABORINANE WITH ALKYL BEING UNDECANYL AND OCTYL

B. C. PERRY AND J. L. KOENIG

1. Abstract

The polysiloxane side-chain liquid crystalline polymer containing 2-[4-(2S)-methyl-butoxy)-phenyl]-5-(ω-alkyl)-1,3,2-dioxaborinane with alkyl being undecanyl and octyl was studied using solid-state ^{13}C NMR. The spin-lattice relaxation time, T_1 was measured as a function of temperature at two distinct magnetic fields. The T_1's suggest significant motional increases for the mesogenic unit at the LC transition. The relaxation data indicate that two Tgs exist with the second predominantly involving motion of side chain. The 8 and 11 length spacer units offer adequate decoupling. A greater number of gauche conformers exists in the LC state relative to the glassy state. An S_c mesophase is predicted based on optical microscopy "textures." The hindrance of rotation of the branched methyl carbons in the sec-butyl free substituent indicates that a packing interaction is occurring which could only happen in a more ordered smectic phase.

2. Introduction

The ability of disc-like and rod-like molecules to form ordered structures under certain conditions has been known for many years. Only recently, has it been discovered that polymers can also form liquid crystalline (LC) structures where the

---

B. C. PERRY AND J. L. KOENIG - Case Western Reserve University, Cleveland, OH.

Solid State NMR of Polymers, Edited by L. Mathias
Plenum Press, New York, 1991

polymeric properties co-exist with the independent mesogenic properties. There are two basic structural types of liquid crystals: in one, the main chain LC mesogenic group is connected in a linear fashion along the main chain; and in the other, the side-chain LC mesogenic group is attached to a monomer which is subsequently polymerized.

The presence of the rigid mesogenic groups within the polymer chain does not insure liquid crystalline properties [1]. In the case of side-chain LCs, the backbone must deform to allow the LC structure to exist. The stiffness of the groups may increase the transition temperatures of thermotropic systems to values above the decomposition temperatures. The transition from the glassy amorphous phase to a liquid crystalline phase should be accompanied by substantial changes in mobility of the mesogenic group to allow "repacking" of the mesogens. The most common synthetic technique used to decouple the motions of the mesogen and the backbone is by the insertion of a flexible spacer between the main chain and mesogen. Generally, longer spacers favor smectic formation and lower transition temperatures.

In order to optimize decoupling, it is desirable to be able to view each carbon in the polymer individually to determine specific motions that are occurring. Solid-state ^{13}C NMR can resolve most of the carbons in LCs. Also, the T_1 value are quite sensitive to the LC transition. It is the purpose of this paper to use ^{13}C NMR techniques to study the temperature dependence of the carbon T_1 values of certain side chain polymer LCs and to study the motional changes induced at the transition temperatures.

3. Experimental

The polymer LCs used in this study were synthesized by Hahn and Percec [2,3] (Fig. 1). The LCs are specified on the basis of spacer length, C=11 and C=8. DSC and optical microscopy experiments are reported in the literature. There were no thermal pretreatments of the polymers prior to the NMR studies.

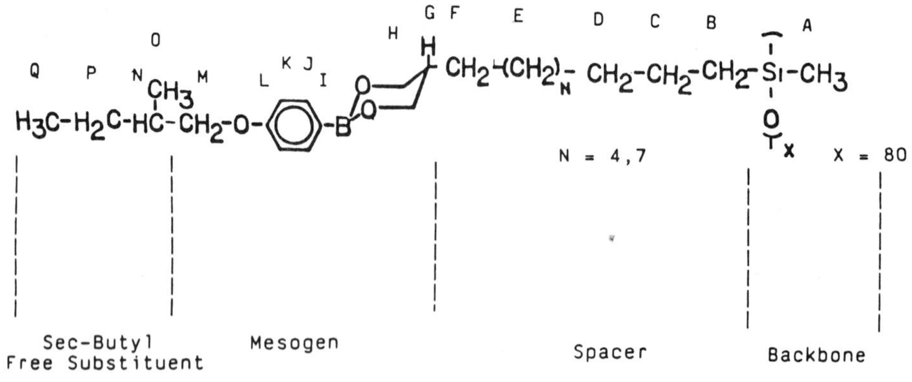

Figure 1. The structure and chemical shift assignments of the C=11 and C=8 LCs.

Table I TRANSITION TEMPERATURES (°C) OF THE C=11 AND C=8 LCs DETERMINED BY DSC [2,3]

	C=11	C=8
Tg_1	---	---
Tg_2	-7	4
T_m	19	---
T_i	85	74
The DSC scans were taken at a rate of 20°C/min.		

The transition temperatures as measured by DSC are listed in Table I [2,3]. The polymers show a smectic texture after the T_g.

The ^{13}C NMR spectra were recorded on a Nicolet NT 150 and a Bruker MSL 300 spectrometer at 37.72 and 75.47 MHZ, respectively. Magic-angle spinning (MAS) [4], cross-polarization (CP) [5], and dipolar decoupling (DD) [6] are used simultaneously for the cross-polarization experiments. The radiofrequency fields were typically 60 to 70 KHz. The contact time was 1 ms with delays of 2 and 3 s between successive transients. Rotors were supplied by Doty Scientific for the NT 150 and by Bruker Instruments, Inc. for the MSL 300. The rotors were machined from Al_2O_3. End caps of Delrin and Kel-F were used depending on the temperature range studied. Markers were not used in every experiment due to peak overlap so the chemical shift reference was set using the adamantane offset (29.5 ppm) prior to the experiments. The Delrin shift of 89.1 ppm was used as a reference when possible. Spinning frequencies were between 1.0 and 3.3 kHz. Liquid nitrogen gas was used for the low temperature experiments. The FID was collected in a 512W or 1K word data size and zero-filled to 8K before Fourier transformation. The magic angle was set on the NT 150 by maximizing the ratio of carbonyl to methylene peaks of glycine. The magic angle was set on the MSL 300 by maximizing the sidebands of KBr. The Hartman-Hahn match was set by maximizing the peak intensities of adamantane. The 90° pulse was determined by finding the 360° and 720° pulses via the nulling out of the adamantane signal. The rf field was calculated directly from the 90° pulse.

The T_1 inversion-recovery experiments using cross-polarization were performed with the pulse sequence of Maciel and Melchior [7,8]. Delay times of 500 μs to 12 s were used while the number of transients were approximately 1000 on the MSL 300 and 2000 on the NT 150 for the majority of the samples. Each T_1 versus temperature sequence from low to high temperature was run consecutively to avoid relaxation effects. Standard pulse sequences were used to obtain the cross-polarization and gated high power decoupling experiments with typical contact times of 1 ms and delays of 2 s. The T_1 inversion-recovery intensities were plotted as a function of time and curve-fitted to the inversion-recovery equation on a microvax. Proton saturation was not applied during the inversion-recovery delay time. Transient NOE effects were not considered in the analysis.

4. Results and Discussion

The DSC and optical microscopy work done by Hahn [2,3] indicates that the polymers begin to flow and exhibit a smectic texture above T_{g2}. The thermal history of the polymers has little effect on the transition temperatures. Two separate T_gs are observed for the siloxane copolymers: one at low temperatures T_{g1} which is assigned to backbone motions and a T_{g2} which is assigned to side-chain motions. The homopolymer used in this study exhibited a T_{g2}. The T_{g1} may not appear because the weight percent of the siloxane backbone is small compared to the side chains. Boron is used in the cyclohexane ring to avoid mixtures of cis and trans conformers that have plagued other similar systems [9,10]. Qualitatively, when the spacer is lengthened one observes the expected variations in the glass transition temperatures. Increasing the flexibility decreases the temperature of the observed glass transitions. Odd-even effects have been observed in other systems [11-13] for a homologous series but consecutive methylene lengths are needed to probe this effect.

NMR measurements are performed on the samples without any previous thermal treatment. The DSC results indicate that the LCs are relatively insensitive to thermal history. The polymer is a LC at room temperature. Orientation of the LC in the magnetic field was not considered a problem because orienting smectic polymers is difficult [14] and all measurements are done with magic angle spinning which inherently removes orientation effects.

The observed chemical shifts for the LCs are listed in Table II for the 11 and 8 methylene spacer LCs. The observed assignments were made on the basis of the calculated shifts in combination with cross-polarization, gated high power decoupling, and dipolar dephasing experiments on a number of LCs with similar structures [15]. The Theoretical chemical shifts were calculated using the assignments from Silverstein *et al.* [16] and Maciel [17] (Table II).

Qualitatively, some observations can be made for the C=11 LC using conventional CP experiments. A comparison of the CP/MAS/DD experiments for the C=11 LC at three temperatures is shown in Fig. 2. At -15° C one sees a broadening of the methyl peaks due to dipolar broadening from motions at the decoupling frequency. The intensities of the aromatic carbons are relatively small due to long relaxation times and to a lack of bonded protons. At 20° C, the methyl peaks sharpen and the aromatic peaks grow in relative intensity. The methylene carbons at 68 and 73 ppm, which are separated at -15°C, become broader at 20° C. A further increase in the temperature to 50° C, near the T_m, shows a decrease in the methyl peaks due to poorer CP efficiency. Also, at this temperature the resolution of the carbons at 68 and 73 ppm is enhanced with a reversal in relative intensities, and there is a narrowing and an increased relative intensity of the aromatic carbons. The C=8 has similar trends with the exception that no broadening of the carbons at 68 and 73 ppm is observed.

Preliminary relaxation experiments on these LCs using T_1 and $T_{1\rho}$ indicated that the T_1 is more sensitive to the motional changes. Also, the MHz motions are

Table II CALCULATED AND OBSERVED CHEMICAL SHIFTS OF THE C=11 AND C=8 LCs IN PPM AT 25° C. ADAMANTANE REFERENCE (29.5 PPM)

Carbon	Calculated	C=11	C=8
A	0.4	1.0	0.97
B	18	19.1	20.1
C	25	24.3	26.1
D	33	---	---
E	31.5	31.7	31.4
F	32-40	---	---
G	---	---	---
H	65.9	67.2	68.0
I	130	137	137
J	129.5	137	137
K	114	115	114
L	158	160	159
M	66-70	73	73.8
N	36.7	35.7	36.6
O	18.6	17.6	17.9
P	29.3	27.3	27.8
Q	11.3	12.9	13.1

localized so individual carbons can be probed. The experiments are performed at two magnetic fields to determine on which side of the T_1 minimum the relaxation times occur. The questions to be addressed are as follows. What types of motion are occurring at the T_{g2} that allow "repacking" of the side chains? Are the motions decoupled from the main chain? Finally, are any differences observed between C=11 and C=8?

It should be noted that the T_g is a dynamic process so it depends on the frequency at which it is being measured. DSC is a comparatively static technique whereas the NMR experiments are probing the MHz region. Thus, the T_g will appear a higher temperatures [18]. The T_m is not a dynamic process thus the NMR and DSC will coincide.

Cross-polarization is chosen to study the inversion-recovery in order to shorten the repetition times and increase the signal to noise. The CP efficiency decreases at the higher temperatures. It is difficult to obtain reliable data on the aromatic carbons at the lower temperatures so relaxation times are reported only for the methylene and methyl carbons. Qualitative observation can be made for the aromatic carbons.

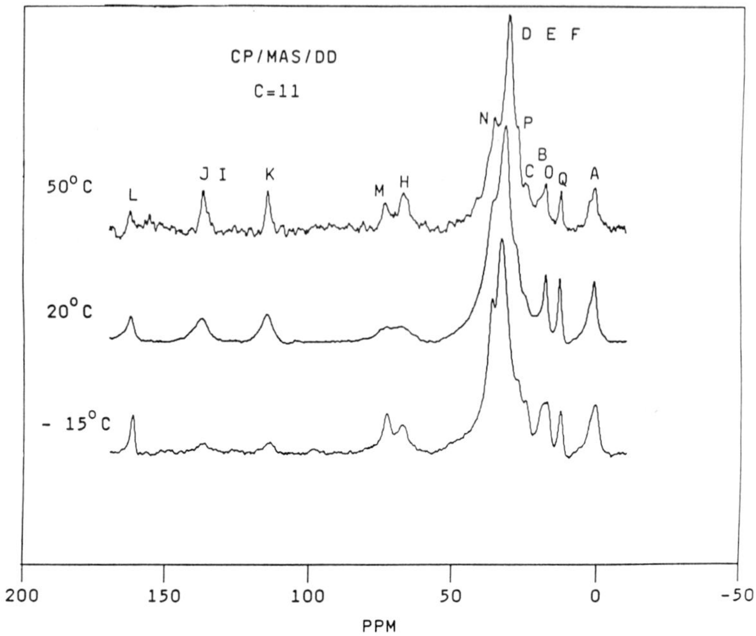

Figure 2. The CP/MAS/DD spectra of the C=11 LC at -15° C, 20° C, and 50° C at 37.7 MHz.

The inversion-recovery spectra for the C=11 and C=8 polymers below and above T_g are shown in Figs. 3 and 4, respectively. The typical T_1 inversion-recovery curve is shown in Fig. 5 with the solid line being the fitted curve. Carbon T_1 values for the C=11 and C=8 polymers at 37.7 and 75.5 MHz as a function of temperature are shown in Tables III and IV, respectively. A plot of the field dependence of the spacer carbons (31 ppm) is shown in Fig. 6. A plot of the temperature dependence of the T_1 values at 75.5 MHz for C=11 of the spacer (31 ppm) and ring (68 ppm) carbons is shown in Fig. 7. An expanded view of the relaxation behavior of the inversion point is shown in Figs. 8A and 8B in order to emphasize the s-shaped inversion of the peak at 31 ppm.

4.1. Methyl Carbons in the C=11 LC

Three methyl carbons are present in the C=11 polymer. One is associated with the polysiloxane backbone and two are assigned to the free substituent (sec-butyl) attached to the mesogenic unit. The backbone methyl carbons T_1 values are considerably longer and relatively invariant to the temperature and field strength. The sec-butyl carbons' T_1s increase with temperature and are invariant to field strength below T_{g2}. The increase of T_1 with temperature infers that the T_1 lies on the fast side of the minimum. The backbone methyl is more mobile relative to the sec-butyl methyl carbons although the T_1 differences decrease at elevated temperatures. Within the sec-butyl group, one interesting trend is observed. At both fields, the terminal methyl carbons (13 ppm) have a larger T_1 value than the branched methyl (17 ppm) by a factor of approximately two. There may be some hindrance of rotation as a result of packing constraints as the ratio of the T_1 values (T_1(13 ppm)/T_1(17 ppm)) becomes larger above T_{g2} (Fig. 9).

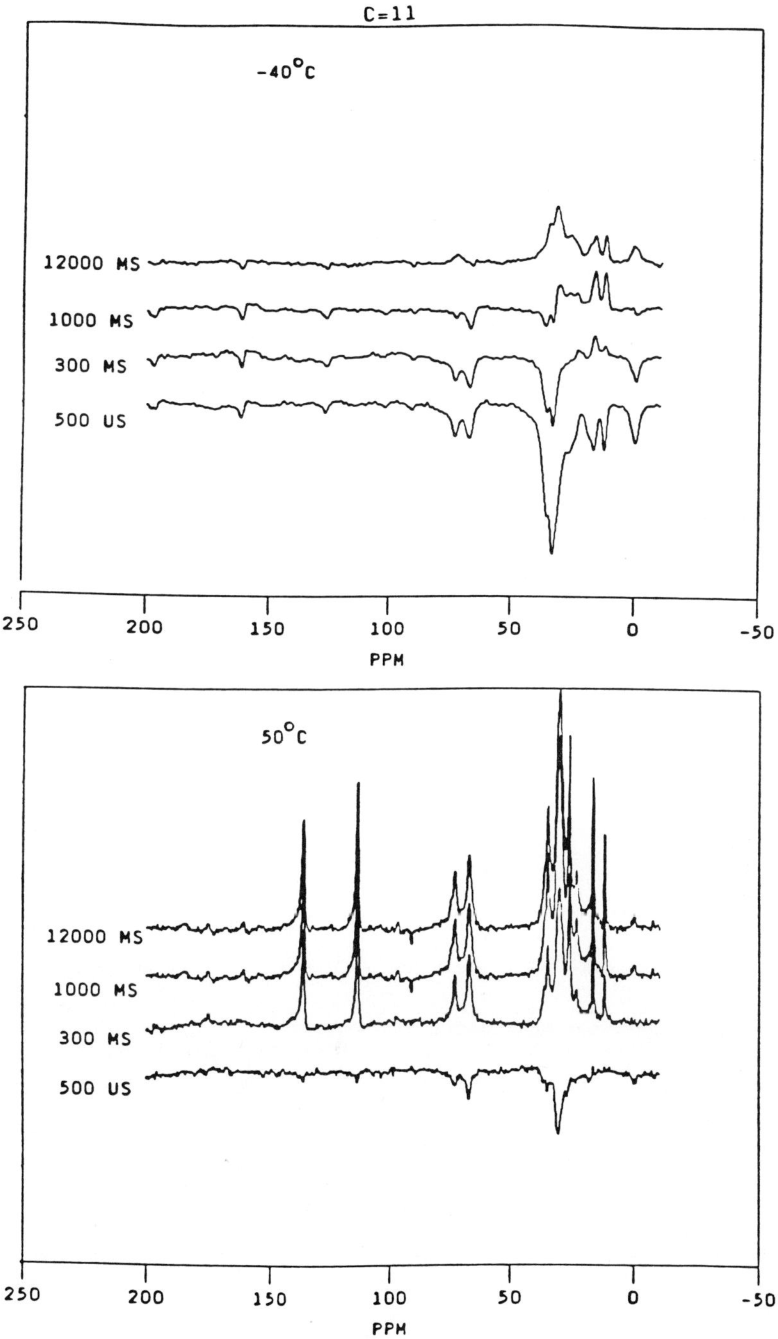

Figure 3. The inversion-recovery spectra for the C=11 LC at -40° C and 50° C at 75.5 MHz.

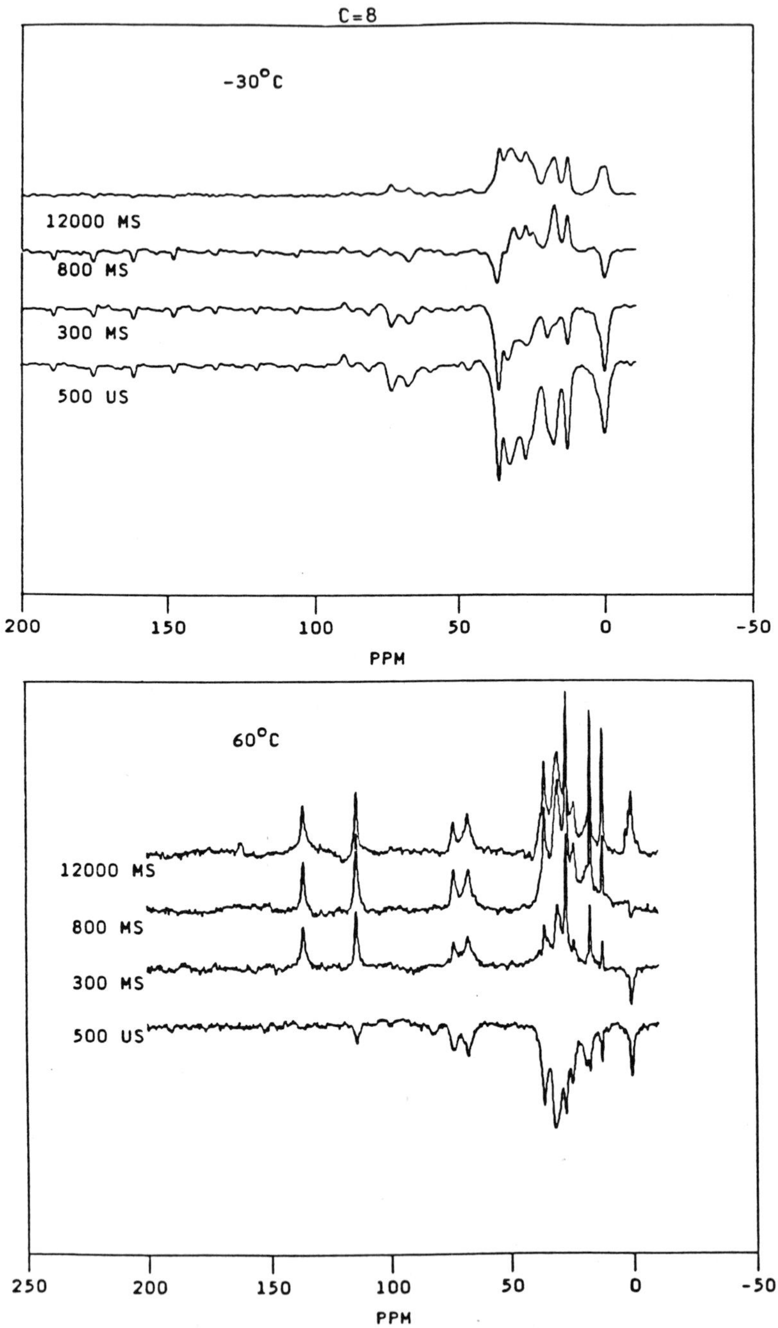

Figure 4. The inversion-recovery spectra for the C=8 LC at -30° C and 60° C at 75.5 MHz.

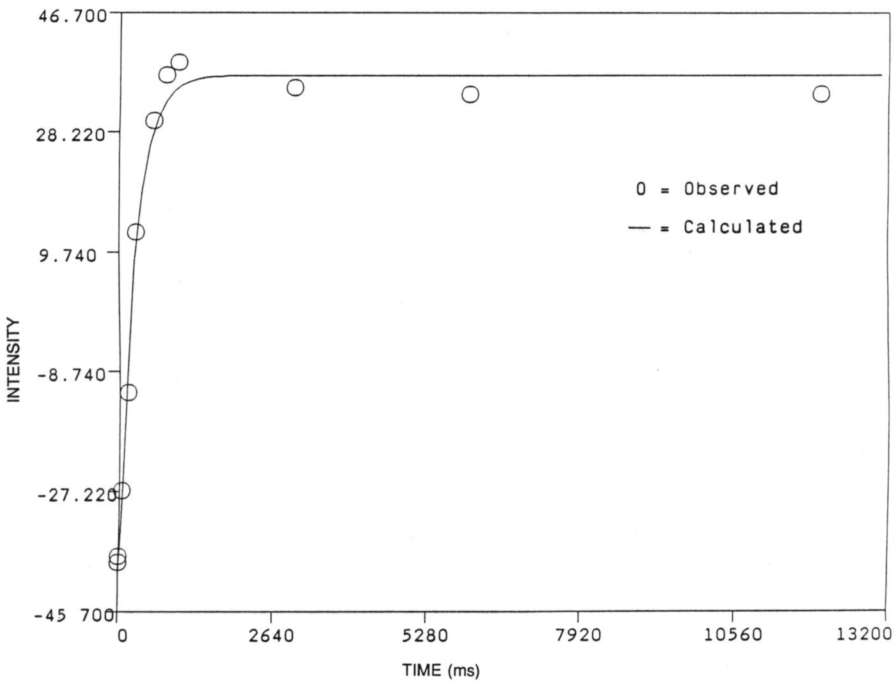

Figure 5. The observed and calculated inversion-recovery of the central spacer carbons, E (31.5 ppm), at a temperature of 25° C at 75.5 MHz.

4.2. Methylene Carbons in the C=11 LC

There are six resolvable methylene peaks. Three separate portions of the LC can be observed. The peaks at 21, 25, and 31 ppm are associated with the spacer. The peak at 68 is associated with the mesogen while the peak at 73 ppm is assigned to the sec-butyl group. The methylene at resonance 28 ppm is a composite resonance which is assigned to the flexible free substituent and spacer carbons.

Generally, the field dependence of the methylene carbons indicate that they lie on the slow side of the minimum (Fig. 6). The general tendency of the T_1 values is to decrease with increasing temperature until the T_{g2} is reached at 20° C (37.7 MHz) or 35° C (75.5 MHz); then the values increase as the T_m is approached indicating that a local minimum may exist.

A comparison of the spacer carbons at 31 ppm and 25 ppm indicates that from -15° C to 50° C the T_1 values are the same. However, there is a large discrepancy at -40° C with the T_1 of the spacer carbons at 31 ppm being twice as long in value as the 25 ppm carbons. The LC is well above T_{g1} but well below T_{g2} at this temperature so there is a decoupling of motion within the spacer at low temperatures; but as the T_{g2} is approached the motions become cooperative. The type of motion involved is most likely a trans-gauche (T-G) isomerization [19]. At low temperature a slight chemical

Table III C-13 T_1 VALUES OF THE METHYL, METHYLENE, AND METHINE CARBONS OF THE C=11 LC AS A FUNCTION OF TEMPERATURE AT 37.7 AND 75.5 MHz

Methyl Carbons										
	37.7 MHz				75.5 MHz					
T(C)	-15	20	40	50	-40	-20	0	25	35	50
ppm										
1.0	1.0	1.1	1.6	1.8	1.2	.98	.96	1.5	1.6	---
13	.29	.42	.88	1.3	.23	.27	.32	.50	.48	---
17	.18	.15	.36	.55	.18	.19	.21	.26	.24	---
Methylene Carbons										
	37.7 MHz				75.5 MHz					
T(C)	-15	20	40	50	-40	-20	0	25	35	50
ppm										
73	.44	.05	.09	.09	1.1	.52	.36	.29	.14	.30
68	.92	.03	.07	.08	2.8	.83	.53	.30	.12	.26
28	.12	.08	.16	.20	.49	.22	.23	.19	.13	.32
31	.17	.08	.19	.22	.78	.30	.26	.27	.19	.40
25	.13	.09	.19	.20	.36	.21	.17	.22	.17	.31
Methine Carbons										
	37.7 MHz				75.5 MHz					
T(C)	-15	20	40	50	-40	-20	0	25	35	50
ppm										
36	.20	.10	.19	.22	1.0	.51	.31	.27	.19	.40
Experimental error is ±15%.										

shift from 31.5 ppm to 34 ppm is observed for the main spacer peak (Fig. 10). The other spacer carbons are unaffected. The shift is a conformational effect. The general trend for aliphatic chains observed in polyethylene is a downfield shift for the trans methylenes [20]. Lauprêtre [19] and Samulski [21] and others [22,23] have shown that a significant number of gauche conformers must exist in the LC state. In fact the conformation is quite dependent on the position of the carbons in the spacer. There is an odd-even effect for the probability of a trans bond for a spacer between two cores. An odd-numbered spacer will tend to propagate this alternating effect whereas an even-numbered spacer should destructively interfere yielding a relatively constant trans probability [21]. Additionally, ^2H spectroscopy by Samulski [21] and Müller [22] show that for main chain LCs, the central methylenes have a lower number of trans bonds in the LC state but as the temperature is lowered the central methylenes become ordered at a faster rate. The increase in the number of trans conformers is greater for the central carbons. Spiess [23] has seen a large disparity in the micro-order parameter for spacer carbons next to the mesogen as compared to spacer carbons near the backbone in the frozen LC state for side-chain LCs. The shift thus indicates that on average the central methylenes are in a more extended conformation below T_{g2}. The high degree of order and the ability to maintain the order when cooled below T_g for the LCs is quite apparent as the central methylenes contain an increased number of trans bonds below T_{g2}. Figure 8B indicates that at low temperatures at a recovery time of 800 ms, the large spacer has two components at 34.0 and 30.7 ppm. The former relaxes slower. Figure 11 shows a plot of T_1 for the two components as a function of temperature. Above 25° C, inversion-recovery is not able to separate the two components. Below T_{g2}, a larger number of central methylenes exist in trans conformations causing the 2 to 3 ppm shift. Additionally, the carbons which still exist in the less extended conformation possess more mobility.

Table IV C-13 T_1 VALUES OF THE METHYLENE, METHYL, AND METHINE CARBONS OF THE C=8 LC AS A FUNCTION OF TEMPERATURE AT 37.7 AND 75.5 MHz

Methyl Carbons											
	37.7 MHz						75.5 MHz				
T(C)	-40	-10	10	25	40	60	-30	0	25	45	60
ppm											
1.0	.48	.99	1.0	1.3	1.8	1.0	1.6	1.5	2.2	2.3	2.3
13	.05	.31	.31	.49	.44	.74	.32	.54	.65	.87	.76
17	.06	.19	.22	.22	.25	.27	.21	.32	.38	.44	.39
Methylene Carbons											
	37.7 MHz						75.5 MHz				
T(C)	-40	-10	10	25	40	60	-30	0	25	45	60
ppm											
73	4.9	.32	.15	.06	.12	.10	1.1	.43	.30	.29	.21
68	*	.70	.28	.17	.14	.06	4.7	.96	.44	.32	.17
28	.48	.27	.15	.13	.14	.13	.27	.23	.23	.25	.20
31	.60	.30	.18	.15	.16	.14	.29	.28	.37	.36	.27
25	.63	.22	.16	.14	.21	.13	.30	.36	.32	.40	.26
Methine Carbons											
	37.7 MHz						75.5 MHz				
T(C)	-40	-10	10	25	40	60	-30	0	25	45	60
ppm											
36	.91	.50	.35	.18	.16	.18	.92	.62	.39	.41	.34

Experimental error is ±15%. *Very long inversion-recovery.

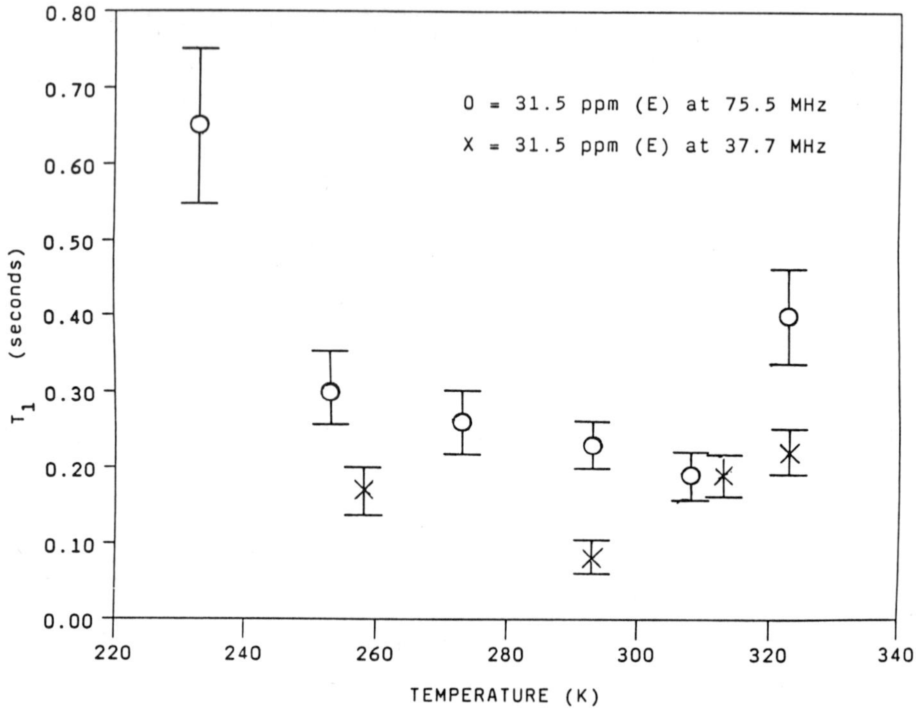

Figure 6. The T_1 values as a function of temperature and field strength for the central spacer carbons E (31.5 ppm).

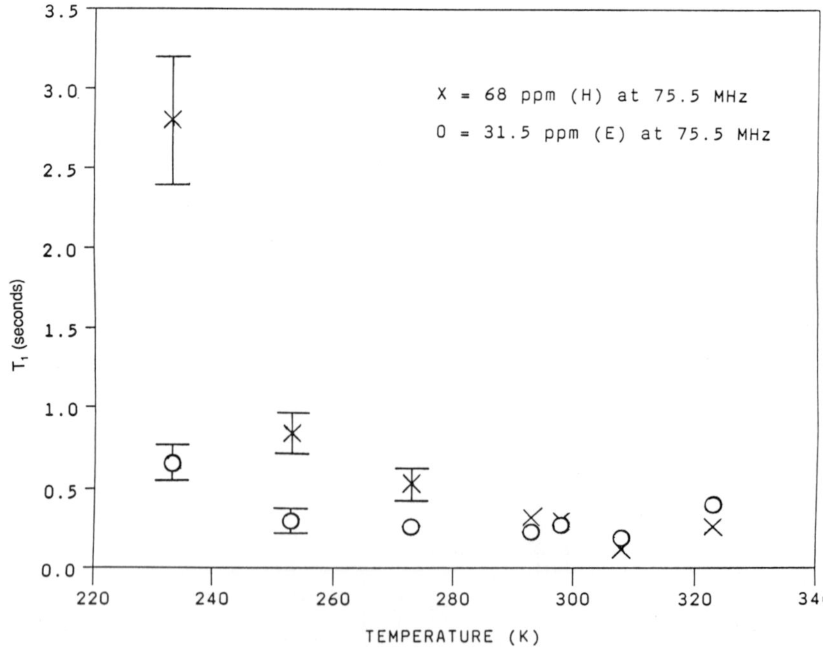

Figure 7. The T_1 values of the mesogenic carbons, H (68 ppm), and the central spacer carbons, E (31.5 ppm), as a function of temperature at 75.5 MHz.

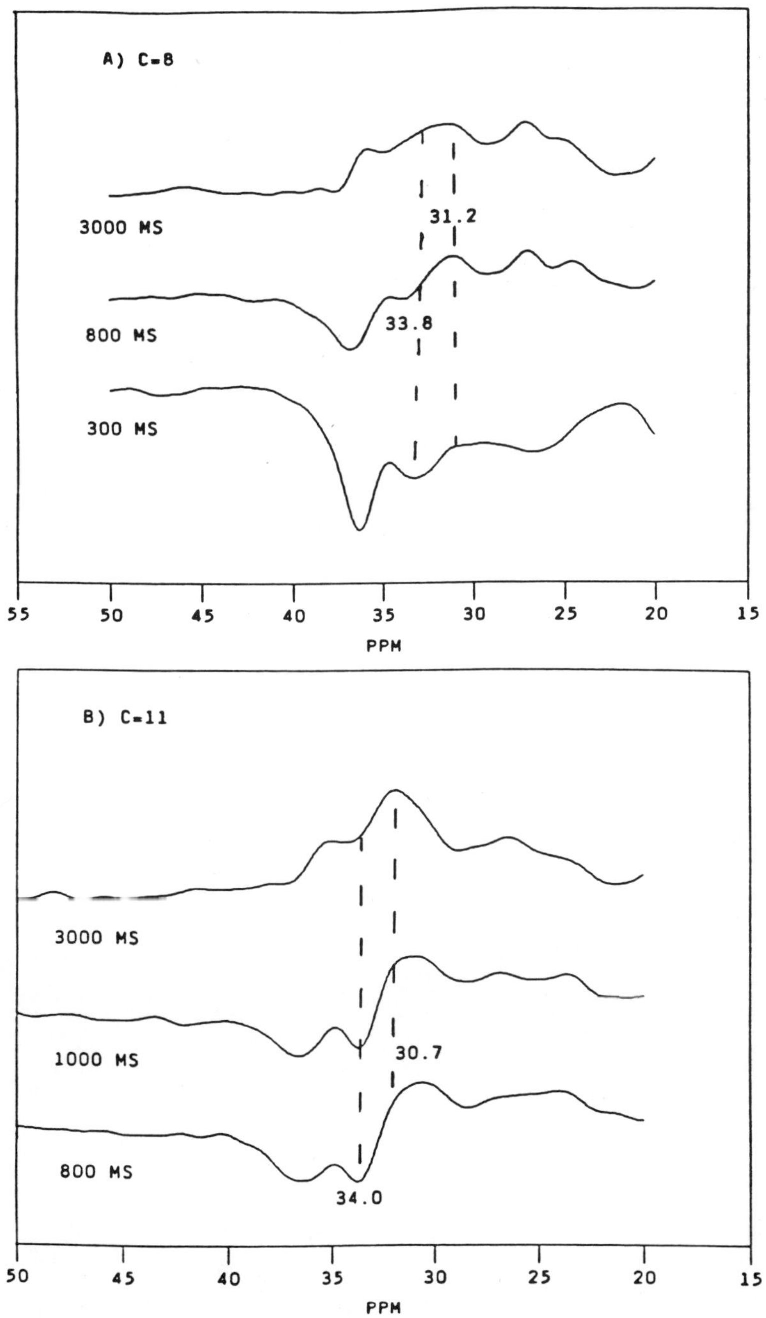

Figure 8. The inversion-recovery spectra of the expanded central methylene region for the C=8 and C=11 LCs at -40° C and 75.5 MHz.

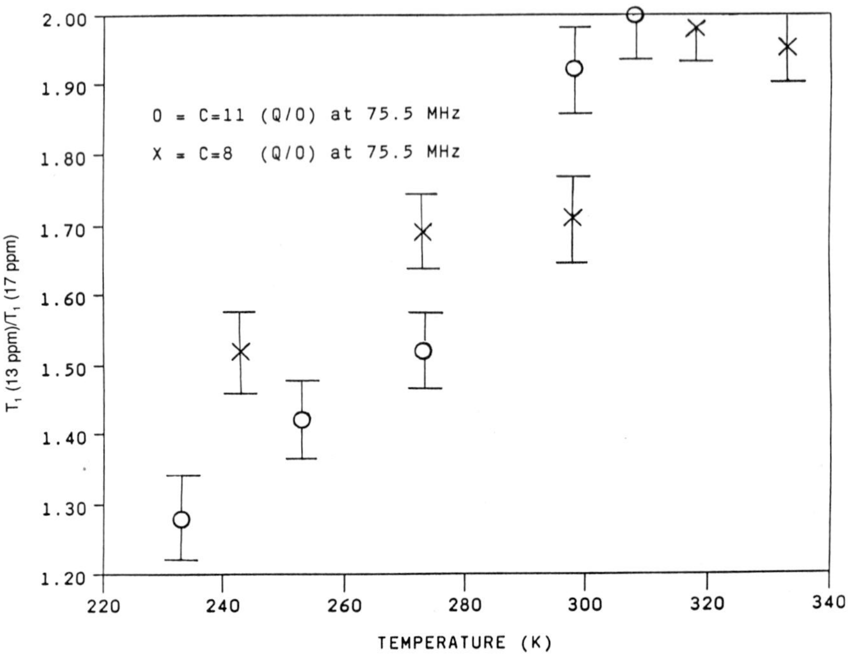

Figure 9. The ratio of T_1 values (Q/O) for the methyl carbons in the sec-butyl group for the C=8 and C=11 LCs at 75.5 MHz.

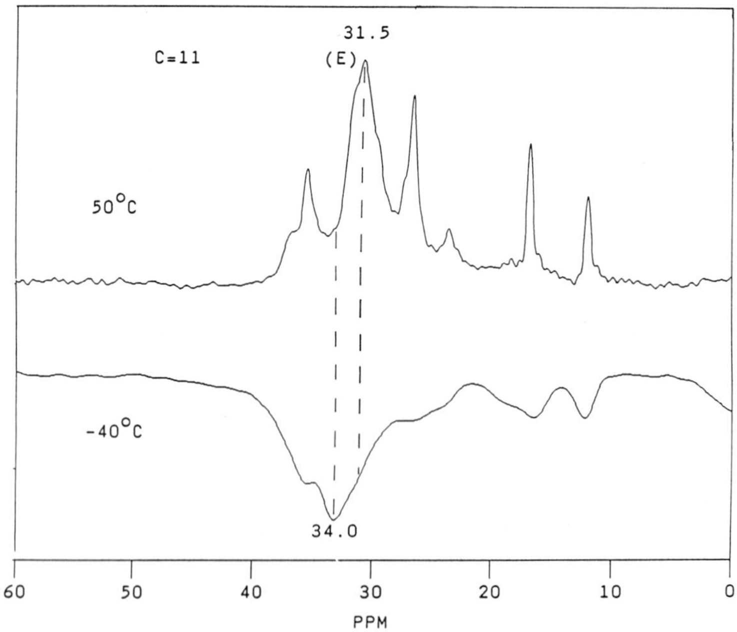

Figure 10. The CP spectra showing the temperature dependent chemical shift of the central spacer methylene carbons, E (31.5 ppm), at 50° C and -40° C for the C=11 LC at 75.5 MHz.

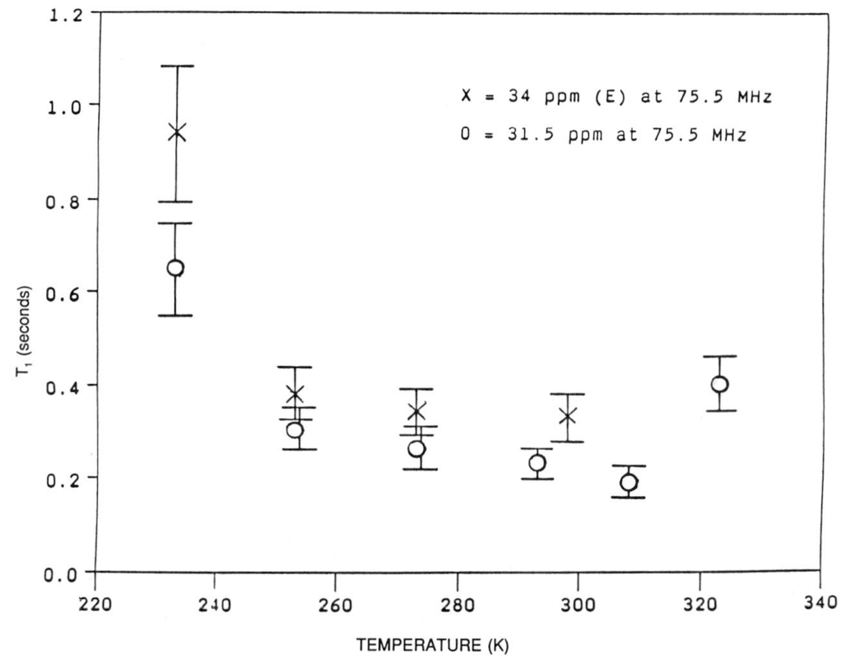

Figure 11. The T_1 values of the two components of the central spacer methylene carbons, E (31.5 ppm), as a function of temperature for the C=11 LC at 75.5 MHz.

The mesogenic methylene carbons become broader when the T_{g2} is approached as observed with CP (Fig. 2). The T_1 values for the 68 and 73 ppm carbons are quite long at low temperatures. There is a decrease by a factor of 8 for the 73 ppm carbons and a decrease by a factor of 23 for the 68 ppm carbons from -40° C to 25° C. This compares with changes by a factor of 2 to 4 for the other methylene carbons. The transition at T_{g2} is associated with a large scale increase in mobility of the rigid mesogenic unit. The T_1 values are not reported for the aromatic carbons, but qualitatively one can see a large increase in relative intensity coupled with rapid inversion-recovery at temperatures above 25° C (Fig. 3). Inversion recovery of the aromatic carbons is not obtained at low temperatures; but at high temperatures, the inversion recovery occurs at approximately 150 ms.

4.3. Methyl Carbons in the C=8 LC

The C=8 LC also has three distinct methyl peaks. There is a field dependence of the measured T_1. The T_1 temperature dependence is similar to the C=11 LC. The backbone methyl exhibits more mobility than those in the sec-butyl group. The same trend within the sec-butyl group is observed, with the terminal methyl being more mobile than the branched methyl including an enhancement above T_{g2} (Fig. 9).

4.4. Methylene Carbons in the C=8 LC

The three methylene carbons in the C=8 spacer exhibit the same trends with field strength and temperature at 37.7 MHz as the previously discussed C=11 LC.

There is a very small temperature dependence at 75.5 MHz contrary to LC I (C=11). There is no evidence of decoupling along the chain at any temperatures using T_1 values. There is, however, a downfield shift of the main spacer peak at lower temperatures. The same shift is observed in the C=11 spacer. The C=8 spacer is an even-numbered chain length but it still appears to have substantial reordering of the central methylenes below T_{g2}. This is the same effect observed by Müller [22] for a ten methylene spacer length chain in a main chain LC. There are two components observed at low temperatures for the 31 ppm spacer peak with the 33 ppm component relaxing slower (Fig. 8A).

The mesogenic and sec-butyl methylenes have the most dramatic change in T_1. The T_1 values decrease when the LC transition temperature is approached with the ring carbons having the largest relative decrease. One also observes a rapid inversion-recovery of the aromatic carbons above the transition temperature (Fig. 4).

4.5. A Comparison of the C=11 and C=8 LC

Overall, the two LC systems exhibit similar trends. In both, the T_{g2} transitions are associated with the large scale increases in the amplitude and/or frequency of the mesogenic units. However, there is evidence of decoupling along the spacer based on the T_1 relaxation times below T_{g2} for the C=11 system but not for the C=8 system. The main spacer peak has a larger number of trans conformers in the glassy state as evidenced by a downfield chemical shift. The two component inversion-recovery of the major spacer peak (C=11, C=8) at low temperatures is observed.

Comparison of the C=11 to the C=8 LC can be done using the same conditions. We chose to use the T_1 values at 0° C and 25° C at 75.5 MHz (Table V). The lower temperature was chosen in order to insure that both systems were below the transition temperature. The carbons associated with the peaks at 25, 31, 68, and 73 ppm are clearly resolved and are in various sections of the LC. At 0° C there are significant differences (greater than 20%) between the mesogenic and sec-butyl carbons. In each case, the more flexible polymer has the shorter relaxation times. The spacer carbons (31 ppm) are identical while the beta spacer carbons (25 ppm) indicate that there is more mobility for the longer spacer.

Table V C-13 T_1 VALUES FOR THE METHYLENE CARBONS OF THE C=11 AND C=8 LCs AT 75.5 MHz, AT 0° C AND 25° C

	0°C		25°C	
ppm	C=8	C=11	C=8	C=11
73	.43	.36	.30	.29
68	.96	.53	.44	.30
31	.28	.26	.37	.27
25	.36	.17	.32	.22

Experimental error is ±15%.

As the T_{g2} is approached (25° C), the mesogenic carbons (68 ppm) are more mobile in the C=11 LC but the differences are smaller. There is a difference in both spacer carbons because the C=11 LC, which is above its T_{g2}, has increased mobility with respect to the C=8 LC.

The differences of the mesogen mobilities at 0° C indicates that lengthening the spacer affects the motions of the mesogenic unit below T_{g2} so that the decoupling of the motions is increased. The enhancement in the mobility of the C=11 spacer at the elevated temperature (25° C) indicates that decoupling is not complete. The differences observed in the beta spacer carbons and the backbone carbons between the C=11 and C=8 LC at 0° C indicates that lengthening the spacer from 8 to 11 has an effect on the mobility of the methylenes near the backbone.

5. Conclusion

Both of the spacers in the C=11 and C=8 LC systems offer significant decoupling of the mesogenic groups and the polymer chain to allow smectic formation. At the DSC transition temperature, large scale increases are observed in amplitude and/or frequency of the motion of the dioxaborinane and aromatic rings in the mesogenic unit which allows "repacking" of the side chains. There is evidence for the decoupling of motions along the spacer in the glassy phase for the C=11 system; but no decoupling is observed in the liquid crystalline state. Below the T_{g2}, a slight chemical shift of the spacer indicates the different conformations which exist above and below the transition. The carbons associated with the LC state have greater relative mobility. A comparison of the two systems at 0° C and 25° C confirms the increase in mobility of the mesogen in the C=11 LC and that the decoupling is not complete.

6. Acknowledgments

I would like to thank Virgil Percec and Bruce Hahn for supplying the polymers that were used in this study. I would also like to acknowledge the MRG for supporting this research under grant (DMB-8614093).

7. References

1. Wendorff, J. H.; Finkelmann, H. and Ringsdorf, H. J., *Polym. Sci., Polym. Symp.*, *63*, 245, 1978.
2. Hahn, B. and Percec, V., *Macromolecules*, *12(20)*, 2961, 1987.
3. Hahn, B. and Percec, V., *Mol. Cryst. Liq. Cryst. Inc. Nonlin. Opt.*, *157*, 125, 1988.
4. Andrew, E. R., *Prog. Nucl. Magn. Reson. Spectrosc.*, *8*, 1, 1972.
5. Pines, A.; Gibby, M. F. and Waugh, J. S., *J. Chem. Phys.*, *56*, 1776, 1972.

6. Schaefer, J., *Topics in Carbon-13 NMR Spectroscopy*, Levy, G. C., Ed., Wiley, New York, 1974; p 149.
7. Sullivan, M. J. and Maciel, G. E., *Anal. Chem.*, *54*, 1606, 1982.
8. Sullivan, M. J. and Maciel, G. E., *Anal. Chem.*, *54*, 1615, 1982.
9. Hsu, C. S.; Rodriguez-Parada, J. M. and Percec, V., *Makromol. Chem.*, *188*, 1017, 1987.
10. Hsu, C.; Rodriguez-Parada, J. M. and Percec, V., *J. Polym. Sci., Part A, Polym. Chem.*, *25*, 2425, 1987.
11. Müller, K. and Kothe, G., *Ber. Bungenges. Chem.*, *89*, 1214, 1985.
12. Samulski, E.; Gauthier, M. M.; Blumstein, R. B. and Blumstein, A., *Macromolecules*, *17*, 479, 1984.
13. Ober, C. K., *Advances in Polymer Science*, *59*, Springer-Verlag, Berlin, 1984; p 104.
14. Shibaev, V. P. and Platé, N. A., *Advances in Polymer Science*, *60/61*, Springer-Verlag, Berlin, 1984; p 175.
15. Perry, B. C., Ph.D. Thesis, Case-Western Reserve University, Cleveland, OH, 1988.
16. Silverstein, R. M.; Bassler, G. C. and Morrill, T. C., *Spectrometric Identification of Organic Compounds*, 4th Ed., John Wiley and Sons, Inc., New York, 1981; p 249.
17. Sindorf, D. W. and Maciel, G. E., *J. Am. Chem. Soc.*, *105*, 1851, 1983.
18. Monnerie, L., *Pure Appl. Chem.*, *57(11)*, 1563, 1985.
19. Oulyadi, H.; Lauprêtre, F.; Mauzac, M.; Richard, H.; Gasparoux, H. and Monnerie, L., *Macromolecules*, submitted for publication.
20. Ando, I.; Yamanobe, T.; Sorita, T.; Komoto, T.; Sato, H.; Deguchi, K. and Imanari, M., *Macromolecules*, *17*, 1955, 1984.
21. Samulski, E. T., *Isr. J. Chem.*, *23*, 329, 1983.
22. Müller, K.; Meier, P. and Kothe, G., *Progress in NMR Spectroscopy*, *17*, 211, 1985.
23. Boeffel, Ch.; Hisgen, B.; Pschorn, U.; Ringsdorf, H. and Spiess, H. W., *Isr. J. Chem.*, *23*, 388, 1983.

Chapter 13

VT-MAS NMR AS A PROBE OF MOLECULAR MOBILITY AND LOCAL COMPOSITION: SOLIDIFIED ACRYLIC DISPERSIONS

R. VOELKEL

Dispersions from emulsion copolymerization of acrylic monomers have been solidified into films. The heterogeneity of these films, molecular composition of the phases and molecular dynamics have been investigated by the use of ^{13}C-CP/MAS-NMR exploiting the $T_{1\rho}(H)$ and $T_{1\rho}(C)$ behavior under variable temperature conditions.

1. Introduction

The copolymerization of an emulsion starts off with a heterogeneous system: the emulsified hydrophobic monomer forms an oil phase whereas the hydrophilic comonomer is dissolved in the water. Polymerization of such an emulsion yields an emulsion of latex particles which can be solidified by evaporating the water. Butyl acrylate and ethyl acrylate homopolymer latices as well as copolymer latices containing 10% (w/w) of acrylic acid or 15% polymerization at 80°C in water. The emulsion containing the monomers and the sodium salt of sulfated alkyl phenoxyethylate as emulsifier was continuously added to the reactor using potassium persulfate to initiate polymerization. The resulting dispersions of 40% of solid material were solidified into films by evaporating the water at room temperature and by finally storing them under vacuum for complete drying. The solid film which is obtained still possesses a considerable amount of heterogeneity [1]. The electron micrograph shown in Fig. 1 was taken from a latex film containing 90% (w/w) of butyl acrylate and 10% of acrylic acid. In this case, acrylic acid is the high-T_g hydrophilic comonomer.

R. VOELKEL - Kunststofflaboratorium, Abt. Polymerphysik, Festkörperphysik, BASF Aktiengesellschaft, D-6700 Ludwigshafen.

Solid State NMR of Polymers, Edited by L. Mathias
Plenum Press, New York, 1991

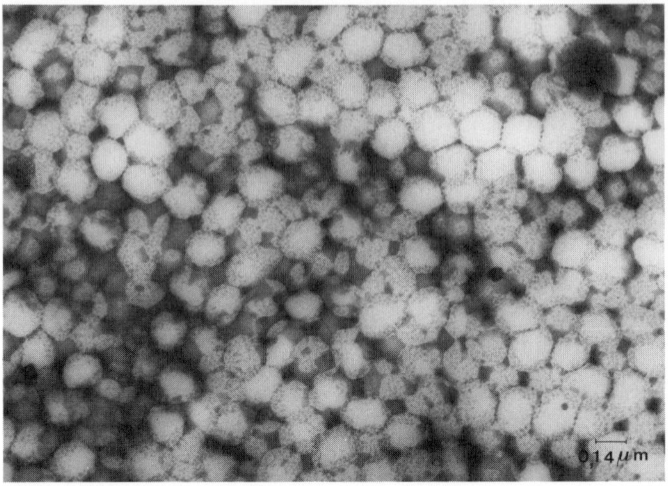

Figure 1. Electron micrograph of a solidified dispersion containing butyl acrylate and 10% (w/w) of acrylic acid. The particle morphology of the dispersion is retained in the film (stained with hydrazine/OsO_4).

The particle morphology of the emulsion is retained in the solid film. It is generally assumed that the particles themselves mainly consist of polymerized hydrophobic monomers (butyl acrylate in this case). The hydrophilic comonomer is preferentially situated among the particles and forms its own phase. Such a two-phase model is a very rough description of the true morphology. We will take a closer look at the system by using solid state NMR spectroscopy and will ask ourselves, how much of the hydrophilic comonomer is contained within the rubbery phase made up primarily by the butyl acrylate in this case. Is part of the hydrophobic monomer built into the hydrophilic polymer? To what extent are the two components segregated? $T_{1\rho}(C)$ and $T_{1\rho}(H)$ are the two quantities which we want to use in order to answer the above questions by means of CP/MAS-NMR.

2. Measuring $T_{1\rho}(H)$ and $T_{1\rho}(C)$

The relaxation time of the carbons in the rotating frame, $T_{1\rho}(C)$, vanes for hydrogen atoms in different positions within the molecule and is governed by the molecular mobility at mid-kilohertz frequency [2]. Contrary to $T_{1\rho}(C)$ the relaxation time $T_{1\rho}(H)$ of the protons is not peculiar to the various positions but is a quantity which is averaged within domains of 1 - 2 nm diameters. $T_{1\rho}(H)$ can therefore be treated as a "phase" quantity, at least in those cases where there is sufficient dipolar coupling among the protons to provide spin diffusion over the range of 1-2 nm [3]. $T_{1\rho}(H)$ was measured via ^{13}C in order to retain the spectral resolution offered by ^{13}C CP/MAS-NMR. We thus correlate the information on chemical structure from ^{13}C-NMR with the phase properties of $T_{1\rho}(H)$ [4].

In order to measure $T_{1\rho}(H)$ we insert a spin-locking period between the proton 90 degree excitation and crosspolarization [5]. It is important to use a fixed time for the ^1H-^{13}C contact for the investigation of heterogeneous systems. Varying the contact-time inevitably also varies the region from which the observed signal originates.

3. Solid Films of Poly(butyl acrylate-co-acrylic acid) Obtained from Emulsion Copolymerization

At 248 K the dipolar interaction among the protons in the poly(butyl acrylate) homopolymer is sufficiently large that all signals experience the same $T_{1\rho}(H)$ which therefore can be called a "phase" quantity (Fig. 2). In heterogeneous systems $T_{1\rho}(H)$ will render us information on the location of a certain molecule which we identify by its chemical shift.

At 248 K the poly(butyl acrylate) homopolymer possesses a $T_{1\rho}(H)$ of 2 ms compared to 5.8 ms of poly(acrylic acid). Both values were measured at a spin-lock field of 50 kHz. The film obtained from copolymerization of an emulsion of butyl acrylate and acrylic acid shows a $T_{1\rho}(H)$ of 3.2 ms (Fig. 3). This value which is

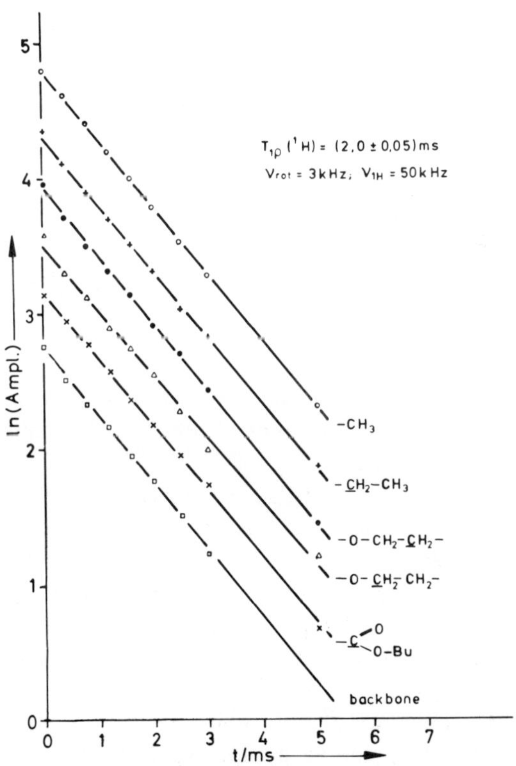

Figure 2. Response of the signal amplitude (arbitrary units) of the various positions in poly(butyl acrylate) to a ^1H prelock [5] of duration *t* prior to a fixed ^1H-^{13}C contact (1 ms). At 248K all positions experience the same $T_{1\rho}(H)$.

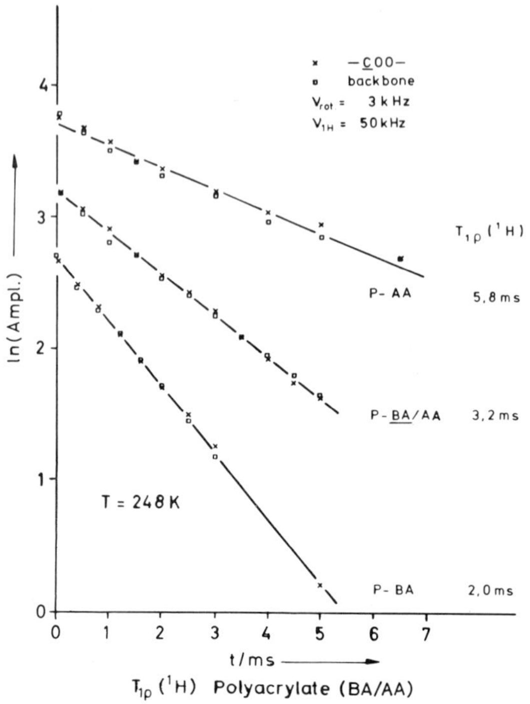

Figure 3. Comparison of the $T_{1\rho}(H)$ behavior of the hompolymers poly(acrylic acid) (P-AA), and poly(butyl acrylate) (P-BA), and a polymer obtained from emulsion copolymerization of butyl acrylate and acrylic acid (P-BA/AA with 10% AA w/w; carboxyl carbon (x) and backbone (□), amplitudes shifted arbitrarily). $T_{1\rho}(H)$ of P-BA/AA lies between those of the homopolymers.

intermediate between the $T_{1\rho}(H)$ relaxation times of the homopolymers indicates that there is in fact some acrylic acid within the phase rich in butyl acrylate.

We now want to determine what effect the vicinity of butyl acrylate and acrylic acid units has on the motion of the butyl acrylate. The relaxation time $T_{1\rho}(C)$ of the ^{13}C in the rotating frame offers us the required information [2,6].

At 283 K, which is above the glass-transition temperature, the $T_{1\rho}(C)$ of the butyl positions in the emulsion copolymer is shorter (Fig. 4, open circles) than in the homopolymer (Fig. 4, crosses). Above the $T_{1\rho}(C)$ minimum a shorter $T_{1\rho}(C)$ indicates an impediment of motion. Copolymerizing butyl acrylate with the much higher T_g acrylic acid should in fact restrict the motion of the butyl acrylate. At 283 K we are indeed above the $T_{1\rho}(C)$ minimum as shown by a variable temperature study (Fig. 5).

$T_{1\rho}(C)$ changes considerably along with the temperature as we would expect for a relaxation which is not governed by spin-spin processes but by motion. What therefore determines the change of $T_{1\rho}(C)$ is the spectral density at the spin-locking frequency, $\omega_{1c}/2\pi = 25$ kHz. At room temperature $T_{1\rho}(C)$ decreases along the butyl chains from the methyl group towards the backbone corresponding to the decrease in spectral density at 25 kHz. An increase in linewidth at the $T_{1\rho}(C)$ minimum [7] furthermore corroborates the interpretation that the changes of $T_{1\rho}(C)$ are caused by molecular motion.

The $T_{1\rho}(C)$ minimum is situated at about 30 K above the glass-transition temperature T_g as measured by differential scanning calorimetry. This difference is due to the mid-kHz frequency employed in the $T_{1\rho}(C)$ NMR experiment. If we compare the position of the $T_{1\rho}(C)$ minimum in the copolymer to that of the poly(butyl acrylate) homopolymer, a shift of about 10 K will be noticed (Fig. 6). This corresponds to a shift in the glass-transition temperature. The dependence of T_g on copolymer composition can easily be measured for non-segregated samples which are obtained from copolymerization in solution. We use this known relation between T_g and copolymer composition to estimate from the shift of the $T_{1\rho}(C)$ minimum what amount of acrylic acid comonomer units is contained within the poly(butyl acrylate) latices, i.e. one region of the partially segregated sample prepared by emulsion copolymerization. The observed shift of 10K of the $T_{1\rho}(C)$ minimum corresponds to a content of about 5% of acrylic acid within the poly(butyl acrylate) latices which is half of the acrylic acid employed in the semi-batch emulsion copolymerization.

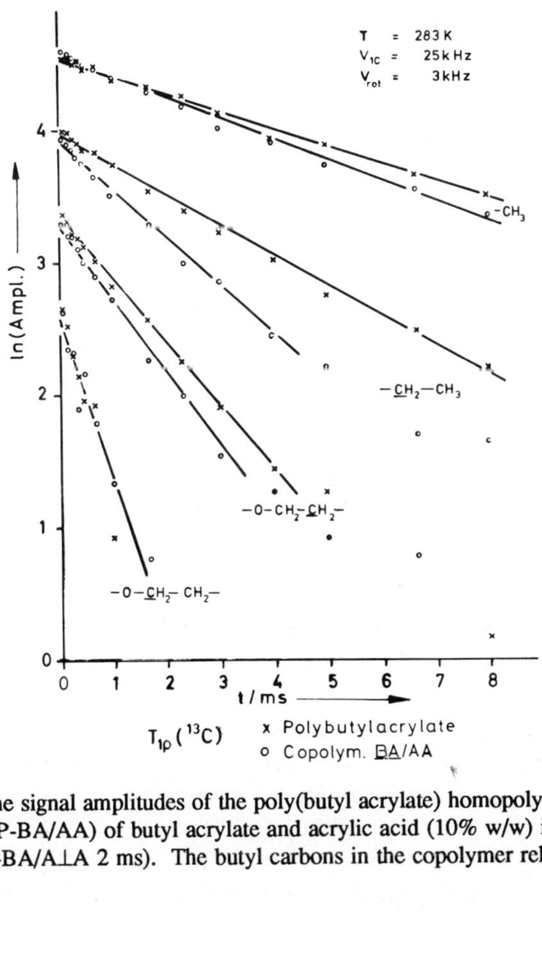

Figure 4. Decay of the signal amplitudes of the poly(butyl acrylate) homopolymer (P-BA, x) and the emulsion copolymer (P-BA/AA) of butyl acrylate and acrylic acid (10% w/w) in a $T_{1\rho}(C)$ experiment (CP P-BA 0.75 ms, P-BA/A⊥A 2 ms). The butyl carbons in the copolymer relax faster than those in the homopolymer.

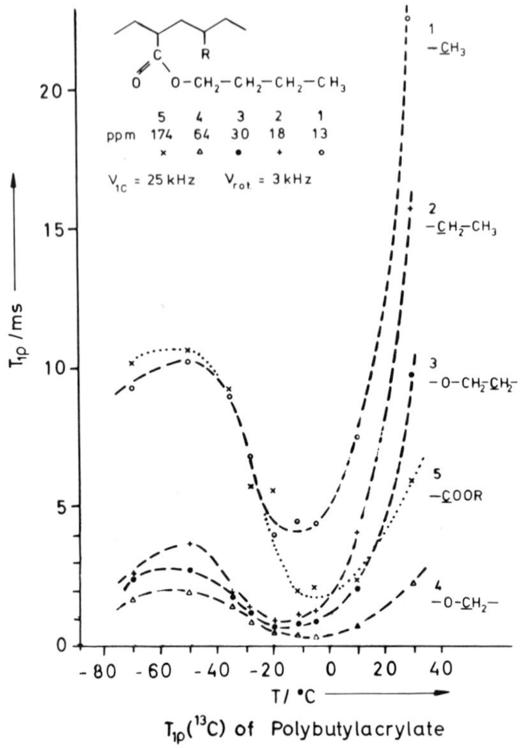

Figure 5. Variable temperature study of $T_{1\rho}(C)$ of the various positions in poly(butyl acryate) showing a $T_{1\rho}(C)$ minimum around -15°C.

The $T_{1\rho}(C)$ of the signal from the carboxylic ester group is longer at the minimum of $T_{1\rho}(C)$ for the copolymer than for the homopolymer (3.2 ms vs. 2 ms). This is not the case for any other position in the copolymer. The explanation may be an increase in anistropy in motion due to hydrogen bonds in the copolymer.

Additionally the decay of $T_{1\rho}(C)$ of the emulsion copolymer is not simply exponential at low temperatures (Fig. 7). As $T_{1\rho}(H)$ stays exponential (Fig. 3), we assume that there is considerable dynamic heterogeneity in the latex phase of the copolymer.

4. Solid Films of Poly(ethyl acrylate/acrylamide) Obtained from Emulsion Copolymerization

How much of the comonomer and whether any of the comonomer at all is contained in the latex phase very much depends on the type of monomers employed. Figure 8 shows data on a latex film from copolymerization of an emulsion of ethyl acrylate and acrylamide (85/15% w/w). $T_{1\rho}(H)$ at -10° C in the surroundings of the ethyl acrylate molecules is practically the same as the one in the homopolymer. This observation indicates that the motions which cause the $T_{1\rho}(H)$ relaxation are the same in the homopolymer as in the emulsion copolymer. The poly(ethyl acrylate) latex particles are more or less pure homopolymeric ones even in the emulsion copolymer. Neither $T_{1\rho}(H)$ nor $T_{1\rho}(C)$ produce much evidence that reasonable amounts of acrylamide are contained within the latex particles formed by poly(ethyl acrylate). This is contrary to the findings for the poly(butyl acrylate-co-acrylic acid) copolymer in which half of the high T_g comonomer employed is located in the latex phase.

So far we have discussed whether there is any content of hydrophilic comonomer within the hydrophobic latex particles. As a last point we now want to reverse the question: is there any hydrophobic monomer within the hydrophilic polymer? Or to be more precise, is there any ethyl acrylate within the poly(acrylamide) regions between the particles?

In order to find out about the content of ethyl acrylate in the glassy interparticulate phase made up primarily of poly(acrylamide), we repeat the measurement of $T_{1\rho}(H)$ at 283 K. At 283 K the $T_{1\rho}(H)$ of the protons as measured via ^{13}C using a fixed contact time clearly shows a two-phase behavior (Fig. 9). There is a very steep decay at short spin-locking periods whereas at longer spin-locking times the decay of magnetization parallels that of the pure poly(acrylamide). A

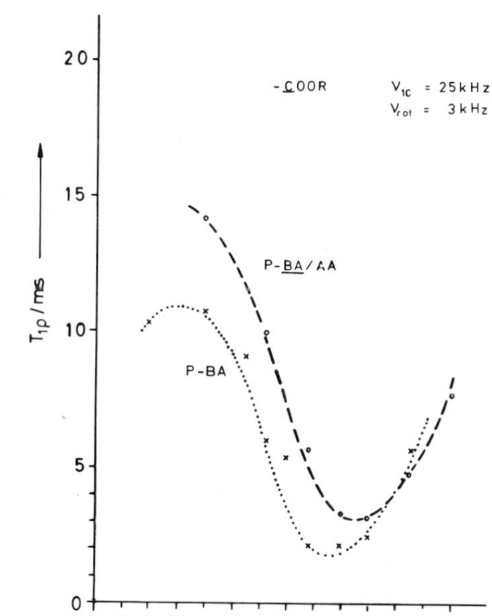

Figure 6. $T_{1\rho}(C)$ of the carboxyl carbon of poly(butyl acrylate-co-acrylic acid) from emulsion copolymerization (P-BA/AA, 10% AA w/w, o) and of the poly(butyl acrylate) homopolymer (P-BA, x) at variable temperature. Note the shift of about 10K in the position of the $T_{1\rho}(C)$ minimum due to the comonomer.

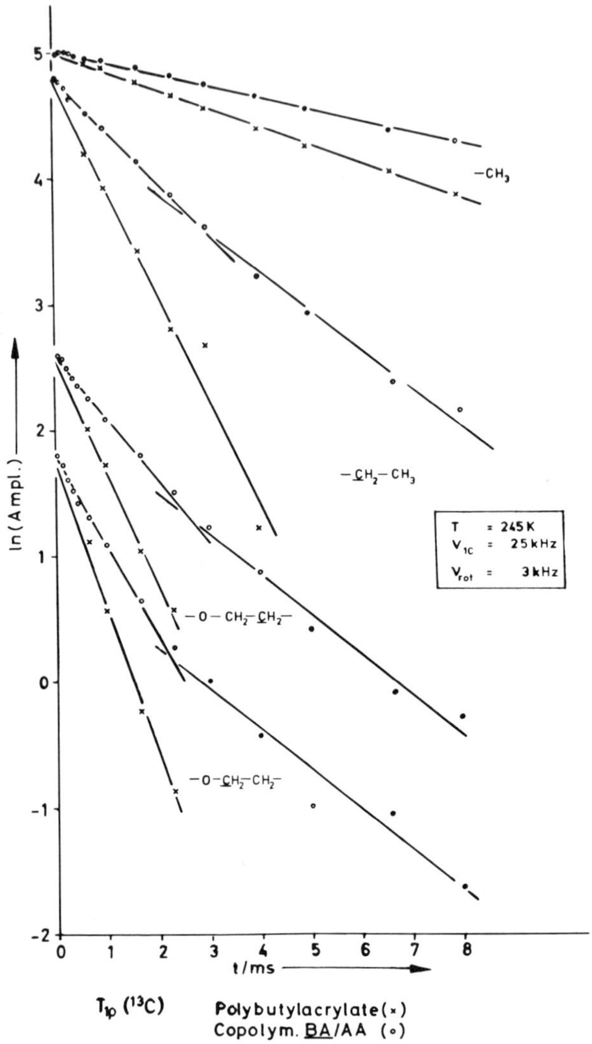

Figure 7. Decay of the signal amplitudes of the various positions in a $T_{1\rho}$(C) experiment at 245K. The copolymer (P-BA/AA; o) from emulsion copolymerization decays non-exponentially unlike to the homopolymer (P-BA, x).

decomposition of the $T_{1\rho}$(H) data into two components yields a short $T_{1\rho}$(H) at short spin-locking times which is perfectly in line with that of the poly(ethyl acrylate) homopolymer. Thus the steep decay at short spin-locking times originates from ethyl acrylate within the latex particles. At longer spin-locking times the decay of $T_{1\rho}$(H) of the ethyl signals in the copolymer follows that of poly(acrylamide).

By extrapolation to zero-time and knowledge of the amount of acrylamide employed we can estimate the amount of ethyl acrylate which is magnetically coupled to the glassy phase. It is only about 10-15 percent of the ethyl acrylate. Nevertheless this small portion is detected at 283 K, because the majority of ethyl acrylate contained in the latex particles cross-polarizes poorly and possesses a very short

$T_{1\rho}(H)$ (1.2 ms at ν_{1H} = 50 kHz). We therefore preferentially observe the signal from those 15 percent of the ethyl acrylate molecules which profit from the much longer $T_{1\rho}(H)$ (4.3 ms) of the glassy phase by spin diffusion.

As $T_{1\rho}(H)$ discriminates at a scale of 1 - 2 nm there must be some ethyl acrylate units within this range towards the glassy interparticulate phase. Either the ethyl acrylate units of the long $T_{1\rho}(H)$ are molecularly mixed into the acrylamide region (we would then expect an influence on the molecular mobility of the ethyl acrylate which should be detectable by $T_{1\rho}(C)$) or the ethyl acrylate molecules of the long $T_{1\rho}(H)$ reside in the outer layers of the latex particles. $T_{1\rho}(C)$ should then be the same as inside the particles themselves. Thus measuring $T_{1\rho}(C)$ of the various regions selectively should allow for distinction between the two possibilities.

The necessary spatial selectivity of the $T_{1\rho}(C)$ measurement is accomplished by an appropriate preparation. Excitation of the ^{13}C signal by a 90 degree pulse will pick up the signal from the highly mobile poly(ethyl acrylate) within the latex particles. A ^1H "pre-lock" of 2 ms realized by cross-polarization of 2 ms before we start the spin-locking of ^{13}C creates the $T_{1\rho}(C)$ of those 10 - 15 percent of ethyl

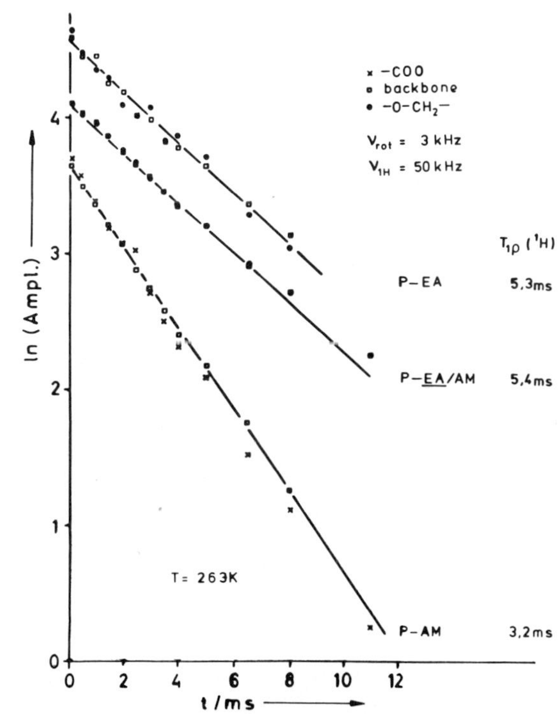

Figure 8. $T_{1\rho}(H)$ decay at 263 K (^1H prelock measurement, CP 1 ms) of poly(ethyl acrylate) (P-EA) and poly(acrylamide) (P-AM) homopolymers and the copolymer latex film (P-DA/AM, 15% w/w AM). No difference can be seen between P-EA/AM and P-EA (see Fig. 3 for comparision with the P-BA/AA system). At 263K the latex particles are frozen in and all EA is detected in the CP experiment, with the latex particles contributing most to the EA signal.

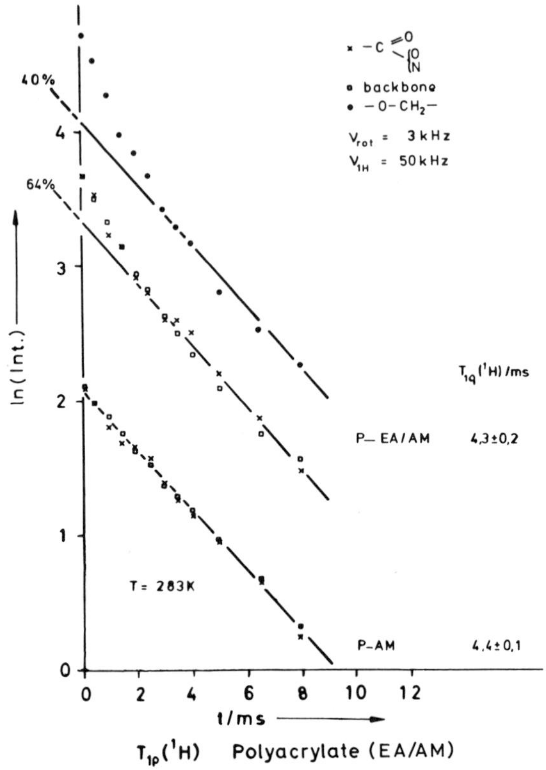

Figure 9. $T_{1\rho}(H)$ of P-EA/AM copolymer and pure P-AM at 283K. 40% of the detected ethyl signal has the $T_{1\rho}(H)$ of the glassy phase. (This 40% of the signal detected by CP corresponds to 15% of *all* ethyl groups, as ethyl within the latex particles at this temperature only shows up at 1/4 the intensity which would correspond to its true amount.)

acrylate units which contribute to the long tail of the $T_{1\rho}(H)$ decay. Results are shown in Figure 10. 90-degree excitation (o) yields a $T_{1\rho}(C)$ decay very similar to that of the poly(ethyl acrylate) homopolymer (□).

Cross-polarization of 2 ms (x) produces a slowly declining curve in the $T_{1\rho}(C)$ plot (Fig. 10). The latter (x) is obviously different from that of the bulk of poly(ethyl acrylate) inside the latex particles (o). The ethyl acrylate units which contribute to the long component of $T_{1\rho}(H)$ are restricted in motion. They are located in a spatially distinct region of different molecular composition. Solidification of a mixture of the two homopolymeric dispersions does not produce a two-component $T_{1\rho}(H)$. We therefore conclude that approximately 10 percent of ethyl acrylate copolymerizes with acryl amide and is located inside the glassy phase.

5. Synopsis

Dispersions from emulsion copolymerization of acrylic monomers have been solidified into films. The heterogeneity of these films, molecular composition of the phases, and molecular dynamics have been investigated by using ^{13}C-CP/MAS-NMR: $T_{1\rho}$(H) of the protons as measured via ^{13}C enables us to decide on the composition of the latex particles. The difference between the $T_{1\rho}$(H) of the latex particles and that of the glassy phase has also been used to select the signals from the interparticulate regions. We thus detect those low T_g monomers selectively which are built into the glassy interparticulate phase. Additional measurements of $T_{1\rho}$(C) of the carbons at variable temperature reflect the influence of the comonomer on molecular mobility. A shift of the minimum of $T_{1\rho}$(C) vs. temperature curve allows for a quantitative determination of the amount of the high-T_g comonomer contained in the latex particles.

In poly(butyl acrylate-co-acrylic acid) copolymers (90/10% w/w) produced by emulsion copolymerization approximately half of the acrylic acid employed is found within the latex particles. The non-exponential decay of $T_{1\rho}$(C) of the copolymer at low temperature indicates dynamic heterogeneity. By use of $T_{1\rho}$(H) it can be shown that this dynamic heterogeneity stems from the latex particles.

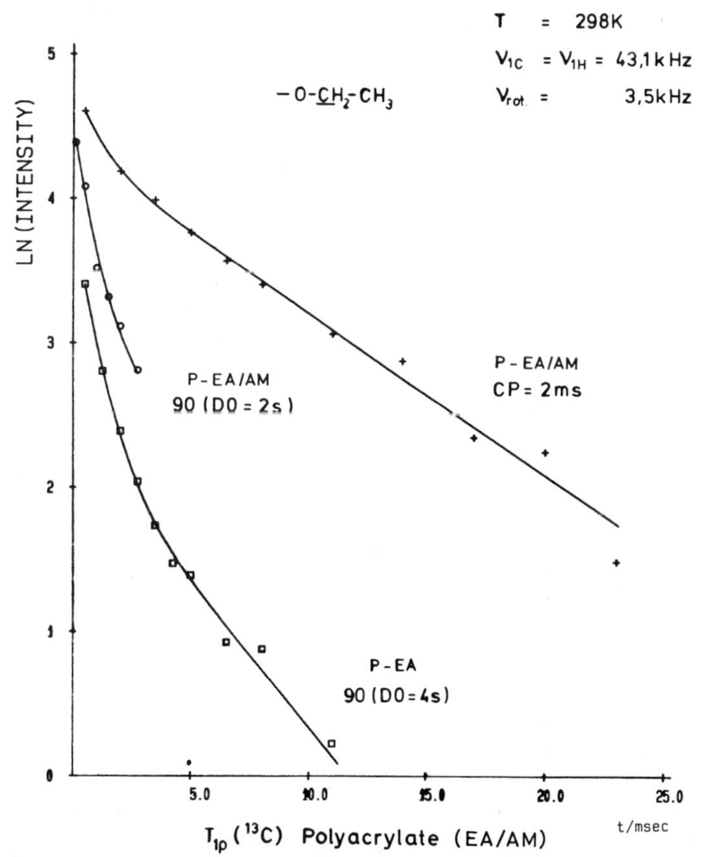

Figure 10. $T_{1\rho}$(C) of the methylene carbon in the ethyl group of P-EA/AM using direct excitation (90°, o) and cross-polarization (CP 2 ms, x) at 298K. $T_{1\rho}$(C) within the latex particles (90°, o) is very similar to the homopolymer (P-EA, □) and much shorter than $T_{1\rho}$(C) of the ethyl groups in the glassy interparticulate regions observed by CP (x).

Neither $T_{1\rho}(C)$ nor $T_{1\rho}(H)$ indicate any incorporation of acrylamide into the latex particles of an ethyl acrylate/acrylamide (85/15% w/w) copolymer from emulsion copolymerization. Yet 15 percent of the ethyl acrylate is contained within the glassy interparticulate phase made up primarily of acrylamide.

These very detailed results demonstrate the utility of CP/MAS-NMR measurements when characterizing latex films of acrylic copolymers obtained by emulsion copolymerization.

6. References

1. Zosel, A; Heckmann, W.; Ley, G. and Mächtle, W., *Coll. Polym. Sci.*, *265*, 113, 1987.
2. Schaefer, J.; Stejskal, E. O. and Buchdahl, R., *Macromolecules*, *10*, 384, 1977.
3. McBrierty, V. J. and Douglas, D. C., *J. Polym. Sci., Macromol. Rev.*, *16*, 295, 1981.
4. Schaefer, J.; Sefcik, M. D.; Stejskal, E. O. and McKay, R. A., *Polym. Prepr.*, *20*, 247, 1979; *Macromolecules*, *14*, 275, 1981.
5. Aujla, R. S.; Harris, R. K.; Packer, K. J.; Parameswaran, M.; Say, B. J.; Bunn, A. and Cudby, M. E. A., *Polym. Bull.*, *8*, 253, 1982.
6. VanDerHart, D. L. and Garroway, A. N. *J. Chem. Phys.*, *71*, 2773, 1979; Schaefer, J.; Stejskal, E. O.; Steger, T. R.; Sefcik, M. D. and McKay, R. A., *Macromolecules*, *13*, 1121, 1980.
7. Laupretre, F.; Monnerie, L. and Virlet, J. *Macromolecules*, *17*, 1397, 1984.

Chapter 14

MOLECULAR MOTION IN CRYSTALLINE AND MESOMORPHOUS PHASES OF LARGE SIZE CYCLOALKANES AS A MODEL FOR POLYETHYLENE

MARTIN MOLLER, GERHARD KOGLER, DIETER OELFIN AND HANSOTTO DROTLOFF

1. Abstract

Cycloalkanes with 22 to 96 units crystallize in lamellae, where two adjacent stems are connected by two tight folds. Before melting, they form mesophases, which can be described as conformationally disordered crystals (Condis crystals). Conformational disorder and molecular mobility in the solid and in the mesomorphic state have been studied using MAS ^{13}C-NMR spectroscopy and ^{2}H-NMR spectroscopy. In the case cyclodoeicosane and cyclotetraeicosane, MAS ^{13}C-NMR spectra show that slow jump type reptation results in an exchange of the carbon atoms between stems and folds already in the crystalline state. Conformational disorder and fast molecular motion is observed for the mesophase. The exchange of methylene groups can occur with rates which are larger than $10^{8}s^{-1}$. The molecular dynamics in the mesophase is described by fluctuation of additional gauche conformations. The passage of a gauche-trans-gauche defect along the chain leads to small translational motion of the chain segment and can thus explain the translational diffusion of methylene groups from the inner part of the lamellae to the folds and vice versa. This observation of fast chain reptation through the lamellae may help to understand the formation of extended chain crystals in polyethylene.

MARTIN MOLLER, GERHARD KOGLER, DIETER OELFIN, AND HANSOTTO DROTLOFF - Institut fur Makromolekulare Chemie, Universitat Freiburg, Hermann-Staudinger-Haus, Stefan Meier Str. 31, 7800 Freiburg, Federal Republic of Germany.

Solid State NMR of Polymers, Edited by L. Mathias
Plenum Press, New York, 1991

2. Introduction

Flexible chain molecules, such as polytetrafluorethylene [1,2], trans-1,4,-polybutadiene [3,4], poly-p-xylylene [5], polydiethylsiloxane [6,7,8], polyethylene [9,10,11], the polyphosphazenes [12], n-alkanes [13,14] and cycloalkanes [15,16,17,18,19] show solid-solid transitions from an ordered crystal to a state of intermediate order. Because of the absence of mesogenic units in the flexible molecules and the fact that the molecules are packed regularly in three dimensions, these mesophases should not be described as liquid crystals. The molecules are flexible and form lamellar crystallites. Typically, the chains within the lamellae are packed hexagonally in the high temperature phase. This implies an axial symmetry around the long axis of the stems, irrespective of the molecular structure.

For the polymers and cycloalkanes, chain folding and the resulting re-entry of the molecules in the lamellae prevents a rotation of the whole molecule, as was proposed for the n-alkanes [13].

A conformational disordered state is proposed for mesophases of flexible chain molecules [20,21]. In that case, the axial symmetry around the stems in the crystallite is the result of complex segmental molecular motions. The formation of a hexagonal packing, caused by dynamic conformational disorder is drawn in Fig. 1.

A special phenomenon, observed for the mesophase of polyethylene is the formation of extended chain crystals of a crystallinity close to 100% as shown schematically in Fig. 2 [22]. Translational motion of the chain segments through the lamellae is required to explain the conversion of folded chain crystals (FCC) to extended chain crystals (ECC) without melting [23].

Cycloalkanes with more than 20 methylene units form lamellae crystals. Because of the ring structure, adjacent stems in the lamellar are connected by two folds. Although the chain folding is certainly much more regular than in polymer crystllites, the cycloalkanes represent well defined model compounds for the investigation of the solid state structure and dynamics of flexible polymers. Recently,

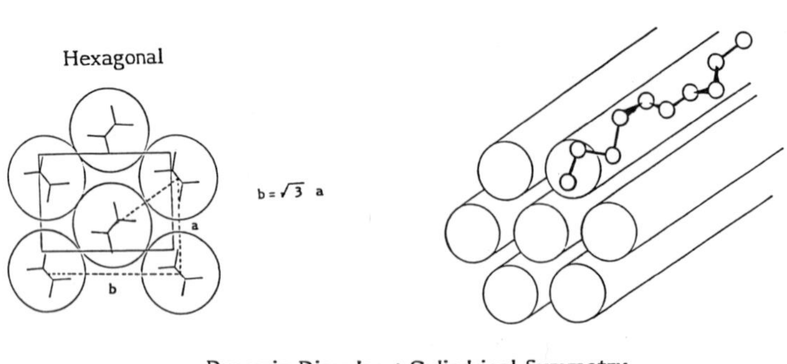

Figure 1. CONDIS CRYSTALS-Hexagonal packing of flexible chain molecules within the condis state.

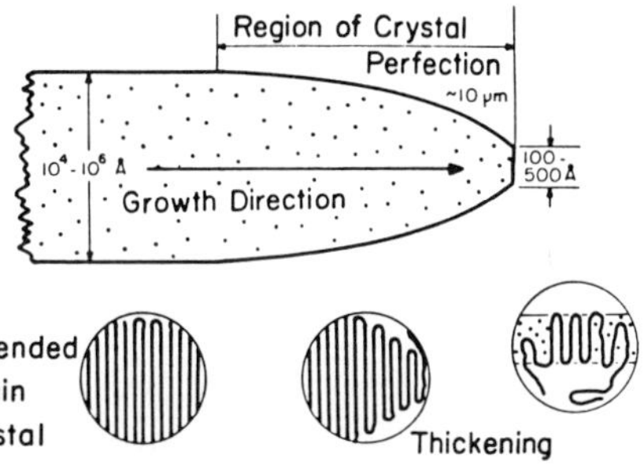

Figure 2. Lamellar thickening of polyethylene when crystallized in the mesophase according to [16].

we reported calorimetric data and variable temperature MAS ^{13}C NMR spectra of cycloalkanes with 12 to 96 methylene groups [16,19]. Cycloalkanes with 12, 14, and 24 C-Atoms show a thermal transition prior to melting. Intermediate size cycloalkanes, e.g. $C_{48}H_{96}$, do not show a phase transition below the melting point. The thermal behavior of large cycloalkanes with 72 and 96 C-Atoms depends on the crystallization conditions. Only melt crystallized samples show a solid-solid transition. Solution crystallized samples exhibit a monoclinic subcell, while within the melt crystallized samples, the methylene units are arranged orthorhombically. The solid-solid transition corresponds to a change from orthogonal to hexagonal packing [24]. In the case of the smaller cycloalkanes, the results clearly showed the onset of conformational interconversions below the disordering transition [25]. In the present paper, we want to discuss variable-temperature, solid state ^2H NMR and MAS ^{13}C NMR experiments on cyclodoeicosane, cyclotetraeicosane, cyclodoheptacontane and cyclohexanonacontane.

3. Experimental

Synthesis and purification of the partially deuterated samples cyclotetraeicosane, cyclooctatetracontane, cyclodoheptacontane and cycloexanonacontane has been achieved according to References 27, 16. Defined crystallization of the large size cycloalkanes was essential to obtain homogeneous samples. Solution crystallized samples of C_nH_{2n} with n= 48, 72, and 96 were prepared by slow (0.35 K/h) cooling of a dilute solution in butyl acetate. Melt crystallized samples with n = 72 and 96 were achieved by slowly cooling from the melt (rate < 0.35 K/h). ($C_{22}H_{44}$ was generously donated by Dr. R. G. Snyder, Berkeley, California.)

DSC data were determined with a Perkin-Elmer Model DSC 7. For experimental details see reference 16. Temperature accuracy was better than 0.3 K.

X-Ray diffraction powder patterns were recorded with a Simon-Guinier camera, Enraf-Nonius FR 533. The well characterized low temperature modification of cyclodoeicosane at -160° C [26] was used as a standard.

High resolution solid-state ^{13}C NMR spectra were recorded at 75.47 MHz on a Bruker CXP 300 spectrometer. For further details see reference 28.

^{2}H NMR spectra were recorded on a Bruker CXP 300 Spectrometer operating at 46.073 MHz. The quadrupole echo sequence was employed. The delay between the two 90° pulses (3μs) was varied from 20 to 100 μs [29]. Recycle time was varied between 30 s at low temperatures and 1 s in the mesophase. Spectra were obtained with a 833 kHz spectra width, accumulating from 500 to 3000 scans according to the signal/noise ratio and folded about the center frequency.

4. Results and Discussions

Table I shows the transition temperatures and transition entropies per methylene unit for C_nH_{2n} with n = 22, 24, 72 and 96 [17,18]. As mentioned above, the thermal behavior of cyclodoheptacontane and cyclohexanonacontane depends on the crystallization conditions. Monoclinic packing is observed for solution crystallized samples. They do not show any phase transition below the melting point. Melt crystallized samples are packed orthorhombically and show a solid-solid transition to a hexagonal high temperature phase [24]. The smaller rings are always packed in a monoclinic lattice. Samples with up to 24 methylene groups also showed a disordering transition below the melting point. Intermediate size cycloalkanes, $C_{36}H_{72}$ and $C_{48}H_{96}$ are packed monoclinically and behave similarly to solution crystallized large size rings.

Figure 3 demonstrates the structure observed for $(CH_2)_{24}$ at -160° C [26]. The molecules adopt a rectangular shape in which two parallel all-anti stems are connected by tight folds of a ggagg conformation. Four molecules are packed in a monoclinic

Table I TRANSITION TEMPERATURES AND TRANSITION ENTROPIES OF A SERIES OF CYCLOALKANES:

1. Solution Crystallized
2. Melt Crystallized

n	T_{tr} K	$\Delta S_{tr}/CH_2$ J/K•mol	T_m K	$\Delta S_m/CH_2$ J/K•mol
22	298.5	5.4	322.0	1.3
24	299.5	5.4	323.0	1.4
48	---	---	362.5	7.9
72[1]	---	---	381.2	8.9
72[2]	352.8	1.1	381.0	4.2
96[1]	---	---	391.2	8.4
96[2]	362.4	1.1	389.5	4.8

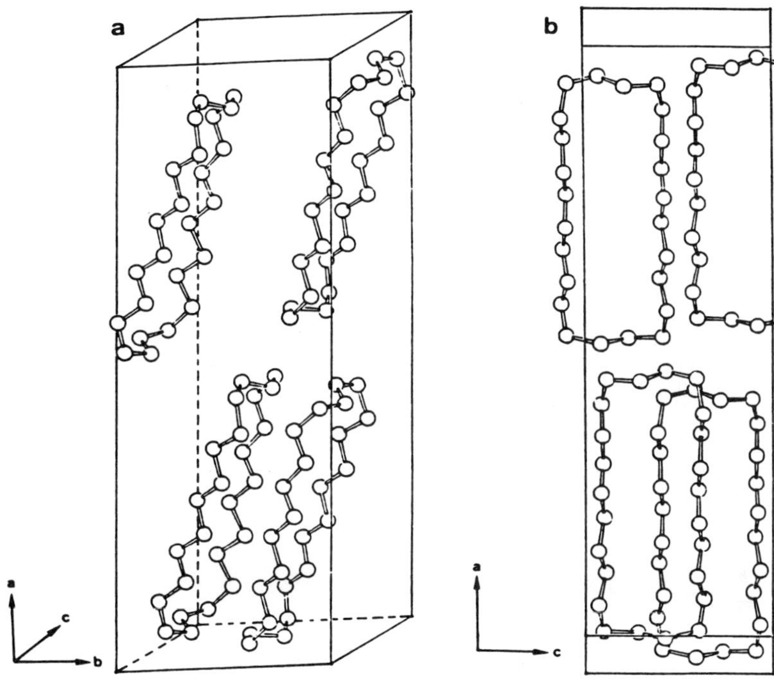

Figure 3. Structure of cyclotetraeicosane below 299 K; a: perspective plot; b: projection on the (010) plane.

sub-cell forming a layered structure in which the stems are inclined towards the normal of the lamellar surface. ($\beta = 98.2°$). In the mesomorphic state above 299 K, two molecules are packed in an orthorhombic unit cell with a = 1.0997 nm, b = 1.2388 nm, and c = 0.9222 nm [30]. Thus, the coupling of the adjacent layers is lost, but the molecular structure is still that of two parallel stems connected by tight folds. The stems form the long axis of the molecules, which is parallel to the b axis. The b-axis is by 0.14 nm shorter than the half value of the a-axis in the monoclinic low temperature cell. This corresponds to a decrease of 5.5% in the length of the stems. A similar decrease in the length of the stems has been observed for $(CH_2)_{22}$ at the transition at 298 K [31]. It could be shown by infrared experiments that the number of gauche bonds in the $(CH_2)_{22}$ structure increases at the solid-solid transition from 8 to 10 [32]. Thus the decrease in the length of the stems is explained by one gauche arranged carbon-carbon bond per stem. This is also shown in the high resolution solid-state ^{13}C NMR spectra [33]. In addition, the NMR spectra indicate the dynamic character of the disordering.

Figure 4 shows the temperature dependence of the ^{13}C NMR spectra of cyclodoeicosane. At low temperatures, conformationally different sites of the methylene carbons can be resolved as sharp signals in the spectrum. The resonances in the 220 K spectrum can be assigned to the different four bond carbon segments (aa.aa, aa.ag, aa.gg, ag.ga, ga.gg) within the crystal conformation [28]. Line

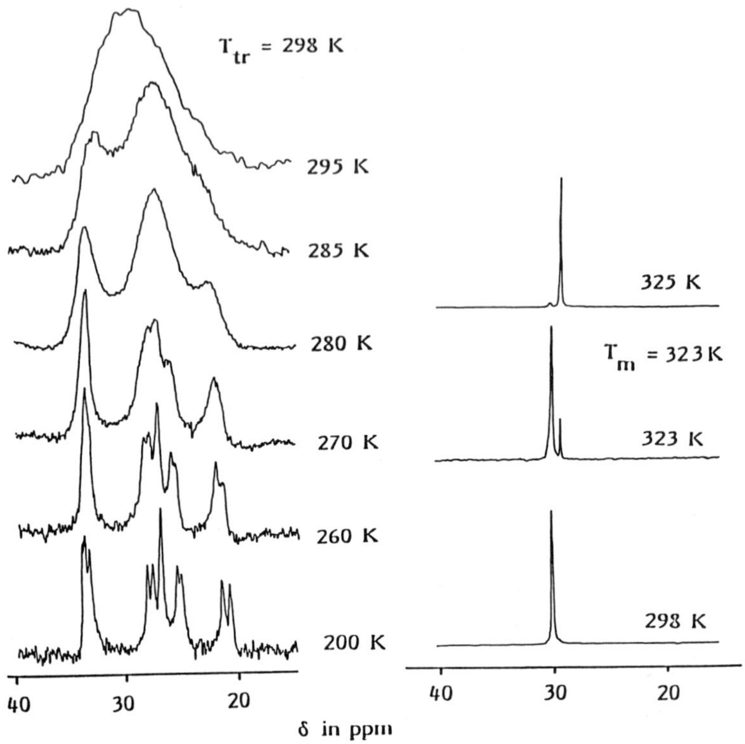

Figure 4. Temperature dependence of MAS ^{13}C NMR spectra of cyclodoeicosane.

broadening and the coalescence to an averaged signal is observed on raising the temperature from below the transition to the mesomorphic phase. Heat capacities below T_d did not indicate any disordering process. Thus, the exchange of the carbons between the conformationally distinguished sites must result from a jump-type motional process via transition states which are little populated [25]. Above the solid-solid transition only one signal is observed, which is narrow as if recorded from the melt.

Simulation of the slow- to fast-exchange transition in the ^{13}C NMR spectra for the exchange between the different conformational sites below the phase transition is presented in Fig. 5: ^{13}C NMR lines were calculated assuming that all carbons exchange with the same rate between the different sites. The points give the experimental data, the dotted lines give the calculated spectra, where the exchange rate is the only fitted parameter. The rates are slow, increasing from a few to about 1000 Hz until T_d is reached. Plotting the rate constant versus T^{-1}, a line with a constant slope is obtained as presented in Fig. 6. From the temperature dependence of the exchange rates, an activation energy of ΔE=62 kJ/mol is calculated.

Due to a jump type increase of the efficiency and the rates of the motional processes, one sharp, liquid type, resonance line is recorded above T_d. The perfect

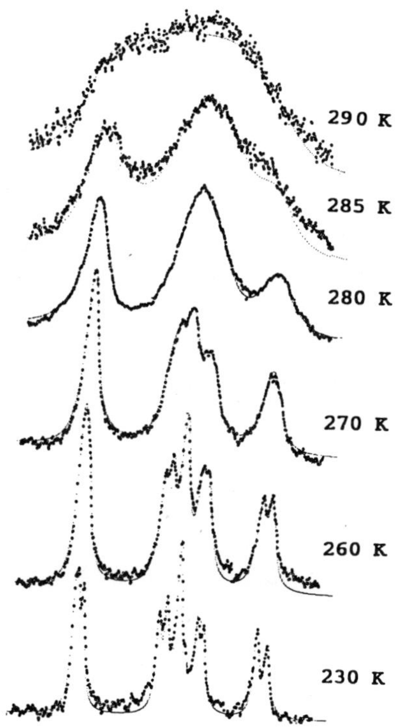

Figure 5. Experimental and calculated ^{13}C NMR spectra of cyclodoeicosane below the solid-solid transition.

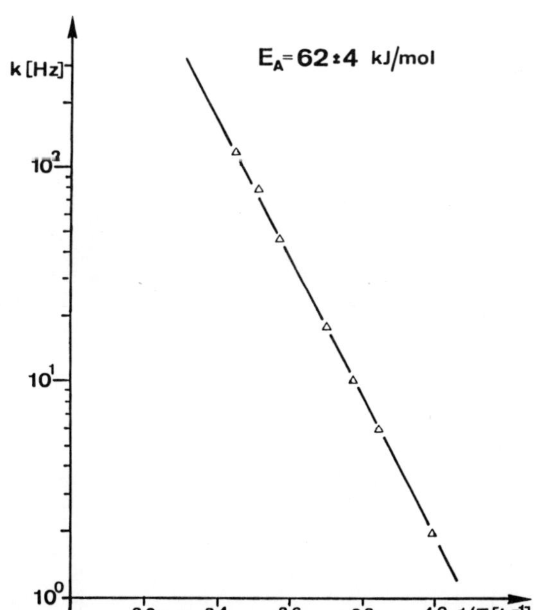

Figure 6. Plot of the rate constants from the simulation of the ^{13}C NMR spectra versus T^{-1}. The slope gives the listed activation energy.

averaging of the ^{13}C resonance above the solid-solid transition indicates fast diffusive reptation of the methylene strings through the folds and stems.

In conclusion, the picture in Fig. 7 emphasize the difference in the molecular mobility below and above the solid-solid transition. Below the solid-solid transition, jump type interconversion processes occur in the fully ordered crystal and result in an exchange of the methylene units between the conformationally different positions of the ring. Above the phase transition, the increase of specific volume and the formation of a partially disordered structure results in additional molecular mobility leading to a jump type increase of the rates of the motional processes.

Above the phase transition, the molecular mobility was also investigated by means of ^2H NMR experiments.

Solid state ^2H NMR of powder samples give the well known Pake spectrum with a quadrupole splitting of $\Delta\upsilon_{xx}$=125 kHz. Molecular motion leads to an averaging of the quadrupole interactions, resulting in decreased line widths. In the region of the fast exchange, $\tau<10^{-8}$ s, line width and line shape of the spectrum depend only on the type of molecular motion. Increasing velocity of the motions do not result in an additional decrease of the line width. Molecular motion with correlation times $\tau<10^{-4}$ s are slow with respect to the time resolution of the ^2H NMR spectrum and can not be observed by a change of the resonance line. In the intermediate range 10^{-4} s > τ > 10^{-8} the averaging is not complete and the ^2H NMR resonances depend on the type of molecular motion as well as on the rate [29].

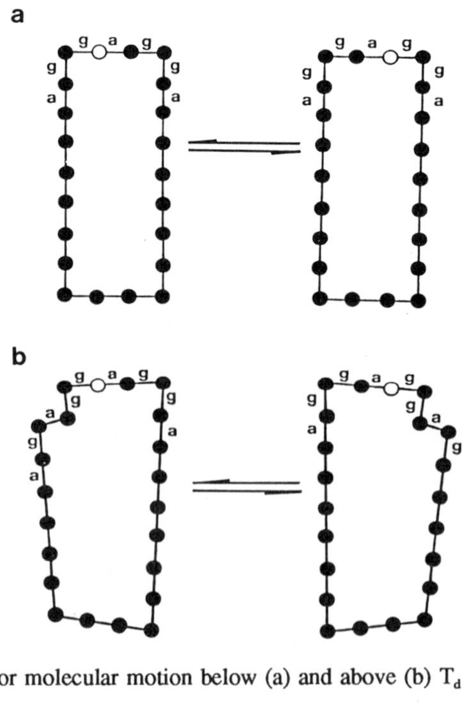

Figure 7. Model for molecular motion below (a) and above (b) T_d for cyclodoeicosane.

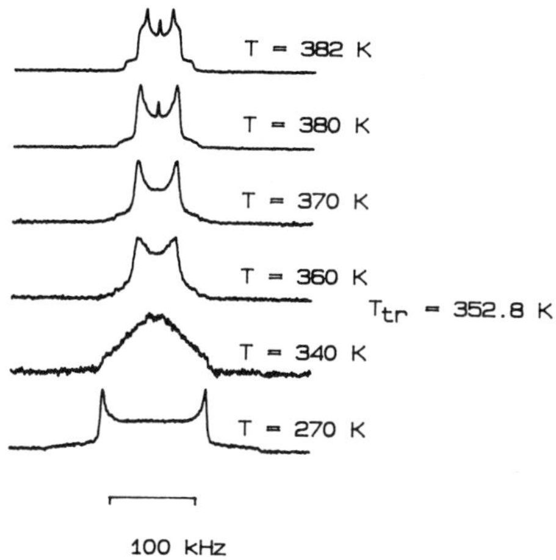

Figure 8. Temperature dependence of the ^{2}H NMR spectrum of cyclodoheptacontane.

Figure 8 shows the ^{2}H NMR spectra of melt crystallized $C_{72}H_{120}D_{24}$ at various temperatures: At 270 K, the ^{2}H NMR quadrupolar splitting is $\Delta\upsilon\perp = 125$ kHz which corresponds to the full Pake spectrum of a rigid system. At 340 K, still below the phase transition, a slight motional narrowing of the ^{2}H NMR line occurs. Above the solid-solid transition until 380 K, the spectrum is axially symmetric with $\Delta\upsilon\| = 90$ kHz and $\Delta\upsilon\perp = 44$ kHz. Raising the temperature above 380 K, the line shape changes to a non-axially symmetric form ($\Delta\upsilon_{yy} = 30.5$ kHz, $\Delta\upsilon_{xx} = 47.5$ kHz, $\Delta\upsilon_{zz} = 78$ kHz, $\eta = 0.21$). The axially symmetric spectrum at 380 K resembles a one-component spectrum in fast exchange. However, dependence of the line shape on the delay time between the two 90° pulses of the spin echo experiment and the additional change of the line shape at 382 K show that the molecular mobility has not yet reached correlation times $\tau < 10^{-8}$ s. The spectrum at 382 K does not depend noticeably on the delay time τ and thus appears to be a fast exchange spectrum.

Simulation of the spectra will be discussed, using simple motional models. According to the condis crystal concept, the hexagonal structure results from disordering of the chain conformation. Trans-gauche isomerization leading to gauche conformers is assumed as the motional mechanism. Corresponding motional models have already been successfully used to describe molecular mobility in biomembranes and polymeric membranes [34,35,36].

Figure 9 shows the geometry of an alkyl chain containing gauche conformers. In principle, only two angles between the C-D bonds and the long axis of the chain appear: C-D bonds are located in a plane at an angle of 90° and on a cone at an angle of 35.3°. Trans-gauche isomerization results thus in an exchange between the C-D bonds in the plane and those on the cone, as well as in a rotational motion of the C-D bonds around the long axis of the stems.

Such motional processes can be described using the 6 site jump model [34,35,36] drawn in Fig. 10. The 6 site jump consists of two axially symmetric three-

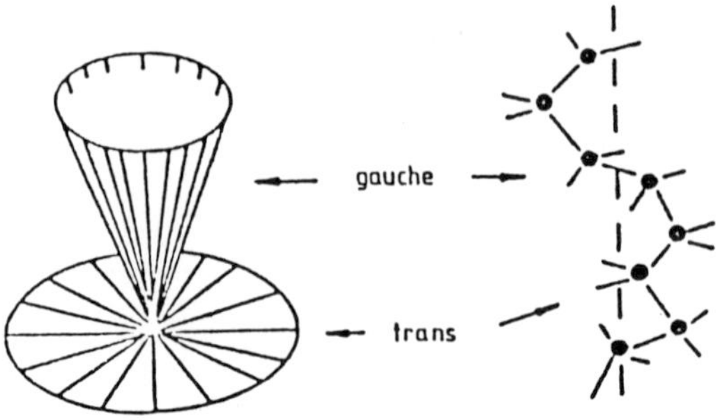

Figure 9. Geometry of an alkyl chain containing gauche rotamers according to [34].

site jumps on two cones with angles to the z axis of 90° and 35.3°. The population of the two cones can be varied according to the assumed trans/gauche ratio. In our calculations, the rotation around the long axis of the stems was assumed to have the same rate constant as the exchange between the plane and the cone.

Calculations of fast exchange ^{2}H NMR spectra using this model lead to motionally narrowed, axially symmetric spectra. Increasing the gauche concentration, the spectrum width decreases, while the line shape does not change. The axially symmetric spectrum in the mesophase of $C_{72}H_{120}D_{24}$ can be calculated, assuming a gauche concentration of 12.5%. Experimental and calculated spectra are drawn in Fig. 11. This simple motional model does not enable us to explain the variation of the line shape on raising the temperature further and the dependence of the spectra on the pulse delay between the two 90° pulses of the spin-echo experiment. Thus additional molecular motions must be included in the motional model.

As mentioned above, ^{13}C NMR spectra give evidence of a diffusion of the methylene units through the ring. Consequently, exchange between the stems and the folds must be included for the calculations of the ^{2}H NMR spectra. A possible mechanism of this diffusion is presented in Fig. 12.

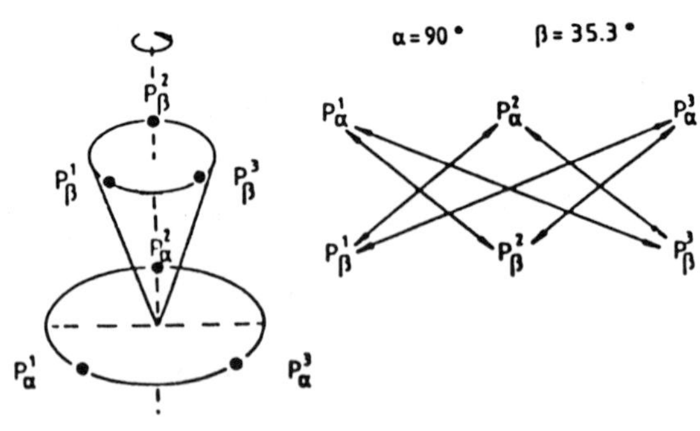

Figure 10. Six site jump model to describe fast trans-gauche isomerization within an alkyl chain according to [34].

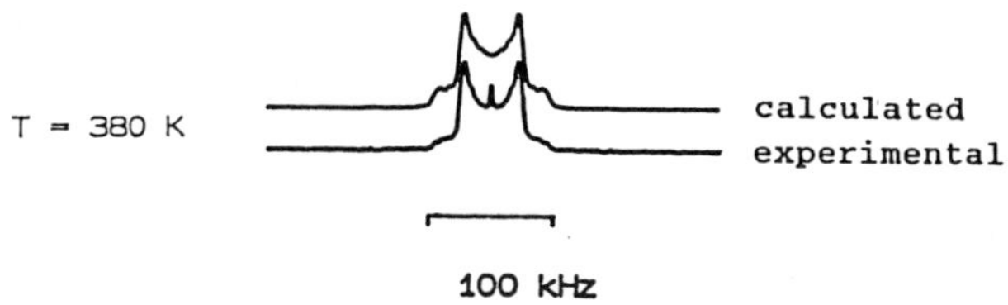

Figure 11. Experimental and calculated fast-exchange ^2H NMR spectra using the 6 site jump model with a ratio of gauche/trans rotamers of 12.5%.

Passage of a gauche defect through a chain segment leads to a small translational displacement of the methylene groups. Several repeated passages of gauche defects may explain a diffusion of the methylene units through the lamellae, as is observed by the chemical exchange in the ^{13}C NMR spectra.

In a first approach to the problem, a ten-site jump model, as schematically drawn in Fig. 13, was used for the line shape calculations. It consists of the above described 6 site jump for the motion within the stems. Four additional C-D orientations for the fold, as they can be calculated from X-ray analysis of the small rings [37], were included in the motional model.

The assumed angles Θ to the z-axis and Φ in the plane of the x-y-axes as used in the calculations are given in Table II.

Population of the four fold-positions was assumed to be 8/72 according to the ratio fold/stem.

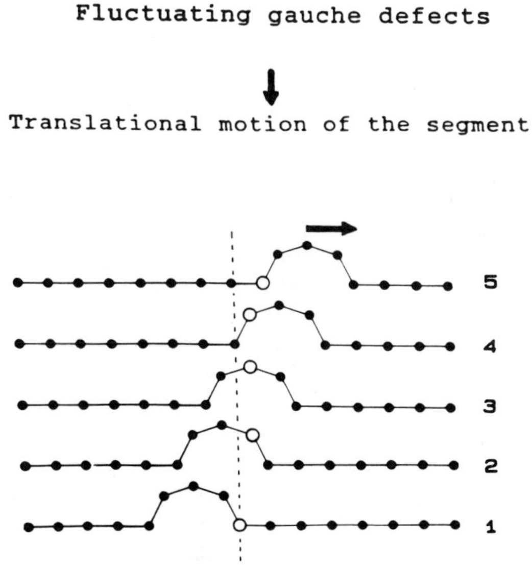

Figure 12. Indication, how the passage of a gauche defect over a chain segment leads to a small translational motion of the segment.

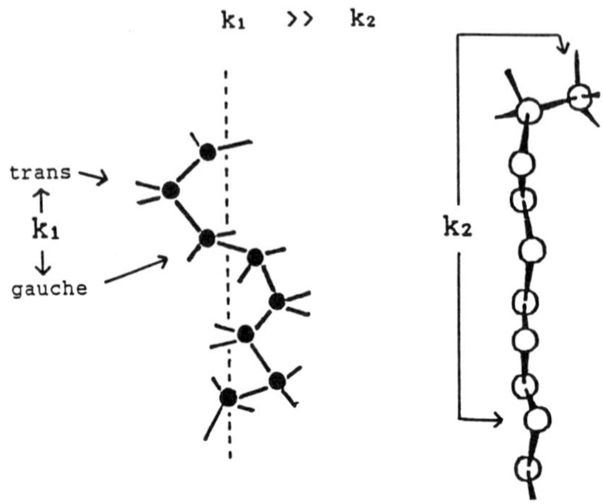

Figure 13. Two rate constants describe molecular motion in cycloalkanes. The rate constant k_1 describes trans-gauche isomerization, k_2 describes the exchange between the stem and the fold.

Two rate constants must be distinguished in this model: k_1 characterizes the rate for the exchange between a gauche and a trans orientation of a methylene unit; k_2 the exchange between the stems and the folds. The value of k_2 must be much smaller than k_1 because several passages of gauche defects are necessary to move a methylene unit from the inner part of the lamella to the folds.

Exchange between the 6 position for the stems on one side and the positions 7 and 8 for the fold on the other side were described by the smaller rate constant k_2. Exchange within the 6 positions for the stems as well as between position 7 and 9, respective 8 and 10 was described by k_1.

Assuming a constant ratio of $k_1/k_2 = 10^4$, the temperature dependent changes of the ^2H NMR spectra (changing from an axially symmetric spectrum at 380 K to a non axially symmetric spectrum at 382 K) can be simulated qualitatively, increasing the rate of exchange. Experimental and calculated spectra are drawn in Fig. 14.

In the case of the axially symmetric spectrum at 380 K, the gauche-trans isomerization is fast enough to be observed in the spectrum, while the exchange

Table II ADDITIONAL JUMP ANGLES FOR THE TEN SITE JUMP MODEL

Position	Θ	Φ
7	78.8°	50.5°
8	38.7°	-76.2°
9	137.0°	102.0°
10	33.0°	102.0°

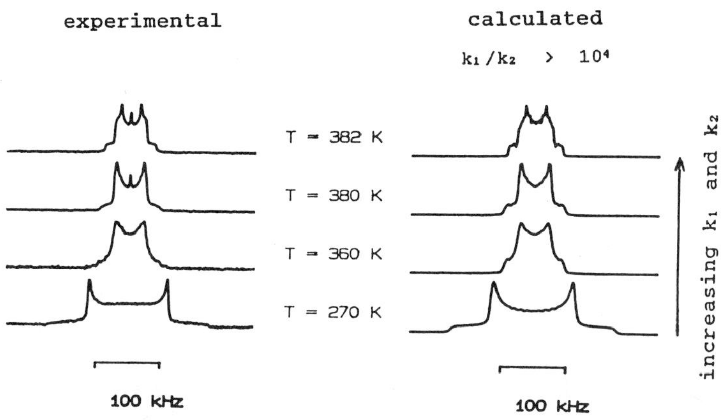

Figure 14. Experimental and calculated ^{2}H NMR spectra of cyclodoheptacontane assuming a ten-site jump model, using a constant ratio $k_1/k_2 = 10^4$ and a constant concentration of gauche defects.

between the stems and the folds is slow and can not be observed. In that case the 10 site jump model corresponds to the 6 site jump model. However, increasing both rate constants, assuming a constant ratio k_1/k_2, the exchange between the stems and the fold becomes also fast enough to be observed by a change of the line shape. Values of $\Delta\nu_{zz}$ = 75 kHz and η = 0.16 were obtained for the calculated fast-exchange spectrum compared to $\Delta\nu_{zz}$ = 78 kHz and η = 0.22 for the experimental spectrum. The small difference may result from the simplified description of the fold region.

For the simulation of the spectra the population of the gauche rotamers could be assumed to be constant in the mesophase. There are 12.5% gauche rotamers necessary to fit the line width, similar to the line shape calculation using the 6 site jump model as described above.

The defined line shapes in the mesophase indicate, that the molecules are in a stable, well defined motional state. A constant ratio of gauche rotamers in the molecules can be assumed, no gradual increase of conformational disorder can be observed.

It must be noted, that this model is a rather rough approach to the problem of molecular mobility in the cycloalkanes. The axially symmetric 6 site jump model for the stems is probably not valid in the region of the fold. In addition, the assumption of the exchange from the stems to fixed positions in a fold with a defined conformation is certainly a simplification which does not represent the real case. Another simplification of the problem is the assumption of a constant ratio k_1/k_2, indicating equal activation energies for the two processes, which is probably not valid. However, the change of the line shape from an axially symmetric spectrum to a non-axially symmetric one with increasing temperature gives evidence of the two molecular processes with a large difference of their rate constants. The temperature dependence of the ^{2}H NMR line shape can be explained consistently and in agreement with the ^{13}C NMR results by the incorporation of the fold segment C-D orientations into the motional model.

We could show, that the motional averaging may be explained on the base of the simple and not uncommon fluctuation of conformational defects along the chain. The results described may indicate a reptation of the chains as a consequence of segmental trans-gauche isomerization which cause the axial symmetry around the stems in the hexagonal phase. Corresponding molecular motion can be used as a model to explain the lamellar thickening observed in the hexagonal mesophase of polyethylene.

5. Acknowledgement

The authors want to thank Prof. Dr. H. Spiess and Dr. V. Macho for their support with the line shape simulation. We also want to thank Dr. R. G. Snyder for valuable discussions and the $C_{22}H_{44}$ sample and Prof. Dr. G. Thiele and Dr. J. Grossmann (Institut fur Anorganische Chemie der Universitat Freiburg) for support with the x-ray diffraction experiments. Financial support was granted by the Deutsche Forschungsgemeinschaft within the SFB-60.

6. References

1. Sperati, C. A. and Starkweather, J. R., *Fortschr. Hochpolym. Forsch.*, *2*, 465, 1961.
2. Lau, S. F. and Wunderlich, B., *J. Polym. Sci., Polym. Phys. Ed.*, *22*, 379, 1984.
3. Corradini, P. and Petracone, V., "Interaction Between Synthetic Polymer Molecules" in Ciba Foundation Symposium, Wolstenholme, G. E. W., O'Connor, M., Ed., p 173, Elsevier, Amsterdam, 1972.
4. Moller, M., *Makromol. Chem., Rapid Comm.*, *9*, 107, 1988.
5. Kirkpatrick, D. E. and Wunderlich, B., *J. Pol. Sci.*, *24*, 931, 1986.
6. Kogler, G.; Loufakis, K.; Bohnert, R. and Moller, M., in *Integration of Fundamental Polymer Science and Technology*, *3* Lemstra, P. J. and Kleintjens, L. A., Ed., Elsevier Applied Publishers, New York, in Press.
7. Papkov, V. S.; Godovsky, Y. K.; Swistunov, W. S.; Litvinov, V. M. and Zadanov, A. A., *J. Pol. Sci., Polym. Chem. Edn.*, *22*, 3617, 1984.
8. Beatty, J.; Pochan, J. M.; Froix, M. F. and Hinan, D. D., *Makromolecules*, *8*, 547, 1975.
9. Yamamoto, T.; Miyaji, H. and Asai, K., *Jap. J. Appl. Phys.*, *16*, 1891, 1977.
10. Hikosaka, M.; Minomura, S. and Seto, M., *Japan J. Appl. Phys.*, *19*, 1763, 1980.
11. Wunderlich, B., *Makromolecular Physics*, Vol. 1, Academic Press, New York, 1973.
12. Schneider, N. S.; Desper, C. R. and Beres, J. J., in "Liquid Crystalline Order in Polymers", Blumstein, A., Ed., Academic Press, New York, 1978, p 299.

13. Broadhurst, M. J., *J. Res. Natl. Bur. Stand, 66a*, 241, 1962.
14. Snyder, R. G.; Maroncelli, M. and Strauss, H. L., *J. Chem. Phys., 82*, 2811, 1985.
15. Drotloff, H. and Moller, M., *Thermochim. Acta., 112*, 57, 1987.
16. Drotloff, H.; Emeis, D.; Waldron, R. F. and Moller, M., *Polymer, 28*, 1199, 1987.
17. Moller, M.; Oelfin, D. and Kogler, G., *Polymer Preprints, 29*, 38, Am. Chem. Soc., 1988.
18. Kogler, G.; Drotloff, H. and Moller, M., *Mol. Cryst., Liqu. Cryst., 153*, 179, 1987.
19. Moller, M.; Waldron, R. F.; Drotloff, H. and Kogler, G., in *Integration of Fundamental Polymer Science and Technology, 2*, Lemstra, P. J. and Kleintjens, L. A., Ed., Elsevier Applied Publishers, New York, 1988.
20. Wunderlich, B. and Grebowicz, J., *Adv. in Pol. Sci., 60,61*, 1, 1984.
21. Wunderlich, B.; Moller, M.; Grebowicz, J. and Baur, H., *Adv. in Pol. Sci., 87*, 1, 1988.
22. Wunderlich, B. and Arakawa, T., *J. Polymer Sci., Part A, 2*, 3703, 1964.
23. Gruner, C. L. and Wunderlich, B., *J. Pol. Sci. Part A2, 7*, 2099, 1969.
24. Trzebiatowski, T.; Drager, M. and Strobl, G. R., *Makromol. Chem., 183*, 731, 1982.
25. Drotloff, H.; Rotter, H.; Emeis, D. and Moller, M., *J. Am. Chem. Soc., 109*, 7797, 1987.
26. Groth, P., *Acta. Chem. Scand., A33*, 199, 1979.
27. Schill, G.; Zurcher, C. and Fritz, H., *Chem. Ber., 111*, 2910, 1978.
28. Moller, M.; Gronski, W.; Cantow, H. -J. and Hocker, H., *J. Am. Chem. Soc., 106*, 5093, 1984.
29. Spiess, H. W. and Sillescu, H., *J. Magn. Reson., 42*, 381, 1980.
30. Moller, M.; Oelfin, D.; Drotloff, H. and Grossmann, J., to be published.
31. Grossmann, H. P., *Polym. Bull., 5*, 137, 1981.
32. Shannon, V., "IR-Spectroscopy Investigations of Medium Size Cycloalkanes", Doctoral Thesis, University of California, Berkeley, CA, 1985.
33. Drotloff, H., "Struktur und Bewegungszustande von Cycloalkanen im kristallinen und mesomorphen Zustand. Eine Studie zum Festkorperverhalten von flexiblen Kettenmolekulen", Doctoral Thesis, Freiburg, Federal Republic of Germany, 1987.
34. Ebelhauser, R. and Spiess, H. W., *Ber Bunsenges. Phys Chem., 89*, 1208, 1985.
35. Huang, T. H.; Sharjune, R. P.; Wittebort, R. J.; Griffin, R. G. and Oldfield, E., *J. Am. Chem. Soc., 102*, 7377, 1980.
36. Blume, A.; Rice, D. M.; Wittebort, R. J. and Griffin, R. G., *Biochemistry, 24*, 6220, 1982.
37. Groth, P., *Acta Chem. Scand., A30*, 155, 1976.

Chapter 15

SOLID STATE NMR RELAXATION STUDY OF LIQUID CRYSTAL POLYMERS EMPLOYING A TWO-DIMENSIONAL TECHNIQUE

A. SCHLEICHER, K. MÜLLER AND G. KOTHE

1. Abstract

Molecular motions of liquid crystal polymers in the solid state cover a broad dynamic range, extending from the fast-rotational to the ultraslow motional regime. Two-dimensional NMR relaxation spectroscopy is designed to follow these motions and to differentiate the various motional modes. In addition, different types of molecular order, modulated by these motions, are discriminated, likewise.

The solid state dynamics of the studied polyesters sensitively depends upon thermal history. Annealed samples show a typical semicrystalline behavior, manifested by two different motional components. Molecular structure is characterized by a high degree of conformational and orientational order, strongly correlating to the exceptional mechanical properties of these systems.

2. Introduction

Liquid crystal polymers (LCPs) have been reported to exhibit a variety of unusual material properties subject to increased research activities in this field [1,2]. One major objective of these studies is the evaluation of a correlation between the macroscopic bulk behavior and the characteristic molecular properties of these

---

A. SCHLEICHER, K. MÜLLER, AND G. KOTHE - Institute of Physical Chemistry, University of Stuttgart, Pfaffenwaldring 55, D-7000 Stuttgart 80, FRG.

Solid State NMR of Polymers, Edited by L. Mathias
Plenum Press, New York, 1991

systems. In this connection, solid state ^2H NMR has successfully been employed to elucidate the molecular characteristics of such polymers, comprising structural and dynamical aspects, likewise [3-5]. By combining various pulse experiments it was possible to follow molecular motions over an extremely wide dynamic range, not available with any other experimental technique [3,6].

In this contribution, we report on a promising extension of solid state dynamic NMR using a two-dimensional (2D) procedure [7,8]. The method employs the pronounced anisotropy of the nuclear spin relaxation times, observed for polycrystalline or multidomain samples. In these systems, the relaxation times vary as a function of the crystal or director orientation, and equivalently as a function of the frequency across the breadth of the powder spectrum. It turns out that this variation is of particular diagnostic importance for the dynamical characterization of complex molecular systems [9,10].

In the first section the experimental and theoretical background is presented. Then typical results are given to demonstrate the applicability of the new NMR techniques. In the main section 2D relaxation experiments of liquid crystal main chain polymers are presented. Computer simulations provide the orientational distributions and conformations of the polymer chains and the correlation times of the various motions. They are related to the exceptional tensile properties of these systems, evaluated by mechanical measurements [5].

3. NMR Techniques

Multipulse dynamic NMR is a time domain technique. The system is subject to a sequence of non-selective rf pulses and the response $S(t_2)$ after the last pulse is recorded. Fourier transformation (FT) of $S(t_2)$ yields the common 1D frequency spectrum $S(\omega_2)$. Powder lineshapes obtained via multipulse sequences may be distorted as compared with those obtained from the free induction decay due to orientation dependent relaxation times [11-13]. Generally, one observes that with increasing relaxation period t_1 significant spectral changes occur, which are different for the different motions. Apparently, the change of the powder lineshape with the pulse separation t_1 is indicative of the type of motion [3,14,15].

A 2D version of such relaxation experiments is achieved by recording the time signals as a function of successive incremented time intervals t_1. A complex FT in both time domains transforms $S(t_1,t_2)$ into a 2D representation $S(\omega_1,\omega_2)$ of the corresponding relaxation experiment [7,8]. Three different pulse sequences, commonly employed in ^2H NMR relaxation studies, are adapted to this general scheme:

i) inversion recovery

$$(180) - t_1 - (90)_x - \Delta - (90)_y - \Delta - t_2$$

ii) quadrupole echo [16] and

$(90)_x - t_1/2 - (90)_y - t_1/2 - t_2$

iii) Jeener-Broekaert sequences [17]

$(90)_x - \Delta_1 - (54)_y - t_1 - (54)_y - \Delta_2 - (90)_y - \Delta_2 - t_2$

Each sequence defines a specific dynamic window, in which molecular motions can be studied. Generally, inversion recovery in a high magnetic field ($B_z \geq 5$ T) may be used for fast-rotational studies (10^{-12} s $< \tau_R < 10^{-8}$ s), while quadrupole echo experiments are applicable in the slower dynamic range (10^{-8} s $< \tau_R < 10^{-4}$ s) [13]. Finally, by employing Jeener-Broekaert sequences, ultraslow motions (10^{-4} s $< \tau_R <$ 10 s) can be detected [18].

The 2D representation of the various relaxation experiments may be regarded as a graph of the relevant natural widths vs the resonance positions of the individual dynamic spin packets that constitute the spectrum [19]. For example, cross sections through the 2D quadrupole echo spectrum along ω_1 provide homogeneous lineshapes associated with the spin-spin relaxation time T_{2E} [9,10]. It is found that both the magnitude of T_{2E} and the way in which T_{2E} changes across the spectrum are very dependent upon the character of the molecular motion, responsible for spin relaxation [3,10,19].

Analysis of these NMR experiments is achieved by employing a comprehensive relaxation model based on the stochastic Liouville equation [3,13,20]. Various adjustable parameters, characterizing molecular order and dynamics, enter the problem [3]. In the specific case of LCPs, four different motional modes, depicted in Fig. 1, are considered. They describe localized conformational changes (jump processes involving phenyl rings or methylene groups) and overall rotations and fluctuations of larger chain segments (diffusive process). The correlation times τ_J, $\tau_{R\parallel}$ and $\tau_{R\perp}$ specify the time scale of these motions.

Collective molecular reorientations, characteristic of liquid crystals, are considered likewise. For this mechanism, known as director order fluctuations, a continuous distribution of correlation times is predicted. However, in contrast to isolated motions of single molecules, these collective reorientations can only be observed in the anisotropic melt of LCPs [21].

Molecular order is described by contributions to three different levels. They account for conformational order of the various chain segments, orientational order of the polymer chains with respect to a preferential local axis (director), and alignment of the director axes in a laboratory frame. The relevant quantities are denoted as segmental ($S_{Z'Z'}$), orientational (S_{ZZ}) and macroorder parameters ($S_{z''z''}$), respectively [3,5,20].

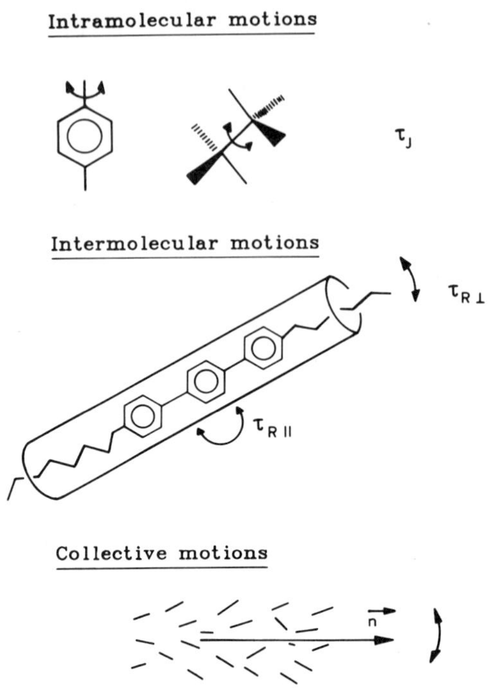

Figure 1. Schematic representation of molecular and collective motions relevant to liquid crystal polymers.

The sensitivity of 2D relaxation spectra to type and time scale of molecular motions is illustrated in Fig. 2. The spectra refer to quadrupole echo sequences and characterize two possible reorientation mechanisms of a methyl group (three-site jump vs continuous diffusion). For simplicity, we have chosen equal correlation times for both motions ($\tau_J = \tau_{R\parallel} = 10^{-6}$ s). Drastic spectral differences are observed. Apparently, these 2D relaxation spectra are highly indicative of the type of motion. The same is true for the corresponding normalized contour plots (right side) [19]. We note that similar 2D relaxation spectra can be obtained from inversion recovery or Jeener-Broekaert sequences. Thus, by applying this 2D technique to different pulse sequences, the various motions can be differentiated over an extremely wide dynamic range, extending from 10^{-12} s in the fast-rotational to 10 s in the ultraslow motional regime. Since the different motions modulate different kinds of molecular order (segmental order, orientational order) these orders can be differentiated, likewise.

4. Experiments and Methods

LCPs of the following chemical structure

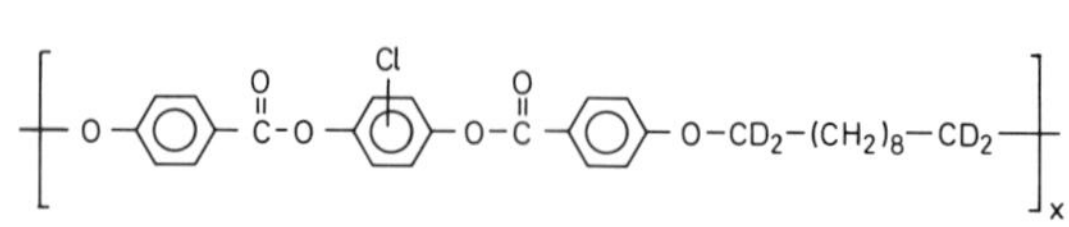

were specifically deuterated at the α-segment of the aliphatic spacer as described elsewhere [22]. The DSC thermograms show a glass transition at T_g = 303 K and a nematic melt, ranging from T_m = 433 K to T_{ni} = 553 K. Typically, 250 mg of the LCP, sealed in a 4 mm NMR tube, were heated to the nematic melt (T = 473 K) outside the NMR magnet and then rapidly cooled to room temperature within a few seconds. Annealed samples were obtained by heat treatment of the quenched samples at 423 K for about 40 minutes.

The ^2H NMR measurements were performed on a Bruker MSL 300 spectrometer at ν = 46 MHz (B_z = 7 T), using quadrature detection and appropriate phase cycling routines. 90° pulse width of the home-built probe was 2.2 s. The number of scans varied between 800 and 1600. The resulting 2D data set consisted of 64 sampling points in the t_1 domain and 2048 sampling points in the t_2 domain. Generally, zero filling for better resolution was employed in t_1. The 2D FT was performed by removing the imaginary data points after the first complex FT in t_2. The second FT, only applied to the real data set, yielded the pure absorption 2D spectrum $S(\omega_1,\omega_2)$ [23]. Normalized contour plots [19] were obtained by dividing the

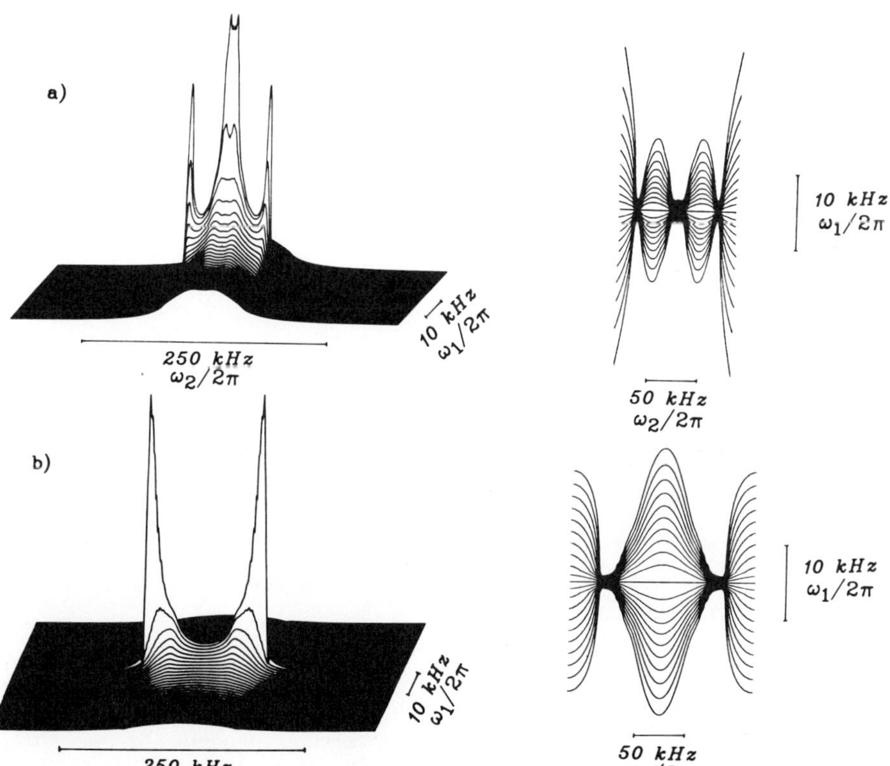

Figure 2. Calculated 2D relaxation spectra (quadrupole echo sequences) of a solid, containing a rotating (deuterated) methyl group. a) Three-site jump, $\tau_J = 10^{-6}$ s. b) Continuous diffusion, $\tau_{R\|} = 10^{-6}$ s. Left side: Stack plots. Right side: Normalized contour plots (contours in units of 5% of maximum amplitude).

ω_2-spectra at the various ω_1 values by the center ω_2-spectrum ($\omega_1 = 0$). From this contour representation the anisotropy of the corresponding relaxation rates can directly be read.

A FORTRAN program package was employed to analyze the ^{2}H NMR measurements [3,20]. The programs simulate 2D NMR relaxation experiments of I=1 spin systems undergoing inter- and intramolecular motion in an anisotropic medium [24]. Numerical solution of the stochastic Liouville equation was achieved using the Lanczos algorithm [25].

5. Results and Discussion

2D NMR relaxation methods have been employed to evaluate the complex solid state behavior of LCPs. The studies were intended to reveal the effect of the thermal history on the molecular properties, determined by the NMR techniques. Only multidomain samples ($S_{z''z''} = 0$) were examined, taking full advantage of the highly informative character of angular dependent relaxation rates.

Figure 3 shows 2D NMR relaxation spectra of solid LCPs, obtained from inversion recovery sequences at T = 363 K. The stack and contour plots refer to a quenched (Fig. 3a) and an annealed sample (Fig. 3b), respectively. Depending on the thermal history, characteristic differences are observed. The quenched sample exhibits a single spectral component with high molecular mobility, characteristic of a liquid crystalline state. In contrast, the 2D spectra of the annealed sample indicate two overlapping components associated with mobile and immobile deuterons in distinct regions of the polymer, which we assign to a liquid crystalline and a crystalline phase [3,5,10]. In that respect, LCPs resemble conventional polymers, which exhibit amorphous and crystalline phases, likewise [5,26].

The assignment is supported by additional 2D relaxation experiments of annealed samples, employing the pulse sequences discussed above. Representative examples are shown in Fig. 4. The 2D spectra refer to quadrupole echo sequences and two selected temperatures - one above (Fig. 4a) and one below the glass transition T_g (Fig. 4b). Note, that only one spectral component is observed for $T < T_g$. Apparently, the mobility of the liquid crystalline fraction is drastically reduced by approaching T_g, whereas the dynamics of the crystalline fraction continues unaffectedly into the glassy state.

Quantitative data concerning molecular order and dynamics of the solid LCPs were obtained by computer simulation of the various 2D NMR experiments. As mentioned above the annealed samples appear to be heterogeneous as two different fractions (liquid crystalline and crystalline) are detected. Diffusive motions of larger chain segments ($\tau_{R\|}$, $\tau_{R\perp}$)is not linear, the apparent activation energies increasing with decreasing temperature. Thus, all intermolecular motions gradually freeze at T_g, possibly due to a loss in free volume [27]. The dynamics of the crystalline component is quite different, exhibiting only restricted conformational motions in the low fre-

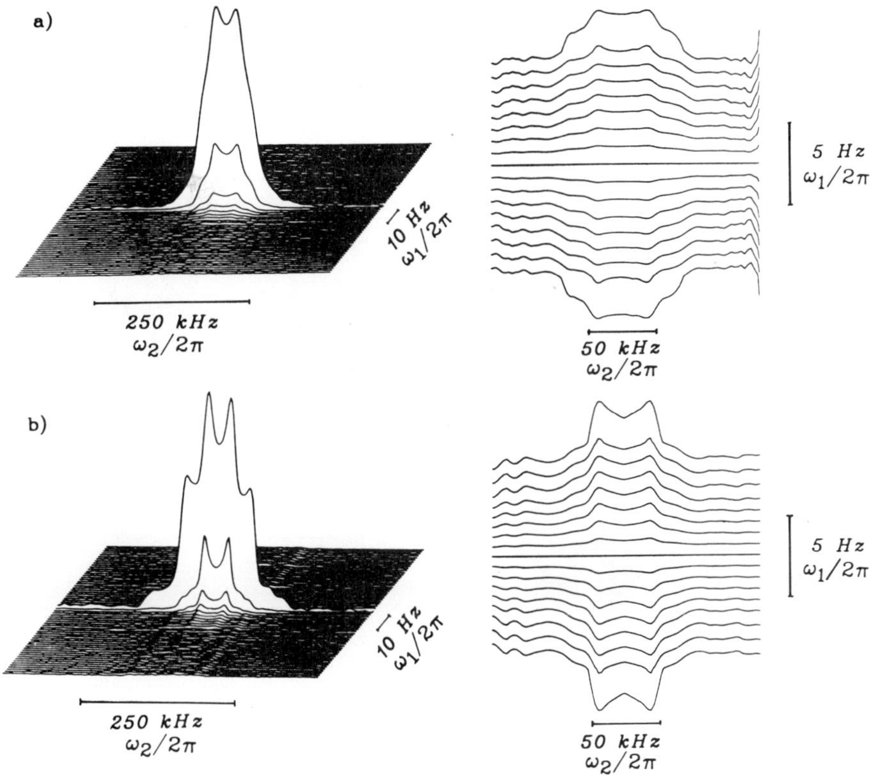

Figure 3. Experimental 2D relaxation spectra (inversion recovery sequences) of the liquid crystal polymer, deuterated at the α-segment of the aliphatic spacer. a) Quenched sample at T = 363 K. b) Annealed sample at T = 363 K. Left side: Stack plots. Right side: Normalized contour plots (contours in units of 10% of maximum amplitude).

quency range (10^{-6} s $< \tau_J <$ 10^{-5} s). This might be explained by a denser packing of this fraction, as observed for conventional crystalline polymers [28].

Simultaneously, the structural parameters of the LCPs could be determined. The most prominent feature concerns the high degree of conformational order [29]. Evidently, highly extended conformers prevail in the liquid crystalline phase, as can be deduced from a segmental order parameter of $S_{Z'Z'} = 0.7$ for this fraction. On the other hand, a conformational order parameter of $S_{Z'Z'} = 0.75$, found for the crystalline component, is far below the limiting value of 1.0, expected for regular crystals [30]. This finding is corroborated by calorimetric studies, indicating non-regular crystalline regions due to conformational disorder [5].

However, a high degree of orientational order ($S_{ZZ} = 0.9$) is found, independent of thermal history and polymer phase [29]. The observed value considerably exceeds that exhibited by low molecular weight nematogens [31]. As predicted [32,33], the polymer chains are highly ordered on a molecular level in agreement with previous ESR studies [34].

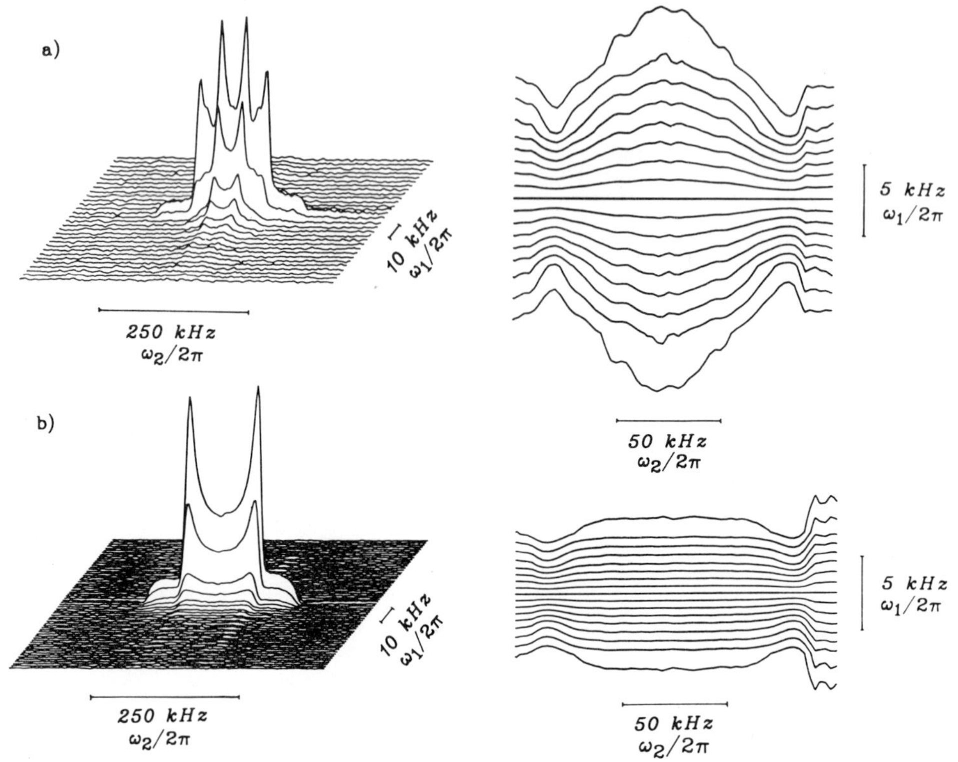

Figure 4. Experimental 2D relaxation spectra (quadrupole echo sequences) of the liquid crystal polymer, deuterated at the α-segment of the aliphatic spacer. a) Annealed sample at T = 363 K (T > Tg). b) Annealed sample at T = 298 K (T < Tg). Left side: Stack plots. Right side: Normalized contour plots (contours in units of 10% of maximum amplitude).

Finally, it should be noted that a complete macroscopic alignment of the director axes $S_{Z''Z''} = 1.0$) is achieved by using strong magnetic or shear fields [3,5,29]. Stress-strain measurements, performed on melt-spun fibres, revealed exceptionally good mechanical properties. The observed tensile moduli (up to 22 GPa) and strengths (up to 0.34 GPa) are at least comparable or even better than those of conventional post-treated polymer fibres [5]. Evidently, the striking mechanical properties of the LCP fibres are closely related to the high degree of molecular order, sorted out by the new NMR techniques.

6. Conclusions

2D NMR relaxation spectroscopy has successfully been employed to study the molecular properties of LCPs in the solid state. The results, referring to quenched and annealed samples, indicate a strong correlation to the exceptional mechanical properties of these systems. It appears that the new NMR techniques present a powerful tool for characterizing complex chemical systems such as LCPs.

7. Acknowledgement

Financial support of this work by the Deutsche Forschungsgemeinschaft and Fonds der Chemischen Industrie is gratefully acknowledged.

8. References

1. Dobb, M. G. and McIntyre, J. E., *Adv. Polym. Sci.*, *60/61*, 61, 1984.
2. Chung, T. S., *Polym. Eng. Sci.*, *26*, 901, 1986.
3. Müller, K.; Meier, P. and Kothe, G., *Prog. Nucl. Magn. Reson. Spectrosc.*, *17*, 211, 1985.
4. Spiess, H. W., *Adv. Polym. Sci.*, *23*, 81, 1985.
5. Müller, K.; Schleicher, A.; Ohmes, E.; Ferrarini, A. and Kothe, G., *Macromolecules*, *20*, 2761, 1987.
6. Spiess, H. W., *J. Polym. Sci.*, *261*, 193, 1983.
7. Aue, W. P.; Bartholdi, E. and Ernst, R. R., *J. Chem. Phys.*, *64*, 2229, 1976.
8. Ernst, R. R.; Bodenhausen, G. and Wokaun, A., *Principles of Nuclear Magnetic Resonance in One and Two Dimensions*, Clarendon Press, Oxford, 1987.
9. Müller, L. and Chan, S. I., *J. Chem. Phys.*, *78*, 4341, 1983.
10. Müller, K.; Schleicher, A. and Kothe, G., *Mol. Cryst. Liq. Cryst.*, *153*, 117, 1987.
11. Jeffrey, K. R., *Bull. Magn. Reson.*, *3*, 69, 1981.
12. Spiess, H. W. and Sillescu, H., *J. Magn. Reson.*, *42*, 381, 1981.
13. Meier, P.; Ohmes, E.; Kothe, G.; Blume, A.; Weidner, J. and Eibl, H. - J., *J. Phys. Chem.*, *87*, 4904, 1983.
14. Torchia, D. A. and Szabo, A., *J. Magn. Reson.*, *49*, 107, 1981.
15. Siminovitch, D. J.; Ruocco, A. J.; Olejniczak, E. T.; Das Gupta, S. K. and Griffin, R. G., *Chem. Phys. Lett.*, *119*, 251, 1985.
16. Powles, J. G. and Strange, J. H., *Proc. Phys. Soc.*, *82*, 6, 1963.
17. Jeener, J. and Broekaert, P., *Phys. Rev.*, *157*, 232, 1967.
18. Spiess, H. W., *J. Chem. Phys.*, *72*, 6755, 1980.
19. Millhauser, G. L. and Freed, J. H., *J. Chem. Phys.*, *81*, 37, 1984.
20. Meier, P.; Ohmes, E. and Kothe, G., *J. Chem. Phys.*, *85*, 3598, 1986.
21. Dippel, T.; Schweikert, K. H.; Müller, K.; Kothe, G. and Noack, F., to be published.
22. Müller, K., Ph.D. Thesis, University of Stuttgart, 1985.
23. Keeler, J. and Neuhaus, D., *J. Magn. Reson.*, *63*, 454, 1985.
24. Schleicher, A.; Müller, K. and Kothe, G., to be published.
25. Moro, G. and Freed, J. H., *J. Chem. Phys.*, *74*, 3757, 1981.
26. Bovey, F. A. and Jelinski, L. W., *J. Phys. Chem.*, *89*, 571, 1985.
27. Wunderlich, B., *Macromolecular Physics*, Vol. 3, Academic Press, New York 1980.

28. McBrierty, V. J. and Douglass, D. C., *J. Polym. Sci. Macromol. Rev.*, *16*, 295, 1981.
29. Müller, K.; Hisgen, B.; Ringsdorf, H.; Lenz, R. W. and Kothe, G., *Mol. Cryst. Liq. Cryst.*, *113*, 167, 1984.
30. Wunderlich, B. and Grebowicz, J., *Adv. Polym. Sci.*, *60/61*, 1, 1984.
31. Hsi, S.; Zimmermann, H. and Luz, Z., *J. Chem. Phys.*, *69*, 4126, 1978.
32. Ronca, G. and Yoon, D. Y., *J. Chem. Phys.*, *76*, 3295, 1982; *80*, 925, 1984.
33. Abe, A., *Macromolecules*, *17*, 2280, 1984.
34. Müller, K.; Wassmer, K. -H.; Lenz, R. W. and Kothe, G., *J. Polym. Sci., Polym. Lett. Ed.*, *21*, 785, 1983.

Chapter 16

DYNAMICS OF SURFACE BOUND POLYMERS AND COUPLING AGENTS

FRANK D. BLUM, ROBERT B. FUNCHESS, AND WIRIYA MEESIRI

1. Introduction

Composite materials have mechanical properties which depend not only upon the properties of each of their components, but also upon the details of the interactions between the components. These interactions take place at the interfacial region. In order to further our understanding of the macroscopic properties, it is necessary to expand our knowledge of the interface at the microscopic level. It is hoped that information about the molecular motion of the surface-adsorbed species in the interfacial region will some day aid in the development of new composite materials and in the understanding of their properties.

Many composite materials do not have the desired physical properties due to a weak interaction of the polymer with the filler or fiber. Improvements in physical properties of composite materials due to enhanced bonding between reinforcements and matrix resin by silane coupling agents are well known. Small amounts of coupling agents at the interface have a profound effect on the performance of composites. These species are active at very low levels (perhaps even monolayers or less) so a microscopic view of their behavior at the interface may help us to understand their macroscopic behavior.

Many forms of chemical spectroscopy are well suited for molecular studies of coupling agents. The molecular structure of aminopropyltriethoxy-silane (APS) on a

---

FRANK D. BLUM, ROBERT B. FUNCHESS, AND WIRIYA MEESIRI - Department of Chemistry and Materials Research Center, University of Missouri - Rolla, Rolla, MO 65401.

Solid State NMR of Polymers, Edited by L. Mathias
Plenum Press, New York, 1991

silica surface studied by infrared (IR) spectroscopy has been reported [1-3]. High-resolution ^{13}C NMR with cross-polarization and magic angle spinning (CP-MAS) were also used to observe structure, orientation, and silane-interface interactions of organosilane coupling agents bound to silica surfaces [4-6]. In addition, the dynamics of alkyl modified silica [7-9] have been studied with CP-MAS. Wideline ^{2}H NMR has been used to probe highly immobile species such as rigid solids or adsorbates attached to a solid surface. Examples include probes of small molecules adsorbed on the surface of alumina [10,11]. Quadrupolar powder patterns are useful in providing information on molecular order and motion [12].

CP-MAS techniques have been used to study interactions between polymers and other materials [13-15], as well as interactions between different polymers [16]. To our knowledge, no deuterium NMR studies of surface-bound polymers have been done. However, wideline ^{2}H NMR has been used to yield information about the dynamics of polymers in bulk [12,17], and this same technique can be applied to surface bound polymers, providing sensitivity is not a serious problem.

In this paper we report some of our preliminary studies on surface bound polymers and coupling agents. We have used IR and NMR to probe the adsorption isotherm and hydrolysis of a series of silane coupling agents. This is followed by investigations of both protonated and deuterated coupling agents by using ^{13}C and deuterium NMR techniques. These are studied as a function of surface coverage. We also report studies of polymers directly adsorbed to a silica surface, specifically acrylate and methacrylate polymers. Some results of CP-MAS NMR studies of poly(isopropyl acrylate) (PIPA) and poly(methyl methacrylate) (PMMA) will be presented below. The objective of the study is to observe the dynamics of the coupling agents and polymers which are attached to the silica surface.

2. Experimental

2.1. Materials

High surface area fumed silicas (Cab-O-Sil grades M-5 and S-17, having nominal surface areas of 200 and 400 m^2/g respectively) were donated by Cabot Corporation. They were dried at a minimum of 105 °C for at least 24 hours prior to use in order to eliminate surface-adsorbed water [18]. Deuterated aminopropyltriethoxysilane (DAPS) and deuterated aminobutyltriethoxysilane (DABS) were prepared by deuterating cyanoethyltriethoxysilane (CES) and cyanopropyltriethoxysilane (CPS), respectively. Aminopropyltrimethoxysilane (APS), CES, and CPS were purchased from Petrarch Systems. The method of deuteration has been described by Stein *et al.* [19]. It involves the catalytic hydrogenation of the cyano group to the amine with deuterium gas. DAPS was hydrolyzed in water at concentrations of 1% and 4% DPS by weight. After 2 hours, an 8 ml aliquot of each solution was added to about 0.15 g of M-5 silica in a 15 ml screw-cap tube. The tube was shaken on a vortex mixer for a brief period, tumbled for 15 minutes, and centrifuged at 3,500 rpm for 45 minutes. The supernatant solution was decanted and saved for FT-IR analysis. The wet treated silica was subjected to i) no treatment, ii) one water wash, iii) two water washes, iv) one acetone wash, or v) one deuterium oxide wash before drying. The S-17 silica samples were also treated similarly by 1% DAPS and 1% APS solutions followed by one water wash. All treated silicas were dried in a vacuum oven at 110 °C for 12 hours prior to NMR studies.

Various concentrations of hydrolyzed coupling agents were prepared and used to create a Beer's law plot. The absorbances of the coupling agent solutions were measured using a Perkin-Elmer 1750 FT-IR spectrometer. The solutions were injected into a demountable liquid cell fitted with calcium fluoride windows with a 0.1 mm Teflon spacer. Spectra of supernatant solutions were collected in single ratio mode at 2 cm^{-1} resolution and 3 scans. A solution of the solvents used was also run. The absorbance values were determined manually by measuring the absorbance at about 2840 cm^{-1} in the difference spectrum. This resonance corresponds to the C-H stretching mode of the methoxy group. The concentrations of coupling agents were determined by comparison with the Beer's law plot. The amount of coupling agent on the surface was taken to be the difference between the concentrations of the initial and supernatant solutions.

The adsorption of isotherm for PIPA was determined by weighing samples (about 50 mg each) of M-5 silica into screw-cap tubes. Toluene solutions of varying PIPA concentration were prepared by dilution of standard solution. 5 ml aliquots of these solutions were pipetted into the tubes, which were then shaken in a water bath for approximately 48 hours. The tubes were removed and centrifuged. A measured quantity of supernatant liquid was drawn off and placed in a tared weighing pan. The toluene was allowed to evaporate and the weight of the PIPA residue was determined. Knowledge of the quantities of solution and silica used and of the concentration of the initial solution permits calculation of the final (equilibrium) solution concentration and of the amount of polymer adsorbed per gram of silica. The amount of PIPA adsorbed per square meter of surface was calculated using the nominal figure of 200 m^2/g silica supplied by Cabot Corporation. The silica with adsorbed PIPA was rinsed with distilled toluene and allowed to dry.

2.2. NMR Studies

A Varian VXR-200 NMR spectrometer equipped with a solid-state MAS probe for ^{13}C and a wideline probe for deuterium was used to collect spectra of the solid materials. ^{13}C CP-MAS spectra of coupling agents on silica surfaces were obtained by dipolar decoupling and the CP-MAS technique at 50.3 MHz with a contact time of 0.5 ms. A 90° pulse width of 10.5 s was used and the cross polarization done at the same power level. Dipolar decoupling of the protons was done at about three times this power level. The chemical shifts were measured relative to the aromatic resonance at 123.1 ppm of hexamethylbenzene. Typically, the wideline deuterium spectra of the solids were collected at 30.7 MHz in a solid-state echo mode with a spectral width of 1.67 and 0.5 MHz, a 2 s 90 ° pulse width, and a repetition time of about 1 s.

In the case of PIPA, spectra were obtained for the bulk polymer and for samples at various levels of surface coverage. The spectra were taken with a cross-polarization time of 2 ms for both the bulk and surface bound samples. Spinning rates were about 3 kHz, a 10 s 90 ° pulse, and a pulse repetition time of 1 s were used. The number of accumulated transients was about 80,000. All spectra were taken at ambient temperatures. Spectra of PMMA (bulk and one monolayer coverage) were taken under similar conditions. $CDCl_3$ solution spectra of PIPA and PMMA were obtained for comparison with the CP-MAS spectra.

3. Results and Discussion

3.1. Coupling Agent Studies

The proposed structures of deuterated coupling agents synthesized in our laboratory are as follows:

$D_2N\text{-}CD_2\text{-}CH_2\text{-}CH_2\text{-}Si(OC_2H_5)_3$ (DAPS)

$D_2N\text{-}CD_2\text{-}CH_2\text{-}CH_2\text{-}CH_2\text{-}Si(OC_2H_5)_3$ (DABS)

where we define carbon #1 as that carbon next to the amino group. The above coupling agents were characterized by IR, ^1H proton and ^{13}C NMR as described earlier [20]. The deuterium NMR spectra of the coupling agents in solution are shown in Fig. 1. The spectra of both DAPS and DABS in chloroform (spiked with deuterochloroform, 7.2 ppm) show that positions 1 and 2 are highly deuterated. At this sensitivity the other positions with natural abundance deuterons are not observed. The intensities of the $-ND_2$ and $-CD_2$ resonances in the unhydrolyzed coupling agents (Fig. 1A, B) are about the same due to about the same isotopic enrichment. Upon addition of water the intensity of the $-ND_2$ resonance in the hydrolyzed DAPS sample (Fig. 1C) is lower than that of $-CD_2$, which is due to deuterium exchange of the amino-deuterons with water. The evidence of this exchange is the appearance of a deuterium oxide resonance at about 4.67 ppm. The resonance at 1.3 ppm from the amine completely disappears at longer hydrolysis times. Deuteration of the coupling agents prepared under our conditions did not introduce significant deuterium exchange on other carbons, e.g. on position 2, as reported by previous authors [19].

^{13}C CP-MAS and wideline solid-state echo spectra of treated S-17 silica are shown in Figs. 2 and 3 respectively. Fig. 2 shows carbon resonances of (A) protonated and (B) deuterated APS. Both spectra show resonances at about 40, 23, and 7 ppm due to carbons at positions 1, 2, and 3, respectively. The intensity of the deuterated carbon resonance is reduced relative to the protonated resonances because the CD_2 carbon does not cross-polarize as efficiently as the protonated species. The resonances shift upfield when the coupling agent is adsorbed on the surface, most pronounced for carbon position 1 and at least for position 3. Similar results were observed by Zaper *et al.* [6].

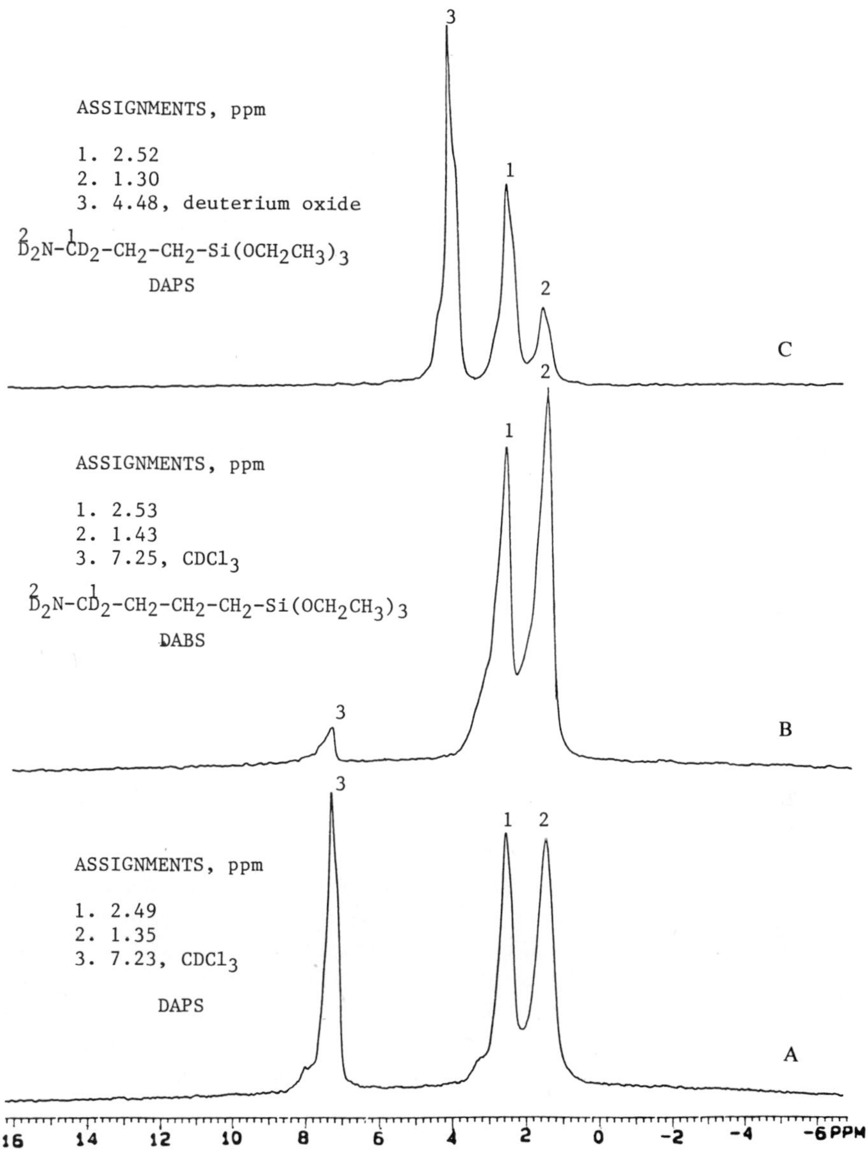

Figure 1. ^2H solution spectra of (A) DABS in chloroform, (B) DAPS in chloroform, and (C) hydrolyzed DAPS in water.

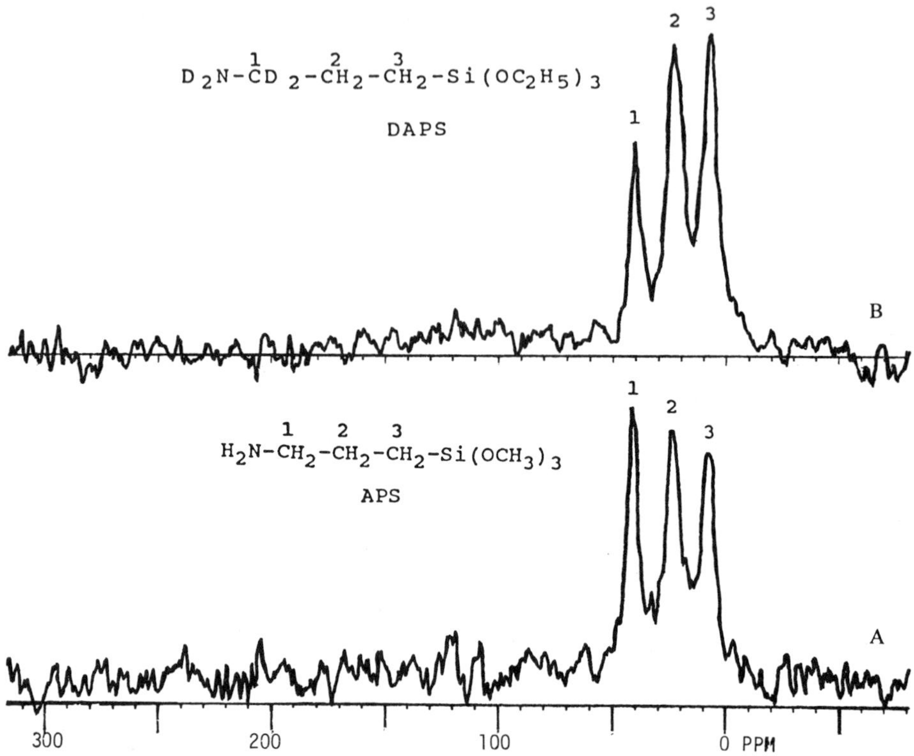

Figure 2. ^{13}C CP-MAS-DD spectra of APS and DAPS on silica. The silanes are shown based on the starting species.

Wideline ^{2}H NMR spectra show broad resonances due to deuterons immobilized on silica surfaces. Additional restricted motion might be due to hydrogen bonding of the amino group with a silanol group on the silica surface. This kind of interaction was reported previously [1,2]. However, the spectra for the DAPS on the surface are not those of a totally immobilized species. Instead an intermediate spectrum is obtained, similar to that previously seen for polymers [12]. The picture is one where the coupling agent is rigid at the Si-O-surface bond but has some motional freedom at the other end of the molecule. The restricted motion results in the broad resonance with a filled in center (Fig. 3A). Upon addition of a small amount of water to the surface, the resonance narrows somewhat and the center fills in further (Fig. 3B) due to faster motion of the deuterons. When excess water is added, the resonance narrows further and shows an additional sharp resonance having a linewidth of about 3.5 kHz (Fig. 3C). The spectrum may be a composite of resonances due to deuterons in regions with different amounts of solvent: the sharp resonance due to deuterons in solvent-rich regions and the broad resonance due to deuterons in regions with less solvent. The above results are, however, not conclusive enough to fully explain the nature of the coupling agent adsorbed to the surface. Work toward this goal is proceeding.

3.2. Polymer Studies

The adsorption isotherm for PIPA was experimentally determined for monolayer coverage in this system to be about 1.3 mg/m^2 (or 260 mg PIPA/g silica). Comparison of the bulk PIPA spectrum with the solution spectrum (Fig. 4) indicates that three carbon-13 peaks do not appear in the bulk PIPA spectrum. By contrast, the bulk PMMA spectrum shows peaks in the same places as the solution spectrum (Fig. 5). Spectra of various PIPA surface coverages (Fig. 6) show the missing PIPA peaks more or less clearly resolved. It should be noted here that the bulk PIPA spectrum presented here is one exhibiting most clearly the lack of certain resonances. Other spectra, taken under different conditions, do show these peaks. This seems to be a function of acquisition parameters (such as spin rate, temperature, decoupler power level, etc.) but the precise nature of this function is unknown. In no case, however, are they resolved as clearly as in the surface spectra.

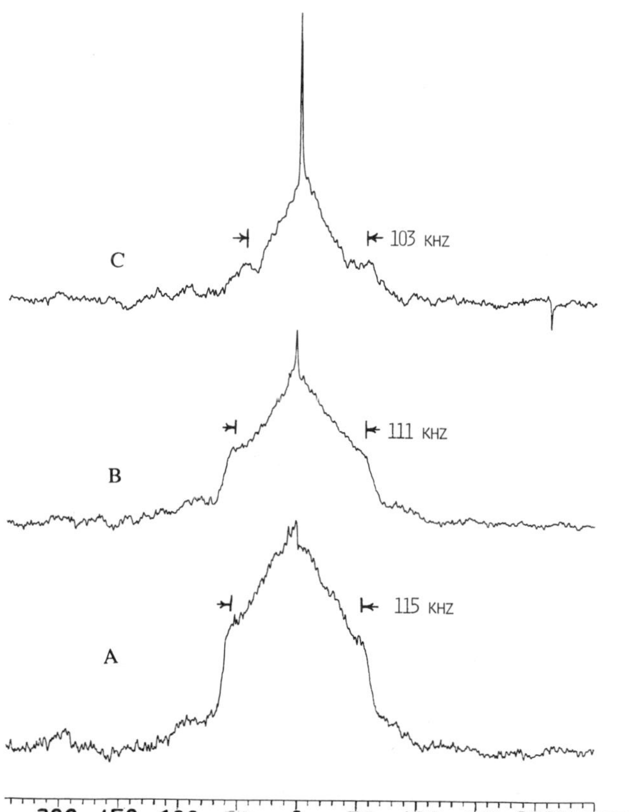

Figure 3. Wideline ^{2}H spectra of DAPS on silica as a function of added solvent: (A) no solvent, (B) small amount of water added, and (C) excess of water added.

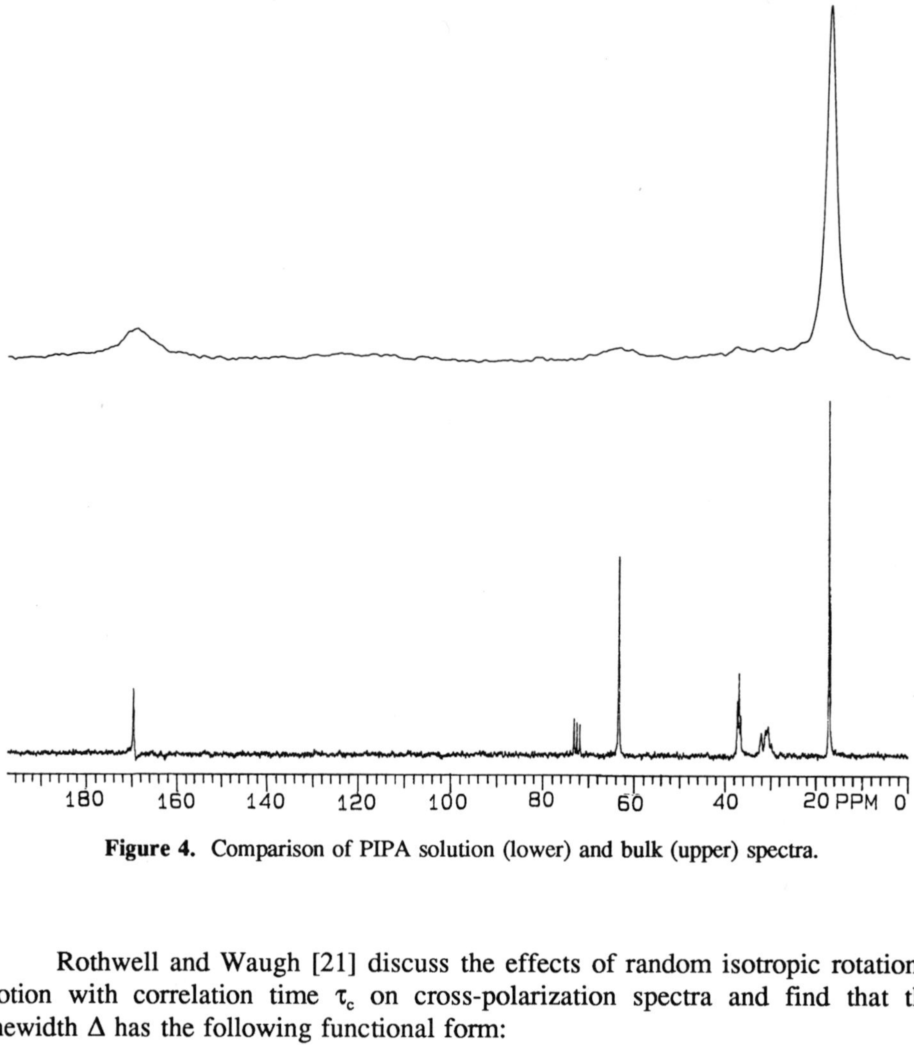

Figure 4. Comparison of PIPA solution (lower) and bulk (upper) spectra.

Rothwell and Waugh [21] discuss the effects of random isotropic rotational motion with correlation time τ_c on cross-polarization spectra and find that the linewidth Δ has the following functional form:

$$\Delta = 2\sqrt{2}\ln 2\ \frac{4\gamma_I^2\gamma_s^2\hbar^2}{15\ r^6}\ I(I+1)\ \frac{\tau_c}{1+\omega_1^2\tau_c^2}$$

where ω_1 is the rf irradiation intensity applied on resonance to the protons. We are only interested in the last term in this equation. This last term (and hence the linewidth) is largest if $\omega_1\tau_c$ is on the order of 1. Since τ_c is dependent on temperature in the simple model

$$\tau_c = \tau_o e^{Ea/RT}$$

the lines will show maximum broadening where

$$\omega_1 = \tau_o^{-1} e^{-Ea/RT}$$

We realize that the backbone motion of the polymer may not be described by simple Arrhenius behavior, but the general trend will be similar to that for small molecules. If we assume that the resonances not appearing in the bulk PIPA spectrum fail to do so because they are broadened by this effect, the presence of these peaks in the adsorbed polymer spectra is evidence that the motion of the surface-adsorbed polymer is significantly different from that of the bulk polymer. By contrast, for PMMA the bulk and surface-adsorbed spectra are quite similar in appearance.

4. Conclusions

^{13}C CP-MAS and ^{2}H wideline NMR have been used to probe adsorbed silane coupling agents on a silica surface. The motion of the adsorbed coupling agent is intermediate between a rigid solid and a small molecule liquid. With the addition of water, a fraction of the coupling agents undergo much faster motion. The present work, while not yet quantitative, shows the potential for the use of deuterium NMR to probe the motion of coupling agents on surfaces.

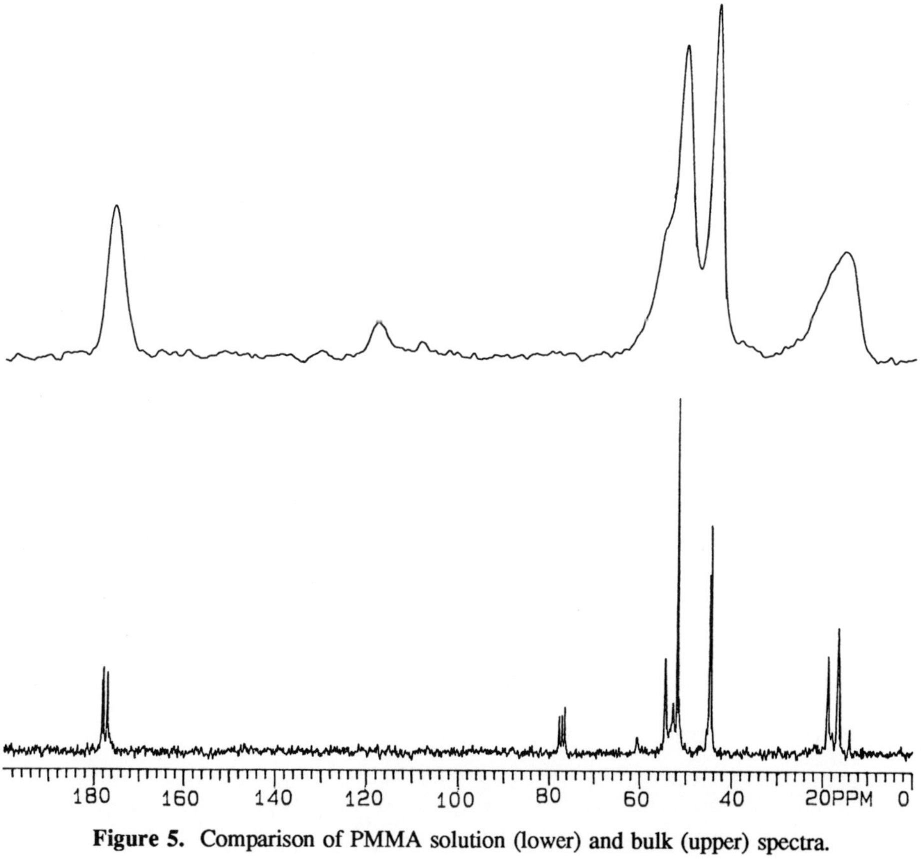

Figure 5. Comparison of PMMA solution (lower) and bulk (upper) spectra.

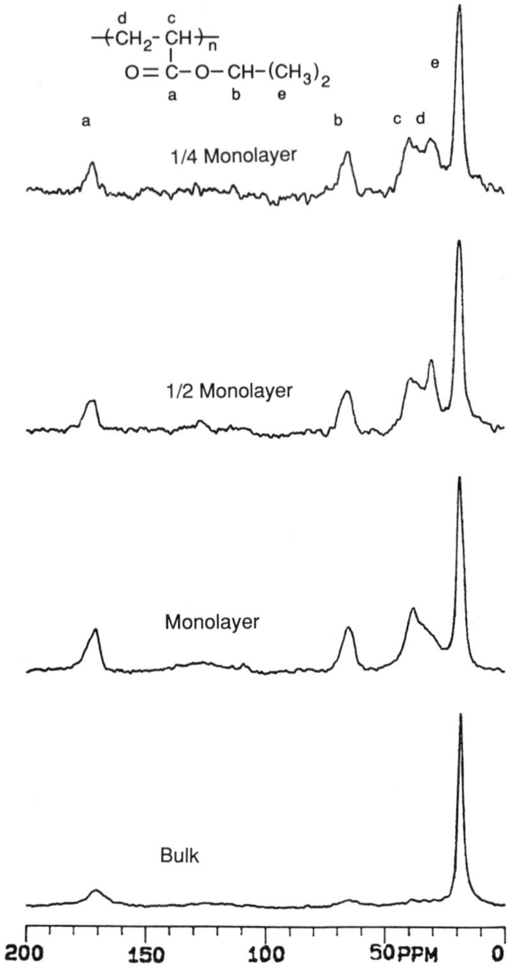

Figure 6. Spectra of PIPA at varying surface concentrations and in bulk.

The CP-MAS spectra of PIPA show the effects of surface adsorption on the ^{13}C linewidths of certain carbons in this polymer. We believe these effects are due to slower motion (longer correlation time τ_c) of the surface-bound material. PIPA has a T_g of 268 K [22], so molecular motion of significant speeds in the bulk polymer at room temperature is not unlikely. PMMA, with a T_g above room temperature [22], has molecular motion in the bulk phase comparable to that in the surface-adsorbed polymer, resulting in similar spectra for both cases.

5. Acknowledgements

The authors wish to thank the donors of the Petroleum Research Fund, administered by the American Chemical Society, and the Office of Naval Research for their financial support.

6. References

1. Naviroj, S.; Koenig, J. L. and Ishida, H., *J. Macromol. Sci. -Phys.*, *B22*, 291, 1983.
2. Naviroj, S.; Culler, S. R.; Koenig, J. L. and Ishida, H., *J. Colloid Interface Sci.*, *97*, 308, 1984.
3. Urban, M. W. and Koenig, J. L., *Appl. Spectros*, *40*, 513, 1986.
4. Leyden, D. E.; Kendall, D. S. and Waddell, T. G., *Anal. Chim. Acta.*, *126*, 207, 1981.
5. Chiang, C.; Liu, N. and Koenig, J. L., *J. Colloid Interface Sci.*, *86*, 26, 1982.
6. Zaper, A. M. and Koenig, J. L., *Polym. Composites*, *6*, 156, 1985.
7. Sindorf, D. W. and Maciel, G. E., *J. Am. Chem. Soc.*, *105*, 1848, 1983.
8. Gandoga, M. E. and Gilpin, R. K., *J. Magn. Reson.*, *53*, 140, 1983.
9. Albert, K.; Evers, B. and Bayer, E., *J. Magn. Reson.*, *62*, 428, 1985.
10. Pearson, R. M., *J. Catalysis*, *46*, 279, 1977.
11. Gottieb, H. E. and Luz, Z., *J. Magn. Reson.*, *54*, 257, 1983.
12. Bovey, F. A. and Jelinski, L. W., *J. Phys. Chem.*, *89*, 571, 1985.
13. Facchini, L. and Legrand, A. P., *Macromolecules*, *17*, 2405, 1984.
14. Chang, J. J.; Pines, A.; Fripiat, J. J. and Resing, H. A., *Surface Sci.*, *47*, 661, 1975.
15. Resing, H. A.; Garroway, A. N.; Weber, D. C.; Ferraris, J. and Slotfeldt-Ellingsen, D., *Pure Appl. Chem.*, *54*, 595, 1982.
16. Douglass, D. C. and McBrierty, V. J., *Polym. Eng. Sci.*, *19*, 1054, 1979.
17. Vega, A. J., *Polym. Preprints*, *22(22)*, 282, 1981.
18. Botham, R. and Thies, C., *J. Colloid Interface Sci.*, *31*, 1, 1969.
19. Stein, J.; Valenty, S. J.; Smith, G. A.; Brezniak, D. V. and Prutzman, L. C., *Macromolecules*, *19*, 2291, 1986.
20. Blum, F. D.; Funchess, R. B. and Meesiri, W., *Interfaces in Polymer, Ceramic and Metal Matrix Composites*, Ishida, H., Ed., Elsevier Science, New York, 1988, p 209.
21. Rothwell, W. P. and Waugh, J. S., *J. Chem. Phys.*, *74*, 2721, 1981.
22. Brandrup, J. and Immergut, E. H., Eds. *Polymer Handbook*, Wiley Interscience, New York, 1975.

Chapter 17

SOLID-STATE ^{2}H AND ^{13}C NMR STUDY OF THE STRUCTURE OF POLYANILINES

S. KAPLAN, E. M. CONWELL, A. F. RICHTER AND A. G. MACDIARMID

1. Introduction

Polyaniline has been the subject of considerable scientific inquiry because of its unique electrical behavior and its potential as an environmentally stable conducting polymer. It is synthesized by the electrochemical or chemical oxidative polymerization of aniline and can exist as a number of unique structures, characterized by the oxidation state, i.e., the ratio of amine to imine nitrogens, and the extent of protonation. These different structures can be interconverted by acid/base or oxidation/reduction treatment. Six basic repeat unit building blocks have been proposed from which the different forms of polyaniline are composed [1]. They are designated by 1 = reduced or 2 = oxidized, followed by A = amine or imine base, S′ = monoprotonated salt or S" = diprotonated salt. The unprotonated forms of polyaniline consist of reduced base units, 1A, and oxidized base units, 2A, represented as follows:

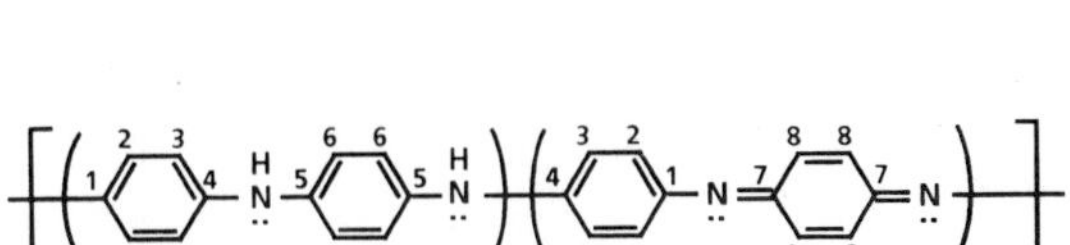

1A 2A (I)

S. KAPLAN AND E. M. CONWELL - Xerox Webster Research Center, 800 Phillips Road 0114-39D, Webster, New York 14580; A. F. RICHTER AND A. G. MACDIARMID - Department of Chemistry, University of Pennsylvania, Philadelphia, Pensylvania 19104.

Solid State NMR of Polymers, Edited by L. Mathias
Plenum Press, New York, 1991

where the oxidation state of the polymer increases with decreasing values of y ($0 \leq y \leq 1$). Claims have been made that polyaniline polymers having the following compositions (and others) have been isolated: the fully reduced leucoemeraldine base (y = 1), the half oxidized emeraldine base (y = 0.5), and the fully oxidized pernigraniline base (y = 0) [1]. Protonation of the base forms leads to polymers whose conductivity depends upon the ratio of reduced and oxidized units as well as the extent of protonation. Of particular interest is the highest conductivity form, the emeraldine salt, which has an idealized composition corresponding to an equal number of 1A and 2S" units:

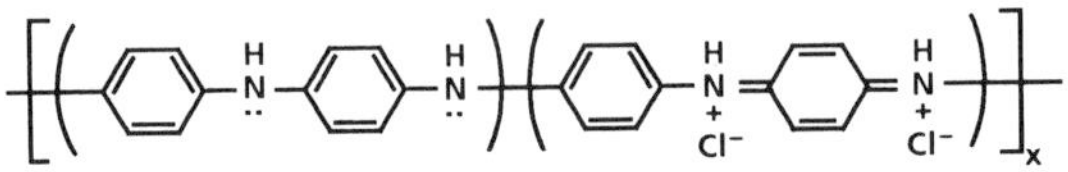

1A 2S" (II)

Structural characterization of polyanilines has been limited because, like most conducting polymers, they are largely insoluble in common organic solvents. X-ray diffraction measurements show that emeraldine base and emeraldine hydrobromide are completely amorphous polymers [2]. Raman studies of films of emeraldine base have identified the presence of para-disubstituted benzene and quinone diimine moieties [3] and, taken together with infrared spectra [3,4], have provided evidence for a head-to-tail polymerization of aniline, with no ortho incorporation of phenylenediamine groups. Cross-polarization magic angle spinning ^{13}C nuclear magnetic resonance (CPMAS NMR) spectra also resolve localized benzenoid and quinoid ring structures [5,6], indicating that extensive electron delocalization along the backbone does not occur. For the conducting salt form, X-ray photoelectron spectroscopy (XPS) N_{1s} and C_{1s} spectra [7], supported by CPMAS NMR [8], show that most of the charge accompanying protonation resides on the nitrogen atoms, with very little on the carbons. In this paper NMR techniques provide further insight into the structure and dynamics of three polymers: leucoemeraldine base, emeraldine base, and the emeraldine HCl salt.

2. Experimental

2.1. Material Preparation

Polyanilines used in the current study were synthesized by methods previously described [1,9,10]. Emeraldine hydrochloride was prepared by oxidizing a 1 M aqueous HCl solution of aniline (or aniline-d_5) with ammonium peroxydisulfate, $(NH_4)_2S_2O_8$, at ~5 °C. The precipitate was washed with 1 M HCl and dried under vacuum. The emeraldine hydrochloride was converted to the base form by stirring with a dilute (0.1 M) aqueous solution of NH_4OH, followed by repeated washing with 0.1 M aqueous NH_4OH and drying under vacuum. Leucoemeraldine base (hereinafter referred to as leucoemeraldine) was prepared by reduction of the emeraldine base with excess neat degassed hydrazine for 48 h under static vacuum (initial pressure: 1 μm

Hg). In order to avoid oxidation, leucoemeraldine samples were handled in an argon atmosphere, and NMR magic angle spinning was accomplished using nitrogen rather than air as the driving gas. Deuterated polymers were weighed and loaded into 5 mm NMR tubes in a glove box, and then attached to a vacuum line for flame sealing.

2.2. NMR Spectroscopy

All solid-state NMR spectra were obtained on a Bruker CXP spectrometer operating at 200 MHz for ^{1}H in a Cryomagnet Systems, Inc. 47 KGauss widebore superconducting magnet system. ^{13}C cross-polarization (CP) measurements at 50.3 MHz were performed in a Doty Scientific probe with magic angle spinning (MAS) at 4.2-5.2 KHz. Spectral assignments are facilitated by using the cross-depolarization (CDP) technique (Fig. 1) to distinguish between protiated and non-protiated carbons [11,12]. (The term protiated is used here to refer to carbons with attached hydrogens so as to avoid confusion with the term protonated, which, in this paper, refers to H^+ addition.) After the cross-polarization mixing period and prior to carbon observation, the 1H rf phase (i.e., ^{1}H spin temperature) is inverted for 50-100 μs. During this period carbon spins cross-polarize in the reverse direction, protiated carbons the fastest and non-protiated carbons the slowest. Therefore, in the CDP spectra non-protiated carbons appear with nearly full intensity while protiated carbons are reduced in intensity or inverted.

Solid-state ^{2}H NMR spectra were measured on ring-deuterated polyanilines at 30.7 MHz in a Cryomagnet Systems, Inc. high power probe. The probe has a Q of ~100 and achieves 90° pulse widths of 2 μs with ~500 watts power in a 5 mm sample tube. Data were acquired with quadrature detection using the quadrupolar-echo pulse sequence, ($90°_x$-τ_1-$90°_y$-τ_1-acquire-RD), with alternate scans added and subtracted in order to minimize distortions from probe acoustic ringing. Typically, the slowest relaxing deuterium component had a T_1 of a few seconds; consequently, recycle delays

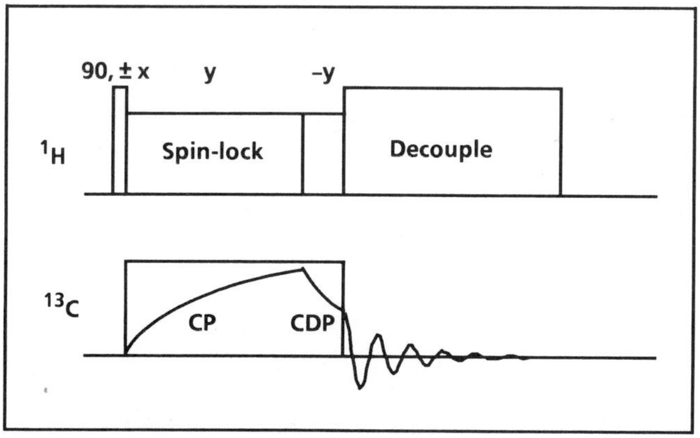

Figure 1. Pulse sequence for the cross-depolarization technique (CDP) used to invert carbons with directly attached hydrogens. The ^{1}H spin temperature reversal (-y) pulse is typically 50-100 μs. The standard CP experiment is identical except this pulse is omitted.

(RD) of 10-20 s, depending upon sample and temperature, were used. Data were acquired with a dwell of 0.2 μs per point (2.5 MHz acquisition bandwidth), and the free-induction decay was left-shifted 12 points (2.4 μs) in order to start at the echo peak. Selective observation of the more quickly relaxing component was achieved with five saturation pulses preceeding the solid-echo sequence as follows: $((90^\circ\text{-}100\ \text{ms})_5\text{-}90^\circ_x\ \text{-}\tau_1\text{-}90^\circ_y\text{-}\tau_1\text{-acquire-RD})$.

3. Results and Discussion

3.1. ^{13}C NMR

Figure 2 shows ^{13}C CP-MAS and CDP-MAS spectra of leucoemeraldine, partially air-oxidized leucoemeraldine and emeraldine base polymers. Peak assignments, determined previously,[13] correspond to the numbering scheme given above in structure I. The C = N quinoid resonance (C-7) observed at 157.3 ppm in

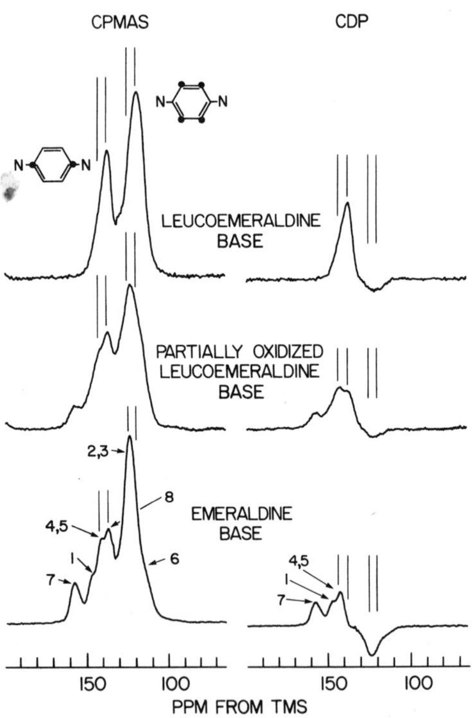

Figure 2. ^{13}C CP and CDP magic angle spinning spectra of (top) leucoemeraldine base, (center) leucoemeraldine oxidized at 80 °C and 100% relative humidity for 2 days and (bottom) emeraldine base polymers. Top and center spectra: spinning speed = 4.2 KHz, no. of scans = 20,000. Bottom spectra: spinning speed = 5.0 KHz, no. of scans = 48,000. All spectra: mixing time = 0.5 ms, recycle delay = 1.0 s. Vertical bars are placed at 142, 136, 123 and 118 ppm. Assignments correspond to the numbering scheme of Structure I.

the emeraldine base spectra is absent from the leucoemeraldine spectra. However, after heating at 80 °C for 2 days in 100% humidity air (Fig. 2, center), the quinoid peak appears, and it is clear that oxidation has occurred, leading to the partial formation of emeraldine base. A similar result has been deduced from Raman and infrared spectroscopies [14]. It is plausible that, in the presence of water acting as a catalyst, oxygen extracts amine hydrogens to convert a 1A unit to a 2A unit. For our sample, oxidation is slow; even under severe conditions of heat and humidity for two days, only about 50% conversion occurred. However, sample color change from the initial light grey of leucoemeraldine to a dark green was evident after only ~10% conversion and to a dark blue shortly thereafter.

The non-protiated carbon resonance (136.3 ppm) of leucoemeraldine, diminishes in intensity in the partially oxidized material and is absent from emeraldine base. The 5.3 ppm shift difference from C-5 of emeraldine suggests that the resonance at 136.3 ppm can be assigned to blocks of 1A units. The absence of spectral intensity at 136.3 ppm in the CDP spectrum of emeraldine base leads to the conclusion that all 1A units have 2A units as neighbors. Furthermore, ^1H NMR spin counting shows that there are nearly equal numbers of 1A and 2A units (y = 0.46) [13]. Taken together, these observations indicate that emeraldine base is primarly an alternating head-to-tail copolymer of 1A and 2A units.

The 8-12 ppm spectral linewidths measured for both emeraldine base and leucoemeraldine are attributed primarily to local fluctuations in conformational and configurational geometries and a distribution in chain packing [13]. The ^{13}C CPMAS spectrum of the HCl salt of emeraldine is a single ~60 ppm broad unresolved resonance. It has been shown that this featureless signal is inhomgeneously broadened, likely as a result of local variations in charge density along the polymer backbone [13].

3.2. ^2H NMR Lineshapes

For polymers with 1,4-disubstituted benzenoid rings in the backbone, 180° ring-flips can occur, and the flipping populations and rates are a probe of local structure [15-20]. Such flips are absent in crystalline regions of semicrystalline polymers [17]. However, in amorphous regions of polymers, deuterium NMR lineshapes characteristic of flipping as well as static rings are observed. This marked motional heterogeneity is taken as a manifestation of differences in packing at the different sites [15,16]. For amorphous polymers there is usually a very broad distribution of correlation times, extending as much as 6-10 orders of magnitude [19]. Furthermore, it is often necessary to incorporate rapid small angle librations as well into simulations to fit the experimental spectra [15-17,20]. It is important to note that ring flipping may be influenced by factors other than local free volume. For example, π-bonding to quinoid moieties will prevent rotation. It has been suggested that for poly (butyleneterephthalate) the partial double bond character of the carbonyl-aromatic bonds constrains flipping [17]. Also, for charged species, such as emeraldine hydrochloride, the electrostatic attraction of counterions may be important.

^2H spectra for the ring-deuterated emeraldine base as a function of temperature are plotted in Fig. 3, which shows the fully relaxed spectra obtained with echo evolution times, τ_1, of 20 μs and 80 μs, as well as partially relaxed spectra with τ_1 = 20 μs. Partial relaxation, achieved by the presaturation experiment, suppresses the static powder pattern, and the spectra are dominated by a lineshape characteristic of rings undergoing 180° flips. The residual static component in the partially relaxed spectra can be entirely accounted for by incomplete saturation of the rigid deuterons. In Fig. 4 the τ_1 = 20 μs data are deconvoluted into rigid and flipping ring components. The deconvolution was accomplished by taking appropriate linear combinations of the simple solid-echo spectra and the spectra obtained by the presaturation experiment. Calculated spectra, using a quadrupole coupling constant ($e^2qQ/\hbar$) of 178 KHz for both the flipping and non-flipping components, are plotted in Fig. 4 for the T = 413 K data.

As seen in Fig. 4, the rigid and flipping component lineshapes are nearly independent of temperature. This observation holds for all three polymers and indicates that polymer dynamics can be conveniently described in terms of rings that undergo rapid flips ($\tau_c<10^{-7}$ s) and rings that flip slowly ($\tau_c>10^{-4}$ s), if at all. This is not to say that there is not a continuous distribution of flip rates. In fact, ^2H and ^{13}C

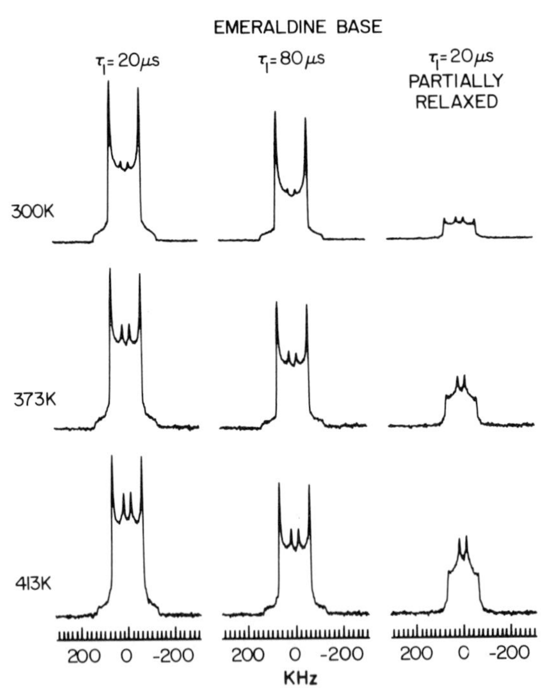

Figure 3. ^2H Spectra of ring-deuterated emeraldine base at T = 300K, 373K and 413K. At each temperature spectra are obtained with the two pulse quadrupole echo sequence using echo delay times of τ_1 = 20 μs and τ_1 = 80 μs and with the presaturation sequence (see text) to obtain partially relaxed spectra. All spectra were measured with a repetition delay of 10 s. For T = 300K, no. of scans = 6600 while for T = 373K and 413K, no. of scans = 2300.

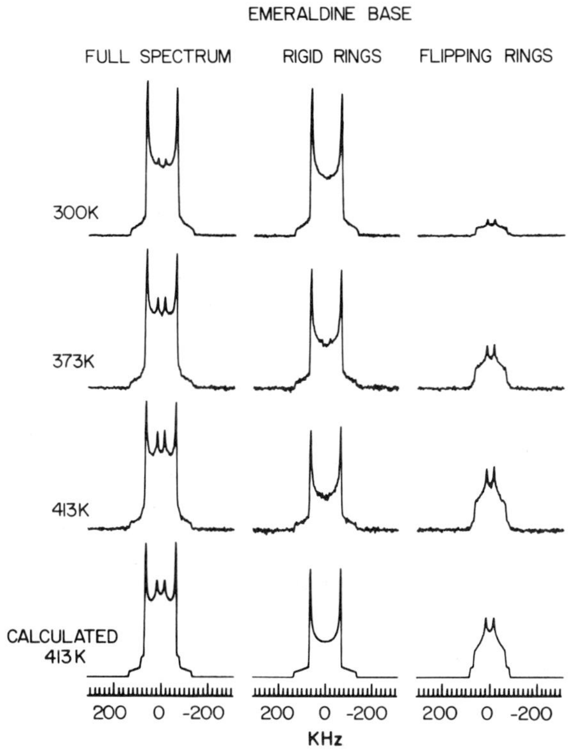

Figure 4. ^2H Spectra of ring-deuterated emeraldine base. Fully relaxed spectra are obtained with the quadrupole echo pulse sequence using the same experimental conditions as described for the left column of Fig. 1. Rigid and flipping ring lineshape components are experimentally determined from appropriate linear combinations of the fully and partially relaxed spectra plotted in Fig. 3. The calculated spectra for the T = 413K data use a quadrupole coupling constant ($e^2qQ/\hbar$) of 178 KHz. Calculated spectra were corrected for both the excitation bandwidth of the finite length radiofrequency pulses [31] as well as for the frequency response of our high Q (= 100) probe. Corrections for frequency-dependent signal loss due to motion during echo formation time, τ_1,[22] were not necessary to achieve good lineshape fits.

relaxation data suggest that there is a very wide distribution [21]. Because of deuteron signal reduction due to intermediate regime motion (10^{-7} s<τ_c<10^{-4} s) during the echo formation time, τ_1 [19,22,23], it is often difficult to distinguish a very broad distribution of correlation times from a bimodal superposition of very rapid and very slow motions. For emeraldine base, signal reduction factors based on total spectral integration have been computed. After accounting for signal loss with increasing temperature due to the Boltzmann factor, it was found that from 303 K to 413 K signal intensity loss was only ~4%. The absolute signal reduction factors may be higher, but since there is little additional signal reduction over the temperature range accessed, the simple determination of the ratio of fast to slow flippers is a good approximation of the polymer dynamics. Deuterium lineshapes of ring-deuterated leucoemeraldine and emeraldine hydrochloride are identical to those of emeraldine base; at a given temperature only the relative contributions of the two components differ. Signal intensity reductions over the measured temperature range were ~12% and ~2% for leucoemeraldine and emeraldine hydrochloride, respectively.

Because the leucoemeraldine and emeraldine base polymers have a distribution in packing densities, it is reasonable that local free volume determines the rate of ring flipping. With increasing temperature, free volume increases, and rings are sterically less confined by neighboring chains. Therefore, flipping rates increase and the number of fast flippers should increase, as is observed in the deuterium spectra of Fig. 4. It has been shown that the free volume fraction, f, for a polymer glass can be expressed by a simple exponential relationship [24,25]:

$$f = Ae^{\bar{\varepsilon}/kT}$$

where A is nearly constant, $\bar{\varepsilon}$ is the mean thermodynamic energy required for hole (i.e., free volume) formation, k is Boltzmann's constant and T is temperature. Under the assumption of proportionality between the fraction of fast flippers and the fractional free volume, the slope of the Arrhenius plots of the fraction of fast flippers vs. inverse temperature (Fig. 5) should give the mean energy required to create sufficient free volume for rotation to occur. For the emeraldine and leucoemeraldine base forms ε derived from these plots is ~3.0 Kcal/mole, which is consistent with hole energies determined from equation of state calculations for typical amorphous polymers [26]. This agreement supports the hypothesized relationship between the

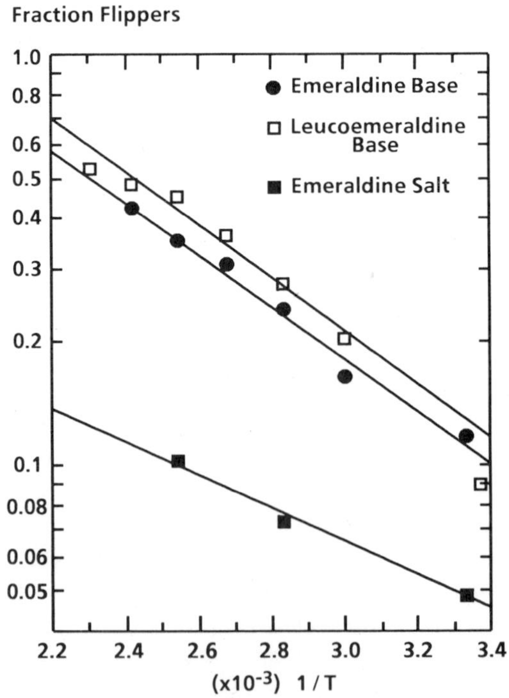

Figure 5. Semi-logarithmic plot of the lineshape fraction attributed to rings undergoing rapid 180° flips as a function of inverse temperature for emeraldine base, leucoemeraldine and emeraldine hydrochloride. Solid lines are linear least square regression fits.

NMR measured number of fast flippers and free volume, at least over the measured temperature range. Since 1 of every 4 rings in the emeraldine base is quinoid and cannot flip, emeraldine base should have fewer flippers than leucoemeraldine, as was observed. A similar fraction of benzenoid rings flips in both polymers, suggesting that local chain packing is similar for the two polymers.

The emeraldine salt, on the other hand, has 30-40% as many flipping rings as the emeraldine base over the measured temperature range. This decrease in the number of flippers is attributable to either a direct steric interaction with the chloride counterions or to partial π-electron delocalization into the ring-nitrogen bonds. X-ray photoelectron valence band spectra show that protonation of the emeraldine base to form emeraldine hydrochloride is accompanied by a change in chain geometry [27]. This structural change is probably a manifestation of the partial π-character of approximately half of the ring-nitrogen bonds.

Emeraldine hydrochloride (II), schematically depicted above, likely rearranges via a proton-assisted spin-unpairing mechanism to form two isolated polarons (semiquinone radical-cations), [4,9] shown in III.

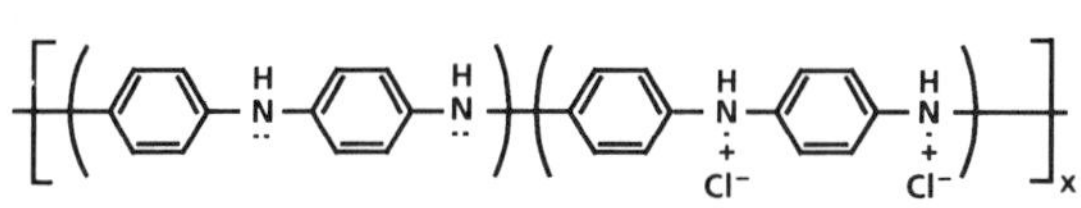

(III)

It is believed that these polaron pairs separate because of colomb repulsion to form the poly (semiquinone radical cation) structure IV, where charge is situated on alternate nitrogens [9].

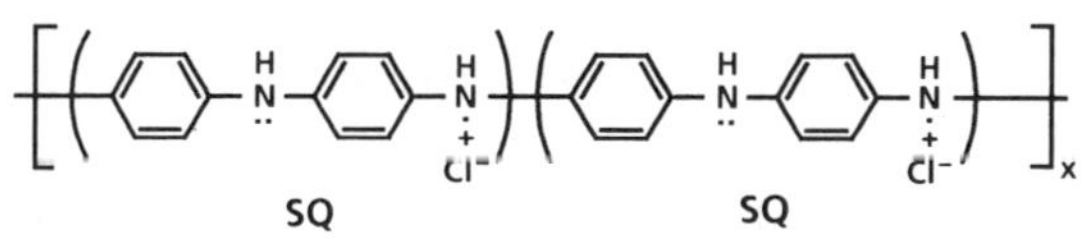

(IV)

In IV, the electron density distribution associated with the polarons will impart some π-character to at least one of the C-N bonds of each ring. This partial π-character and/or the steric interactions with adjacent chloride counterions should inhibit flipping. However, for the unseparated polaron structure of III, only three of every four rings would likely be frozen, while one ring could flip, given sufficient free volume. Therefore, the experimental determination that emeraldine hydrochloride has approximately 1/3 the number of fast flippers of the emeraldine base is consistent with structure III but not with structure IV. This conclusion is dependent upon the assumptions that rings adjacent to polaron centers cannot flip rapidly and that chain packing in the base and salt forms is similar. These assumptions are reasonable since the temperature dependence of the fraction of flippers in the salt form (Fig. 5) has a slope corresponding to ε~2 Kcal/mole, which is within experimental error of that measured for leucoemeraldine and emeraldine base. One would expect this value to be considerably higher if rings adjacent to polaron centers could also flip (because they require additional energy to do so).

Structure III, however, is not a realistic model for emeraldine hydrochloride, because polarons are not expected to remain strictly localized [4,9]. An alternative explanation for the ^{2}H NMR data is derived from recent redox [28] and volumetric [29] titration studies suggesting that protonation of emeraldine base (which has been prepared and handled in air so that y~0.40-0.46 [13,28]) is simultaneously accompanied by reduction in oxidation state (due to hydrolysis) to y~0.6 [28] and that some amine nitrogens are protonated to convert 1A to 1S′ units [29,30]. If reduced groups constitute 60% of the protonated polymer, then redox rearrangement of equal numbers of 1A and 2S" units to form poly (semiquinone radical cation) segments [30] would lead to structure V.

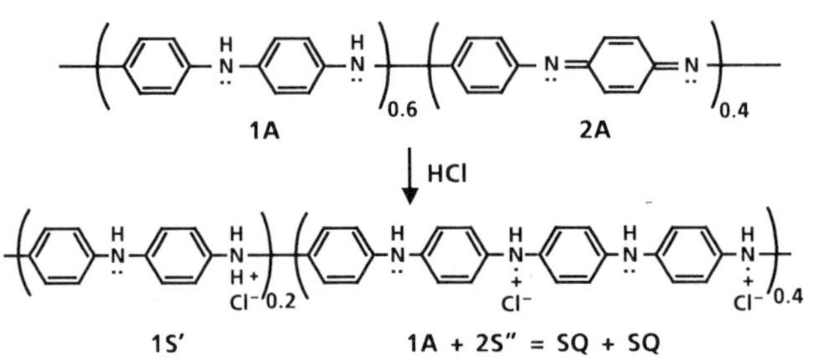

(V)

With the assumption that rings in 1S′ units can flip (i.e., chlorides do not inhibit flipping) while those of the polaron segments cannot, structure V is then expected to have nearly 30% the number of flippers of the emeraldine base. Finally, structures III and V are both consistent with the elemental analysis determination of Cl/N = 0.5 [10] and the NMR measured stoichiometry for the ring-deuterated salt of H/N = 1.0 [13] (for structure III: H/N = 1.0; for structure V: H/N = 1.1). The present data is consistent with but does not prove structure V to be the correct representation of emeraldine hydrochloride. In any case, it has been demonstrated that the simplified picture of IV requires 1A (as in III) or 1S′ (as in V) "defects" to explain the observed population of flippers. The data says nothing about the distribution of such defects along the chains.

4. Conclusions

Carbon and deuterium NMR measurements have been used to characterize the structure and dynamics of insulating and conducting forms of polyaniline. By comparison with the ^{13}C NMR spectra of pristine and partially oxidized leucoemeraldine, emeraldine base is shown to be an alternating copolymer of reduced 1A, [-(C_6H_4)-N(H)-(C_6H_4)-N(H)-], and oxidized 2A, [-(C_6H_4)-N=(C_6H_4)=N-], units. Separation of deuterium spectra into flipping and rigid rings has provided a convenient approximation for the dynamics of polyanilines, although deuteron spin alignment and carbon spin-lattice relaxation [21] demonstrate that there is, in fact, a wide distribution of correlation times for ring flips. Temperature dependence of the fraction of benzenoid rings undergoing rapid 180° flips reveals a wide distribution of local packing densities, consistent with the high degree of structural heterogeneity implicit in the ^{13}C linewidths. Finally, for the conducting emeraldine hydrochloride, the determination that about a third as many rings are fast flippers as compared with the emeraldine base, is consistent with a structure containing unseparated polarons or with a more realistic one containing poly (semiquinone radical cation) segments and residual reduced units.

5. Acknowledgements

This work was supported in part (A.F.R. and A.G.M.) by the Defense Advanced Research Projects Agency through a grant monitored by the Office of Naval Research. Helpful discussions with A. J. Epstein are gratefully acknowledged.

6. References

1. Chiang, J. -C. and MacDiarmid, A. G., *Synth. Met.*, *13*, 193-205, 1986.
2. Annis, B. K.; Narten, A. H.; MacDiarmid, A. G. and Richter, A. F., *Synth. Met.*, *22*, 191, 1987.
3. Furukawa, Y.; Hara, T.; Hyodo, Y. and Harada, I., *Synth. Met.*, *16*, 189-198, 1986.
4. Wudl, F.; Angus, Jr., R. O.; Lu, F. L.; Allemand, P. M.; Vachon, D. J.; Nowak, M.; Liu, Z. X. and Heeger, A. J., *J. Am Chem. Soc.*, *109*, 3677-3684, 1987.
5. Devreux, F.; Bidan, G.; Syed, A. A. and Tsintavis, C., *Journal de Physique*, *46*, 1595-1601, 1985.
6. Hjertberg, T.; Salaneck, W. R.; Lundstrom, I.; Somasiri, N. L. D. and MacDiarmid, A. G., *J. Polym. Sci: Polym Lett. Edn.*, *23*, 503-508, 1985.
7. Salaneck, W. R.; Lundstrom, I.; Hjertberg, T., Duke, C. B.; Conwell, E. M.; Paton, A.; MacDiarmid, A. G.; Somasiri, N. L. D.; Huang, W. S. and Richter, A. F., *Synth. Met.*, *18*, 291-296, 1987.
8. Stafström, S.; Sjögren, B.; Wennerström, O. and Hjertberg, T., *Synth. Met.*, *16*, 31-39, 1986.
9. MacDiarmid, A. G.; Chiang, J. -C.; Richter, A. F. and Epstein, A. J., *Synth. Met.*, *18*, 285-290, 1987.
10. MacDiarmid, A. G.; Chiang, J. -C.; Richter, A. F.; Somasiri, N. L. D. and Epstein, A. J. in *Conducting Polymers*, Alcácer, L., Ed., Reidel Publications, Dordrecht, Holland, 1987; pp 105-120.
11. Melchior, M. T., 22nd Experimental NMR Conference, Asilomar, CA, Poster B29, 1981.
12. Zumbulyadis, N. J., *Chem. Phys.*, *86*, 1162-1166, 1987.
13. Kaplan, S.; Conwell, E. M.; Richter, A. F. and MacDiarmid, A. G., *J. Am. Chem. Soc.*, *110*, 7647-7651, 1988.
14. Ohira, M.; Sakai, T.; Takeuchi, M.; Kobayashi, Y. and Tsuji, M., *Synth. Met.*, *18*, 347-352, 1987.

15. Spiess, H. W., *Colloid and Polym. Sci.*, *261*, 193-209, 1983.
16. Spiess, H. W. in *Advances in Polymer Science*, *66*, Kausch, H. H. and Zachmann, H. G., Eds.; Springer-Verlag, Berlin Heidelberg, 1985.
17. Cholli, A. L.; Dumais, J. J.; Engel, A. K. and Jelinski, L. W., *Macromolecules*, *17*, 2399-2404, 1984.
18. Dumais, J. J.; Cholli, A. L.; Jelinski, L. W.; Hedrick, J. L. and McGrath, J. E., *Macromolecules*, *19*, 1884-1889, 1986.
19. Wehrle, M.; Hellmann, G. P. and Spiess, H. W., *Colloid and Polym. Sci.*, *265*, 815-822, 1987.
20. Roy, K. R.; Jones, A. A. and Inglefield, P. T., *Macromolecules*, *19*, 1356-1362, 1986.
21. Kaplan, S.; Conwell, E. M.; Richter, A. F. and MacDiarmid, A. G., *Macromolecules*, *22*, 1669-1675, 1989.
22. Spiess, H. W. and Sillescu, H., *J. Magn. Reson.*, *42*, 381-389, 1981.
23. Wittebort, R. J.; Olejniczak, E. T. and Griffin, R. G., *J. Chem Phys.*, *86*, 5411-5420, 1987.
24. Chow, T. S., *Advances in Rheology*, Vol. 3, Polymers, Mena, B., et al., Eds., 177-183, 1984.
25. Chow, T. S., *J. Chem. Phys.*, *79*, 4602-4607, 1983.
26. Chow, T. S., *Macromolecules*, *22*, 698-701, 1989.
27. Salaneck, W. R.; Lundstrom, I.; Hjertberg, T.; Duke, C. B.; Conwell, E. M.; Paton, A.; MacDiarmid, A. G.; Somasiri, N. L. D.; Huang, W. S. and Richter, A. F., *Synth. Met.*, *18*, 291-296, 1987.
28. Asturias, G. E.; MacDiarmid, A. G.; McCall, R. P. and Epstein, A. J., *Synth. Met.*, *29*, E151-E156, 1989.
29. Menardo, C.; Genoud, F.; Nechtschein, M.; Travers, J. P. and Hani, P., in *Electronic Properties of Conjugated Polymers*, H. Kuzmany et al., Eds., Springer-Verlag, 244, 1987.
30. Ray, A.; Richter, A. F.; MacDiarmid, A. G. and Epstein, A. J., *Synth. Met.*, *29*, E151-E156, 1989.
31. Bloom, M.; Davis, J. H. and Valic, M. I., *Can. J. Phys.*, *58*, 1510-1517, 1980.

Chapter 18

THE SOLID STATE ^{29}Si AND ^{13}C NMR OF POLY(DI-*n*-ALKYLSILYLENES)

F. A. BOVEY AND F. C. SCHILLING

1. Abstract

Poly(di-*n*-hexylsilylene) is composed of a crystalline phase (I) in which the silicon main chain has an all-*trans* or planar zigzag conformation, together with a less ordered phase (II) in which non-trans conformations are present. The solid state ^{29}Si NMR spectrum exhibits two resonances, the more shielded of which corresponds to Phase II. A I → II transition occurs at 40°-42°C, above which only the more shielded ^{29}Si resonance is observed. The ^{13}C spectra are consistent with the disordering of the sidechains in Phase II. The Phase I ^{13}C resonances exhibit a doubling which is not at present understood. Poly(di-*n*-pentylsilylene) and poly(di-*n*-butylsilylene) exhibit I → II transitions at 70° and 87°, respectively. Their ^{29}Si spectra show strikingly different behavior from that of the poly(di-*n*-hexylsilylene). The crystalline phase (I) shows a more shielded resonance than that of the disordered phase. This is consistent with the X-ray findings of a 7/3 helical backbone bond arrangement. The Phase I ^{13}C spectrum of poly(di-*n*-butylsilylene) shows peak doubling similar to that of the dihexyl polymer.

2. Introduction

The soluble polysilylenes having the novel structure

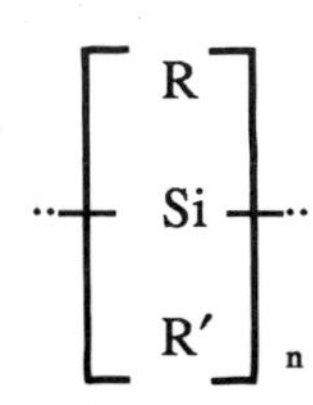

F. A. BOVEY AND F. C. SCHILLING - AT&T Bell Laboratories, Murray Hill, New Jersey 07974.

Solid State NMR of Polymers, Edited by L. Mathias
Plenum Press, New York, 1991

are a class of polymers which have recently generated great interest [1-5]. They exhibit a wide variety of physical properties depending on the nature of R and R′, which may be alkyl or aryl. When R ≠ R′, the silicon atoms become pseudoasymmetric, introducing the possibility of variations in sterochemical configuration, which may be viewed in terms of *meso (m)* or *racemic (r)* dyads:

m *r*

Judged from the configurational sequence probabilities revealed by ^{29}Si resonance intensities in solution spectra [6], tacticity seems to be Bernoullian with a P_m of 0.5 or nearly so. Copolymers with differently substituted silicon atoms are also known. In the work reported here we deal with polysilanes in which R=R′=*n*-alkyl=*n*-butyl, *n*-pentyl, or *n*-hexyl.

The polysilanes have been intensively studied by several groups, employing the entire armamentarium of structural methods, including UV, Raman and IR spectroscopy, X-ray, NMR (both solution and solid state), calorimetry, and light scattering. There have also been several theoretical studies of conformational energies which we shall refer to in due course.

In this discussion we will deal primarily with the solid state magic angle ^{13}C and ^{29}Si NMR spectroscopy of the three polysilanes we have indicated. However, the other properties of these polymers must receive some mention. Of these the most striking is the *thermochromic* behavior [7,8], which has stimulated much of the interest in these materials. This is illustrated in Fig. 1 (C. Walsh, unpublished

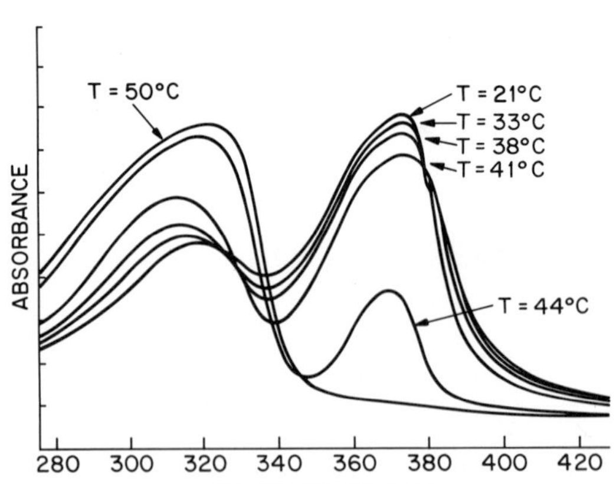

Figure 1. UV absorption spectra of poly(di-*n*-hexylsilylene) film as a function of temperature (C. Walsh, unpublished observations).

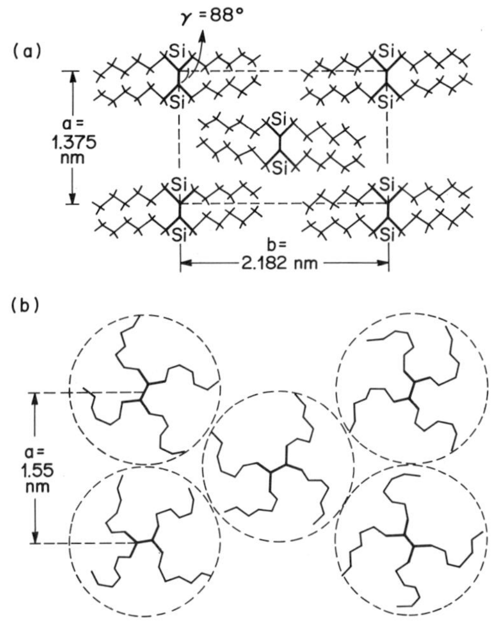

Figure 2. Schematic views of the solid state structures of (a) Phase I and (b) Phase II of poly(di-*n*-hexylsilylene) (PDHS), seen along the main chain direction.

observations), which shows the UV absorption spectrum of a poly(di-*n*-hexylsilylene) (PDHS) film observed between 21° and 50°C. At 50°C a single absorption band at 320 nm is observed, very similar to the absorption maximum for this polymer in solution [7,8]. As the sample is cooled a second absorption band is observed at 374 nm which continues to grow on further cooling. This phenomenon is reversible. It is not observed in the solid-state absorption spectra of poly(di-*n*-butylsilylene) (PDBS) or poly(di-*n*-pentylsilylene) (PDPS). The chain conformation of PDHS below ca. 42°C has been determined by X-ray [9-11] and electron diffraction [11]. The conformation of the crystalline phase, termed Phase I, is found to correspond to an all-*trans* silicon backbone, with side-chains which are probably also all *trans* or nearly so. There is also a substantial disordered phase, Phase II, which in some preparations (particularly those crystallized from solution) may constitute the major share of the sample. Above 42°, this becomes the only phase present. As determined by NMR analysis of Phase II, non-*trans* conformations (vid. inf.) are introduced into the main chain and *gauche* conformations into the side-chains. The chains become more mobile and there is a substantial decrease in density (ca. 10%). Despite this conformational disordering, *inter*molecular order remains very high, the Si chains preserving their original direction and being packed as cylinders in a regular hexagonal array. The structures of Phase I and Phase II have not yet been determined by full crystal-structure analysis, but schematic models, [9-11,20] viewed along the main chain axis, are compared in Fig. 2.

The dramatic red shift shown in Fig. 1 for Phase I is attributed to the existence of "sigma conjugation" in the all-*trans* silicon backbone and its reduction when conformational irregularities are introduced above the disordering transition, in which Phase I disappears in favor of Phase II. The existence of π conjugation is well known in structures with extended series of conjugated double bonds, as is the fact that in such systems the first excited electronic state is lower in energy than for isolated double bonds, resulting in a marked red shift of the UV absorption maximum, even into the visible. In the silane backbone the silicon atoms are in sp^3 states, just as in a paraffinic hydrocarbon chain, but the sigma electrons appear to be capable of delocalization in the same manner as the π electrons in conjugated double bond systems. This sigma conjugation, together with the inherent flexibility of the silicon chain, results in a strong coupling of the electronic transitions with the chain conformation. We shall consider this matter further in the discussion of the conformations of poly(di-*n*-butylsilylene) (PDBS) and poly(di-*n*-pentylsilylene) (PDPS).

3. The Structure and NMR of PDHS in the Solid State

The molecular conformation of PDHS below the transition temperature has been determined by X-ray [9-11] and electron diffraction [11]. From uniaxially drawn samples at room temperature, molecular repeats of 0.400 nm [11] or 0.407 nm [9] were observed and interpreted as reflecting an all-*trans* silicon backbone. When PDHS is heated through its thermochromic transition (ca. 38-43°C), the changes observed in the X-ray diffraction patterns indicate progressive disappearance of all reflections from Phase I, accompanied by a large increase in the intermolecular diffraction peak of Phase II. This reflection becomes stronger and narrower as the temperature is raised well above the transition, even as high as 180°--a remarkable observation.

In Fig. 3 the solution and solid state ^{13}C NMR spectra of PDHS are compared. In Fig. 3(a) is shown the 50.3 MHz spectrum in toluene-d_8 solution. The resonance assignments have been made by 2D NMR experiments described elsewhere [6]. The progressive broadening of the resonances of carbons C-6 to C-1 corresponds to increasingly restricted motion near the sterically crowded main chain, making averaging of carbon-proton dipole-dipole interactions less complete. In Fig. 3(b) is shown the solid state PDHS ^{13}C spectrum [14] under conditions of magic angle spinning with relatively weak proton decoupling, sufficient only to collapse the multiplicity arising from scalar ^{13}C-^{1}H coupling but having no effect on the dipole-dipole interactions. Under these conditions, only Phase II (and any amorphous fraction) are mobile enough to be visible. Peak broadening is much accentuated compared to the solution spectrum, but the pattern is similar. Chemical shifts are virtually unchanged, showing that the average chain conformation in the mobile phase is the same as in solution.

In Fig. 4(a) both magic angle spinning and high power proton decoupling are applied. In addition, proton-carbon cross-polarization is employed to enhance the ^{13}C signal strength. (The temperature is 23°C). A contact time of 1 ms is allowed between the protons and carbons, sufficient to permit resonances of both Phase I and Phase II to appear [14]. The designations of the crystalline Phase I resonances are underlined while those of the mobile Phase II are not underlined. We observe that the Phase II resonance for each is more shielded than the corresponding Phase I resonance, reflecting the presence of *gauche* conformations in the disordered side chains of Phase II. (It should be emphasized that the peaks of Phase I and Phase II cannot be quantitatively compared because of differences in cross-polarization efficiency, the proportion of Phase II being actually higher than Fig. 4(a) suggests).

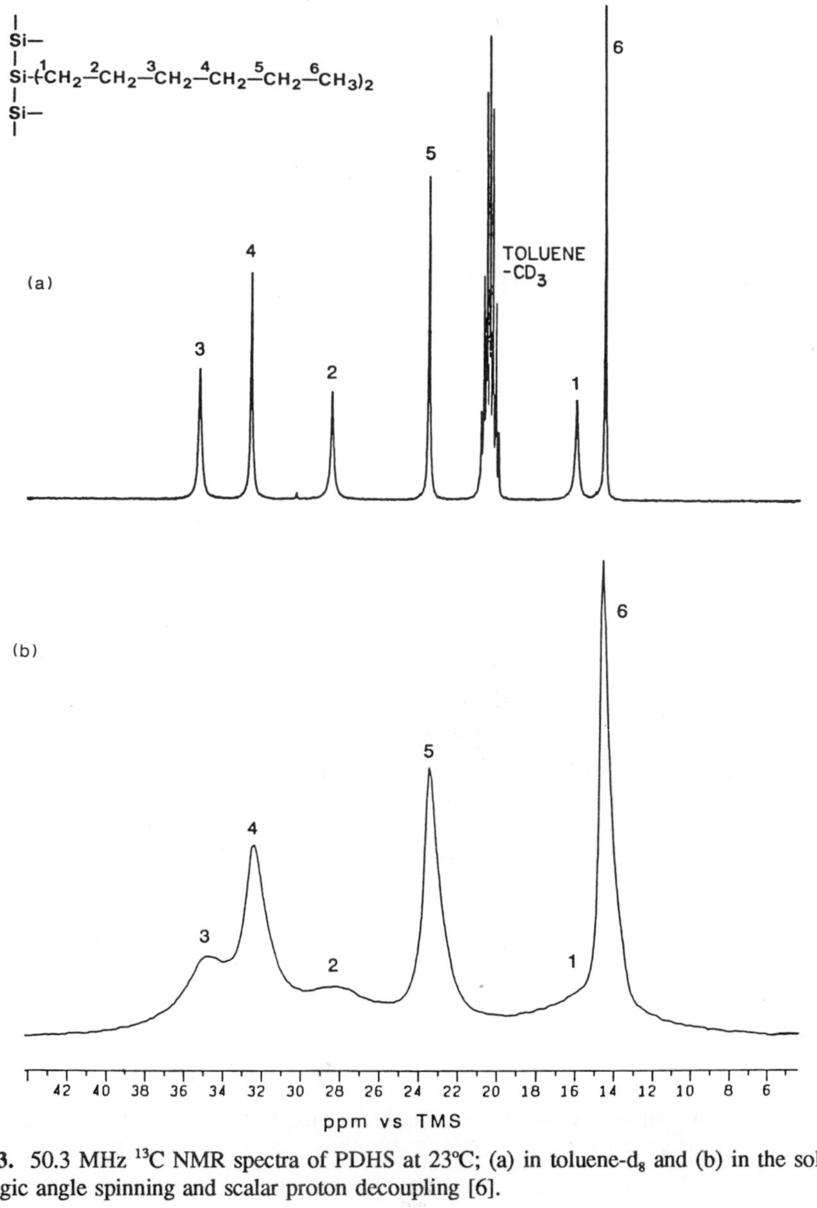

Figure 3. 50.3 MHz ^{13}C NMR spectra of PDHS at 23°C; (a) in toluene-d_8 and (b) in the solid state with magic angle spinning and scalar proton decoupling [6].

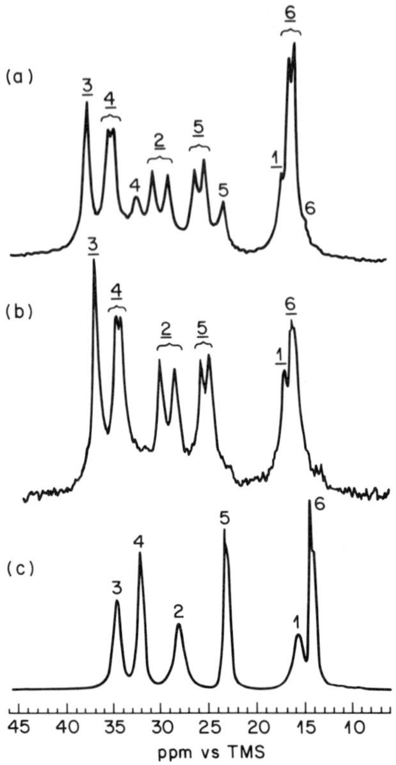

Figure 4. 50.3 MHz ^{13}C solid state spectra of PDHS at 23°; (a) with magic angle spinning, dipolar decoupling, and cross-polarization with a contact time of 1 millisecond; (b) same conditions but with a contact time of 0.2 millisecond; (c) same as (a) but at 50° [14].

In Fig. 4(b) is shown a spectrum obtained under the same conditions as Fig. 4(a) but with a contact time of only 0.20 ms. One now sees only the resonances of Phase I. It is notable that in both spectra the Phase I resonances are split into approximately 1:1 doublets. The origin of this doubling is at present unknown. It may arise from two slightly different modes of side chain packing within the polymer crystal, although as yet this suggestion has no X-ray support.

The solid state ^{13}C spectrum of PDHS at 50°C, above the disordering transition, is shown in Fig. 4(c), under conditions of magic angle spinning with high power proton decoupling. We now of course see only the resonances of Phase II. Measurement of ^{13}C spin-lattice relaxation times were made on Phase II at 61°C and Phase I at 23°C. For Phase II values were obtained ranging from 0.15 s for C_1 to 2.2 s for C_6, indicated greater mobility as one proceeds away from the main chain. For Phase I the T_1 values increase from 110 s for C_5 (C_6 was not measured) to 160 s for C_1; under the much greater motional restriction prevailing in Phase I, this trend again indicates decreasing mobility as one proceeds inward toward the silicon backbone.

The ^{29}Si NMR spectrum is of particular interest since it permits a direct measurement of the conformation and motion of the backbone. Such spectra for

PDHS have been reported by Gobbi et al. [15] and Schilling et al [14]. The ^{29}Si spectra at 39.75 MHz with proton cross-polarization, magic angle spinning and dipolar proton decoupling are shown in Fig. 5 below and above the 42°C disordering temperature [16]. (The vertical scaling is constant for all spectra.) In the spectrum at 21°C (a) we observe two resonances. As the temperature is increased the more shielded peak increases in intensity, becoming the entire spectrum at 45°C. It is therefore evident that it corresponds to Phase II. As with ^{13}C, ^{29}Si-proton cross-polarization is less effective for the more mobile Phase II, which therefore is somewhat underrepresented in these spectra.

Measurement of silicon spin-lattice relaxtion times also strongly demonstrates the large difference in mobility of the two phases. For Phase II at 61°C a T_1 of 5.5 s is observed, whereas for Phase I a T_1 of 3.2 h is found, one of the longest on record.

It should be mentioned that, in agreement with the fact that Phase II of PDHS is of ca. 10% lower density than Phase I, it has been found [17] that under 1000 atm. pressure the disordering transition of PDHS is increased by 22°, as judged by the optical density at 370 nm. This corresponds to a heat of transition of ca. 7 kcal. -mole^{-1}, in approximate agreement with direct thermal measurements of the transition endotherm. At any given temperature below the transition, the effect of pressure in favoring Phase II is found to be reversible upon removing the pressure.

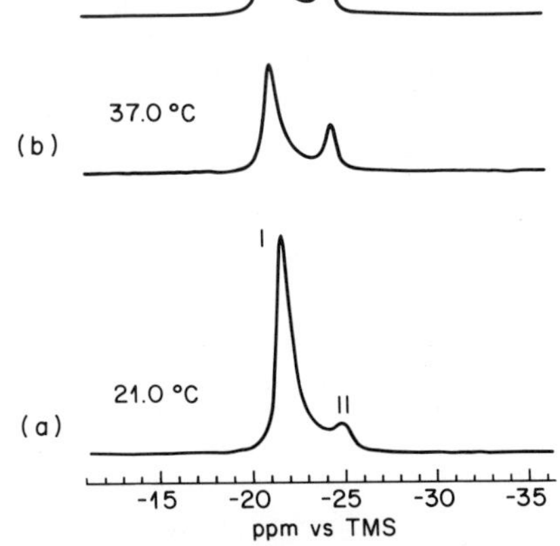

Figure 5. 39.75 MHz ^{29}Si magic angle spinning spectra with dipolar decoupling and proton cross polarization of PDHS in solid state at (a) 21.0°C; (b) 37.0°C; (c) 37.5°C; (d) 38.0°C; 45.0°C. (F. C. Schilling, unpublished observations).

4. The Solid State Structure and NMR of Poly(di-*n*-pentylsilylene) (PDPS) and Poly(di-*n*-butylsilylene) (PDBS)

When the alkyl side chains are shortened by one or two carbon atoms the conformational behavior of the polysilanes becomes more complicated. The structure of the dipentyl polymer, PDPS, has been determined by Miller et al. [18] using Raman spectroscopy and X-ray diffraction. The main chain conformation is not all-*trans* (Fig. 6(a)) as in Phase I of PDHS but instead is a 7/3 helix in which every Si-Si bond is rotated 30° in the same direction throughout the chain. Fig. 6(b) shows a left-handed helix generated in this manner, so far as is known at present right-handed helices are equally probable. According to the conformational energy calculations of Farmer et al. [19] this represents a lower energy state than the all-*trans* planar zigzag not only for PDPS but also for PDHS and PDBS (vid. inf.). It is further believed that the *gauche* state (Fig. 6(c)), which in hydrocarbon chains is of almost equal probability to the *trans* state, is of very high energy owing to side chain interactions and plays no part. Transition to the disordered Phase II occurs at ca. 70°, compared to 42° for PDHS, and is accompanied by a broadening but only minimal bathochromic shift (<10 nm) of the UV absorption maximum [16].

It has been found that Phase I of the dibutyl polymer (PDBS) also has a 7/3 helical structure [20]. The Phase II transition occurs at ca. 86°C [18] and is accompanied by UV absorption effects similar to those reported for PDPS. It appears that a rotation of 30° away from *trans* is sufficient to destroy or substantially diminish the σ-conjugation believed responsible for the bathochromic UV absorption shift in all-*trans* PDHS.

The effect of the 7/3 helical conformation on the solid state ^{29}Si NMR spectrum is striking. In Fig. 7 are shown 39.75 MHz ^{29}Si spectra of PDHS, PDPS,

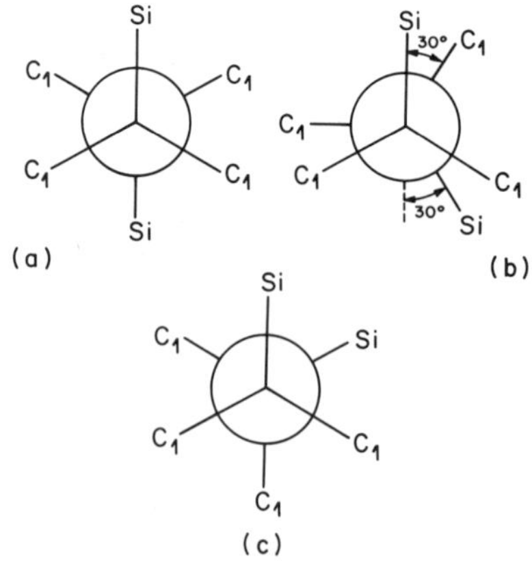

Figure 6. Newman projections of main-chain conformations of a poly(dialkylsilylene) chain; C_1 represents the first carbon of the alkyl side-chain. (a) All-*trans* planar zig-zag conformation; (b) 7/3 helical conformation; (c) *gauche* conformation (energetically forbidden).

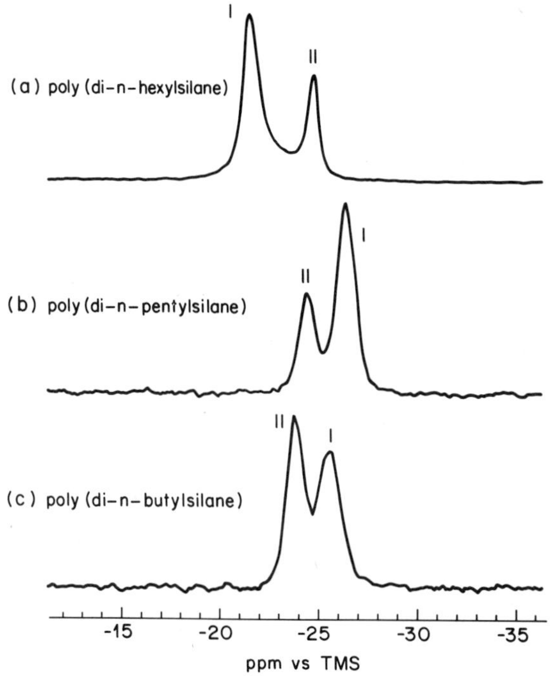

Figure 7. 39.75 MHz ^{29}Si magic angle spinning spectra (same conditions as in Fig. 5) taken near the midpoint of the Phase I-Phase II disordering transition for (a) PDHS); (b) PDPS); and (c) PDBS [14].

and PDBS at temperatures near the midpoint of their Phase I-Phase II transitions, so that resonances of both forms may be seen. It will be observed that Phase I resonances of PDPS and PDBS are more shielded than those of Phase II, in contrast to what is observed for PDHS, in which the Phase I resonance is *less* shielded than that of Phase II. Referring again to Fig. 6, one must conclude that in the 7/3 helical conformation the rotation of one side chain C_1 carbon away from the 60° dihedral angle of the *trans* conformation is more than compensated, with respect to shielding of ^{29}Si, by the rotation of the other C_1 carbon 30° closer to the silicon atom. (It is well known that in hydrocarbon chains maximum shielding of carbons occurs when the dihedral angle is at or near the eclipsed state.) The Phase II ^{29}Si resonance is at nearly the same chemical shift for all three polymers. It seems likely that in this phase the backbone conformation represents mainly a rapidly averaging nearly equimolar mixture of local trans (0°) and 7/3 helical (± 30°) conformations, these angles representing the Si-Si-Si-Si dihedral. (Contributions from changes in side chain conformations are not considered in this analysis).

The effect of pressure on the conformation of PDBS in the solid state is of great interest, but has not yet been explored by NMR. Optical [17] and X-ray observations [21] show that under 2500 atmospheres pressure the helical form is partially transformed into an all-*trans* form like that of PDHS. Upon release of the pressure this transformation persists, but upon heating ca. 53°, the polymer reverts to the 7/3 helical form.

5. References

1. Wesson, J. P. and Williams, T. C., *J. Polym. Sci., Polym. Chem. Ed.*, *18*, 959, 1980.
2. Trujillo, R. E., *J. Organomet. Chem.*, *198*, C27, 1980.
3. West, R.; David, L. D.; Djurovich, P. I.; Stearly, K. L; Srinivasan, K. S. V. and Yu, H., *J. Am. Chem. Soc.*, *103*, 7352, 1981.
4. Trefonas, III, P.; Djurovich, P. I; Zhang, X. M.; West, R.; Miller, R. D. and Hofer, D., *J. Polym. Sci., Polym. Lett. Ed.*, *21*, 819, 1983.
5. Zeigler, J. M.; Harrah, L. A. and Johnson, A. W., *SPIE Adv. Resist Technol.*, *539*, 166, 1985.
6. Schilling, F. C.; Bovey, F. A. and Zeigler J. M., *Macromolecules*, *19*, 2309, 1986.
7. Harrah, L. A. and Zeigler, J. M., *J. Polymer Sci., Poly. Lett Ed.*, *23*, 209, 1985.
8. Trefonas, III, P.; Damewood, J. R.; West, R. and Miller, R. D., *Organometallics*, *4*, 1318, 1985.
9. Kuzmany, H.; Rabolt, J. F.; Farmer, B. L. and Miller, R. D., *J. Chem. Phys.*, *85*, 7413, 1986.
10. Rabolt, J. F.; Hofer, D.; Miller, R. D. and Fickes, G. N., *Macromolecules*, *19*, 611, 1986.
11. Lovinger, A. J.; Schilling, F. C.; Bovey, F. A. and Zeigler, J. M., *Macromolecules*, *19*, 2657, 1986.
12. Klingensmith, K. A.; Downing, J. W.; Miller, R. D. and Michl, J., *J. Am. Chem. Soc.*, *108*, 7438, 1986.
13. Schweizer, K. S.; Harrah, L. A. and Zeigler, J. M., ACS Advances in Chemistry Series, in press.
14. Schilling, F. C.; Bovey, F. A.; Lovinger, A. J. and Zeigler, J. M., *Macromolecules*, *19*, 2660, 1986.
15. Gobbi, G. C.; Fleming, W. W.; Sooriyakumaran, R. and Miller, R. D., *J. Am. Chem. Soc.*, *108*, 5624, 1986.
16. Schilling, F. C., unpublished observations.
17. McGregor, R. and Walsh, C., unpublished observations.
18. Miller, R. D.; Farmer, B. L.; Fleming, W. W.; Sooriyakumaran, R. and Rabolt, J. F., *J. Am. Chem. Soc.*, *109*, 2509, 1987.
19. Farmer, B. L.; Rabolt, J. and Miller, R. D., *Macromolecules*, *20*, 1167, 1987.
20. Schilling, F. C.; Lovinger, A. J.; Bovey, F. A. and Zeigler, J. M., in *Adv. in Chem.*, Vol xxx, Zeigler; J. M. and Fearon, F. G. Eds., American Chemical Society, Washington, D. C., in press.
21. Lovinger, A. J., unpublished observations.

Chapter 19

^{29}Si AND ^{13}C MAGIC ANGLE SAMPLE SPINNING NUCLEAR MAGNETIC RESONANCE SPECTROSCOPY OF CERAMIC FIBERS PREPARED BY PYROLYSIS OF ORGANOSILICON POLYMERS

JONATHAN LIPOWITZ AND GARY L. TURNER

1. Abstract

Silicon oxycarbide (Si-C-O) and oxycarbonitride (Si-N-C-O) fibers, prepared by pyrolysis of organosilicon polymers, have been characterized by ^{29}Si and ^{13}C magic angle spinning(MAS) NMR. The Si-C-O fibers are partially amorphous and the Si-N-C-O fibers are totally amorphous. A single-pulse FT-NMR technique using ≥ 60s pulse delay with phase cycling and with multiple signal acquisitions (>200) was used. A spinning rate > 4 KHz is required to remove spinning sideband interference. Boron nitride rotors were used for sample spinning in ^{13}C NMR experiments. ^{29}Si chemical shifts of the five possible tetrahedral species about silicon (SiC_4, SiC_3O, SiC_2O_2, $SiCO_3$, SiO_4) in silicon oxycarbides (NILCALON™ fibers) can be partially resolved at 6.3 Tesla (^{29}Si resonance frequency of 53.8 MHz). These Si-C-O compositions were previously shown to contain β-SiC nanocrystallites, 1 to 3 nm in size, and a carbon nanophase distributed throughout the continuous amorphous phase. Chemical shift assignments, relative to external TMS, are based on analogous organosilicon compounds and on the various silicon carbide polytypes.

Signal broadening can be attributed to the solid-state effect and to chemical shift variations caused by a distribution of bond angles and bond lengths in these amorphous materials. In addition, second order chemical shifts (e.g., next nearest neighbor effects) may also contribute to the broadening. Signal deconvolution of the

---

JONATHAN LIPOWITZ - Dow Corning Corporation, Midland, MI 48640 and
GARY L. TURNER - Spectral Data Services, Inc., Champaign, IL 61820.

Solid State NMR of Polymers, Edited by L. Mathias
Plenum Press, New York, 1991

five tetrahedral species gives structural information which is quantitatively consistent with elemental analysis data.

^{13}C MAS NMR data shows the presence of carbide carbon and a graphitic-like carbon phase in all these ceramic compositions. Quantitative analysis of carbide to carbon ratios is not possible because the elemental carbon nanophase contains ~10^{20} spins/g (ESR data). The ESR signals exhibit no fine structure and resemble pyrolytic carbon spectra. Thus, carbon atoms in the vicinity of unpaired electrons are expected to have extremely broad signals which are difficult to quantitate. Spin-echo techniques are unable to give quantitative measurement of carbide carbon to elemental carbon ratios.

In the silicon oxycarbonitrides, the ^{29}Si chemical shift range is too small to give separation of the 15 possible tetrahedral structures about silicon. However, peak shape and peak maxima are consistent with a random structure of C, N and O bonding to silicon. ^{13}C chemical shifts show carbide carbon and elemental (graphitic-like) carbon. The carbon phases contain ~10^{20} spins/g which prevents quantitative measurement of carbide C to elemental C ratios.

2. Introduction

There has been much recent interest in preparation of ceramics by pyrolysis of polymeric precursors [1-8]. Characterization of chemical bonding in such materials is difficult if they are largely amorphous, as is often the case when pyrolyzed to 1000 to 1400 C [9], and little progress was made until solid state magic angle spinning(MAS) NMR techniques were recently employed [10-12]. At higher pyrolysis temperatures, some crystallization of these metastable glassy structures often occurs and conventional X-ray diffraction techniques can be used for partial structural characterization [12]. There are experimental difficulties in obtaining good NMR spectra since spin-lattice relaxation times (T_1) for ^{29}Si and ^{13}C nuclei in these materials are long. T_1 relaxation times of minutes to hours have been recently reported for ^{29}Si and ^{13}C nuclei in various silicon carbide polytypes [13,14] and in silicon nitrides [15]. Furthermore, protons or other nuclei which would permit cross polarization are not present [10-12].

The intent of this work was to gain insight into the chemical bonding and phase composition of these "inorganic amorphous polymers" and to determine if MAS NMR techniques could be used for quantitative analysis of composition and phase structure. Previous work using ^{29}Si MAS NMR and other techniques [12] had demonstrated in a qualitative manner the chemical bonding and phase structure of Si-C-O and Si-C-N-O ceramic compositions. The Si-C-O compositions consist of an amorphous, continuous silicon oxycarbide [$Si(C,O)_4$] phase containing nanometer-scale crystallites of β-SiC and a nanometer-scale phase which resembles pyrolytic carbons (polyaromatic carbon with two-dimensional turbostratic order) [16]. The Si-C-N-O compositions consist of a silicon oxycarbonitride [$Si(N,C,O)_4$] amorphous phase containing a similar C nanophase resembling pyrolytic carbon.

3. Experimental

Fibers characterized include a standard grade of commercial Si-C-O fiber (NICALON™ fiber, Nippon Carbon Co., Ltd.) containing ~15 wt% oxygen, referred to as (SG, NICALON™), and a recent ceramic grade NICALON™ fiber containing ~10 wt% oxygen, referred to as (CG, NICALON™) [17-19]. A Si-C-N-O ceramic fiber derived from methylpolydisilylazane (MPDZ) polymer, referred to as MPDZ fiber [7,20] and a Si-N-C ceramic fiber derived from hydridopolysilazane (HPZ) polymer, referred to as HPZ fiber [21,22] were also characterized. A Si-C-O fiber derived from methylpolysilane polymer, referred to as MPS fiber [23-25] was characterized as well. The latter three fibers were developed at Dow Corning under a Government (DARPA) sponsored contract "Advanced Ceramics Based on Polymer Processing" (Air Force Contract F33615-83-C-5006).

Elemental analyses and rule-of-mixtures compositions, calculated using free energies of formation for stable silicon-containing species [12], are shown in Tables I and II, respectively.

^{13}C and ^{29}Si MAS NMR spectra of finely ground powders were obtained on an FT NMR spectrometer operating at 68.1 and 53.8 MHz, respectively, using an Oxford Instruments (Osney Mead, Oxford, U.K.) 6.3 Tesla, 39 mm bore superconducting solenoid magnet. A Nicolet Instrument Corp. (Madison, WI) Model 1280 computer system for data acquisition and an Ehrhorn Technological Operations, Inc. (Canon City, CO) Model Alpha PA-30/80 amplifier for final rf pulse generation were used. Spectra were obtained using Doty Scientific, Inc. (Columbus, SC) or Probe Systems, Inc. (Champaign, IL) probes. For both ^{13}C and ^{29}Si, 90° pulses were 5-6 μs. Recycle delays of 60 s were typically used. Chemical shifts are reported in parts per million (ppm) from external Me_4Si, where more positive values correspond to low-field, high-frequency, paramagnetic, deshielded values (IUPAC δ scale). Line broadenings due to exponential multiplication were typically 25-50 Hz. The

Table I. TYPICAL ELEMENTAL ANALYSES IN WT%

Ceramic Fiber	Si	C	N	O	Total[a]
MPS	68.3	27.1	0	0.9	96.3
PMS[b]	59.1	29.3	0.1	6.3	98.7[c]
CGN	58.4	31.2	0.1	10.0	99.7
SGN	56.2	30.4	0.1	14.2	100.9
MPDZ-PhVi	47.0	28.9	14.5	7.5	97.9
HPZ	59.4	10.1	28.9	3.1	101.5

[a] Also H, usually < 0.1 wt%; Cl, usually < 0.1 wt% (not in SGN or CGN).
[b] This MPS ceramic fiber sample was used for MAS NMR experiments. It also contained 3.4 wt% Cl and 0.5 wt% Al [25].
[c] Total includes Cl and Al content.

Table II. TYPICAL RULE OF MIXTURES COMPOSITIONS IN MOLE % (WT%)

	MPDZ-PhVi	HPZ	SGN	CGN	MPS[a]
SiO_2	8.0 (14.3)	6.8 (5.8)	15.1 (26.8)	10.9 (19.1)	1.1 (1.7)
Si_3N_4	8.9 (37.1)	35.5 (71.3)	0	0	0
SiC	22.9 (27.2)	32.3 (18.5)	52.1 (61.6)	60.6 (70.9)	92.9 (94.0)
C	60.2 (21.4)	25.5 (4.4)	32.8 (11.6)	28.5 (10.0)	0
Si	0	0	0	0	6.0 (4.3)

[a] The MPS sample used for MAS NMR experiments had a rule of mixtures composition, ignoring Cl, Al and N content, of 7.5 mol% (12.5 wt%) SiO_2, 72.4 mol% (80.7 wt%) SiC and 20.2 atom% (6.8 wt%) C.

experimental method [26] and a review of MAS NMR spectroscopy of ceramics [27] have been published.

^{29}Si MAS NMR spectra were obtained utilizing a "one-pulse" experiment with phase cycling. For ^{13}C MAS NMR spectra, in addition to the one-pulse experiment, spin echo experiments were performed to obtain undistorted spectra of the very broad amorphous carbon component. The two echo experiments used were modifications of the basic Hahn echo [28]; 90_x - τ - 180y -τ - echo; the first modification was using the echo sequence with a 16-step phase cycling [29], and the second was to use a 90_x - τ - 90_y - τ - echo sequence [30].

At least 200, and generally ~250, signal acquisitions were obtained to give signal to noise ratios as shown in the NMR spectra. A sample spinning rate of >4 KHz was required to minimize spinning sideband interference. Boron nitride sample spinning rotors were specifically fabricated for ^{13}C NMR experiments to eliminate interference from the rotors. A recycle delay of 60s was used. Spectral deconvolutions shown in Figs 1. and 2 were performed using the Nicolet software program "NMRCAP".

ESR spectra were obtained on a Bruker 200 spectrometer. Powdered ceramic samples and freshly recrystallized DPPH were appropriately diluted with powdered KCl so that ESR signals of similar signal strength could be obtained on samples run in quartz tubes at comparable filling factors (~0.15 g sample). All samples were run at the same instrumental settings. Signal strength was obtained by performing a double integration on the first derivative spectra using IBM software provided with the instrument. Signal strength and sample weight could then be normalized to the known spin density and signal strength of the DPPH sample. All samples exhibited Lorentzian line shapes except for HPZ which gave a Gaussian shape. Calculated spin concentrations and g-values are shown in Table VIII.

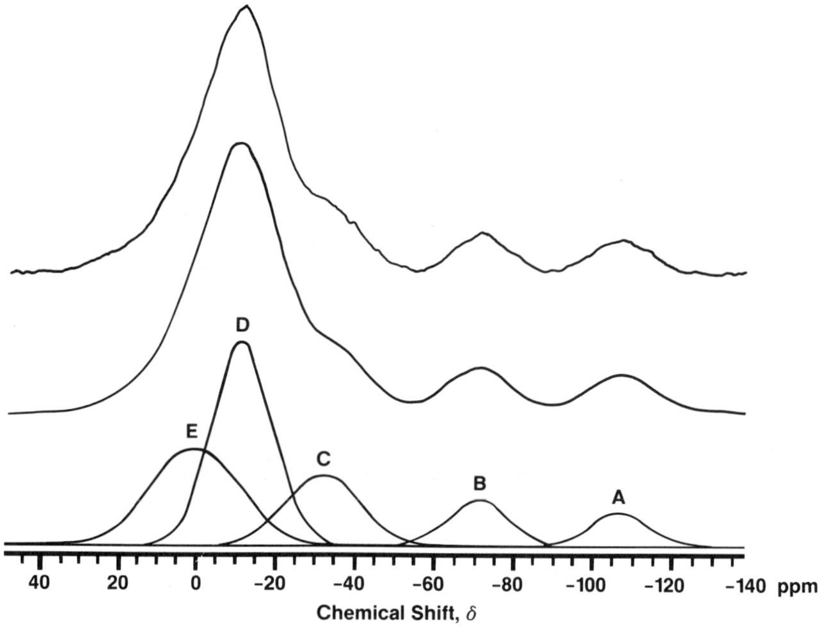

Figure 1. ^{29}Si MAS NMR Spectrum of Powdered SGN Fiber, deconvoluted, 4.5 kHz spin rate.

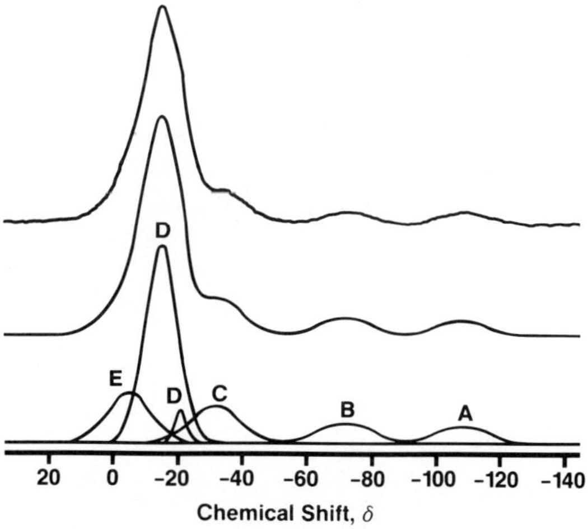

Figure 2. ^{29}Si MAS NMR Spectrum of Powdered CGN Fiber, deconvoluted, 4.5 kHz spin rate.

4. Results and Discussion

Figures 1 to 5 show ^{29}Si MAS NMR spectra of the ceramic fibers. In the Si-C-O fibers (SGN, CGN, MPS) there are five possible tetrahedral structures for the silicon atoms if only Si-C and Si-O bonds are present, as suggested by the infrared spectra [12]. The tetrahedral structures are SiC_4, SiC_3O, SiC_2O_2, $SiCO_3$, and SiO_4. All appear to be present in the NICALON™ (SGN, CGN) fibers and their deconvolutions are shown in Figs. 1 and 2. Oxygenated tetrahedra are more prevalent in SGN than in CGN, reflecting the higher oxygen concentration in SGN (Tables I and II). Chemical shifts are assigned based on known shifts for the various silicon carbide polytypes [13,14] and for organosilicon compounds with mixed $[Si(C,O)_4]$ tetrahedral structures [31]. Model compounds used and ^{29}Si NICALON™ chemical shifts are shown in Table III. Note that the most intense signal (at δ~-12 to 15 ppm) in Si-C-O fibers occurs close to the chemical shift of β-SiC (δ=-18.3 ppm), indicating mainly SiC, as predicted by rule-of-mixtures calculations (see below) showing SiC to be the dominant species. X-ray diffraction and transmission electron microscopy data shows that β-SiC nanocrystallites are present [12]. The ^{29}Si MAS NMR spectrum for MPS fiber (Fig. 3) is rich in SiC_4 structure (and β-SiC nanocrystallites by X-ray diffraction) but contains no highly oxygenated $SiCO_3$ and SiO_4 structures because this MPS fiber sample contains only 6 wt% 0 [25].

The ^{29}Si MAS NMR signal of each possible tetrahedral structure about silicon in all these ceramic materials is broadened by the solid-state effect and by the

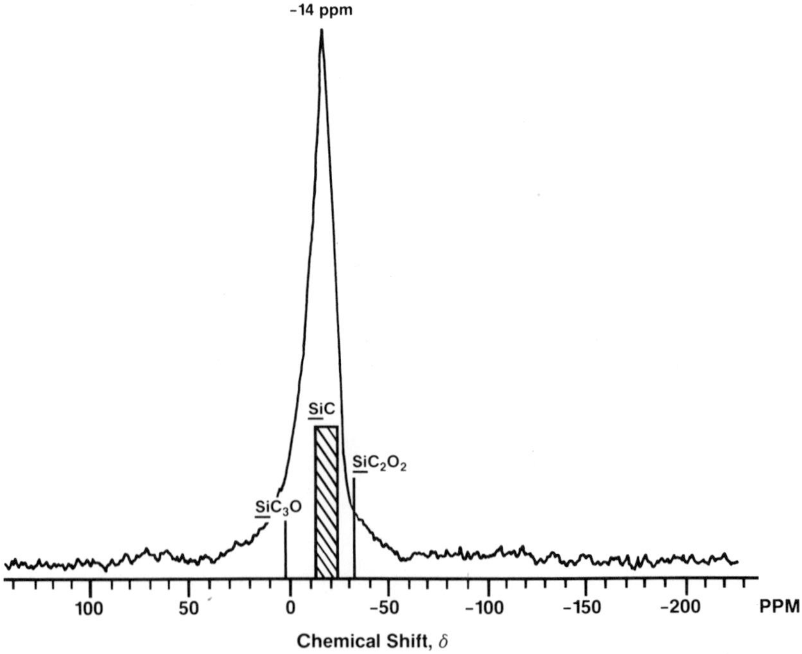

Figure 3. ^{29}Si MAS NMR Spectrum of Powdered MPS Ceramic Fiber.

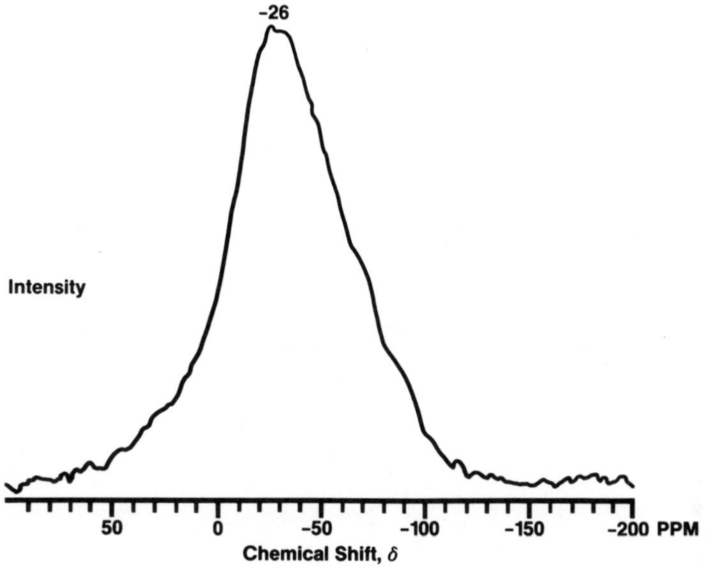

Figure 4. ^{29}Si MAS NMR Spectrum of Powdered MPDZ-PhVi Ceramic Fiber.

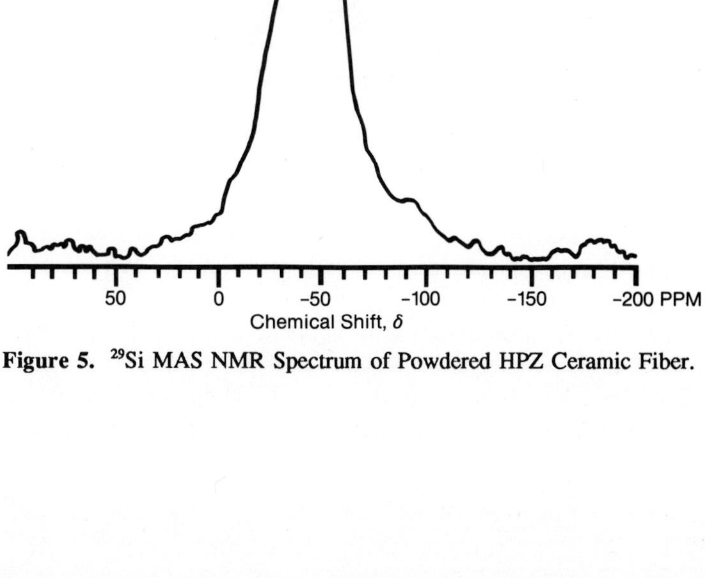

Figure 5. ^{29}Si MAS NMR Spectrum of Powdered HPZ Ceramic Fiber.

Table III. MODEL COMPOUNDS FOR ^{29}Si NMR CHEMICAL SHIFTS OF Si-C-O CERAMICS[a]

			Assignment[b]	
Tetrahedral Structure	Compound	δ, ppm	SGN	CGN
SiO_4	$Si(OSiMe_3)_4$	-105	-106.6	-108.6
	SiO_2(1)	-114.0		
SiO_3C	$MeSi(OSiMe_3)_3$	-65.0	-71.5	-72.1
	$MeSi(OMe)_3$	-41.5		
SiO_2C_2	$Me_2Si(OSiMe_3)_2$	-22.1	-32.8	-32.6
	$Me_2Si(OMe)_2$	-2.5		
$SiOC_3$	Me_3SiOMe	+17.2		
	$(Me_3Si)_2O$	+6.7	-0.7	-5.8
SiC_4	6H-SiC β-SiC Me_4Si	-13.9, -20.2, -24.5[c] -18.3[c] 0	-12.2	-15.3

[a] [31]
[b] see Table VI and Figs. 1 and 2
[c] [13, 14]

distribution in bond angles and bond lengths in amorphous structures as well as by second nearest neighbor effects. There are 15 possible tetrahedral structures $[Si(N,C,O)_4]$ in the Si-C-N-O amorphous ceramic system represented by MPDZ-PhVi and HPZ. Broad, unresolved envelopes covering this chemical shift region are thus expected and observed in Figs. 4 and 5. The ^{29}Si MAS NMR spectrum for MPDZ-PhVi is enriched in tetrahedral species in the highfield (SiO_4, $SiCO_3$) and lowfield (SiC_3O) chemical shift regions because of its higher oxygen and carbon content. Chemical shifts of the oxygenated species in Table III cover a broader range than do the $Si(C,N)_4$ model compounds shown in Table IV. Chemical shifts of compounds with C, N, and O bonded to silicon have not been reported.

Note that the ^{29}Si resonance peak for HPZ ceramic shown in Fig. 5 (δ=-46 ppm) is at the chemical shift recently reported [15] for amorphous silicon nitride (Table IV) since the predominant rule-of-mixtures component in HPZ is Si_3N_4. The MPDZ-PhVi NMR peak (δ=-26 ppm) occurs close to the SiC chemical shift region (δ=-14 to -24 ppm) since the predominant rule-of-mixtures component is SiC.

Rule-of-mixtures compositions in Table II were calculated by assigning all oxygen to silicon as SiO_2; then all nitrogen (if any) to silicon as Si_3N_4; then carbon to silicon as SiC; excess carbon or silicon (if any) is considered to be in the elemental standard state. These rules are obtained using increasing Gibbs free energies of formation per equivalent of silicon for the species SiO_2, Si_3N_4 and SiC. These calculated compositions do not actually exist in the metastable fiber but represent the thermodynamically stable composition and are useful in interpreting characterization data [12].

In the following equations, A through E are the deconvoluted peak areas in area %/100 from Table VI.

$$Si = A + B + C + D + E = x$$

$$Carbon = \frac{B}{4} + \frac{C}{2} + \frac{3E}{4} + D = y$$

$$Oxygen = 2A + \frac{3B}{2} + C + \frac{E}{2} = z$$

Empirical formula is $Si_xC_yO_z$, not including excess C.

Empirical formula from rule-of-mixtures calculation is obtained from relative moles of SiC and SiO_2 using molar ratios from Table II.

$$\begin{array}{ll} & SiC \\ + & \underline{uSiO_2} \\ & Si_{1+u}C_1O_{2u} \text{ (not including excess C)} \end{array}$$

Table V shows the method used to calculate an empirical formula from the deconvoluted ^{29}Si MAS NMR spectra of the SGN and CGN ceramic fibers for comparison with the rule of mixtures empirical formula (not including elemental

Table IV. ^{29}Si NMR CHEMICAL SHIFTS OF MODEL Si-C-N COMPOUNDS[a]

Tetrahedral Structure	Compound	δ, ppm
SiC_3N	Me_3SiNMe_2	+5.9
SiC_2N_2	$Me_2Si(NMe_2)_2$	-1.7
$SiCN_3$	$MeSi(NMe_2)_3$	-17.5
SiN_4	$Si(NMe_2)_4$	-28.1
	α-Si_3N_4	-46.8, -48.9[b]
	β-Si_3N_4	-48.7[b]
	amorphous-Si_3N_4	-46.4[b]

[a] [31]
[b] [15]

Table V. COMPOSITION FROM ^{29}Si MAS NMR SIGNAL DECONVOLUTION

Peak	Assignment
A	SiO_4
B	SiO_3C
C	SiO_2C_2
D	$SiOC_3$
E	SiC_4

carbon). Comparison of the NMR and rule of mixtures empirical formulae are shown in Table VI. Through agreement is not precise, it is reasonable in view of the experimental errors inherent in the spectral deconvolution, errors in the rule of mixtures calculation based on elemental analyses, and perhaps most important, the possibility of differential saturation of the various resonances.

^{13}C MAS NMR spectroscopy of these ceramics provides information on the chemical environment about the carbon atoms (Figs. 6-9). Two types of carbon are seen; carbide-like carbon which occurs in the chemical shift region typical of various SiC polytypes (δ=+15 ppm to +25 ppm) [13,14] and a graphitic-like carbon which is found near the chemical shift region reported for graphite (δ=+155 to 200 ppm, broad) [32]. Graphitic carbon was not found in HPZ ceramic fibers in which its calculated composition is low (data not shown). The carbon phase in these ceramics is similar to the pyrolytic carbons in structure, based on Raman spectroscopy [12] and electron spin resonance data (see below).

Table VI. COMPARISON OF THE EMPIRICAL FORMULAE OF SGN AND CGN (WITHOUT ELEMENTAL C) BASED ON ^{29}Si NMR AND RULE OF MIXTURES DATA

		SGN		CGN	
Line	**Assignment**	**δ, ppm**	**Area %**	**δ, ppm**	**Area %**
A	SiO_4	-106.6	6.5	-108.6	5.9
B	SiO_3C	-71.5	8.8	-72.1	7.9
C	SiO_2C_2	-32.8	17.0	-32.6	14.4
D	SiC_4	-12.2	39.4	-15.3	54.8
E	$SiOC_3$	-0.7	28.3	-5.8	17.0

For SGN: $Si_1C_{0.712}O_{0.574}$ from NMR data
$Si_1C_{0.760}O_{0.480}$ from rule of mixtures calculation

For CGN: $Si_1C_{0.767}O_{0.465}$ from NMR data
$Si_1C_{0.760}O_{0.354}$ from rule of mixtures calculation

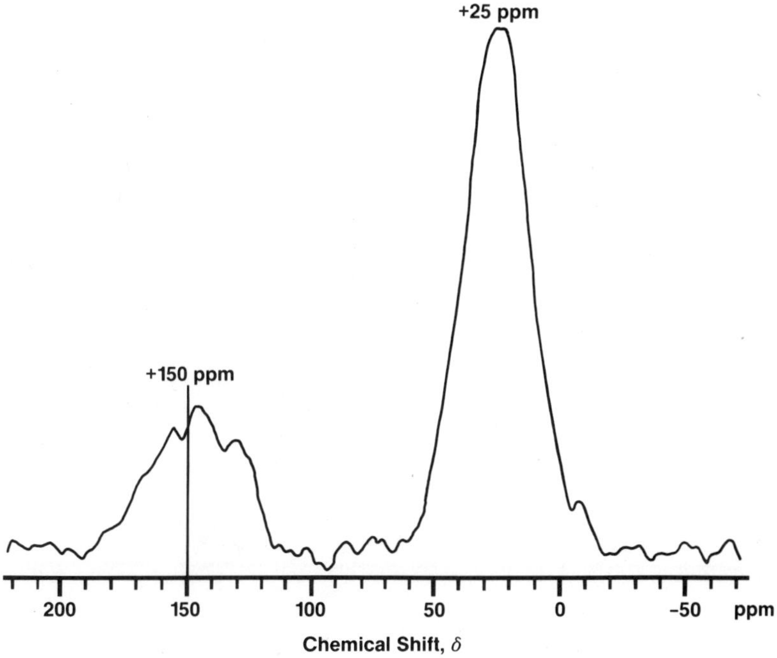

Figure 6. ^{13}C MAS NMR Spectrum of Powdered SGN Ceramic Fiber.

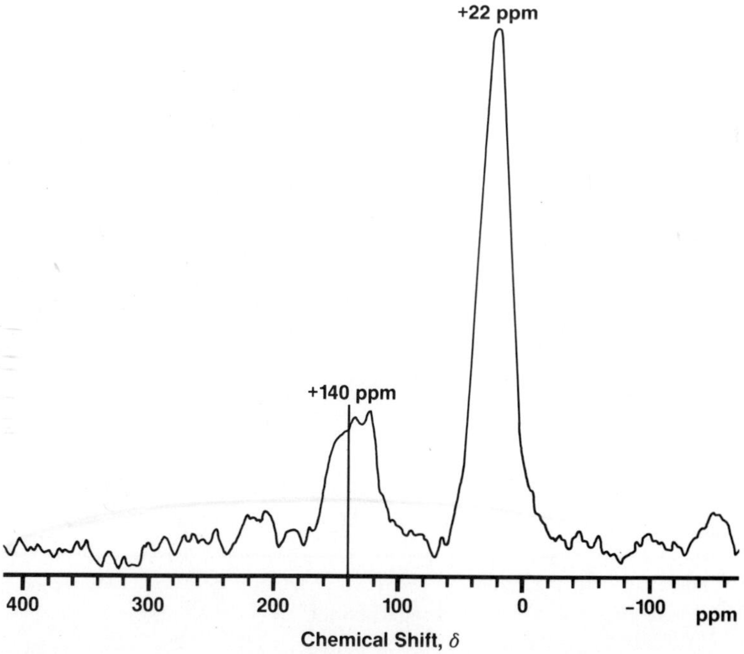

Figure 7. ^{13}C MAS NMR Spectrum of Powdered CGN Ceramic Fiber.

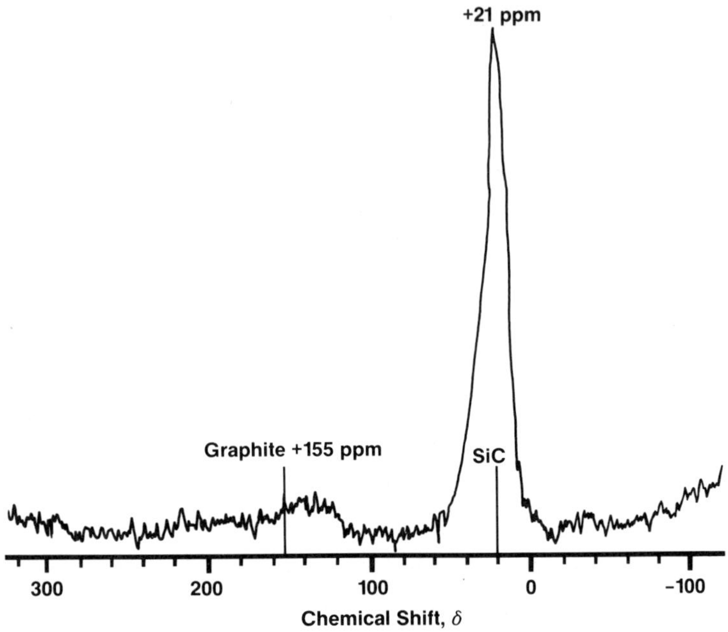

Figure 8. ^{13}C MAS NMR Spectrum of Powdered MPS Ceramic Fiber.

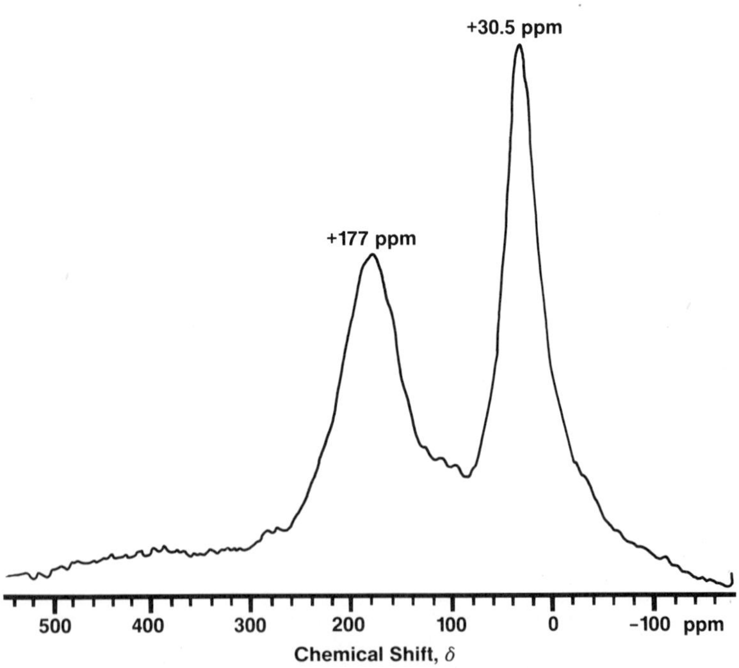

Figure 9. ^{13}C MAS NMR Spectrum of Powdered MPDZ-PhVi Ceramic Fiber.

Table VII. ANALYSIS OF ^{13}C MAS NMR SPECTRUM OF SGN BY THE SPIN ECHO TECHNIQUE

$90°_x$ - τ - $90°_y$ Experiment, τ = 50 ms	
Integration	C/SiC = 0.346
Deconvoluted spectrum	C/SiC = 0.321
$90°_x$ - τ - $90°_y$ Experiment, τ = 25 ms	
Integration	C/SiC = 0.427
Deconvoluted spectrum	C/SiC = 0.384
Rule of Mixtures	
C/SiC = 0.597	

A ratio of elemental carbon to carbide carbon can be obtained directly from the ^{13}C NMR MAS spectrum by integration of the graphitic carbon signal and the carbide carbon signal. However, this experimental ratio is always lower than the calculated ratio because carbon atoms in proximity to unpaired spins are expected to have very short relaxation times. These carbon atoms will not be included in the integration. Carbide carbon atoms, by analogy with literature values for T_1 [13,14], and from our own estimates, have T_1 relaxation times of several minutes. Calculated and observed ratios are shown in Table VII. Use of spin-echo techniques [28-30] improves agreement. Use of a shorter delay time (τ) in the spin echo pulse sequence further improves agreement with the rule of mixtures ratio.

Electron spin resonance data [33], Table VIII, show that all of these ceramic fibers exhibit strong signals (10^{19} to 10^{20} spins/gram) without fine structure, similar to those found for pyrolytic carbons which typically contain highly delocalized unpaired electrons [34]. That is, a derivative of a simple broad resonance line without fine structure is obtained (Fig. 10). G-values close to that for a free spin, g = 2.0023, are found as expected for pyrolytic carbons.

Table VIII. ESR PARAMETERS FOR CERAMIC FIBERS

Fiber	Spins/g x 10^{19} [a]	g-value
SGN	1.56 ± 0.47	2.0035
HPZ	1.57 ± 0.24	2.0019
MPDZ-PhVi	5.64 ± 0.78	2.0032

[a] average of 2 independent sample preparations and ESR measurements.

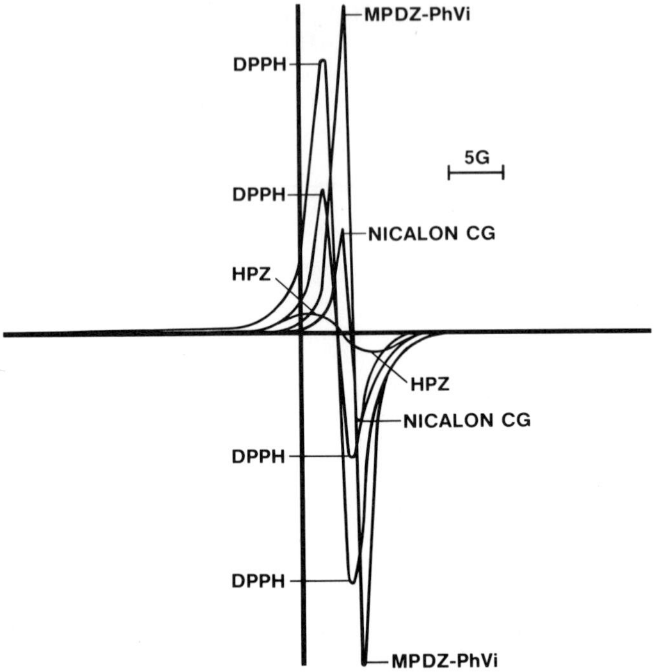

Figure 10. ESR Spectra of Powdered Ceramic Fibers.

5. Acknowledgments

This work was performed under Air Force Contract F33615-83-C-5006 sponsored by the Material Sciences Division of the Defense Advanced Research Projects Agency and administered by Air Force Systems Command, Aeronautical Systems Division/PMRRA, Wright-Patterson AFB, OH, 45433. We thank the program managers, Alan P. Katz and Kay A. Hardman-Rhyne, for their support and encouragement. The authors also thank G. Eastland for providing the ESR data.

6. References

1. Wynne, K. J. and Rice, R. W., *Ann. Rev. Mater. Sci.*, *14*, 297-334, 1984.
2. Rice, R. W., *Am. Ceram. Soc. Bull.*, *62*, 889-92, 1983.
3. Yajima, S., *ibid.*, *62*, 893-98, 903, 1983.
4. West, R.; David, L. D.; Durovich, P. I.; Yu, H. and Sinclair, R., *ibid.*, *62*, 899-903, 1983.
5. Wills, R.; Markle, R. A. and Mukherjee, S. P., *ibid.*, *62*, 904-11, 915, 1983.
6. Schilling, C. L., Jr.; Wesson, J. P. and Williams, T. C., *ibid.*, *62*, 912-15, 1983.

7. Baney, R. H., "Some Organometallic Routes to Ceramics," Ch 20 in Ultrastructure Processing of Ceramics, Glasses and Composites, Hench, L. L. and Ulrich, D. R., Eds., John Wiley and Sons, New York, 1984.
8. Mah, T.; Mendiratta, M. G.; Katz, A. P., et al., *Am. Ceram. Soc. Bull.*, *66*, 304-8, 317, 1987.
9. Bender, B. A.; Rice, R. W. and Spann, J. R., *J. Am. Ceram. Soc.*, *70*, C-58-60, 1987.
10. Inkrott, K. E.; Wharry, S. M. and O'Donnell, D. J., *Mat. Res. Symp. Proc.*, Vol 73, 165-71, 1986.
11. Lipowitz, J.; Freeman, H. A.; Harget, P.; Goldberg, H. and Prack, E., *ibid.*, 489-94, 1986.
12. Lipowitz, J.; Freeman, H. A.; Chen, R. T. and Prack, E. R., *Adv. Ceram. Mater.*, *2*, 121-8, 1987.
13. Hartman, J. S.; Richardson, M. F.; Sheriff, B. L. and Winsborrow, B. G., *J. Am. Chem. Soc.*, *109*, 6059-67, 1987.
14. Guth, J. R. and Petuskey, W. T., *J. Phys. Chem.*, *91*, 5361-64, 1987.
15. Carduner, K. R.; Carter, R. O., III; Milberg, M. E. and Crosbie, G. M., *Anal. Chem.*, *59*, 2794-97, 1987.
16. Ishikawa, T. and Magaoki, T., Eds., *Recent Carbon Technology*, JEC Press, Cleveland, OH, 1983, Section 2.3.
17. Yajima, S.; Hasegawa, Y.; Hayashi, J. and Imura, M., *J. Mater. Sci.*, *13*, 2569-76, 1978.
18. Hasegawa, Y.; Imura, M. and Yajima, S., *ibid.*, *15*, 720-8, 1980.
19. Yajima, S.; Hayashi, J. and Omori, M., U. S. Patent 4,100,233, July 11, 1978, assigned to Tohoku University.
20. Gaul, J. H., Jr., U. S. Patent 4,340,619, July 20, 1982, assigned to Dow Corning Corporation.
21. Cannady, J. P., U. S. Patent 4,535,007, August 13, 1985, U.S. Patent 4,543,344, September 24, 1985, assigned to Dow Corning Corporation.
22. LeGrow, G. E.; Lim, T. F.; Lipowitz, J. and Reaoch, R. S., *Am. Ceram. Soc. Bull.*, *66*, 363-7, 1987.
23. Baney, R. H. and Gaul, J. H. Jr., U.S. Patent 4,298,559, November 3, 1981, assigned to Dow Corning Corporation.
24. Baney, R. H.; Gaul, J. H. Jr. and Hilty, T. K., *Organometallics*, *2*, 859-64, 1983.
25. Lipowitz, J.; LeGrow, G. E.; Lim, T. F. and Langley, N. R., *Ceram. Eng. Sci. Proc.*, *9* [7-8], 931-42, 1988.
26. Smith, K. A.; Kirkpatrick, R. J.; Oldfield, E. and Henderson, D. M., *Am. Mineral.*, *68*, 1206-15, 1983.
27. Turner, G. L.; Kirkpatrick, R. J.; Risbud, S. H. and Oldfield, E., *Am. Ceram. Soc. Bull.*, *66*, 656-63, 1987.
28. Hahn, E. L., *Phys. Rev.*, *80*, 580-594, 1950.
29. Kunwar, A. C.; Turner, G. L.; and Oldfield, E., *J. Magn. Reson.*, *69*, 124-27, 1986.
30. Fukushima, E. and Roeder, S. B. W., *Experimental Pulse NMR: A*

Nuts and Bolts Approach, Addison-Wesley Publishing Company, 1981, pp 251-256.

31. Marsmann, H., in *NMR, Basic Principles and Progress*, Diehl, P., Fluck, E. and Kosfeld, R., Eds., New York, Springer-Verlag, 1981, pp 65-235.
32. Haworth, D. T. and Wilkie, C. A., *J. Inorg. Nucl. Chem.*, *40*, 1689-90, 1978.
33. We thank G. Eastland for the ESR data.
34. Singer, L. S. and Lewis, I. C., *Appl. Spectrosc.*, *36*, 52-7, 1982.

Chapter 20

VARIABLE-TEMPERATURE SOLID-STATE NMR OF POLYPHOSPHAZENES

JAMES F. HAW AND RICHARD C. CROSBY

1. Introduction

When studying polymers in the solid state, questions frequently arise concerning structure, morphology, reactivity, or the motion of chains and functional groups. Since many of these properties are temperature dependent in typical polymers, variable-temperature (VT) solid-state NMR is often a valuable tool for polymer characterization. Several general articles on solid-state NMR have been published elsewhere [1,2], and the subject has also been introduced earlier in this volume. Applications of the VT experiment have also been reviewed [3,4].

This chapter describes some of our VT NMR studies of phosphazene polymers, with emphasis upon studies of solid materials and gels. Polyphosphazenes are inorganic polymers which are based on chains of alternating phosphorus and nitrogen atoms [5,6]. The development of these materials is the direct result of the discoveries by Allcock and co-workers of methods for converting the hexachlorocyclotriphosphazene (**1**) into the soluble, linear chloropolymer (**2**) and substitution reactions to convert **2** into a wide variety of hydrolytically stable polymers with diverse properties. These polymers are of considerable interest because of their mechanical and thermal properties, solvent resistance, and potential for providing biocombatible materials.

We are using VT high-resolution solid-state NMR [7] to study the synthesis, morphology, reactivity, and molecular dynamics of polyphosphazenes [8]. In this

---

JAMES F. HAW AND RICHARD C. CROSBY - Department of Chemistry, Texas A&M University, College Station, TX 77843

Solid State NMR of Polymers, Edited by L. Mathias
Plenum Press, New York, 1991

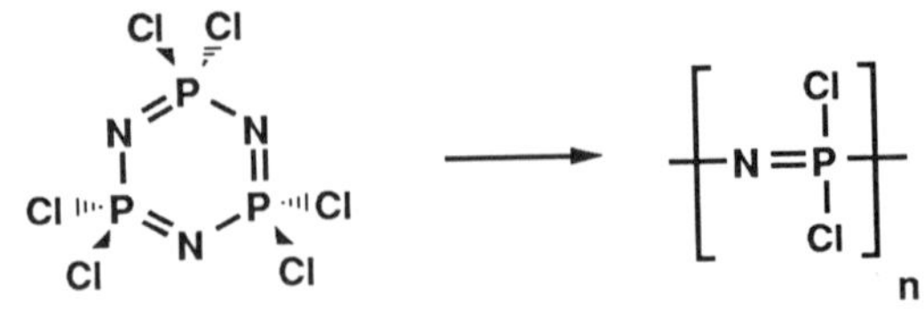

1 2

chapter we report preliminary results on the molecular dynamics and morphology of two solid phosphazene polymers, poly[bis(ethoxy)phosphazene] (PBEP) and poly[bis(3-methylphenoxy)phosphazene] (PBMP). We also report a study involving both solid-state and solution-state NMR methods which is directed toward an understanding of the roles of water in the polymerization of **1** to give **2**.

All solid-state spectra were obtained on a Chemagnetics M-100S solid-state NMR spectrometer equipped with a 2.35-T superconducting magnet with a 70-mm air bore and a home-built VT CP/MAS probe. Torlon and Teflon were used in place of Delrin and other standard plastics in the fabrication of some of the structural components and insulators in this probe. Observation frequencies between 25 MHz (^{13}C) and 40 MHz (^{31}P) are obtained by replacing the tuning capacitor of the observe channel resonant circuit. The diameter of the probe is 60 mm; since the magnet is not equipped with room-temperature shims, this leaves a 5-mm air gap between the probe and the bore surface of the magnet. During prolonged high-temperature operation a stream of cool nitrogen gas is passed through this air gap to minimize heating of the magnet. A purge stream of cool nitrogen gas is also used to maintain the probe electronics near room temperature; this keeps probe returning to a minimum.

The CP/MAS NMR probe has two independent spinning-gas streams: the drive gas, which is used to propel the rotor; and the bearing gas, which is used to stabilize the rotor. For high temperature operation the gas streams are heated in two stages. The first stage uses independent heaters and controllers for the two gas channels while the second stage is provided by a single aluminum block heater. This arrangement allows the temperature of the bearing- drive-gas streams to be independently controlled and is important for accurate sample temperature measurement and control [7].

Before discussing the solid-state NMR of polyphosphazenes, it is instructive to consider a much simpler application of VT solid-state NMR which will serve a pedagogical purpose as well as providing a basis for an understanding of some of the polyphosphazene spectra. Dramatic changes in spectral parameters are often observed during solid-solid phase transitions. For example, 1,4-diazabicyclo-[2,2,2]-octane (DABCO) undergoes a transition at 351 K to a high temperature plastic crystalline phase in which rapid reorientation is possible about all axes. Figure 1 shows ^{13}C CP/MAS spectra of DABCO obtained at several temperatures in the vicinity of the transition [9]. In the low temperature phase, the line shape is dominated by an incompletely averaged dipolar coupling to the quadrupolar ^{14}N nucleus [10]. In the high temperature phase, molecular reorientation is sufficiently rapid to average the

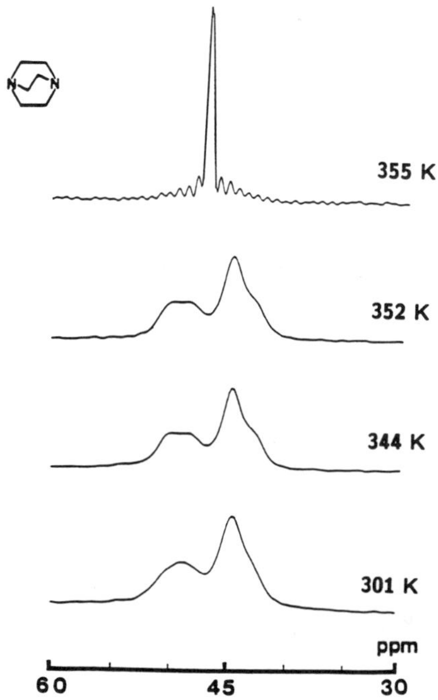

Figure 1. ^{13}C CP/MAS NMR spectra of 1,4-diazabicyclo-[2,2,2]-octane (DABCO) at several temperatures in the vicinity of the solid-solid phase transition temperature (351 K).

^{14}N-^{13}C dipolar coupling to zero, and the ^{13}C resonance is a single sharp line. Since polyphosphazenes contain phosphorous atoms which are adjacent to two ^{14}N atoms, we might expect to observe ^{31}P line shapes that reflect the degree to which molecular motion averages this "^{14}N effect."

An illustration of how we are using ^{31}P VT MAS and ^{13}C VT CP/MAS NMR to study the molecular motion and morphology of polyphosphazenes is seen in Fig. 2 which shows VT ^{13}C CP/MAS and ^{31}P mas spectra of PBEP at various temperatures. The line shapes and intensities of the variable-temperature spectra, in conjunction with relaxation measurements (not shown) provide information about molecular motion and the morphology of the solid polymer. The ^{13}C spectrum at 298 K has a poor signal-to-noise ratio, due to the fact that the polymer is so mobile at this temperature that the ^{1}H - ^{13}C dipolar interaction responsible for cross polarization is greatly reduced. At 173 K, a temperature well below the glass transition temperature (Tg) of this material (Tg of PBEP is 189 K), the signal-to-noise ratio of the ^{13}C CP/MAS spectrum is greatly increased due to the decrease in molecular motion.

Inspecting the ^{13}C spectra in Fig. 2, one notes that the line width of the CH_2 resonance is strongly temperature dependent. The resonance is sharp at both high and low temperatures, but is broad at intermediate temperatures. At 203 K, for example, the CH_2 carbon signal is so broad that it is not observed. Rothwell and Waugh have shown that molecular motion at the frequency of the ^{1}H decoupling field (48 kHz in this case) can account for this type of behavior [11].

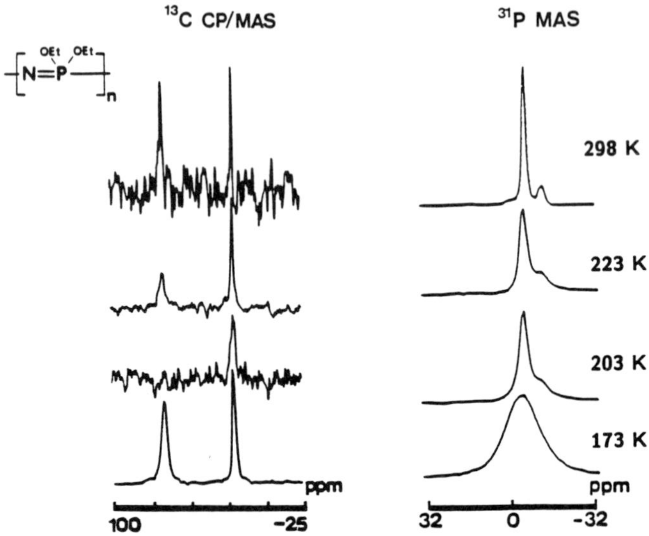

Figure 2. Variable-temperature ^{13}C CP/MAS (left) and ^{31}P MAS spectra (right) of the inorganic polymer poly[bis(ethoxy)phosphazene] (PBEP).

To further investigate molecular motion in solid PBEP, ^{31}P MAS spectra, which are sensitive to main-chain motions in phosphazene polymers, were acquired over the same temperature range used for the ^{13}C data. In the ^{31}P spectra in Fig. 2, one sees a gradual increase in the line width as the temperature is decreased. This indicates that the main-chain motion is not isotropic and rapid in the temperature range studied, otherwise the dipolar coupling to ^{14}N (the dominant line-broadening interaction for ^{31}P with these experimental conditions) would be completely averaged. Upon dropping below Tg, there is a significant further increase in line width. These data indicate that the main chains of PBEP are undergoing large-amplitude anisotropic motions above Tg. This main-chain motion could account for the relaxation behavior observed for the CH_2 carbon resonance since the side groups will share some of the main-chain motions.

Many polyphosphazenes are semicrystalline and undergo two first-order phase transitions. The first transition, which occurs at a temperature designated T(1) is a conversion to a mesomorphic phase that appears to involve a pseudohexagonal structure [12]. Depending on the type of side group present this mesomorphic state can persist over a temperature range of 150 - 250 K [13]. The second transition is the true melting point (T_m). Alexander and coworkers have used wide-line NMR to study the mesomorphic transition in poly[bis(2,2,2-trifluoroethoxy)phosphazene] and have concluded that the polymer backbone is undergoing large amplitude motions at temperatures above T(1) [14]. Using differential scanning calorimetry Singler and coworkers have observed this type of transition behavior for PBMP [15]. Figure 3 reports ^{31}P CP/MAS spectra of PBMP obtained at temperatures ranging from 223 K to 453 K. The spectrum of PBMP obtained at 223 K is a 600 Hz-wide resonance with a maximum at -13 ppm and hints of two shoulders at -8 ppm and -18 ppm. We have recently reported that the ^{31}P MAS NMR spectra of the elastomeric polymers poly[bis(methoxy)phosphazene] and poly[bis(ethoxy)phosphazene] broaden significantly at temperatures below the glass transitions due to dipolar coupling to ^{14}N [10]. A similar effect is operative here, and it is clear from Fig. 3 that all of the PBMP is in a rigid state at 223 K, presumably a semicrystalline phase. At higher temperatures, however, an additional, sharp resonance is observed at -18.4 ppm. Relaxation measurements (not shown) reveal this resonance to be due to a highly mobile component which is almost certainly the mesomorphic phase. The T(1) transition is essentially complete by 363 K, and no further changes are observed with further increases in temperature to 453 K.

The spectra of PBMP were obtained with a MAS spinning system based on Torlon, a high temperature engineering polymer. Although this material has

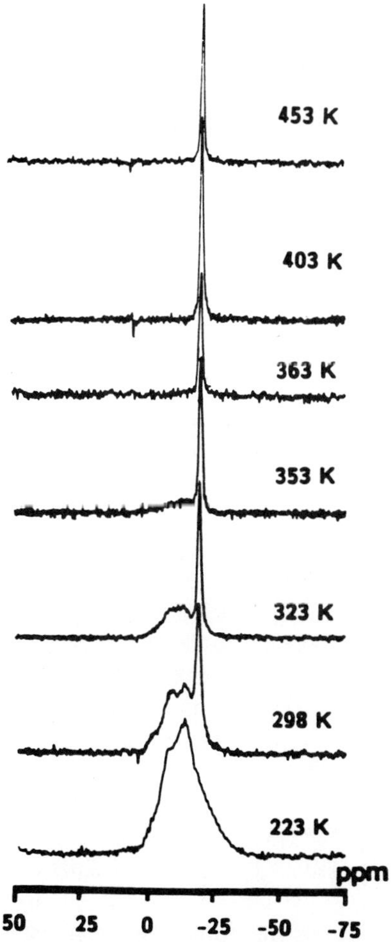

Figure 3. Variable-temperature ^{31}P CP/MAS NMR spectra of poly[bis(3-methylphenoxy)phosphazene]. 2-ms contact times were used and each spectrum is a result of 200 transients except for the 223 K spectrum which is a result of 400 transients.

unacceptable ^{13}C, ^{15}N and ^{1}H background signals, it is completely free of other resonances. This spinning system was much easier to machine than analogous ceramic-based systems and is very forgiving of rotor crashes. We have used the Torlon system to obtain high-quality ^{31}P MAS spectra at temperatures as high as 523 K with a spinning speed of 4.5 kHz [16].

We are investigating the use of "solution-state" NMR spectroscopy to study the polymerization mechanism(s) of **1** at actual operating temperatures. In a typical experiment, **1** and a catalyst (usually BCl_3) or impurity such as water are prepared in a sealed glass reaction tube which is then loaded into a high temperature multinuclear solution-state NMR probe. We are able to observe spectra during the course of polymerization, even after the polymer has gelled due to hydrolysis and cross-linking reactions. We are curious about the effects on the polymerization reaction of low concentrations of species derived from partial hydrolysis of **1**. This curiosity was motivated in part by the detection of such species in even highly purified samples of **1** (vide infra). We therefore prepared a sample of **1** with an intentionally large concentration of partially hydrolyzed impurities.

Fifteen grams of **1** were dissolved in a mixture of 75 mL of diethyl ether, 1 mL of pyridine, and 4 mL of distilled water. The resulting solution was shaken for approximately 3 hrs, washed once with a 5% HCl solution to remove excess pyridine and once with a saturated NaCl to remove excess HCl. The solvent was then evaporated, yielding a white solid. A solution-state ^{31}P NMR spectrum confirmed that the product was primarily unchanged **1** along with the small amounts of **1** which had been hydrolyzed to varying degrees. A 4 g sample of this material was sealed in a thick-walled 10 mm NMR tube which was inserted into a Varian XL-200 NMR spectrometer equipped with a VT probe. The probe was heated to 140 C (slightly above the melting point of **1**) and a ^{31}P spectrum of the molten starting material was obtained. The probe temperature was then raised to 180 C and spectra were acquired at various times over the next 7 hrs. These spectra are shown in Fig. 4. The first spectrum (0.4 hrs) consists primarily of the resonance due to **1** (20 ppm) and several small peaks on the low field shoulder of that peak which we attribute to various hydrolyzed species. As the reaction time increases several new peaks appear; for example, at 2.4 hrs there are four new peaks. Although we have not as yet definitely assigned these peaks we feel that the three at 5.0 ppm, 0.5 ppm, and -4.0 ppm are probably due to linear hydrolysis products of **1** of the type proposed by Stokes [17]. The peak at -16.0 ppm can be attributed to the phosphorus atoms in the linear portion of a reactive intermediate of the type which has been proposed as the propagation step in one possible reaction mechanism of the melt polymerization [15]. This assignment is consistent with the work of Moran [18], in which the compound $[Cl(PCl_2N)_6PCl_3]^+$ $[PCl_6]^-$ was characterized by ^{31}P NMR. At 2.8 hrs a peak appears at -19.0 ppm and continues to grow as time passes. The chemical shift of this peak is similar to that observed for the linear chloropolymer in solution [19], and it is, therefore, reasonable to assume that we are observing the formation of linear chloropolymer in the melt polymerization. After 5.2 hrs a new peak is evident at -17.0 ppm. This peak continues to grow and at 6.6 hrs it surpasses in intensity the peak at -16.0 ppm. At this time we are unsure as to the origin of this peak, but feel that it could possibly be

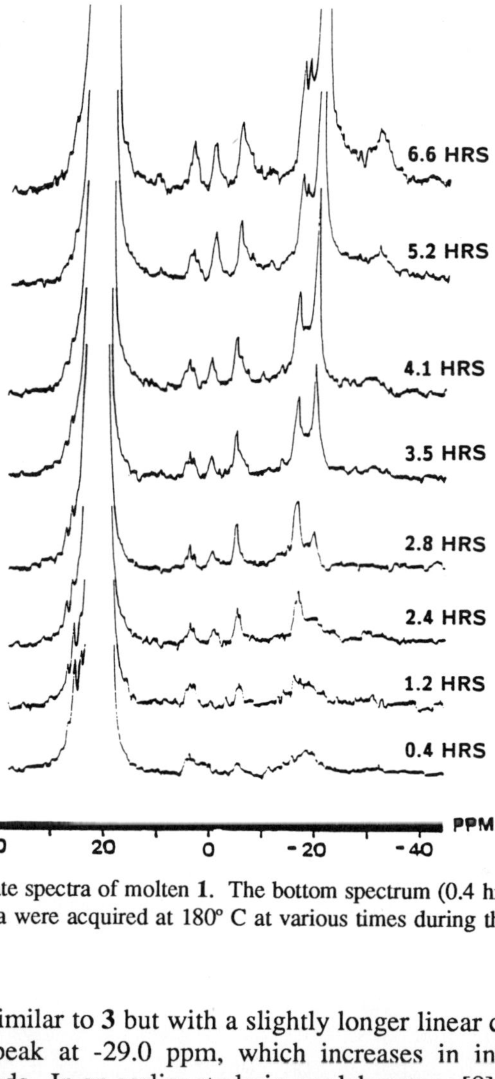

Figure 4. ^{31}P solution-state spectra of molten **1**. The bottom spectrum (0.4 hrs) was acquired at 140° C. The rest of the spectra were acquired at 180° C at various times during the melt polymerization.

due to some species similar to **3** but with a slightly longer linear chain length. At 5.2 hrs there is a new peak at -29.0 ppm, which increases in intensity as the melt polymerization proceeds. In an earlier study in our laboratory [8] we observed P-O-P cross-links at this same chemical shift, so we suspect that we are seeing this type of cross-linking during the melt polymerization. Synthetic and spectroscopic efforts are in progress to verify the identities of the peaks in the above experiment.

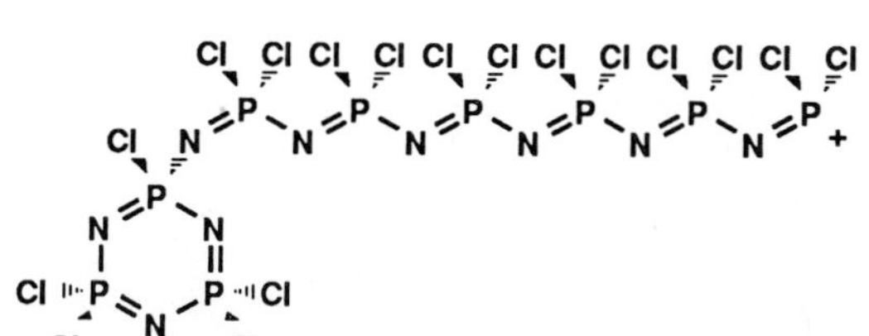

3

It is well known that the rate of melt polymerization of **1** and the degree of cross-linking are profoundly influenced by the exposure of **1** to even trace amounts of water. Indeed, it has even been proposed that completely pure **1** does not polymerize at all reasonable temperatures. Given the high reactivity of **1** with water, it is unlikely that molecular water will exist under melt polymerization conditions. Rather, it seems likely that one or more hydrolyzed species are responsible for initiation and cross-linking reactions under melt polymerization conditions. It is, therefore, crucial to an understanding of the polymerization mechanism(s) that we identify the hydrolyzed species of **1** and study their chemistry. We have taken the first steps in that direction.

At first glance, it might seem reasonable to study protonated functional groups by dissolving **1** in a suitable solvent and applying solution state NMR methods. OH and NH groups, however, readily undergo chemical exchange in solution, and a single (uninformative) resonance is generally observed at the weighted-average chemical shift. A better approach is to perform solid state NMR experiments. Proton NMR is normally a very difficult experiment in the solid state as a result of homonuclear dipole-dipole couplings, and the technically sophisticated CRAMPS experiment is generally required to obtain satisfactory results. In samples of **1**, however, protons are a chemically dilute spin, and proton-proton couplings are negligible. As a result, it is possible to obtain good-quality solid state proton spectra using magic angle spinning alone.

Representative spectra for three samples of **1** are reported in Fig. 5. The lower-most spectrum in that figure is of a highly purified sample of **1** which was prepared using the multiple recrystallization and sublimation procedure of Allcock and coworkers [20]. It is obvious that there are easily detectable quantities of several (as yet unassigned) proton containing functional groups in even this rigorously purified sample. The proton concentration of a commercial sample of **1** (Aldrich 99+%) was found to be even higher and to involve different functionality. Finally, a sample was prepared under intentional hydrolysis conditions; its spectrum is reported at the top of Fig. 5. Inspecting that spectrum, one notes a very high signal-to-noise ratio indicative of the mich higher proton concentration and a diversity of chemical functionality. Further experiments are in progress to assign these resonances and to elucidate the roles of various species in the catalysis of polymerization and cross-linking reactions.

In conclusion, these preliminary results show that multinuclear, high resolution, VT solid-state NMR is a powerful tool for the study of molecular motion and morphology of polyphosphazenes. These results also illustrate that solid-state ^{1}H NMR might make possible the identification of the actual species involved in "water"-catalyzed polymerization and crosslinking. This is a result which would have important implications for the technology and economics of polyphosphazenes, including the discovery of new catalysts and, possibly, reagents that inhibit undesirable cross-linking reactions.

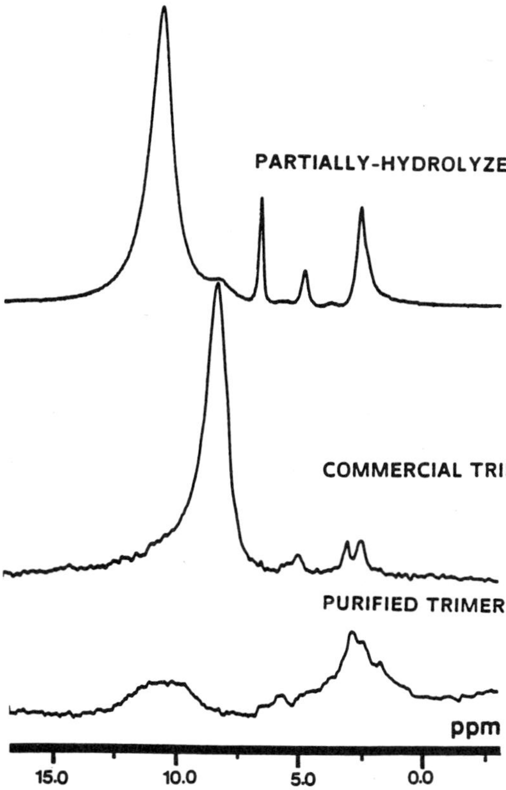

Figure 5. Solid-state ^{1}H MAS spectra of various samples of **1**.

2. References

1. Yannoni, C. S., *Acc. Chem. Res.*, *15*, 201-208, 1982.
2. Maciel, G. E., *Science (Washington, DC)*, *226*, 282-288, 1984.
3. Lyerla, J. R.; Yannoni, C. S. and Fyfe, C. A., *Acc. Chem. Res.*, *15*, 208-216, 1982.
4. Haw, James F., *Anal. Chem.*, *60*, 1988, in press.
5. Allcock, H. R., *Science*, *193*, 1214-1219, 1976.
6. Allcock, H. R., *Angew. Chem. Int. Ed. Engl.*, *16*, 147-156, 1977.
7. Haw, James F.; Campbell, Gordon C. and Crosby, Richard C., *Anal. Chem.*, *58*, 3172-3177, 1986.
8. Crosby, R. C. and Haw, J. F., *Macromolecules*, *20*, 2324-2326, 1987.
9. Haw, J. F.; Crook, R. A. and Crosby, R. C., *J. Magn. Reson.*, *66*, 551-554, 1986.
10. Hexam, J. G.; Frey, M. H. and Opella, S. J., *J. Chem. Phys.*, *77*, 3847-3855, 1982.
11. Rothwell, W. P. and Waugh, J. S., *J. Chem. Phys.*, *74*, 2721-2732, 1981.

12. Schneider, N. S.; Desper, C. R.; Singler, R. E.; Alexander, M. N. and Sagalyn, P. L. in "Organometallic Polymers" Carraher, C. E.; Sheats, J. E. and Pittman, C. U., Ed., Academic Press, New York, 1978.
13. Desper, C. R.; Schneider, N. S. and Higgenbotham, E., *J. Polym. Sci., Polym. Lett. Ed., 15*, 457-461, 1977.
14. Alexander, M. N.; Desper, C. R.; Sagalyn, P. L. and Schneider, N. S., *Macromolecules, 10*, 721-723, 1977.
15. Singler, R. E.; Schneider, N. S. and Hagnauer, G. L., *Polym. Eng. Sci., 15*, 321-338, 1975.
16. Crosby, R. C.; Haw, J. F. and Lewis, D., manuscript in preparation.
17. Stokes, H. N., *Inorg. Nucl. Chem., 30*, 1405-1413, 1968.
18. Moran, E. F., *Inorg. Nucl. Chem., 30*, 1405-1413, 1968.
19. Allcock, H. R.; Kugel, R. L. and Valan, K. G., *Inorg. Chem., 5*, 1709-1715, 1966.
20. Allcock, H. R.; Gardner, J. E. and Smeltz, K. M., *Macromolecules, 8*, 36-42, 1975.

Chapter 21

HYDROGEN CYANIDE POLYMERS: SOLID STATE NMR (^{15}N, ^{13}C) INVESTIGATIONS

C. N. MATTHEWS AND R. LUDICKY

1. Abstract

Liquid hydrogen cyanide polymerizes spontaneously to a black solid from which a yellow-brown powder can be extracted by water. CPMAS NMR spectra have been used to determine the composition of both kinds of polymers. In the water-soluble case the unambiguous presence of secondary amide groups (as in peptide links) has been established by double cross-polarization studies on polymers synthesized from equimolar amounts of $H^{13}CN$ and $HC^{15}N$. These investigations, and the results of electric discharge experiments which produce HCN from methane and ammonia, are consistent with the hypothesis that the original heteropolypeptides on Earth were synthesized directly from hydrogen cyanide and water without the intervening formation of α-amino acids. The black crust on the nucleus of Halley's comet may also consist largely of HCN polymers.

2. Introduction

Hydrogen cyanide polymers--heterogeneous solid mixtures--have resisted unambiguous characterization ever since they were discovered over 180 years ago [1]. In our time these ubiquitous HCN products have been assigned significant roles in prebiotic and extraterrestrial chemistry [2-5], making it all the more necessary to establish their structures in a convincing way [6].

C. N. MATTHEWS - Department of Chemistry, University of Illinois at Chicago, Chicago, IL 60680 and R. LUDICKY - Morton International, Specialty Chemicals Division, 1275 Lake Avenue, Woodstock, IL 60098.

Solid State NMR of Polymers, Edited by L. Mathias
Plenum Press, New York, 1991

To this end we have been collaborating with Jacob Schaefer and his Monsanto coworkers on CPMAS solid state ^{15}N NMR studies of these intriguing macromolecules [7-10]. We discuss here the synthesis and analysis of HCN polymers with particular reference to this NMR work. We then relate our HCN research to the chemistry now proceeding on other solar system bodies such as Halley's comet, and conclude with a cyanide model for the origin of life on Earth.

3. Hydrogen Cyanide Polymers: Synthesis and Analysis

Liquid HCN polymerizes spontaneously to a brown-black solid at room temperature in the presence of a base such as an amine or ammonia [3,8]. Polymerization also occurs readily in non-aqueous solvents or in water [1-3]. Two types of structural units appear to be present in these solid materials. Most stable are the ladder polymers [1] shown in Fig. 1, formally derived from the olefinic tetramer of HCN, diaminomaleonitrile (A), which is usually found among the products. Most probably, however, polymerization to the substituted polymethylene B proceeds by way of an HCN dimer [1]. Cyclization to C and D then leads to ladder structures possessing conjugated >C=N- bonds, as proposed by Völker [1] on the basis of extensive physical and chemical investigations which have been recently supplemented by the work of Umemoto et al [11].

More controversial is the existence of the polyamidine structures shown in Fig. 2. Although polyaminomalonitrile (F) can be formally derived by addition

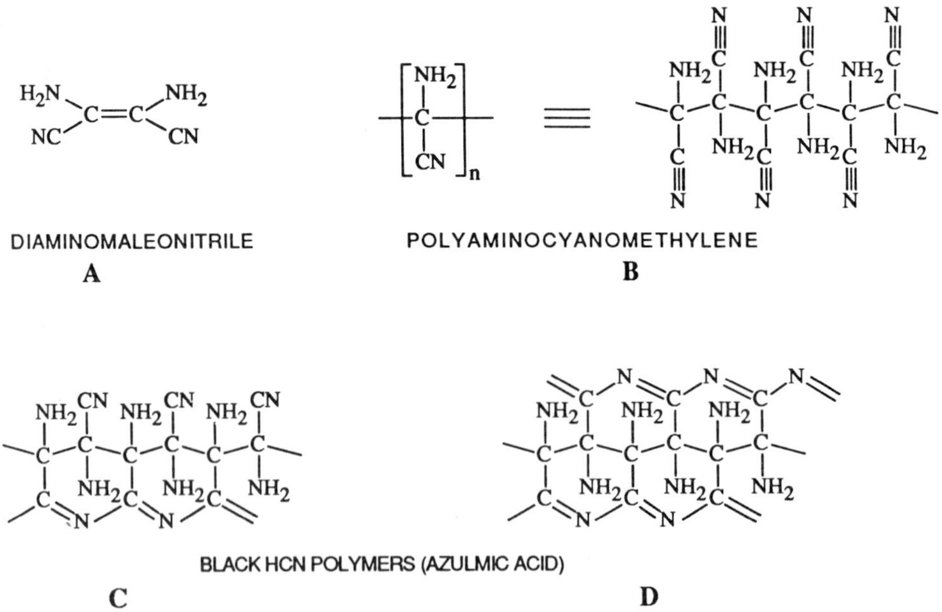

DIAMINOMALEONITRILE
A

POLYAMINOCYANOMETHYLENE
B

BLACK HCN POLYMERS (AZULMIC ACID)

C

D

Figure 1. HCN polymers formally derived from HCN tetramer (diaminomaleonitrile).

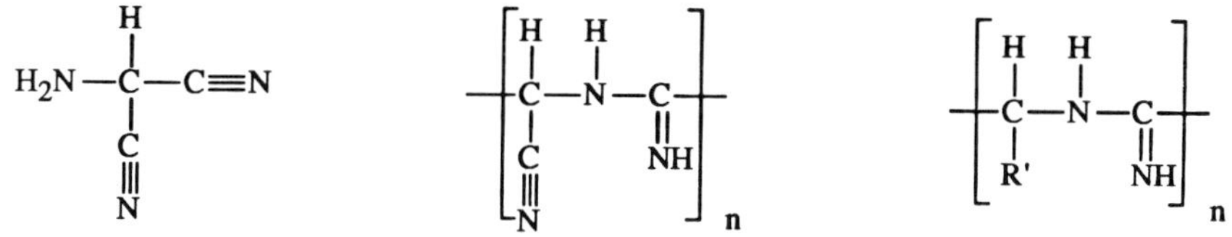

AMINOMALONONITRILE **E** POLYAMINOMALONONITRILE **F** HETEROPOLYAMIDINES **G**

Figure 2. HCN polymers formally derived from HCN trimer (aminomalononitrile).

polymerization of the reactive trimer aminomalononitrile (E), again it seems more likely that polymerization occurs through an HCN dimer [2,3,12]. Further reactions of HCN with the highly activated nitrile groups of F then yield the heteropolyamidines G which are readily converted by water to heteropolypeptides [2,3]. Overall, this series of reactions constitutes a route for the direct synthesis of polypeptides without the intervening formation of α-amino acids.

Several kinds of experiments have provided results consistent with this polyamidine model. In general, water-soluble, yellow-brown solids can be extracted from the products of each of the following types of reactions:

1. base-catalyzed polymerization of liquid HCN, alone, in water or in non-aqueous solvents [3];

2. electric discharge experiments producing HCN from methane-ammonia mixtures [2];

3. alkaline hydrolysis of aminoacetonitrile [13], aminomalononitrile (E,HCN trimer) [14], and diaminomaleonitrile (A, HCN tetramer) [15], all of which are ready sources of HCN at high pH;

4. HCN modification of the reactive nitrile side chains of poly-α-cyanoglycine, a polyamide analog of polyaminomalononitrile (F) synthesized from the N-carboxyanhydride [16,17] related to α-cyanoglycine.

As predicted, acid hydrolysis of these yellow-brown polymers yields not just glycine, the major product, but other α-amino acids as well, such as alanine, aspartic acid, glutamic acid, serine, threonine and leucine, together with some α-amino acids not found in proteins. Also, as predicted, the glycine is perdeuterated when D_2O/DCl is used for hydrolysis instead of H_2O/HCl [18], an indication of the highly acidic nature of the methine carbons of F and of the great reactivity of its nitrile groups brought about by its delocalized structure (Fig. 3).

Comparable polypeptides with molecular weights of 4000-20000 daltons, obtained by Draganic, Niketic et al. [19, 20] in their studies of ^{60}Co irradiation of

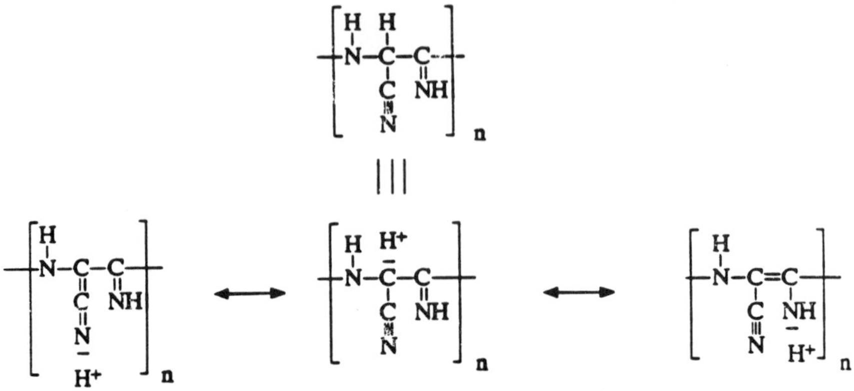

Figure 3. Delocalized structure of polyaminomalononitrile (F).

aqueous cyanide solutions, yielded up to 20% amino acids on hydrolysis, could be cleaved with proteolytic enzymes, gave positive biuret tests, and showed the expected infrared peaks for hydrocarbon (~3μ) and amide (~6μ) groups. Radical mechanisms have been proposed for this γ-radiation chemistry. Whatever the detailed pathway, it seems reasonable to suppose that polyaminomalononitrile (F) was a necessary intermediate arising from cyanide generated in these high energy reactions.

4. Hydrogen Cyanide Polymers: Solid State NMR Investigations

With the development of cross-polarization magic-angle spinning ^{15}N NMR it became possible to detect specific nitrogen functionalities within these complex mixtures [7]. Natural abundance 9 MHz cross-polarization ^{15}N NMR spectra for three model compounds [7] are shown in Fig. 4. The dominant resonance from a typical protein (a), due to the main chain peptide nitrogen, appears at about 95 ppm downfield from solid ammonium sulfate. For poly-α-cyanoglycine (b), the polyamide analog of polyaminomalononitrile (F), an additional broad peak arising from the nitrile is present. The spectrum of diaminomaleonitrile (c) shows the expected resonances due to nitrile (δ_N~220 ppm) and amine nitrogen (δ_N~30 ppm).

Compare these spectra with those of HCN polymer mixtures (Fig. 5,d-f). These samples were enriched with labelled hydrogen cyanide and the spectra obtained at 20 MHz. Sample d, prepared from an equimolar mixture of $H^{13}C^{15}N$ and $H^{12}C^{14}N$ and treated with water for 30 days, shows peaks at 220 ppm and 30 ppm corresponding to nitrile and amine groups, perhaps belonging to A, B, and C. Further resonances in the 200-240 ppm region can be ascribed to azomethine functions (C=N bonds) present in the black polymers C and D. Peaks at ~95 ppm could indicate the presence of polypeptides. Ammonia is present (δ_N=0) as expected following hydrolysis of polyamidines to polypeptides. Evidence for such a change is provided by spectra e and f obtained from a sample prepared from $H^{13}C^{14}N$ and $H^{12}C^{15}N$. These spectra indicate a partial loss of azomethine functionalities--amidines?--and show that peptide/amide linkages (δ_N~95 ppm) appear upon treatment with water. Presumably ammonia, diaminomaleonitrile (A), and low molecular weight polypeptides dissolve out of the water insoluble fraction (f), which very likely consists of azulmic acid (δ_N~200-240 ppm) together with polypeptides.

Some ambiguity remains, however, particularly in determining whether the 95 ppm resonances are due to the nitrogen of primary or secondary amides. In this respect, further clarification was achieved by using double cross-polarization NMR involving the sequential transfer of spin polarization from ^1H to ^{15}N to ^{13}C, followed by detection of ^{15}N under magic-angle spinning conditions [8,9]. The $^{15}N \rightarrow {}^{13}C$ transfer drains some of the polarization of ^{15}N bonded to ^{13}C. The double cross-polarization spectrum is thus a measure of all ^{15}N-^{12}C nitrogen and a fraction of the $^{15}N^{13}C$ nitrogen. A cross-polarization spectrum shows the total ^{15}N. The subtraction of a double cross-polarization spectrum from a cross-polarization spectrum will result in a difference spectrum showing the ^{15}N-^{13}C drained through the $^{15}N \rightarrow {}^{13}C$ transfer. This spectrum will show the formation of new $^{15}N^{13}C$ bonds if different sources of labelled carbon and nitrogen, such as $H^{13}CN$ and $HC^{15}N$, are used in a mixed label experiment.

Figure 5 shows single cross-polarization spectra compared with double cross-polarization difference spectra of HCN polymers synthesized from equimolar mixtures of $H^{13}C^{14}N$ and $H^{12}C^{15}N$ [8,9]. We have already interpreted the single cross-

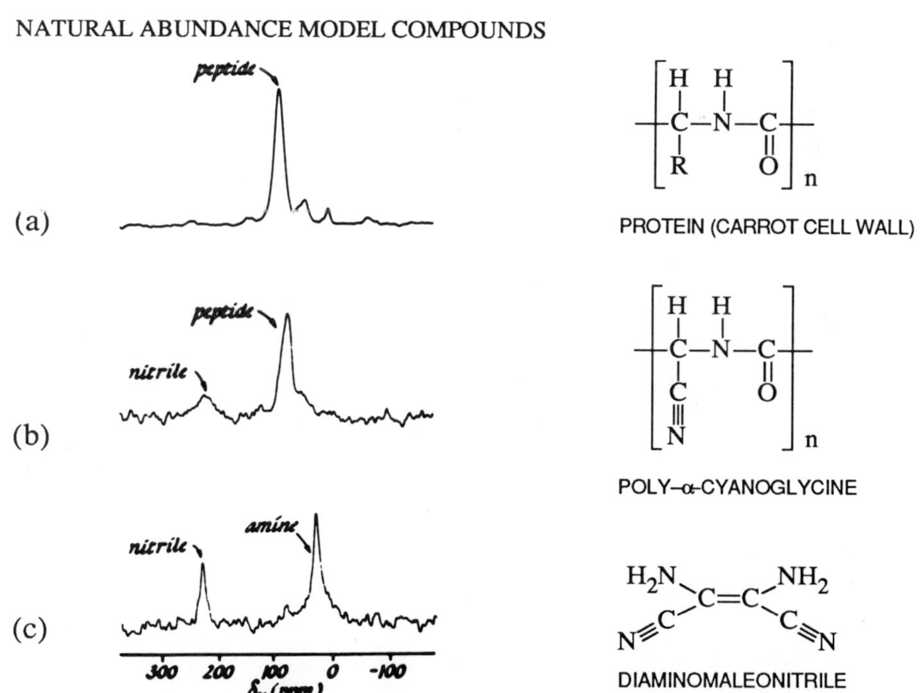

Figure 4. Solid state cross-polarization ^{15}N NMR spectra (natural abundance).

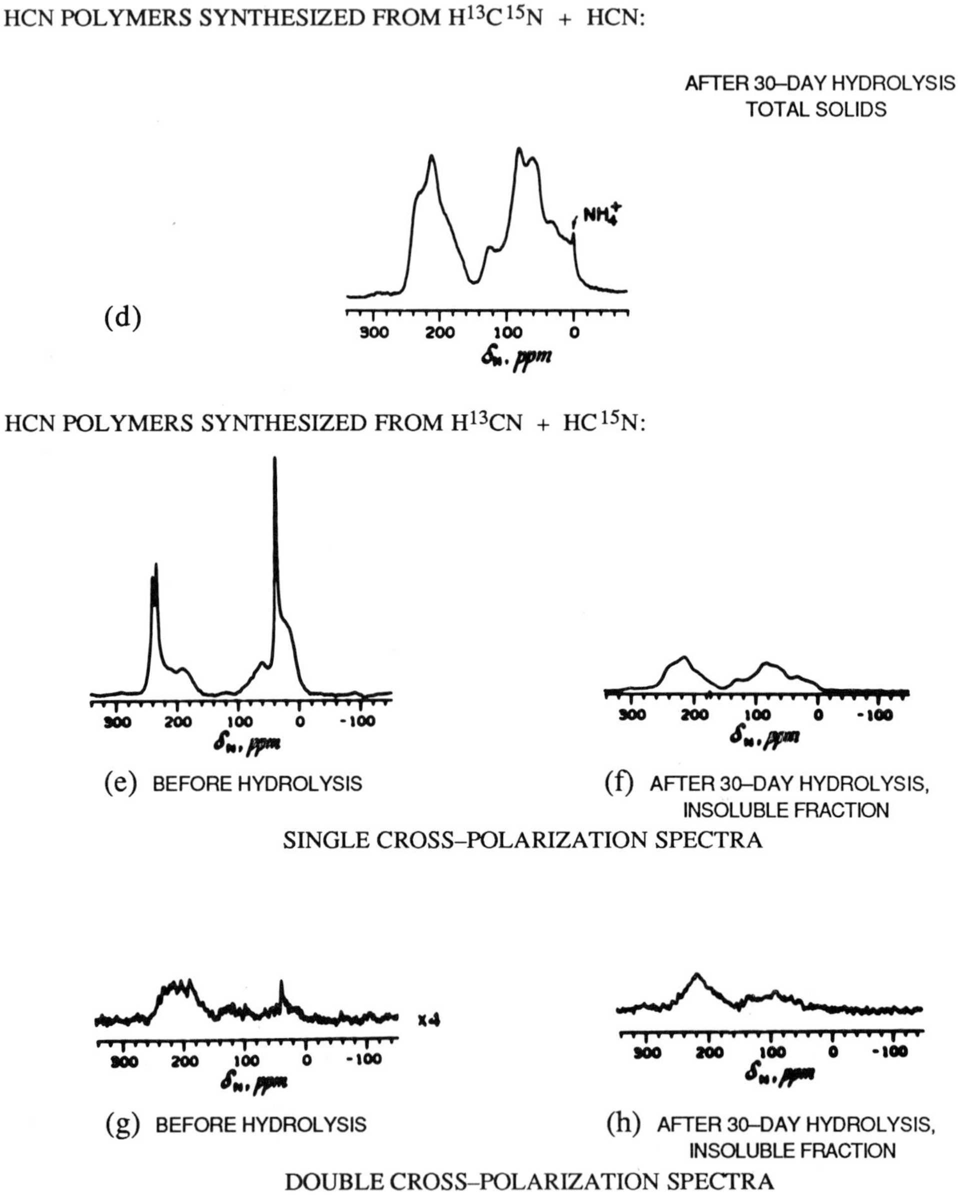

Figure 5. Solid state cross-polarization ^{15}N NMR spectra (labelled).

polarization spectra (e,f). The difference spectra (g,h) supply further information since their very existence depends on the formation of new ^{13}C-^{15}N bonds. Possible C-N bond scrambling was ruled out by cross-polarization experiments on polymers formed from $H^{13}C^{15}N$ diluted with $H^{12}C^{14}N$. Therefore, every ^{15}N which has formed a new ^{13}C-^{15}N bond must still be bonded to its original carbon atom. This implies that the amide peak (δ_N~95 ppm) in the difference spectrum (h) must be at least secondary in nature, as in a peptide link. This evidence, along with the release of ammonia (δ_N=0) from HCN polymers treated with water, is consistent with the proposed pathway:

$$H^{13}C^{14}N + H^{12}C^{15}N \xrightarrow{H_2O} \text{—}^{12}CH(R)\text{—}^{15}NH\text{—}^{13}C(O)\text{—}$$

A further technique--CPMAS dipolar rotational spin-echo ^{15}N NMR [10]--was introduced to measure directly the average number of protons attached to the amide nitrogens. The nitrogen chemical shift region selected restricts the structures examined to amidines and amides, and the ^{1}H-^{15}N dipolar coupling identifies the amidine or amide as primary or secondary. The majority of the amidine or amide nitrogens in the base-catalyzed HCN polymers examined were secondary and hence peptide-like. Questions remain. Why are none of the observed amidine or amide nitrogens bonded to aliphatic carbons, as established by double cross-polarization NMR (^{13}C and ^{15}N)[10]? Because of the delocalized structure and resulting high acidity of the methine group in polyaminomalononitrile? Further investigations are in order.

Taken together, these NMR studies point to the presence of both peptide and ladder structures in HCN polymers exposed to water. NMR detection of substantial peptide presence in water-insoluble material as well as in the aqueous extracts suggests that some peptides are attached to the ladder polymers, probably through their amino groups, i.e. hybrid polymers may be present.

5. Hydrogen Cyanide Polymers in the Solar System

Comparable cyanide polymers are synthesized when methane and ammonia mixtures are converted to hydrogen cyanide by electric discharges [2]. The expected predominance on cometary nuclei of frozen volatiles such as methane, ammonia, and water subjected to high energy sources [21] thus makes them ideal sites for the formation and condensed-phase polymerization of HCN. Dust emanating from the nucleus, contributing to the coma and tail, would also arise partly from the polymer [22]. Results of the recent Halley missions support this view:

1. For the first time, the nucleus of a comet was detected unambiguously. Following Vega observations of a dark surface largely masked by clouds [23], the Giotto multicolor camera showed the presence of a potato-shaped core (~15 x 10 km) with a very low albedo (2-4%) [24,25]. We propose that this non-volatile black crust is largely

composed of hydrogen cyanide polymers and related compounds [22]. Most of the cometary activity appeared to originate from a few discrete sources on the nucleus, notably two major, bright dust jets.

2. Giottos' neutral mass spectrometer confirmed that Halley's exhalations consisted of water vapor (at least 80% by volume) together with methane, ammonia, and carbon dioxide [26]. The infrared observations of Vega 1 showed the surface temperature to be above 300°K, suggesting that the frozen volatiles were covered with a layer of debris perhaps 1 cm thick [27].

3. Among the most prominent emissions in the near ultraviolet and visible spectrum of Halley were those of the CN radical [28,29]. Ground-based studies of Halley showed the presence also of HCN [30]. From their investigations of CN jets in the inner coma A'Hearn et al conclude that while some of the CN radicals of Halley might be split off from HCN, other CN sources must exist tied to very small grains [31].

4. Mass spectra from Halley's coma sent back by the Vega and Giotto missions showed most of the cometary dust grains to be rich in H, C, O, and N, indicating the presence of polymeric organic material [32,33]. The close approach of Giotto to Halley's nucleus enabled particle types to be distinguished. In addition to silicate grains and particles of mixed composition, as well as H, C and H, C, O grains, there appear to be populations consisting only of H, C, O, N, and others only of H, C, N [34]. Both kinds could be closely related to HCN polymers.

It has been argued that comets contributed substantially to early chemical processes on Earth, supplying a large fraction of the volatiles necessary for the emergence of life [35]. Cometary transport of cyanide-related polymers, polypeptides in particular, could certainly have enriched any primordial pool of organic compounds. Whether they would have survived the drastic conditions of impact is an open question.

However, the possibility that cyanide chemistry as described above might be proceeding on Halley and other comets does suggest that HCN polymerization is a preferred pathway within the solar system. HCN polymers may account for the dark color of asteroids and other outer solar system bodies [36]. Carbonaceous chondrites from the asteroid belt might therefore also possess HCN products. Studies of the Murchison meteorite have shown that free α-amino acids are present together with acid-labile amino acid precursors of undefined structure [37,38]. We have shown that these water-soluble yellow-brown solid extracts in many ways resemble peptide segments derived from HCN polymers [39]. Hydrogen cyanide polymerization could account, too, for much of the yellow-brown-orange coloration of Jupiter and Saturn [4,40]. Indeed, HCN has been found in Jupiter's reducing atmosphere and in the atmosphere of Titan, the largest moon of Saturn [41]. Most intriguing is an orange haze high in Titan's stratosphere that may consist of cyanide polymers [40]. After settling on the frozen surface of Titan, the polyamidines would be converted by water to polypeptides.

6. Hydrogen Cyanide Polymers and the Origin of Life

While life would not be expected to arise in such environments, the visible cyanide chemistry on these solar system bodies is a continuing reminder that hydrides of the elements oxygen, carbon, and nitrogen might have been a ready source of prebiotic molecules on Earth, as Urey had originally suggested [42]. In particular, the ubiquitous presence of HCN polymers invites the reexamination and possible reinterpretation of almost all studies concerned with the origin of proteins. Which came first (Fig. 6), amino acids or their polymers?

In Miller-Urey experiments [42], for example, it seems clear from our reinvestigations that the primary products obtained from methane, ammonia and water

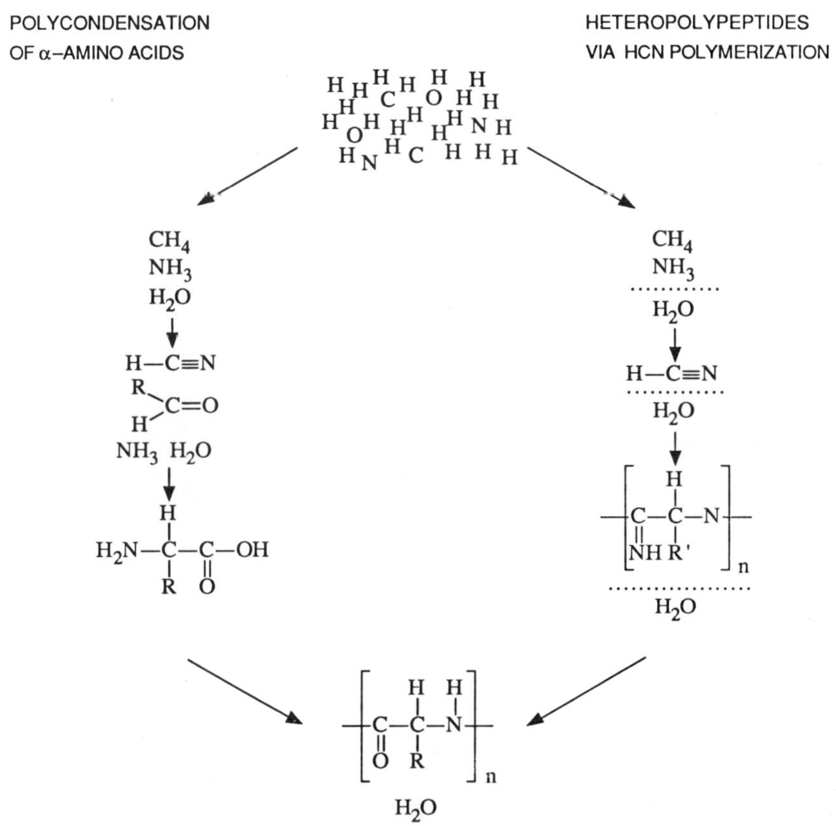

Figure 6. Two opposing models for the origin of proteins. Which came first, amino acids or their polymers?

were not α-amino acids, as claimed, but rather HCN polymers. HCN would be readily formed form methane and ammonia by electric discharge reactions [2] and by elimination from intermediates such as aminoacetonitriles and HCN oligomers [13-15]. The polymers then became hydrolyzed to amino acids either during reflux in the reaction flask, or later during the working-up procedure. The same conclusion applies to virtually all simulations of the chemistry of primitive atmospheres, as well as to studies of aqueous cyanide chemistry by Oró [43], Ferris [44] and others. In our view, these investigations ostensibly yielding α-amino acids actually supply evidence for the abundant prebiotic existence of polymeric protein ancestors--heteropolypeptides synthesized directly from hydrogen cyanide and water [2-4].

Optimum conditions might well have existed on primitive Earth with its methane-ammonia atmosphere, a ready source of hydrogen cyanide. As polyamidines formed within these clouds of HCN, eventually settling onto land and sea together with other products of atmospheric photochemistry, a proteinaceous matrix developed able to take part in and promote the interactions leading to the emergence of life [4,45].

7. Conclusions

The hypotheses that the dark crust of comet Halley consists largely of HCN polymers and that the original polypeptides on primitive Earth were synthesized directly from hydrogen cyanide and water are based on the assumption that hydrogen cyanide polymers possess two types of delocalized structures, conjugated >C=N- ladders and polyamidines. CPMAS solid state ^{15}N NMR studies support this view, although at present neither structure can be considered well-established. It is quite remarkable that these suggestive results have been obtained from such crude mixtures. As more efficient separation techniques become available, both for the synthetic polymers and the extracts obtained from meteorites and, eventually, comets, further elucidation by NMR of this unconventional but universal chemistry will undoubtedly become possible.

8. Acknowledgement

The NMR experiments described were initiated and carried out on our HCN material by Jake Schaefer, Ed Stejskal, Bob McKay and Joel Garbow in the Physical Sciences Center of the Monsanto Company, St. Louis, Missouri. We are grateful indeed for their imaginative, subtle and elegant contribution to this long term investigation of HCN polymers.

9. References

1. Völker, T, *Angew. Chem.*, *72*, 379, 1960.
2. Matthews, C. N. and Moser, R. E., *Proc. Natl. Acad. Sci. U.S.*, *56*, 1087, 1966.

3. Matthews, C. N. and Moser, R. E., *Nature, 215*, 1230, 1967.
4. Matthews, C. N., *Proc. Roy. Inst. Gt. Britain, 55*, 199, 1984.
5. Matthews, C. N., in *The Search for Extraterrestrial Life: Recent Developments*, Papagiannis, M. D., Ed., Reidel, Dordrecht, Holland, 151, 1985.
6. Ferris, J. P., *Science, 203*, 1135, 1979.
7. Schaefer, J.; Stejskal, E. O.; Jacob, G. S. and McKay, R. M., *Appl. Spectr., 36*, 179, 1982.
8. Matthews, C. N.; Ludicky, R. A.; Schaefer, J.; Stejskal, E. O. and McKay, R. A., *Origins of Life, 14*, 243, 1984.
9. McKay, R. A.; Schaefer, J.; Stejskal, E. O.; Ludicky, R. A. and Matthews, C. N., *Macromolecules, 17*, 1124, 1984.
10. Garbow, J. R.; Schaefer, J.; Ludicky, R. A. and Matthews, C. N., *Macromolecules, 20*, 305, 1987.
11. Umemoto, K.; Takahashi, M. and Yokota, K., *Origins of Life, 17*, 283, 1987.
12. Kliss, R. M. and Matthews, C. N., *Proc. Natl. Acad. Sci. U. S., 48*, 1300, 1962.
13. Moser, R. E. and Matthews, C. N., *Experientia, 24*, 658, 1968.
14. Moser, R. E.; Claggett, A. R. and Matthews, C. N., *Tetrahedron Lett.*, 1599, 1968.
15. Moser, R. E.; Claggett, A. R. and Matthews, C. N., *Tetrahedron Lett.*, 1605, 1968.
16. Minard, R. D.; Yang, W.; Varma, P.; Nelson, J. and Matthews, C. N., *Science, 190*, 387, 1975.
17. Warren, C. B.; Minard, R. D. and Matthews, C. N., *J. Org. Chem., 39*, 3375, 1974.
18. Matthews, C. N.; Nelson, J.; Varma, P. and Minard, R. D., *Science, 198*, 622, 1977.
19. Draganic, Z.; Niketic, V.; Jovanic, S. and Draganic, I. G., *J. Mol. Evoln., 15*, 239, 1980.
20. Niketic, V., *Origins of Life, 14*, 251, 1984.
21. Whipple, F. L., *Scientific American, 230*, February, 48, 1974.
22. Matthews, C. N. and Ludicky, R. A., Proc. 20th ESLAB Symposium on the Exploration of Halley's Comet, ESA SP-250, 273, 1986.
23. Sagdeev, R. Z. et al., *Nature, 321*, 259, 1986.
24. Reinhard, R., *Nature, 321,313*, 1986.
25. Keller, H. U. et al., *Nature, 321*, 320, 1986.
26. Krankowsky, D. et al., *Nature, 321*, 326, 1986.
27. Combes, M. et al., *Nature, 321*, 266, 1986.
28. Moreels, G. et al., *Nature, 321*, 271, 1986.
29. Levasseur-Regourd, A. C. et al., *Nature, 321*, 341, 1986.
30. Schloerb, F. P.; Kinzel, W. M.; Swade, D. A. and Irvine, W. M., Proc. 20th ESLAB Symposium on the Exploration of Halley's Comet., ESA SP-250, 583, 1986.
31. A'Hearn, M. F.; Hoban, S.; Birch, P. V.; Bowers, C.; Martin, R. and Klinglesmith, D. A., *Nature, 324*, 649, 1986.

32. Kissel, J. et al., *Nature*, *321*, 280, 1986.
33. Kissel, J. et al., *Nature*, *321*, 336, 1986.
34. Clark, B. C.; Mason, L. W. and Kissel, J., Proc. 20th ESLAB Symposium on the Exploration of Halley's Comet, ESA SP-250, 466, 1986.
35. Oró, J., *Nature*, *190*, 389, 1961.
36. Cruikshank, D. P.; Hartmann, W. K. and Tholen, D. J., *Nature*, *315*, 122, 1985.
37. Kvenvolden, K. A.; Lawless, J. G. and Folsome, C. E., *Scientific American*, *227*, June, 38, 1973.
38. Cronin, J. R., *Origins of Life*, *7*, 337, 343, 1976.
39. Matthews, C. N.; Nelson, J. E. and Minard, R. D., Abstracts 6th International Conference on the Origins of Life, 100, 1980.
40. Matthews, C. N., *Origins of Life*, *12*, 281, 1982.
41. Owen, T., *Scientific American*, *246*, February, 98, 1982.
42. Miller, S. L. and Orgel, L. E., *The Origins of Life on the Earth*, Prentice-Hall, Englewood Cliffs, NJ, 1974.
43. Oró, J. and Lazcano-Araujo, A., in *Cyanide in Biology*, Vennesland, B.; Conn, E. E.; Knowles, C. J.; Westley, J. and Wissing, F., Eds., Academic Press, New York, NY, 517, 1980.
44. Ferris, J. P. and Hagan, W. J., *Tetrahedron*, *40*, 1093, 1984.
45. Matthews, C. N., *Origins of Life*, *16*, 500, 1986.

Chapter 22

^{1}H NMR IMAGING OF SOLIDS WITH MAGIC ANGLE SPINNING

D. G. CORY*, A. M. REICHWEIN, J. C. DE BOER, J. W. M. VAN OS, AND W. S. VEEMAN

1. Introduction

NMR imaging of materials holds the potential of combining chemical selection with high spatial resolution deep inside of a solid sample. This field has progressed slowly compared to the related field of medical imaging [1,2] however, since to obtain useful images, studies must overcome the large natural linewidths of most solids while retaining high spatial resolution. Here we suggest a few methods of ^{1}H NMR imaging of solids at high resolution with chemical selection. As with most areas of NMR, there is no "best" experiment in general, and we will attempt to show the flexibility of these methods and how they may be employed for a given sample or situation.

It will be seen that for structurally rigid samples whose homonuclear dipolar couplings (HDC) may be at least partially averaged by magic angle spinning (MAS), very straightforward high resolution imaging techniques exist. Techniques also exist for molecularly rigid samples with larger dipolar couplings, but these are not yet as successful and further developments are needed. One area under development is of a new class of multiple pulse cycles which run synchronously with an oscillating magnetic field gradient. These allow stronger gradients to be used and may achieve higher resolution.

D. G. CORY,* A. M. REICHWEIN, J. C. DE BOER, J. W. M. VAN OS, AND W. S. VEEMAN - Laboratory of Physical Chemistry II, Faculty of Science, University of Nijmegen, Toernooiveld, 6525 ED Nijmegen, The Netherlands. *Present address: Code 6122, Chemistry Division, Naval Research Laboratory, Washington, DC 20375-5000.

Solid State NMR of Polymers, Edited by L. Mathias
Plenum Press, New York, 1991

2. Chemical Selection

It is clearly important in materials science to employ methods which allow chemical selection to be included in the imaging process. Often a material is composed of several chemical species, and the desired image is of only one component. A simple method of combining chemical selection with imaging is to precede the imaging experiment with a selective 90° pulse followed by a nonselective 90° pulse to place the selected magnetization back along the static magnetic field. An image acquired with a second nonselective pulse after the gradient has been switched on will originate from only the selected spins. Selection may be on the basis of chemical shift, or relaxation times. Most chemical shift selection utilizes the isotropic chemical shift (ICS), and must be performed in the absence of the chemical shift anistropy (CSA) which is often larger than the full range of isotropic shifts in ^{1}H NMR [3]. The simplest method of eliminating the CSA while retaining the ICS is magic angle sample spinning (MAS) [4,5]. Mechanical hopping procedures also exist [6,7], but have not found wide spread application. Selective excitation sequences have been constructed by a combination of Dante [8] and multiple pulse [9] methods.

Selective excitation based on relaxation times is also important, and methods exist for a wide variety of relaxation mechanisms [3-5,10] including, but not limited to T_1, T_2, $T_{1\rho}$, and T_{1x}. These are particularly useful where differences in molecular mobility are important.

3. MAS Imaging

It has been stated in the preceding section that MAS is one of the simplest methods of maintaining chemical shift selection while imaging solids. Since the sample is rotating during the imaging experiment, there are at least three choices of applying the magnetic field gradient (MFG):

1. The gradient may be rotated synchronously with the sample [11,12]. This results in each spin packet in the sample experiencing a constant magnetic field.

2. The gradient may be left stationary, and the acquisition synchronized to the spinner position [13-15].

3. The gradient may be rotated asynchronously but with a known angular velocity.

The first approach, synchronizing the gradient rotation to that of the spinner, is conceptually the simplest, and is the only method that we will consider here. In practice, this is accomplished by the hardware diagramed in Fig. 1 and 2.

We currently employ a Doty Scientific 3.5 mm i.d. MAS spinner, which is capable of rotating at up to 17 kHz, in a homemade probe. The rotation rate is

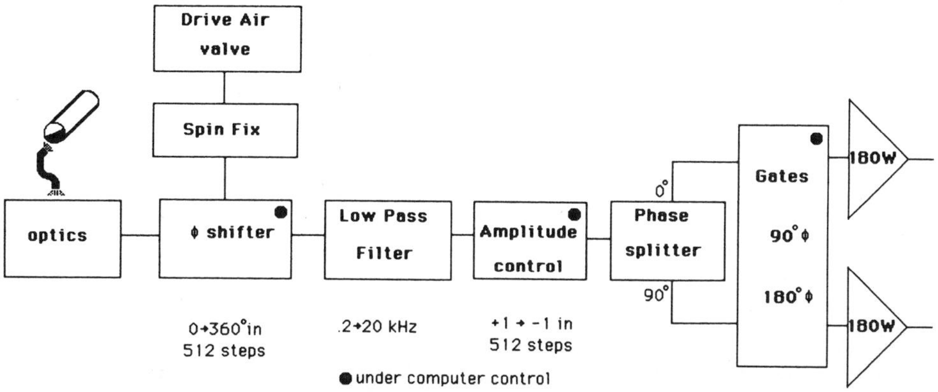

Figure 1. Bloch diagram of the audio circuitry used to control the MFG, and to synchronize it to the spinner position. The spinner frequency is detected optically. The output of a phototransducer is used for long term regulation of the spinner frequency via the Spin-Fix, and is phase locked to a sinusoid whose phase and amplitude are adjustable. This is split into two quadrature signals which are gated and amplified up to 180 W prior to being sent to the gradient coils. Fast 90° and 180° phase shifters just before the final audio amplifiers are used in 2DFI experiments.

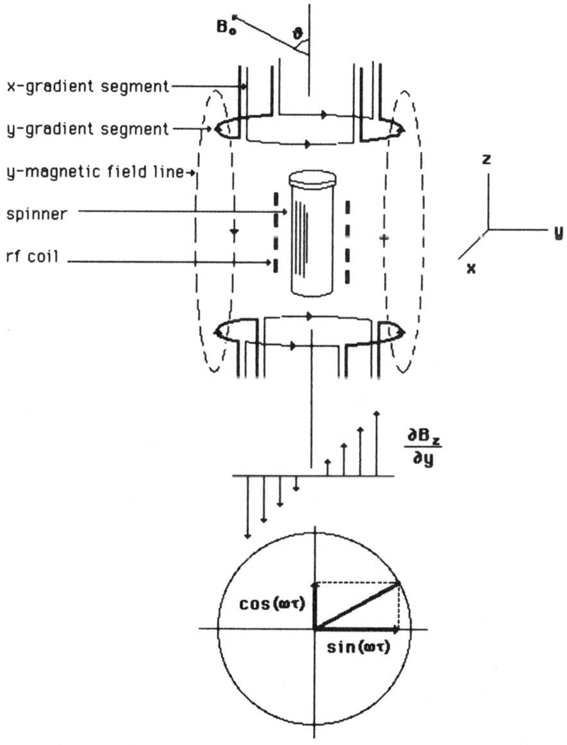

Figure 2. The arrangement of coils which are used to generate a rotating MFG. The y gradient coil segments are connected in series such that current flows in the direction of the arrows. Two magnetic field lines are shown with the direction of the field indicated. The strength of the y MFG ($\partial B_z/\partial y$) is shown schematically below the diagram of the coils. A MFG in the x direction ($\partial B_z/\partial x$) is generated in exactly the same fashion. To rotate the MFG, the voltages applied to the x and y coil sets are modulated by sin and cos functions respectively. The MFG is the sum of these two components which follows a circular path.

detected optically from markings on the base of the spinner, and this signal is used by a "Spin-Fix" module which regulates the spinning speed to within a few Hz for long term stability. The rest of the circuitry is needed to generate two quadrature sinusoidal signals with an adjustable but known phase relation to the spinner signal, and an amplitude which may be varied in evenly spaced steps. The two sinusoids are amplified up to a potential 180 W each and sent to the gradient coils of the probe.

The geometry of the gradient coils are shown schematically in Fig. 2. They consist of two Golay [16,17] coil sets wrapped orthogonally around the spinner housing. The figure describes how this generates a rotating MFG. Golay coils were originally intended to be used in a geometry such that the straight connecting wire segments ran parallel to the static magnetic field. By tilting these at the magic angle, the gradient field is now not generated along the static field, but only that component which is along H_0 contributes. This arrangement scales the applied MFG by $\cos\vartheta_M$ (≈ 0.58). Since the connecting wires are no longer along H_0, they also contribute a distorting field in this application. It may be shown, however, that MAS averages most angular distortions of the MFG to zero.

There are a number of choices of imaging methods which may be employed with a rotating sample, so far the most useful have been:

1. Back projection reconstruction (BP), in which a 2D image is reconstructed from a set of 1D images acquired at incremented orientations of the MFG to the sample [18]. (see Fig. 3a).

2. 2-Dimensional Fourier Imaging (2DFI), in which the MFG is switched by 90° after an initial evolution period, t_1, to acquire a data set suitable for 2DFT processing [19] (see Fig. 3b and c).

Fast echo planar methods have also been described [14,15], but they have yet to prove useful.

BP is more sensitive than 2DFI, since T_2 decays during t_1 reduce the S/N. BP methods also do not normally require gradient switching which often lead to large S/N losses in solids, where the lifetime of the fid may be short compared to the gradient settling time. Methods of storing the magnetization in a state where it does not evolve by T_2 processes during gradient switching are known [20,21], but these also result in losses of sensitivity since only one quadrature component of the fid may be saved at a time.

Figure 4 shows a ^{1}H NMR image [11] of two small pieces of silicon rubber acquired with the 2DFI sequence in Fig. 2b. The ^{1}H NMR spectrum of this material consists of a single sharp resonance, and therefore hardly any line narrowing is necessary to acquire an image. No distortions from the gradient switching of 2DFI are observed since the settling time is short compared to T_2.

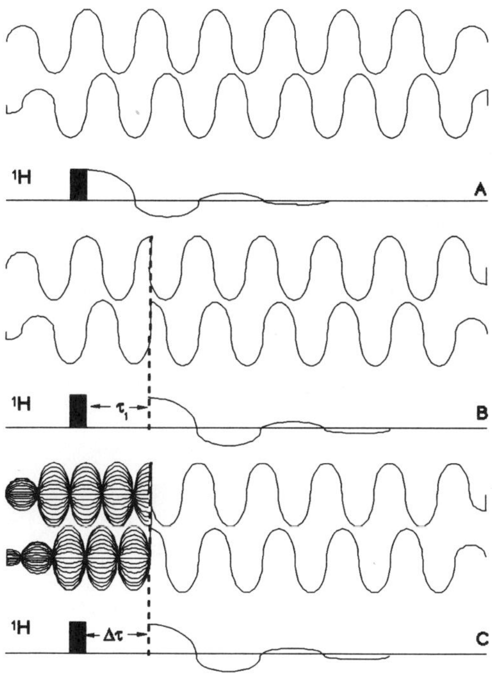

Figure 3. Pulse sequences used to record ^{1}H NMR images of rotating samples. Scheme A is used for BP, while B and C are two versions of the 2DFI method. In each method the gradient is turned on and allowed to stabilize prior to the excitation pulse. For BP the fid is acquired in this MFG, and a complete image is calculated from a set of fids acquired by varying the spinner to gradient phase over 180°. In the 2DFI method (B and C), the phase is rapidly switched by 90° after an evolution time, and then the fid is collected. The evolution time may be incremented (t_1 in B) or kept constant (Δt in C) in which case the strength of the MFG is incremented.

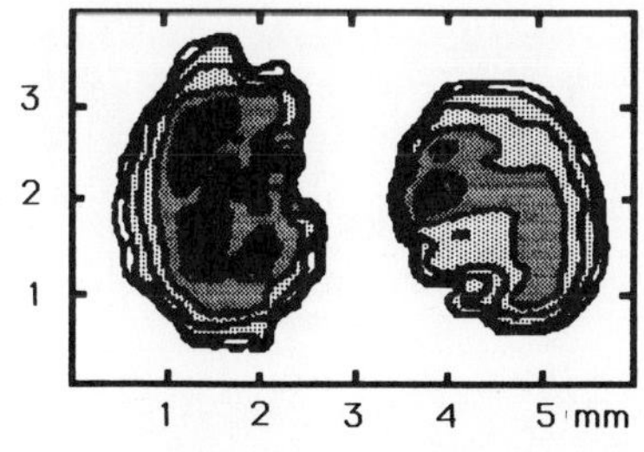

Figure 4. Solid state image of two irregularly shaped pieces of silicone rubber contained in 2 mm holes 1 mm apart. The image was collected with the 2DFI method of Fig. 3b. while the sample was spinning at 2 kHz. (The figure is adapted with permission from ref. 11).

4. The Isotropic Chemical Shift

MAS averages the CSA and possibly the HDC to zero leaving the isotropic chemical shift (ICS) which may distort an NMR image [22-26]. When more than one resonance line is observable in an NMR spectrum, the resultant 2DFI image is a superposition of the images from each resonance line. Each individual image is centered at its resonance frequency in the absence of a magnetic field gradient, and the resulting image is smeared. BP images are distorted in a slightly more complex manner (see for example Fig. 5). Individual resonances in a spectrum may or may not have corresponding distinct spatial distributions, and therefore distinct images. If all resonances arise from the same molecule though, they necessarily have identical spatial distributions and images. In either case a correct image may always be

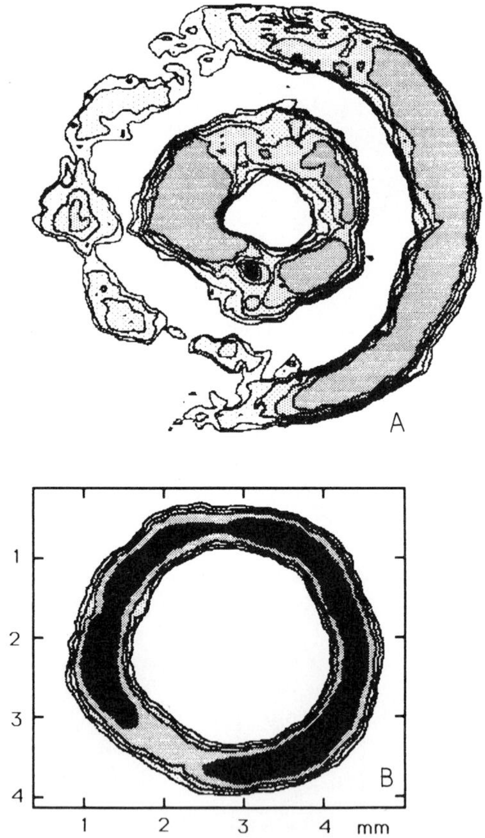

Figure 5. The images in A and B were calculated by the filtered back projection method [18] from the same data set for a polybutadiene (PB) phantom in the shape of a ring, which is described in the text. Image A was calculated from the raw data, and B from data which was first deconvoluted with the spectrum of PB. The extra ring that is observed in A results from chemical shift effects. Since the sample is circular, all 1D images are identical, and the BP algorithm interprets any pair or resonances at constant offsets throughout the experiment as a ring. (The figure is adapted with permission from ref. 25.)

obtained by performing a selective excitation scheme [22] prior to imaging, or by encoding the chemical shift on an axis orthogonal to the image in a multi-dimensional experiment [23]. In effect, these approaches record separate images for each resonance line. When all of the resonances possess the same spatial distribution this is accompanied by a needless sacrifice of sensitivity.

There are currently at least three proven means of removing the ICS in a ^{1}H NMR image of a solid:

1. deconvolution of the gradient broadened image with the spectrum acquired in the absence of a MFG [25].

2. constant time phase encoding schemes of collecting the fid with a stepped MFG [21,26,27].

3. refocused gradient Carr-Purcell sequences [24].

We will discuss the first two of these here, and give some examples. The general topic of refocused gradient imaging will be reserved for Section 7.

4.1. Fourier Deconvolution

The Fourier deconvolution method [25] is valid for BP or 2DFI, is completely independent of our choice of MAS imaging, and versions of this procedure are also employed in ESR imaging [28]. The idea is to deconvolute the image by the NMR spectrum. Such an approach, however, would be very demanding of computer time if carried out in the frequency domain, therefore we perform the corresponding division in the time domain [29].

The fid of a sample in the absence of a magnetic field gradient consists of a set of homogeneously broadened resonances

$$G_0(t) = \Sigma\, p_n g_n(t) \tag{1}$$

where p_n represents the intensity of each of the n homogeneous lines, and $g_n(t)$ is the corresponding lineshape which, for simplicity, is assumed to be Lorentzian.

$$g_n(t) = e^{\,i\omega_n t} e^{-t/T_{2n}} \tag{2}$$

where ω_n is the offset frequency in radiens of line n, and T_{2n} is the static part of its spin-spin relaxation time.

4.2. Back Projection

When a magnetic field gradient is applied across the sample, the intensity function becomes spatially dependent, and each resonance is shifted. In the stabilized linear gradient of the BP method this shift, ω_g, is

$$\omega_g = 2\pi r B \cos\Phi \tag{3}$$

where r and Φ are the spatial coordinates of the spin of interest, relative to the gradient direction, and B is the strength of the gradient field expressed in Hz/unit length. Provided that each resonance has the same spatial distribution, the full fid may be written as

$$G_+(t) = \Sigma\, p_n \int \rho(r,\Phi)\, e^{i(\omega_n+\omega_g(r,\Phi))t} e^{-t/T_{2n}} d\Omega \tag{4}$$

The integration is over the entire sample. As seen, the gradient induced shift adds to the offset frequency, ω_n, and a normalized spatial dependent intensity term, $\rho(r,\Phi)$ has been added. This is independent of the resonance line, and contains the information one desires from an image. For the ideal case of an infinitely sharp line on resonance, the fid in the presence of a field gradient is

$$G_g(t) = \int \rho(r,\Phi)\, e^{i\omega_g(r,\Phi)t} d\Omega \tag{5}$$

The offset terms in Eq. (4) may be eliminated by noting that the sign of the gradient induced shift is changed by rotating the direction of the applied field 180°, $G_-(t)$. It is most informative to separate the fids into the real and imaginary quadrature components, as they are normally acquired. The ideal response, $G_g(t)$, being

$$Re\{G_g(t)\} = \frac{Re\{/G_+(t) + Re\{G_-(t)\}}{2Re\{G_0(t)\}}$$

$$Im\{G_g(t)\} = \frac{Im\{G_+(t)\} - Im\{G_-(t)\}}{2Re\{G_0(t)\}} \tag{6}$$

There is a second equivalent set of expressions for $G_g(t)$ [25]. The data must be acquired with simultaneous acquisition of the real and imaginary points for the imaginary part of $G_g(t)$ to be calculated correctly.

From the integral representation of $G_g(t)$, Eq. (5), it is seen that there are no relaxation terms present. Within the limits of our assumptions, this approach yields infinitely sharp lines regardless of the width of the resonances in the absence of a MFG.

4.3. 2-Dimensional Fourier Imaging

Extension to the 2DFI technique is straightforward, and only the results are listed. The fid of an ideal 2DFI experiment is

$$G_{gg}(t_1,t_2) = \int \rho(r,\Phi)\, e^{i\omega_x(r,\Phi)t_1} e\ ^{i}\omega_y(r,\Phi)t_2\, d\Omega \tag{7}$$

where

$$\omega_x(r,\Phi) = 2\pi r B_x \cos\Phi$$

and

$$\omega_y(r,\Phi) = 2\pi r B_y \cos\Phi$$

2DFT yields the desired image.

To remove distortions in both domains 4 sets of spectra must be acquired ($G_{++}(t_1,t_2)$, $B_x = +$, $B_y = +$; $G_{+-}(t_1,t_2)$, $B_x = +$, $B_y = -$; $G_{-+}(t_1,t_2)$, $B_x = -$, $B_y = +$; $G_{--}(t_1,t_2)$, $B_x = -$, $B_y = -$). The negative sign corresponds to rotating the gradient by 180°. In this nomenclature, the desired result is

$$\mathrm{Re}\{G_{gg}(t_1,t_2)\} = \frac{\mathrm{Re}\{G_{++}(t_1,t_2)\}+\mathrm{Re}\{G_{+-}(t_1,t_2)\}+\mathrm{Re}\{G_{-+}(t_1,t_2)\}+\mathrm{Re}\{G_{--}t_1,t_2)\}}{4\mathrm{Re}\{G_0(t_1+t_2)\}}$$

$$\mathrm{Im}\{G_{gg}(t_1,t_2)\} = \frac{-\mathrm{Im}\{G_{++}(t_1,t_2)\}+\mathrm{Im}\{G_{+-}(t_1,t_2)\}+\mathrm{Im}\{G_{-+}(t_1,t_2)\}-\mathrm{Im}\{G_{--}(t_1,t_2)\}}{4\mathrm{Re}\{G_0(t_1+t_2)\}} \tag{8}$$

Notice that the divisor, G_o, has a time interval t_1+t_2, and must therefore, be measured out to twice as long as the other fids.

In practice, the Fourier deconvolution procedure must be performed with care since noise greatly distorts the results when the divisor is near zero. There are many possible methods of avoiding this, one of which takes advantage of the two equivalent forms of Eq. (6).

The proton image of a polybutadiene phantom [25] in Fig. 5 demonstrates the application of the deconvolution technique to BP imaging. The spectrum of polybutadiene consists of two well resolved resonances from the vinyl and methylene protons. The phantom was made by spinning a small piece of the rubber between two Teflon spacers at about 5 kHz causing the rubber to redistribute itself as a ring around the edge of the spinner.

Figure 5a is the image that was calculated from the raw data, and Fig. 5b was calculated from the deconvoluted data. Distortions are clearly present in the image processed from the raw data, while the image obtained from the deconvoluted data is a quite acceptable representation of the distribution of rubber in the spinner.

4.4. Phase Encoding

The phase encoding pulse sequence [26,27] is shown in Figure 6. The idea here is to acquire each point of the fid after a constant time with a stepped gradient so that all changes arise solely from the gradient. The sequence, therefore, avoids distortions due to chemical shifts, and spin-spin relaxation (T_2). In addition, since the spinner rotates an integral number of revolutions, n, during the encoding time, distortions due to variations in the gradient field with spinner position are not observed. The phase angle of the magnetization in the transverse plane, at time t, has the form

$$\vartheta(t) = |G|\, n \int_0^{t_p} \vec{r}(t') \cdot \hat{G}(t')\, dt' + \int_0^{t} \omega_0 dt' \qquad (9)$$

where $\vec{r}$ is the position of a spin packet, $\vec{G}$ is the gradient vector, $\hat{G}$ is a unit vector in the same direction as $\vec{G}$, and ω_o is an offset frequency. Only the magnitude of $\vec{G}$, $|G|$, is varied between points in the fid, (in steps of ΔG) therefore the details of the path of the magnetization vector over a period, t_p, contained in the integral, are not observed. This is obviously only true of the angular dependence of $\vartheta(t)$ and axial distortions in the field gradient still influence the image.

The experimental parameters are determined by the following two applications of the Nyquist relation

$$2\pi\, \Delta G\, l_s\, nt_p = 2\pi \qquad (10)$$

and

$$2\pi\, G_{max}\, \Delta l\, nt_p = \pi \qquad (11)$$

where l_s and Δl are the sample width and the resolution respectively. In general, it is advantageous to spin as fast as possible to achieve the sharpest resonance. To improve the S/N, n should be as small as possible, but it is limited by the 90° pulse width, t_{90}.

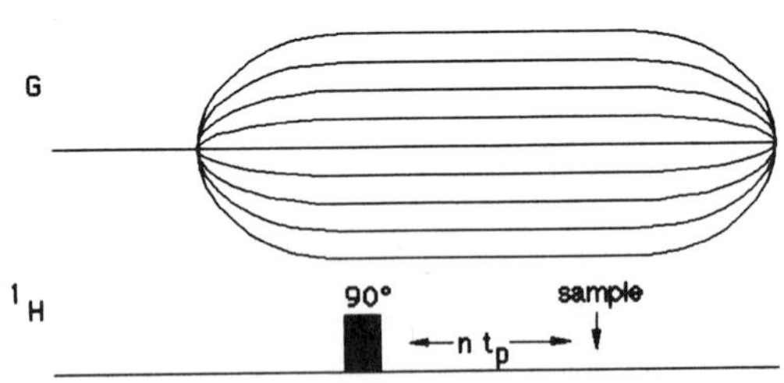

Figure 6. A diagram of the experiment used to obtain the images in figures 4.3 and 4.4. In this experiment, the fids are mapped out one point at a time. First the gradient is turned on and allowed to stabilize, followed by a 90° pulse and an encoding time equal to a multiple of the spinner period. At this time one data point is sampled. Subsequent data points are collected in an identical fashion, with the strength of the gradient field incremented by ΔG. A full set of 1D images suitable for BP reconstruction are acquired by varying the phase of the gradient field to the spinner over 180° in equal steps.

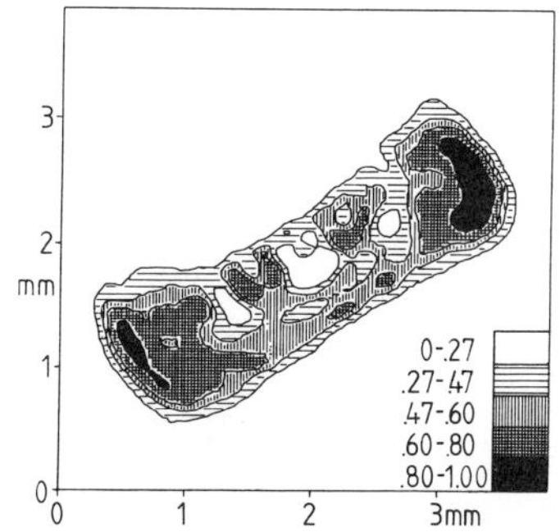

Figure 7. ^{1}H MAS image of a polyisoprene phantom spinning at 5 kHz. The static dimensions of the piece of rubber are 3.3 mm long x 1.1 mm wide x 0.75 mm thick. It has five 150-200 μ holes evenly spaced with centers 500 μ apart. The sample is a soft rubber which deforms due to MAS, but returns to its original shape afterwards.

For a straightforward image reconstruction to work, the excitation pulse must cover the full frequency width of the gradient broadened spectrum, which limits the number of observable volume elements across a sample, NV = $l_s/\Delta l$, to

$$NV \leq \frac{n\ t_p}{2\ t_{90}} \tag{12}$$

For small samples and reasonable 90° pulse widths, this is rarely an important consideration.

Figures 7 and 8 are BP images [26] which were acquired with this sequence. The polyisoprene phantom was studied to investigate possible distortions. At 5 kHz, the sample is stretched and presses against the spinner walls. Though it is cut to be approximately the same thickness along its entire length, there is obviously more intensity where the rubber has flowed against the spinner walls. Five 150 - 200 μ holes were burned into the sample and are still clearly visible. The three central holes are well defined, and have zero intensity inside of them. The central hole is stretched by the spinning, and the two outermost holes are partially closed.

The ends of the piece of rubber in the image are broader than the middle. This is at least partially due to the deformation of the sample, but jitter in the phase lock loop of the gradient to the spinner produces an artifact which has a similar appearance. We have attempted to make this small by strongly locking the gradient and using very stable spinning, however we can not rule out the possibility that the jitter contributes to this shape. Overall, we are led to conclude that the image is an accurate representation of the phantom as it appears under MAS conditions.

Figure 8a and b are ^{1}H NMR images of the polybutadiene fraction of two polybutadiene/polystyrene blends. Both are approximately 50/50 mixtures. The sample corresponding to the image in Fig. 8a is mechanically blended and that of Fig. 8b is cast from toluene. If the polybutadiene was uniformly dispersed through the

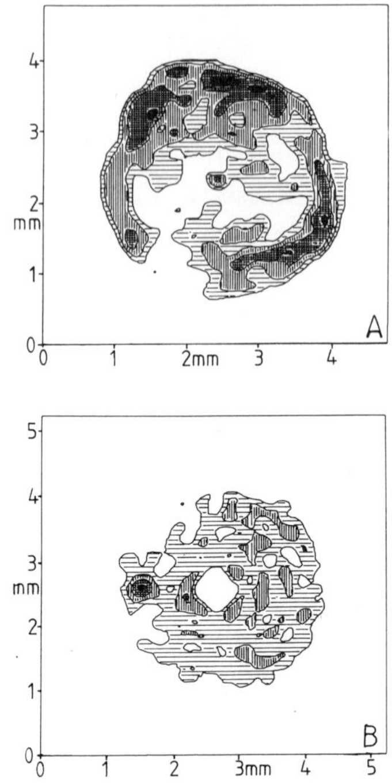

Figure 8. Two ^{1}H MAS images of polybutadiene in blends with polystyrene. A is a mechanical blend, and B is cast from toluene.

sample, a constant level would be observed in the images. The highest levels, black, are areas where the polybutadiene extends through almost the entire thickness of the sample, and the white areas correspond to areas devoid of polybutadiene. Since the sample is a solid disk with no obvious voids, these regions contain predominantly polystyrene. The other levels correspond to mixture of polystyrene and polybutadiene throughout the thickness of the sample.

Clearly very small features are observed in the images, and the samples are obviously incompatible. Some idea of the degree of mixing and the domain size of the rubber particles may be seen.

5. Multiple Pulse Methods

In the previous two sections it has been seen that ^{1}H NMR images may be obtained of materials for which MAS effectively averages the homonuclear dipolar coupling. But, even with the present potential of spinning at up to 23 kHz [30], this is a relatively small subset of the materials one might wish to image. For samples with large dipolar couplings, multiple pulse (MP) methods may be employed to obtain an image [12,21,31,37].

A variety of MP schemes exist for obtaining high resolution ^{1}H spectra of solids [3,4,32]. Their important characteristics for imaging purposes will be explored in section 6, here we will only consider one of the most widely employed, the MREV-8 sequences. MP sequences toggle the spin magnetization between a number of states, the average of which has the desired properties. Sampling of the data, therefore, is restricted to a window in which the averaging is completed, and in practice takes place between closely spaced pulses. MREV-8 coherently averages dipolar interactions to zero, and scales the offset, chemical and gradient shifts by $2/\sqrt{3}$, assuming δ function pulses.

Figure 9 is an image of two irregularly shaped pieces of low density polyethylene (LDPE) [12]. The MREV-8 spectrum of LDPE consists of a single relatively sharp ($\approx$ 1 ppm) resonance. The image was acquired with the BP pulse scheme shown in Fig. 1a coupled with MREV-8 during data acquisition.

LDPE is somewhat too simple a sample since it contains only one rather easily narrowed resonance. A far more demanding sample polystyrene consists of three resonances which are difficult to narrow. Figure 10 displays an image of PS powder which was placed in a 3.5 mm i.d. spinner and spun at 5 kHz. The powder distributed itself in a ring around the edge of the spinner as seen in the image.

It is difficult to avoid lines at zero frequency and rotor lines in CRAMPS spectra, and these artifacts preclude employing Fourier deconvolution to eliminate the ICS. The PS image was therefore obtained with the constant time phase encoding sequence discussed in section 4.4. A constant number of MREV-8 cycles were inserted into the encoding time, after which a single data point was collected in an incremented MFG.

The PS image still displays a distortion which results from acquiring only 16 significant points across the sample and subsequently zero filling to 64 points. The near symmetry of the image is a result of this, as are the intensity distortions and the very good circular shape. Since the sample is indeed circular, all projections are the same and with only 16 points, very few are non-zero, so that small deviations which should appear in the image are lost. Zero filling BP images tends to smooth them in a symmetrical fashion.

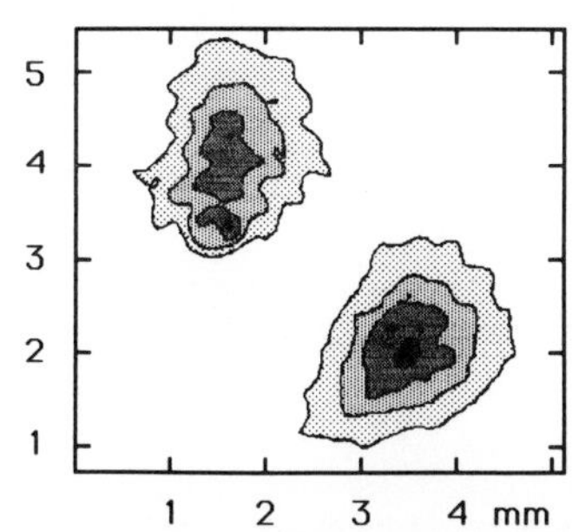

Figure 9. The solid state image of two irregularly shaped pieces of low density polyethylene spinning at 1850 Hz. The image was collected by the BP method using MREV-8 coherent averaging to eliminate homonuclear dipolar couplings. (The figure is adapted with permission from ref. 12.)

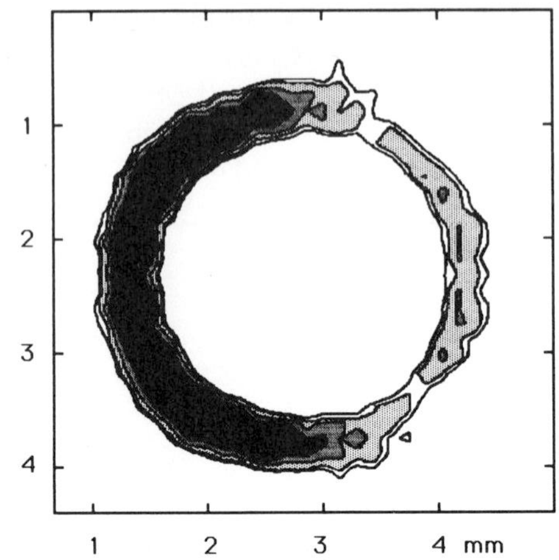

Figure 10. The solid state image of a ring of polystyrene powder spinning at 5 kHz. A constant time phase encoding BP method was employed (see Fig. 6) with MREV 8 coherent averaging to lengthen the fid. Distortions which appear in this image are discussed in the text.

6. Limitations of Multiple Pulse Methods

Line narrowing is not always accompanied by increased sensitivity and improved resolution in solids imaging, and in fact it may have the exact opposite effect. In this section the influences of line narrowing on imaging will be explored [33]. The discussion will consider the constant time phase encoding scheme discussed in sections 4.4 and 5 and reach conclusions based on both the experimental time, and on the maximum number of volume elements (NV) which may be observed across a sample of width l_s. For other types of imaging experiments the results are the same. Throughout, many common imaging variables will be ignored and attention will be focused on those which are pertinent to line narrowing. Detailed discussions of others may be found in the literature [1,2,34].

For simplicity, the S/N is considered normalized to a single data point, and all effects other than the decay of the signal due to spin-spin relaxation are taken to be constant. With an appropriate filter, the noise is dependent on the square root of the bandwidth of the spectrum, and the S/N is

$$S/N \propto \frac{e^{-t/T_2}}{(l_s G_{max})^{1/2}} \tag{13}$$

The shortest time to acquire one data point, $tp \propto (S/N)^{-2}$, may be shown to be

$$tp \propto \frac{l_s}{\Delta l T_2} = \frac{NV}{T_2} \tag{14}$$

where Δl is the resolution. Eq. (14) implies that given sufficient time any arbitrary resolution may be obtained. This is not true for a wide variety of reasons most of which we will ignore. It is sufficient for our purposes to require that the bandwidth of the excitation pulse, t_{90}, approximately cover the full range of frequencies in the image. As was seen in section 4, this places a limit on NV of

$$NV \leq \frac{T_2}{t_{90}}$$

NV may be large for samples with narrow lines, but it is always advantageous to increase T_2. For very rigid samples T_2 may be only a few μ sec and NV is quite small.

One method of increasing T_2 is with MAS. The above analysis holds with T_2 replaced by the effective spin-spin relaxation time under MAS, $T_2^{MAS} \geq T_2$. Provided that MAS narrows the resonance and does not deform the sample, it should always lead to improved images.

Now consider the case of multiple pulse averaging to lengthen T_2 to T_2^{MP}. Here the scaling of I_z interactions by the factor sc, and the offset dependence of MP methods, $\nu \leq \nu_{OS}$, limit the size of G_{MAX}, and NV.

$$NV \leq T^{MP}_2 \, sc \, \nu_{os}$$

Typical values of sc and ν_{os} are given in Table I [35,36]. With MP, NV is generally quite small.

To complete the comparison, the reduction in S/N with MP is also considered. The bandwidth of the receiver must be opened much wider than the spectrum width for MP since the magnetization is observed only a few μ sec after the pulse. This decreases the S/N of the image greatly. Typically the recovery time of the receiver must be approximately 1 μ sec which demands a bandwidth of at least 1 MHz. In other words, MP methods reduce the S/N by a factor of $(b\omega^{mp}/ l_s G_{max})^{1/2}$, or on the order of 10 for a typical experiment. Where $b\omega^{mp}$ is the effective receiver band width under MP. Table II summarizes this comparison of line narrowing approaches to ^1H NMR imaging.

Table I. SCALING FACTORS AND FREQUENCY RANGE OF SELECTED MULTIPLE PULSE SEQUENCES

	sc[a]	ν^b_{os}
MREV-8	√2/3	1.5 kHz
BR-24	2/3√3	4 kHz
BLEW-48	4/3π	4 kHz

a. finite pulse width effects are ignored.
b. approximated from the figures in ref. 36. These numbers are very sensitive to instrument setup, and are only approximate.

Table II. CHARACTERISTICS OF LINE NARROWING METHODS

Line Narrowing	NV_{max}	When to use for equal NV
None	$\frac{T_2}{t_{90}}$	$T_2 \geq T_2^{MAS}$ and $T_2^{MP} \leq \frac{NV}{sc\nu_{OS}[1-\ln((T_2 b\omega^{MP})/NV)]}$
MAS	$\frac{T_2^{MAS}}{t_{90}}$	$T_2^{MAS} > T_2$ and $T_2^{MP} \leq \frac{NV}{sc\nu_{OS}[1-\ln((T_2^{MAS} b\omega^{MP})/NV)]}$
MP	$T_2^{MP}\ sc\ \nu_{os}$	$T_2^{MP} > \frac{NV}{sc\nu_{OS}[1-\ln((T_2 b\omega^{MP})/NV)]}$ and $T_2^{MP} > \frac{NV}{sc\nu_{OS}[1-\ln((T_2^{MAS} b\omega^{MP})/NV)]}$

The time per point for multiple pulse sequences, tp^{mp}, may be calculated in analogy to tp as

$$tp^{mp} \propto bw^{mp}\, e^{2t/T_2^{mp}} \tag{15}$$

which may be compared to tp without line narrowing

$$\frac{tp^{MP}}{tp} = \frac{T_2 bw^{MP}}{NV} \exp\left\{ \frac{NV}{sc\ \nu_{OS}\ T_2^{MP}} - 1 \right\} \tag{16}$$

Multiple pulse sequences are advantageous when this ratio is less than unity. It is immediately apparent that regardless of how well MP methods narrow resonances, T_2 must be less than $(NV/bw^{MP})e^1$ for them to compete with MAS. Setting Eq. 16 equal to one, yields the following equality,

$$\frac{T_2}{T_2^{MP}} = \frac{sc\ \nu_{OS}[\ln[(NV/T_2)(1/bw^{MP})] + 1]}{[NV/T_2]}\ . \tag{17}$$

Equation 17 is plotted in Fig. 5 for a variety of MP sequences. The areas below the curves are where MP leads to improve images.

For very rigid polymers which may have linewidths of up to 100 kHz, MP methods must be employed, but many samples may be more successfully imaged with high speed MAS.

7. Refocused Gradient MP

It has been shown that MP methods suffer from serious drawbacks related to their offset dependence, and their low sensitivity. A solution to the offset problem may be approached by designing new MP sequences exclusively for imaging. At the same time, chemical shift effects may be averaged to zero. Miller and Garroway [24] have demonstrated how this can be accomplished by synchronizing a Carr-Purcell spin echo train to an oscillating MFG. Their sequence eliminates chemical shift, offsets and field inhomogeneities, while retaining the shift from the MFG. The sequence was intended for liquids imaging, and has no effect on the dipolar coupling.

A logical extension of these ideas to MP solid echo sequences is shown in Fig. 11 [37]. Here the dipolar coupling is averaged to zero while retaining the gradient shift and eliminating the ICS and CSA. Since all pulses are at the zero crossing of the MFG, one important source of offset dependence is greatly reduced. Average Hamiltonian theory may be applied to evaluate the sequences dependence of gradient strength [38].

The displayed sequence is one of many which may allow higher resolution images of molecularly rigid solids to be obtained. Since chemical shifts are also

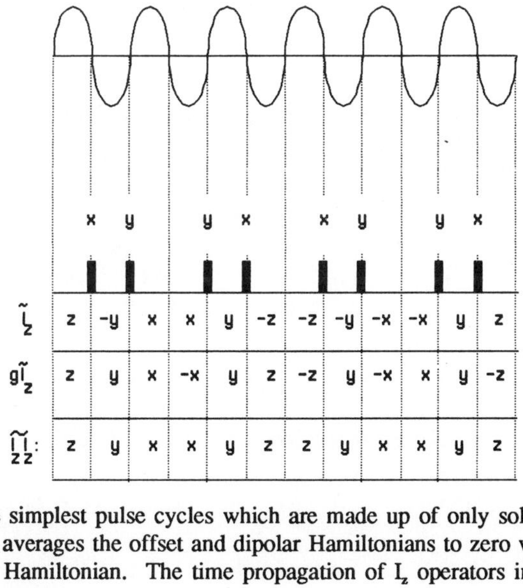

Figure 11. One of the simplest pulse cycles which are made up of only solid echo sequences in a oscillating MFG which averages the offset and dipolar Hamiltonians to zero while retaining a scaled version of the gradient Hamiltonian. The time propagation of I_z operators in the toggling frame is shown below the pulse cycle.

removed, more efficient data collection than constant time phase encoding may be employed.

8. Conclusions

We have attempted to describe the state of the art of ^{1}H NMR imaging of solids as it stands today, and point to where it is likely to go in the near future. Since chemical selection is often necessary in materials imaging MAS should be employed for structurally rigid samples. Provided MAS at least partially averages the homonuclear dipolar coupling, high resolution ($\leq$ 50 μ) images may be routinely obtained.

Molecularly rigid materials for which MAS does not narrow the dipolarly broadened resonance may still be imaged with multiple pulse techniques, but not at the same level of resolution. The limitations are implicit in the MP sequences, and a class of more appropriate MP sequences which incorporate rapidly oscillating gradients are suggested.

9. Acknowledgement

We thank Dirk van Benthum and Henk de Wit for constructing the audio circuitry. Financial support has been generously provided by the Koninklijke/Shell Laboratory (Amsterdam).

10. References

1. Mansfield, P. and Morris, P. G., in "Advances in Magnetic Resonance," Vol. 12, Suppl. 2, Waugh, J. S., Academic Press, New York, 1982.
2. Morris, P. G., "Nuclear Magnetic Resonance Imaging in Medicine and Biology," Claredon Press, Oxford, 1986.
3. Haeberlen, U., in "Advances in Magnetic Resonance," Suppl. 1, Waugh, J. S., ed., Academic Press, New York, 1976.
4. Mehring, M., "High Resolution NMR in Solids," Springer-Verlag, Heidelberg, 1981.
5. Fyfe, C. A., "Solid-State NMR for Chemists," CFC Press, Guelph, Ontario, 1983.
6. Bax, A.; Szeverenyi, N. M. and Maciel, G., *J. Magn. Reson.*, *55*, 494, 1983.
7. Bax, A., Szeverenyi, N. M. and Maciel, G., *J. Magn. Reson.*, *52*, 147, 1983.
8. Morris, G. A., and Freeman, R., *J. Magn. Reson.*, *29*, 433, 1978.

9. Caravatti, P.; Levitt, M. H. and Ernst, R. R., *J. Magn. Reson.*, *68*, 323, 1986.
10. Ernst, R. R.; Bodenhausen, G. and Wokaun, A., "Principles of Nuclear Magnetic Resonance in One and Two Dimensions," Claredon Press, Oxford University Press, London, 1987.
11. Cory, D. G.; van Os, J. W. M. and Veeman, W. S., *J. Magn. Reson.*, *76*, 543, 1988.
12. Cory, D. G.; Reichwein, A. M.; van Os, J. W. M. and Veeman, W. S., *Chem. Phys. Lett.*, *143*, 467, 1988.
13. Matsui, S. and Kohno, H., *J. Magn. Reson.*, *70*, 157, 1986.
14. Matsui, S.; Sekihara, K.; Shiono, H. and Kohno, H., *J. Magn. Reson.*, *77*, 182, 1988.
15. Cory, D. G., *J. Magn. Reson.*, *82*, 337, 1989.
16. Golay, M. J. E., *Rev. Sci. Instrum.*, *29*, 313, 1958.
17. Hoult, D. I. and Richards, R. E., *Proc. Roy. Soc.*, *A344*, 311, 1975.
18. Brooks, R. A. and DiChiro, G., *Radiology*, *117*, 561, 1975.
19. Kumar, A.; Welti, D. and Ernst, R. R., *J. Magn. Reson.*, *18*, 69, 1975.
20. Szeverenyi, N. H. and Maciel, G., *J. Magn. Reson.*, *60*, 460, 1984.
21. Chingas, G. C.; Miller, J. B. and Garroway, A. N., *J. Magn. Reson.*, *66*, 530, 1986.
22. Hall, L. D.; Sukumar, S. and Talagola, S. L., *J. Magn. Reson.*, *56*, 275, 1984.
23. Hall, L. D. and Sukumar, S., *J. Magn. Reson.*, *56*, 314, 1984.
24. Miller, J. B. and Garroway, A. N., *J. Magn. Reson.*, *67*, 575, 1986.
25. Cory, D. G.; Reichwein, A. M. and Veeman, W. S., *J. Magn. Reson.*, *80*, 259, 1988.
26. Cory, D. G.; de Boer, J. C. and Veeman, W. S., *Macromolecules*, *22*, 1618, 1989.
27. Emid, S. and Creyghton, J. H. N., *Physica*, *B125*, 81, 1985.
28. Ohno, K., *J. Magn. Reson.*, *64*, 109, 1985.
29. Champeny, D. C., "Fourier Transforms and Their Physical Applications," Academic Press, London, 1973.
30. Dec, S. F.; Wind, R. A. and Maciel, G. E., *J. Magn. Reson.*, *70*, 355, 1986.
31. McDonald, P. J.; Attard, J. J. and Taylor, D. G., *J. Magn. Reson.*, *72*, 224, 1987.
32. Gerstein, B. C. and Dybowski, C. R., "Transient Techniques in NMR of Solids," Academic Press, London, 1985.
33. Cory, D. G. and Veeman, W. S., *J. Magn. Reson.*, in press.
34. Callahan, P. T. and Eccles, C. D., *J. Magn. Reson.*, *71*, 426, 1987.
35. Burum, D. P. and Rhim, W. K., *J. Chem. Phys.*, *71*, 944, 1979.
36. Burum, D. P.; Linder, M. and Ernst, R. R., *J. Magn. Reson.*, *44*, 173, 1981.
37. Mansfield, P. and Grannell, P. K., *Phys. Rev.*, *B12*, 3618, 1975.
38. Cory, D. G. and Veeman, W. S., unpublished results.

Chapter 23

NMR IMAGING: APPLICATION TO ELASTOMERIC MATERIALS

C. CHANG AND R. A. KOMOROSKI

1. Abstract

We have examined a number of elastomeric systems of similar glass transition temperature T_g but differing chain segmental mobilities. Usable hydrogen images were obtained at ambient temperatures in reasonable times (one half hour to overnight) for bulk *cis*-polybutadiene, natural rubber and a cured, carbon-black-filled *cis*-polybutadiene. Image quality is, as expected, highly dependent on the relative values of T_2, the spin-spin relaxation time, and TE, the time to echo. Thus, under comparable conditions pure *cis*-polybutadiene gave a better image than natural rubber, which in turn gave a better image than the cured, filled *cis*-polybutadiene. Polyisobutylene, a rubber with the same T_g as natural rubber, but a $T_2 < 1$ ms, gave no image under the same conditions. The polymers were imaged using a spin-echo sequence, a gradient-echo sequence, and 3-D thick-slice sequence. The gradient-echo sequence may be better than the spin-echo sequence for detection of macroscopic defects such as voids, gaps, or small foreign particles in elastomers. A crude resolution test was performed on sheets of cured, filled cis-polybutadiene separated by spacers of variable width. Features as small as 0.07 mm could be detected even though the pixel size was considerably larger.

2. Introduction

Nuclear magnetic resonance imaging has evolved into an important diagnostic modality in clinical medicine over the last ten years [1,2]. Whereas the clinical

---

C. CHANG AND R. A. KOMOROSKI - Departments of Radiology, Pathology, and Biochemistry, University of Arkansas for Medical Sciences, Little Rock, AR 72205.

Solid State NMR of Polymers, Edited by L. Mathias
Plenum Press, New York, 1991

applications of NMR imaging have developed quite rapidly, applications in materials and polymer science have been slow to appear. This is due in part to the relative nonavailability and expense of commercial NMR imaging equipment suitable for such studies. It also arises from the belief that most important applications in materials science involve rigid solids, which cannot be studied using standard equipment. Nevertheless, a number of nonmedical applications of NMR imaging has appeared [3]. The majority of these reports deals with diffusion of liquids through rock, wood, or polymers.

Efforts are currently underway on the development of NMR imaging techniques for rigid solids. Solids pose formidable problems for standard NMR imaging techniques. Typical solid-state proton linewidths are on the order of 20 kHz, a value that precludes NMR imaging of rigid solids using standard gradient strengths of about 1-2 G/cm. Although considerable progress has been made, additional development work is necessary before practical techniques are available for imaging truly rigid solids in polymer science.

High resolution NMR spectra, both of ^{1}H and ^{13}C, can be obtained for bulk polymers well above their glass transition temperature T_g [4]. The quality, as reflected in the resonance line widths, of the high resolution spectrum of a bulk, rubbery polymer depends to a large extent on the rate, amplitude, and anisotropy of chain segmental mobility. Typical proton NMR line widths for rubbery polymers at about 100° above T_g are in the range of about 100-2000 Hz. Thus molecular mobility has averaged the spin dipolar interactions significantly from the rigid case, but not to values typical for liquids in an imaging setting (10-50 Hz). The degree of line narrowing can be sufficient for imaging by standard techniques.

Rothwell and Gentempo [5] first demonstrated this by obtaining an image of an unspecified polymer in a composite "sandwich" at a temperature about 80 °C above T_g, which was given as 100 °C. The image of a sample of synthetic *cis*-polyisoprene at room temperature has been reported [6]. In their work applying multiple-pulse line narrowing to imaging, Chingas et al [7] reported an NMR image of a section of neoprene (polychloroprene) hose. No details on NMR imaging of elastomers were explored in any of the above reports, which were directed to other matters.

It is likely that the initial applications of NMR imaging in the science of polymeric materials will be found in heterogeneous, liquid-solid systems (in which case the liquid components will be imaged), in elastomeric systems (filled and/or cured) and composites, and perhaps in semicrystalline polymers where T_g is relatively low. Here we have examined at ambient temperature a number of elastomeric systems of similar T_g's but differing chain segmental mobilities. Both pure amorphous resins and crosslinked, filled materials are studied. Several imaging sequences which are sensitive to different NMR parameters are used. A crude resolution test is performed and used to demonstrate the level of information readily obtainable on available equipment. The present study demonstrates some apparent difficulties and limitations of NMR imaging to elastomeric materials.

3. Experimental Section

Samples of *cis*-polybutadiene, polyisobutylene, and smoked natural rubber were obtained from K. Schur of the BF Goodrich Company. The cured, carbon-black-filled *cis*-polybutadiene was the same as that examined previously using ^{13}C NMR with magic angle spinning [8].

Proton NMR images were obtained at ambient temperature at 200.1 MHz on a General Electric CSI-4.7 system with 33 cm room temperature bore. A GE 3-inch diameter imaging coil was used. The maximum field-gradient strength was 2.0-2.5 G/cm. Spin-warp [9], 2D FT images were obtained using software available with the spectrometer. The radiofrequency carrier was placed midway between the peaks in the spectrum of cis-polybutadiene, and on-resonance with the major ^{1}H peaks for the other polymers. Data from 256 individual phase-encode views were acquired as either a spin echo or gradient echo, and sampled as 256 complex points in the frequency-encode or readout direction. Details are given elsewhere [3].

Figure 1A shows the basic spin-echo NMR imaging sequence used in this work. In the spin-echo method, the minimum time to echo TE is determined by the length of several time periods, including the selective pulse width, the phase-encode time (3 ms), and the acquisition time. These latter two determine the field of view (FOV) in the image. For our FOV of 120 mm, slice thickness of 2 mm and gradient strength, the minimum TE allowed is about 12 ms.

Figure 1B shows the gradient-echo pulse sequence used here. Given our conditions, the minimum TE allowed for the gradient-echo sequence is 6.4 ms. The two sequences, spin-echo and gradient-echo, can produce substantially different images for bulk polymers, as demonstrated below.

Images were also taken using a 3-D thick slice technique. Here, a relatively thick slice is initially excited in the desired plane [2]. Simultaneously to the normal phase encoding (x), spatial information is also phase encoded along z in a small number (8-32) of steps to provide relatively thin, contiguous slices.

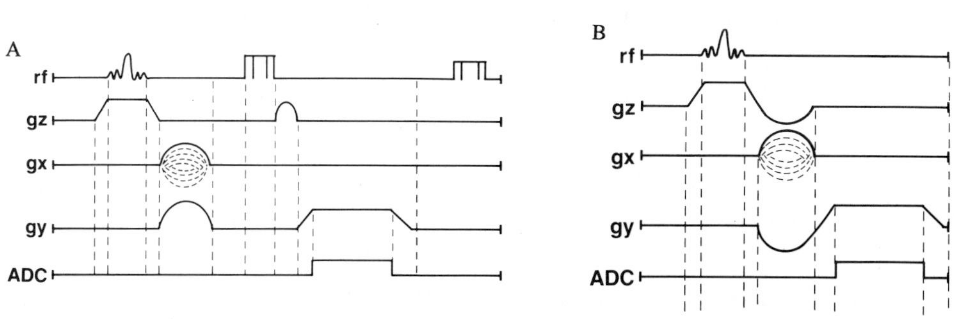

Figure 1. A) Spin-warp, spin-echo sequence for 2D FT NMR imaging; B) Gradient-echo sequence for 2D FT NMR imaging.

Several images were obtained on a GE Signa clinical NMR imaging system at 64 MHz (1.5 T) using the extremity coil provided. Conditions were relaxed relative to those at 4.7 T. The maximum gradient strength was about 1 G/cm, a 3 mm slice was taken and the field of view was 160 mm. All of this serves to increase the signal-to-noise ratio in the 1.5 T image relative to that at 4.7 T.

4. Results

The three totally amorphous bulk polymers, *cis*-polybutadiene, cis-polyisoprene, and polyisobutylene, have low T_g's that are not drastically different from one another, and a range of over three orders of magnitude in backbone segmental mobility at ambient temperature [3,10]. They have approximate proton T_2s of 13, 9, and <1 ms, and T_2^*s of 2.1, 1.0, and 0.2 ms, respectively, at 4.7 T [3]. These quantities were not measured at 1.5 T.

In Fig. 2A is the spin-echo image (TE=12 ms) at 200 MHz of a 2-mm slice of a block of bulk *cis*-polybutadiene. The actual polybutadiene block is 55x110x20 mm in size with irregularly shaped sides. The bottom end of the block contains a high density of trapped air bubbles or voids of about 0.1-1 mm average diameter, and

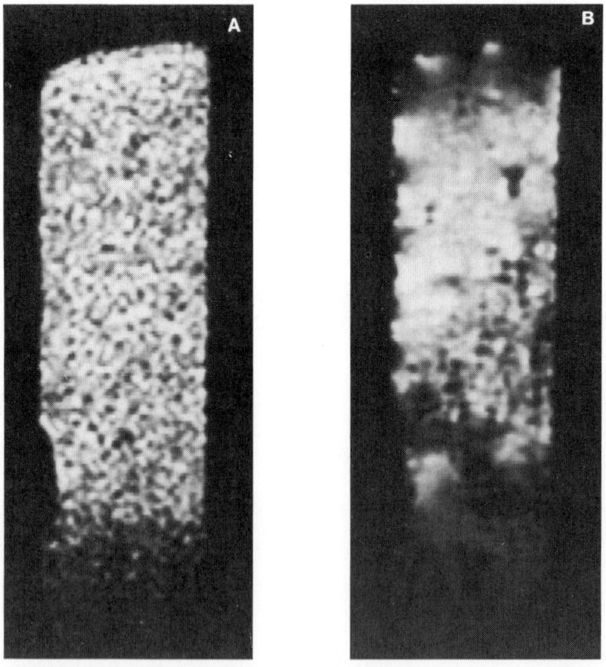

Figure 2. Slices of a *cis*-polybutadiene block at 4.7T; FOV=120x123 mm; resolution, 256x256; shown is 20x110 mm face; A) Spin-echo; slice thickness, 2 mm; no. of transients, 48; TE, 12 ms. B) Gradient echo; slice thickness 2 mm; no. of transients, 72; TE, 6.4 ms.

hence visually appears translucent and somewhat milky. The top end contains few voids, and is somewhat more transparent. Hence the block is a reasonable model of a manufactured rubber product with internal "defects." At the very bottom the signal intensity in the spin-echo image drops off due to the fact that the end of the block is near the end of the rf coil. However, the image is fairly uniform over most of the block.

In Fig. 2B is the gradient-echo (TE=6.4 ms) image at 200 MHz through the same slice as the spin-echo image of Fig. 2A. Although the slice cross section is the same, the appearance of the gradient-echo image is considerably different from that of the spin-echo image. The gradient-echo image shows considerably more structure, that is, the presence of air bubbles or voids which in many instances appear reasonably well defined in the image. Bright areas are observed toward the center of the block. The origins of the bright areas were not visually obvious, but are evidently void-free areas. The intensity in the gradient-echo image of Fig. 2B crudely reflects the variation of void density along the length of the block. Even in the top, void-free end, regions of high and low intensity are seen with no obvious, corresponding features visually apparent.

Good images of *cis*-polybutadiene were obtained on a Signa clinical imager using spin-echo and gradient-echo sequences. Fig. 3 shows images of *cis*-polybutadiene taken at 64 MHz. In Fig. 3A is the spin-echo (TE=14ms) image of the block. The image is relatively uniform in intensity, except for the appearance of some voids in this slice. As expected, the image intensity drops at a TE of 35 ms (Fig. 3B) but the uniformity remains, indicating no T_2 variations in the slice. The gradient-echo image (Fig. 3C) does show the increased void density at the bottom. Moreover, voids appearing in the spin-echo image appear enlarged in the gradient-echo image. The edge is also less well defined in the gradient-echo image.

In Fig. 4A is the spin-echo image of a 2-mm slice (TE=11 ms) of a sample of natural rubber. The sample is a 50x50x35 mm opaque block formed from 11 irregular vertical sheets of rubber fused together, but with some irregularly spaced gaps between several of the sheets. The spin-echo image has the mottled appearance seen for *cis*-polybutadiene, but with less uniformity of intensity across the image. Little evidence of the sheet structure is seen. Figure 4B is the gradient-echo image of the same slice of the same sample. As for polybutadiene, the image is considerably different than the spin-echo image. The darkest regions of the spin-echo image correspond approximately to regions of essentially no signal in the gradient-echo image. The sheet structure is evident in places in the gradient-echo image. The bright, cross-like feature at top center of the block is a zero-frequency artifact at the exact center of the entire image. Because of the shorter T_2 and T_2^* of natural rubber relative to *cis*-polybutadiene, it was necessary to take more averages to obtain a similar signal-to-noise ratio for the natural rubber.

Using the same conditions as for the samples above, we were not able to obtain an image for polyisobutylene. This is not surprising given the short T_2 for this

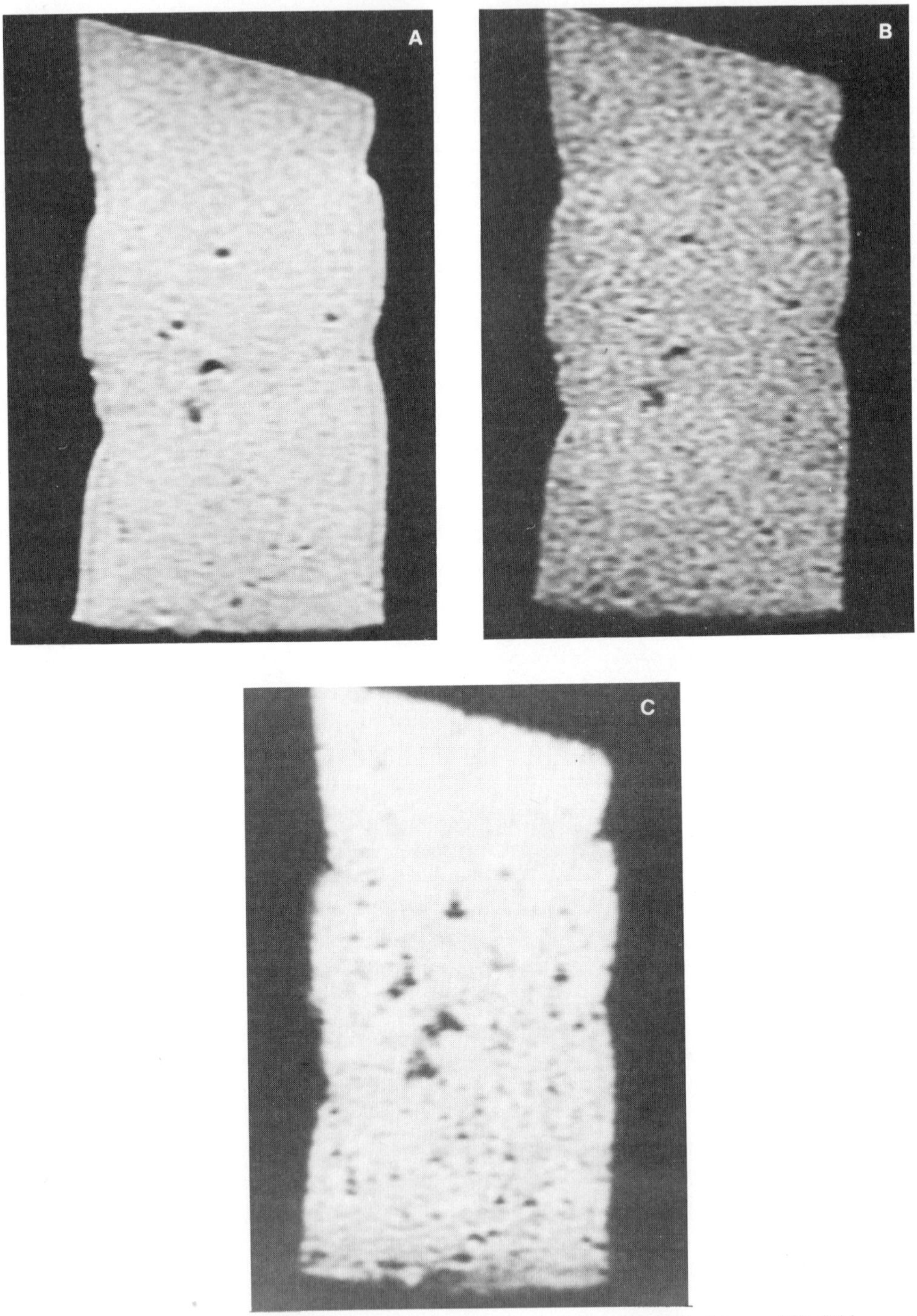

Figure 3. NMR images of *cis*-polybutadiene at 1.5 T. A) Spin-echo; TE, 14 ms; 128x256; no. of transients 200, FOV, 16 cm. B) Spin-echo; TE, 35 ms; 128x256; no. of transients, 200; FOV, 16 cm. C) Gradient-echo; TE, 10 ms; 128x256; no. of transients, 200, FOV, 16 cm.

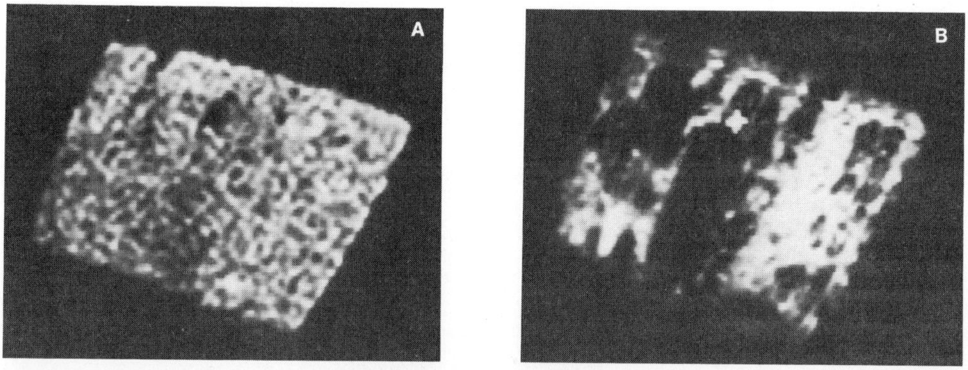

Figure 4. NMR images of a natural rubber sample at 4.7T. FOV, 120x123 mm; resolution, 256x256; shown is 35x50 mm face; A) Spin-echo image with TE of 12 ms; slice thickness, 2 mm; no. of transients, 120. B) Gradient-echo image with TE of 6.4 ms; slice thickness, 2 mm; no. of transients, 240.

rubber. By substantially decreasing the resolution, we were able to shorten TE to 9 ms, allowing us to obtain a very weak, uninformative spin-echo image after considerable signal averaging.

Figure 5 shows the spin-echo image of 10 sheets, each 2 mm thick, of a cured, carbon-black-filled *cis*-polybutadiene separated by glass spacers of variable thickness. The smallest spacer is a sheet of paper approximately 0.07 mm in thickness. Even though the T_2 of this sample is about 4 ms, we are still able to obtain a usable image in an overnight run with a TE of 12 ms. Figure 6 shows a horizontal profile (row # 130) of the data in Fig. 5. This profile demonstrates the typical signal-to-noise ratio of the image, as well as the degrees of resolution between pairs of the various slices. Figure 7 shows another image of the resolution test sample, but at smaller FOV and larger slice thickness.

We also obtained *cis*-polybutadiene images at 4.7 T using the 3-D thick slice technique. Figure 8 shows the set of images taken with eight phase-encode steps in the slice-select direction. Each slice is 1 mm thick, and the slices are contiguous. Spin-spin relaxation time variations are seen within each slice at this longer TE of 26 ms, particularly at the surface and near a hole just above center on the left. The variation in overall intensity among the slices is due to nonuniform excitation over the thick-slice profile (8mm), which was set equal to the field of view in the slice-select direction.

5. Discussion

Our results show that currently available NMR imaging techniques can be used to produce informative images of a range of elastomeric materials. Highly rubbery, bulk polymers of low T_g and high chain segmental mobility are easiest to image using echo techniques. As chain motion slows and T_2 decreases, the images are of lower

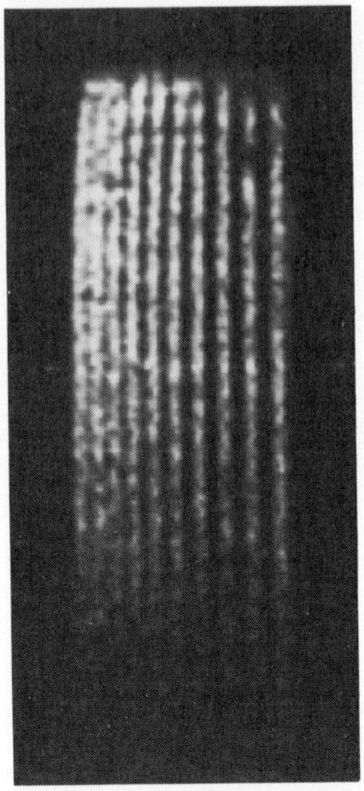

Figure 5. Spin-echo image at 4.7T of 10 sheets of a cured carbon-black-filled *cis*-polybutadiene (2.0 mm sheet thickness) separated by spacers of various thicknesses. The thicknesses of the spacers used are, from left to right in mm: 0.07, 0.15, 0.30, 0.45, 0.60, 0.80, 1.0, 1.15, 1.30. The sample was placed with long dimension alone the coil axis. The slice is 1.6 mm thick. Field of view, 120 mm x 123 mm; resolution, 256x256, number of transients, 240; TE, 12 ms. The signal intensity falls off at the very bottom because the sample extends just to the end of the rf coil.

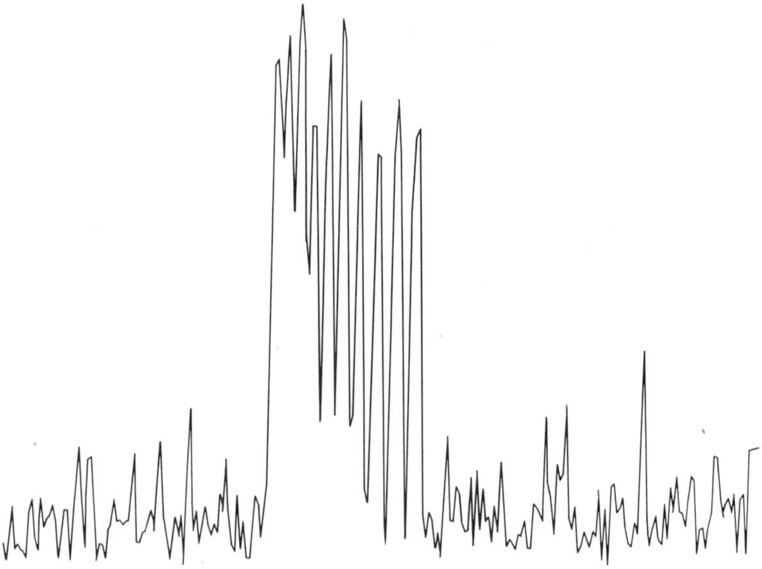

Figure 6. Horizontal profile (row # 130) of the image in Fig. 5. The profile demonstrates the signal-to-noise ratio obtained as well as the degree of resolution of the various rubber sheets at that point.

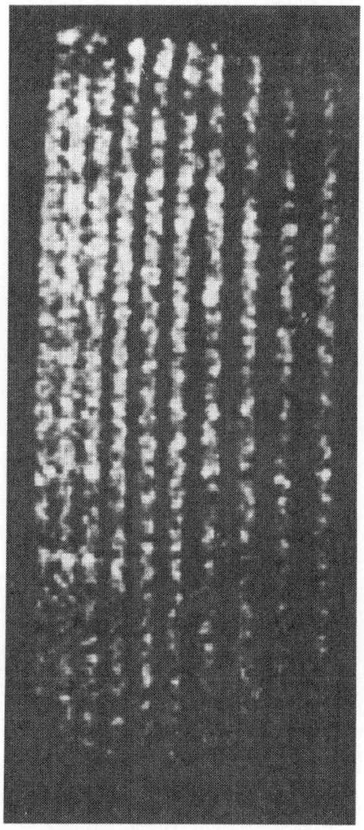

Figure 7. Spin-echo image at 4.7T of same sample as in Fig. 5, but at smaller FOV. The slice is 6 mm thick. Resolution, 256x256; TE, 15 ms; FOV, 80 mm; no. of transients, 240.

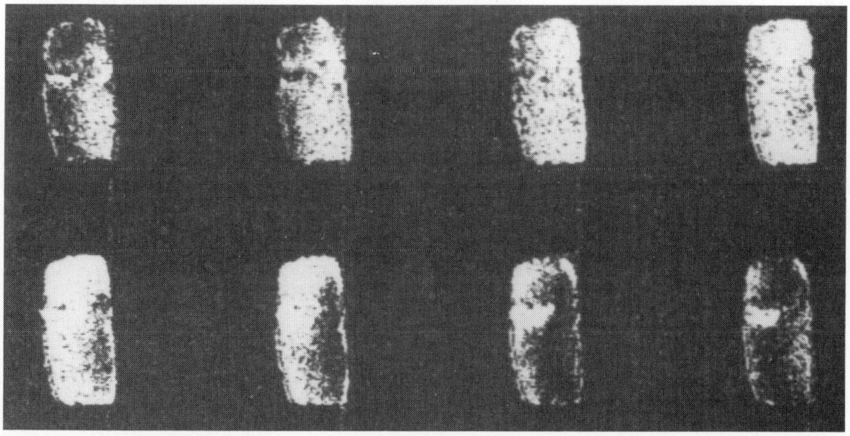

Figure 8. 3-D thick slice images of *cis*-polybutadiene. The eight slices correspond to eight phase-encode steps in the slice-selection direction. TE, 26 ms; 128x256; FOV, 120 mm; no of transients, 32; excitation bandwith of thick slice, 8 mm; FOV in slice-relection direction, 8 mm.

intensity and signal-to-noise ratio. At the TE of 12-14 ms used here, polymers with T_2's of about 4 ms and above can be imaged with varying degrees of difficulty. Polymers of T_2 on the order of 1 ms could not be imaged under our conditions.

We have demonstrated that cured, filled elastomers can be imaged successfully, as long as T_2 has not been shortened too much by the curing or filling process. It is known that a sizable portion of the line width in filled systems can be due to nonmotional factors such as variations in magnetic susceptibility [4]. Such nonmotional factors will be refocused in a spin-echo image and will not contribute to a loss of image intensity, although they will contribute to a loss of resolution. A more systematic study of the effects of curing and filling on elastomer images is necessary.

The gradient-echo images of Figs. 2B, 3C, and 4B are relatively rich in detail. This detail cannot be due exclusively to T_2 effects or to direct absence of spin density due to voids, because these would also be reflected in the spin-echo image. The detail probably arises from magnetic susceptibility variations in polymer at or near void surfaces, or in regions of differing void density. Hence regions of the rubber near the surface of the defect may have shorter T_2^*s than the bulk rubber. In a gradient echo image these surface regions would appear darker, enlarging the apparent size of the defect. The effect would be more pronounced at higher magnetic fields. Such an effect is observed at tissue-air boundaries in medical imaging using gradient-echo sequences [11].

The results in Fig. 3 confirm the above interpretation. All the voids in the spin-echo image (Fig. 3A) appear larger in the gradient-echo image (Fig. 3C). Moreover, additional voids are seen in the gradient-echo image, particularly at the bottom of the block. These voids were apparently too small to observe in the spin-echo image.

The spread of frequencies caused by the application of a linear field gradient G across a thickness Δx is $\gamma G \cdot \Delta x$. To obtain a resolved image along x, it is necessary that the gradient-imposed spread in frequencies dominate the spread in frequencies due to the line width:

$$\gamma G \cdot \Delta x > \Delta \upsilon_{1/2}$$

Alternatively, one can calculate the size of the effect in the image that would be observed at a given line width.

Using pure polybutadiene as an example, we can examine the various factors contributing to a loss of resolution. The largest is the chemical shift effect. The aliphatic and olefinic peaks of *cis*-polybutadiene are separated by about 660 Hz at 4.7 T, and each is about 150 Hz wide. The chemical shift separation results in a chemical shift artifact of .78 mm. In Fig. 2A no chemical shift artifact is seen at the top edge of the block in the frequency-encode (vertical) direction. A clear chemical shift artifact at the top edge is seen in the image at 1.5 T (Fig. 3A).

The line width also influences the resolution of the images. The value of about 150 Hz for *cis*-polybutadiene corresponds to a spatial uncertainty or "fuzziness" of 0.19 mm. Here this is less than the pixel size of 0.47 mm.

It is of interest to consider the resolution test in Figs. 5 and 6 in more detail. Fig. 5 has a 120 mm FOV in 256 points, giving a pixel size of 0.47 mm. Yet all nine spacers can be detected visually, even though the four smallest spacers are less than the pixel size. This seemingly anomalous behavior is readily explained by appeal to Fig. 6, a horizontal slice. Generally it is assumed that, to resolve a particular feature, it must be larger than the pixel size. In our case that would mean resolving the spacer truly as a region of no NMR signal, that is, *to baseline in the slice.* If the spacer is less than the pixel size, then the effect of the spacer is to reduce the intensity of the pixel at that point. Thus the two rubber sheets are resolved in the profile, but not to baseline, for the smallest 7-8 pixels. If the image is displayed at a sufficiently high intensity threshold, the spacer is resolved. This result is reasonable if we consider that the pixel size is 0.47 mm and the fuzziness due to the cured rubber line width is about 0.86 mm.

When features smaller than the pixel size are detected, they of course appear to be of the pixel size. The result in Fig. 7, with a pixel size of 0.31 mm, confirms this. The smallest spacer is still visible, but it is now smaller relative to the thickness of the adjacent rubber sheets.

The implications of the above results need to be carefully considered as more attempts are made at resolving microscopic features in solid polymers by NMR imaging. One important implication is that the effect of features considerably smaller than the pixel dimensions can be "seen," although the features cannot be well characterized. In Figs. 5 and 7 the 0.07 mm paper spacer is resolved as a slightly irregular band of reduced intensity. This smallest spacer is recognized in large part due to its regular features, ie, it is a straight line. In general, one can say little about features at and below the pixel size. This is because features of reduced intensity may be reduced by defects such as voids smaller than the pixel size, by a reduction of spin density over the whole pixel, or by T_1 or T_2 effects. This raises the possibility that a region of reduced intensity in a relaxation-time-weighted image may arise from the presence of defects smaller than the pixel size throughout the region, and not from a difference in the relaxation time for that region. The magnification effect of bulk susceptibility variations will also contribute in the gradient-echo image.

Although it is clear that the images seen here for elastomers do contain macroscopic morphological information, interpretation of that information may not be as straightforward as it is in medical imaging by NMR. The spin-echo images in Figs. 2A and 3A illustrate another complication. For these slices it was expected that the two halves of the polybutadiene block would yield visibly different images. The half with a higher density of voids should have produced a more mottled image than the end of lower void density. This was not the case, and the block appeared relatively uniform. This suggests that factors other than the void density are contributing to the image. Apparently signal-to-noise ratio and resolution are insufficient to distinguish the two halves under those conditions. This is not the case for the gradient-echo images in Figs. 2B and 3C. Here the region of high void density is seen to start at mid-block with decreasing intensity as the block end is approached.

The results in Figs. 5-7 illustrate a dilemma of NMR imaging of materials. A decision must be made as to intensity and contrast of the image. It may often be difficult or impossible to decide which combination of intensity and contrast yields an image "truly" representative of the morphology of the material. In our resolution test of known characteristics, examination of the profile at various points combined with the image itself yielded sufficient information for correct interpretation.

Most of our results have been obtained using single-slice methods. In medical NMR imaging, multislice methods are generally used, resulting in considerable time savings when a large volume must be scanned. Multislice methods can also be used in materials imaging. One disadvantage of multislice imaging is that slice thickness is limited in part by the gradient strength. Longer pulse widths can produce thinner slices, but at the expense of a longer TE. Another disadvantage with multislice methods is the difficulty in getting contiguous slices. Typically two datasets of interleaved slices must be acquired to avoid interslice effects. The 3-D thick slab technique overcomes these problems. By phase encoding in the slice selection direction, slice thickness can be decreased by increasing the number of phase encode steps. The slices are automatically contiguous. The disadvantage of long data acquisition time is not as serious in materials image as in medical imaging. The results in Fig. 8 demonstrate that the technique can produce useful information for elastomeric materials.

For the echo methods employed here, it will be necessary to shorten TE so as to recover most of the intensity of the echo signal and increase signal to noise. The minimum TE available in a particular situation is limited by factors such as the selective rf pulse width, gradient ramp times, the phase encode time, and the acquisition time. The value of TE can be lowered significantly only by decreasing the resolution or increasing the gradient strength. This assumes that the FID has not been lengthened by line-narrowing techniques.

In conclusion, we have demonstrated that NMR imaging techniques can yield information concerning macroscopic defects in elastomeric systems. More work is necessary to define to what extent the image characteristics are influenced by factors such as polymer motion, magnetic susceptibility, relaxation-time variations, and chemical shift effects.

6. Acknowledgement

We thank Dr. Ralph Hurd of GE NMR Instruments for invaluable assistance in obtaining some preliminary data at the GE applications lab. We also thank Ken Schur of BF Goodrich for providing the *cis*-polybutadiene, polyisobutylene and natural rubber samples.

7. References

1. Mansfield, P. and Morris, P. G. "NMR Imaging in Biomedicine," *Advan. Magn. Reson.*, Suppl. 2, Waugh, J. S., Ed., Academic Press, New York, 1982.
2. Morris, P. G. "Nuclear Magnetic Resonance Imaging in Medicine and Biology," Clarendon Press, Oxford, 1986.
3. Chang, C. and Komoroski, R. A., *Macromolecules*, *22*, 600, 1989, and references therein.
4. Komoroski, R. A. in "High Resolution NMR Spectroscopy of Synthetic Polymers in Bulk," Komoroski, R. A., Ed., Ch 4, p 121, VCH Publishers, Deerfield Beach, 1986.
5. Rothwell, W. P. and Gentempo, P. P., *Bruker Reports*, *1*, 46, 1985.
6. Rothwell, W. P.; Tutunjian, P. W. and Vinegar, H. J., *Proc. Industry-University Cooperative Chemistry Program*, Texas A & M Univ. Press, College Station, TX, 1985.
7. Chingas, G. C.; Miller, J. B. and Garroway, A. N., *J. Magn. Reson.*, *66*, 530, 1986.
8. Komoroski, R. A., *Rubber Chem. Technol.*, *56*, 959, 1983.
9. Edelstein, W. A.; Hutchinson, J. M. S.; Johnson, G. and Redpath, T., *Phys. Med. Biol.*, *25*, 751, 1980.
10. Komoroski, R. A. and Mandelkern, L., in "Applications of Polymer Spectroscopy," Brame, E. G., Ed., Academic Press, New York, 1978, Ch 5.
11. Haacke, E. M. and Bellon, E. M., in *Magnetic Resonance Imaging*, Stark, D. D. and Bradley, W. G., Eds., C. V. Mosby, St. Louis, 1988, Ch 8, p 138.

Chapter 24

NMR IMAGING INVESTIGATIONS OF CASE II DIFFUSION IN POLYMERS

L. A. WEISENBERGER AND J. L. KOENIG

1. Abstract

Case II diffusion in glassy polymers is studied extensively using techniques that require destruction of the sample or interruption of the diffusion processes to make proper measurements. NMR imaging is a new technique that does not require the destruction of the sample or interruption of the diffusion processes. This investigation shows that NMR imaging is applicable to studying diffusion processes in polymers. FLASH images provide an accurate representation of the concentration profiles from diffusion of methanol into PMMA and show the major aspects of Case II diffusion. A special diffusion imaging pulse sequence provides information on the self-diffusion of the methanol in the PMMA. Images based on NMR relaxation times show the differences in molecular motions of the methanol in the bulk solvent, at the bulk-rubber interface and at the rubber-glass interface.

2. Introduction

The diffusion of organic solvents into glassy polymers is typically non-Fickian and in the extreme case is known as Case II. Case II diffusion exhibits three characteristics. 1) The solvent concentration increases rapidly at the interface where the swollen polymer and glassy polymer meet. This is known as the sharp front. 2) The solvent concentration is constant from the front out to the bulk solvent. 3) The

---

L. A. WEISENBERGER and J. L. KOENIG - Department of Macromolecular Science, Case Western Reserve University, Cleveland, OH 44106.

Solid State NMR of Polymers, Edited by L. Mathias
Plenum Press, New York, 1991

front advances through the glassy polymer at a constant rate. In recent theoretical treatments [1,2], a Fickian concentration profile preceding the front has been suggested as the fourth characteristic of Case II diffusion. Many of the techniques currently used to investigate diffusion phenomena in polymers are destructive or rely on indirect detection techniques. NMR can examine the diffusion processes and mechanisms without interruption. Also, the sample shape for NMR imaging is not restricted to thin films or rods.

Nuclear magnetic resonance (NMR) is used extensively to characterize the structures and motions of polymers. NMR imaging extends the versatility of NMR to include spatial characterization. A linear magnetic field gradient applied across the sample causes each chemically identical nuclei to spin at slightly different frequencies. If only one type of nuclei exists, then the Fourier transform of the signal reflects the distribution of spins across the sample (when relaxation effects are negligible). The proper application of other orthogonal gradients produces two or three dimensional images. In the presence of relaxation phenomena, the images can be acquired so that they reflect the spatial distributions of relaxation times.

NMR imaging is extensively developed for medical diagnosis and is now being directed towards materials science. With its ability to show the spatial dependence of concentrations and motions, nmr imaging is well suited for investigating diffusion processes. The applicability of imaging to studying diffusion processes is demonstrated in this work through the imaging the diffusion of methanol into PMMA, a system which exhibits Case II behavior and has been extensively investigated.

3. Experimental

3.1 Sample Preparation

The imaged samples are poly(methyl methacrylate) rods or spheres from Commercial Plastics and Supply, Corp., Cleveland, Ohio. The cast and polished spheres are used as supplied in 30 mm and 19 mm diameters. The extruded and polished rods are supplied in 6 ft lengths of 19 mm diameter and are cut to 7 cm in length.

The samples are submersed in hydroxy-deuterated methanol (CH_3OD) from Aldrich Chemical Company, Inc., Milwaukee, WI at either 30 or 60°C. The samples in the methanol are imaged at regular time intervals depending on the temperature.

3.2 NMR Imaging

NMR imaging is done on a Bruker MSL 300 system. The superconducting magnet is 7 T with a 150 mm diameter bore. The gradient coils are housed with the room temperature shim coils. Pre- and post-emphasis are used on the gradient pulses to correct for eddy currents. Shaped pulses are generated in a special Bruker selective

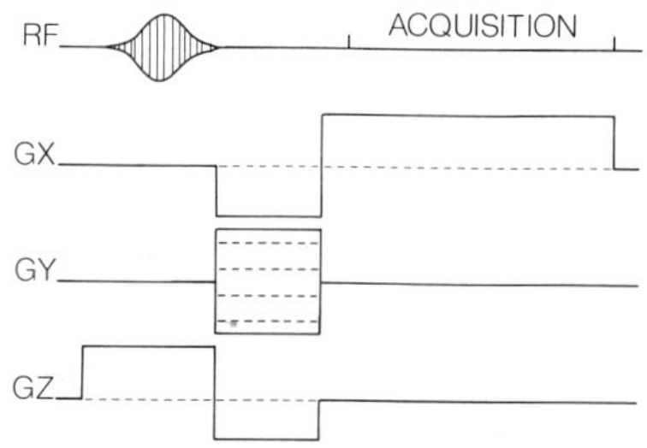

Figure 1. The FLASH imaging pulse sequence.

excitation unit and amplified through an ENI 550L 50 watt linear amplifier. The sample chamber is 65 mm in diameter with an imagable length of about 100 mm.

Data are acquired and Fourier transformed on a Bruker Aspect 3000 computer with an array processor and 256k bytes memory. Images are 256 x 256 pixels. Distance is measured in hertz in the Bruker DIS mode with the twin cursors. Further data processing including noisefiltering and scale expansion, done on a MicroVax II with software written by J. Liu of this group. Images are displayed on a Tektronix 4111 color graphics terminal.

The images are acquired using the FLASH [3] (Fast Low Angle SHot) imaging pulse sequence shown in Fig. 1 unless otherwise noted. FLASH uses a short echo time (TE = 12 msec) and repetition time (TR = 50 msec). The slice-selection pulse is 2 msec which corresponds to about a 20° tip angle at the power level used. The combination of the small tip angle and the repetition time generates a steady state between the relaxation process and the rf excitation allowing signal averaging of several acquisitions. The reversal of the read gradient produces a gradient or field echo rather than the common spin-echo. With the short TE and TR, the resulting image has little T_1 or T_2 dependence and thus accurately reflects the spin density distributions in the sample.

The diffusion pulse sequence in Fig. 2 probes the random motions of the solvent. The sequence is described for fully by Le Bihan, et al. [4] and is an imaging version of the Stejskal and Tanner [5] pulsed gradient technique for studying diffusion.

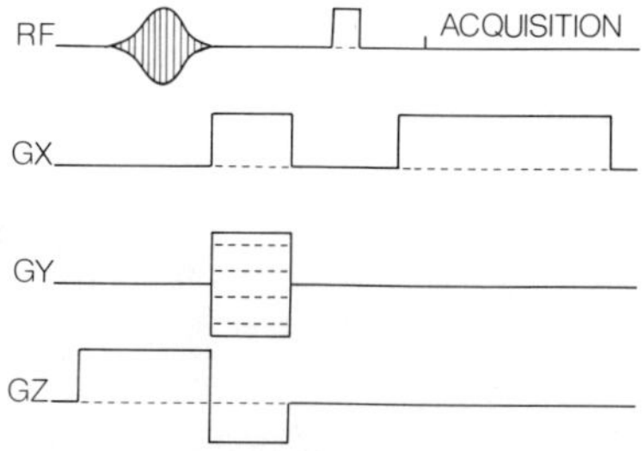

Figure 2. The diffusion imaging pulse sequence.

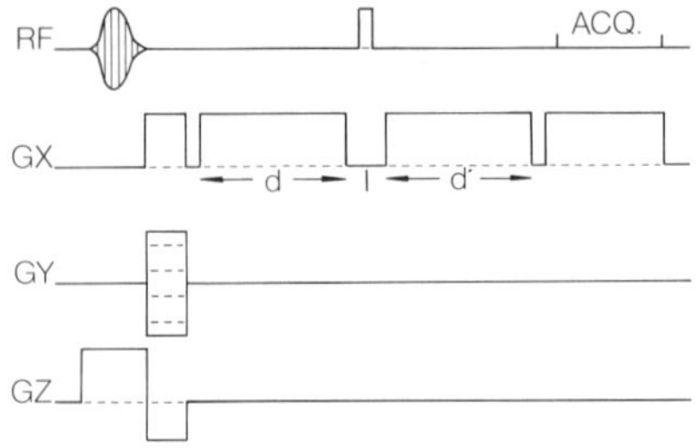

Figure 3. A standard spin-echo imaging pulse sequence.

The standard spin-echo sequence in Fig. 3 is used to acquire images for the T_1, T_2 and spin-density calculations. A slice-selective 90° pulse is used in conjunction with a non-selective 180° pulse. The variable echo times used are 12, 24, 36 and 48 msec with a 4 sec repetition time. The variable repetition times are 0.5, 2, 8, 16 sec with a 12 msec echo time. The T_1, T_2 and spin-density images are generated using software by J. Liu [6] of this group based on the method of combining ratios and least-squares proposed by In den Kleef and Cuppen [7]. The software uses the individual images with different TEs and TRs to generate the "pure" T_1, T_2 and spin-density images.

4. Results

Figure 4a is the FLASH image of a PMMA sphere submersed in methanol at 30°C. Figure 4b is the FLASH image of the same PMMA sphere after 48 hours exposure to the methanol. The resolution is 200 μm with a read gradient (Gx) of 1 G/cm and a spectral width of 20 kHz. The intensity profile of image 2b is shown in Fig. 5. The profile features are (a) the bulk methanol, (b) the imbibed methanol, (c) the diffusion front and (d) the glassy polymer core. The velocity profile is shown in Fig. 6. The circles represent the position of the diffusion front as a function of time. The line is the least-squares linear regression fit to the data. The slope gives a front velocity of 4.9 ± 0.8 nm/sec.

A smaller rod (19 mm) is submersed in methanol at 60°C. The resolution is 150 μm with a read gradient (Gx) of 2 G/cm and a spectral width of 31 kHz. The plot of distance versus the square-root of time is shown in Fig. 7. The circles represent the position of the equilibrium concentration of methanol as a function of the square-root of time. The line represents the linear least-squares fit to the data. The slope is 14.3 $\mu m/\sqrt{sec}$ which gives a diffusion constant of 2.0×10^{-8} cm^2/sec.

The diffusion pulse sequence image (Fig. 8a) is compared with a normal spin-echo image (Fig. 8b). The imbibed methanol is present in the image whereas the bulk methanol is totally absent from the diffusion image.

The results of the T_1 and T_2 calculations are shown as images in Fig. 9. The higher intensities in the relaxation images represent longer values of T_1s and T_2s, respectively. The relaxation times for various regions are given in Table I. The T_1 image shows the bulk methanol to have a longer T_1 than the imbibed methanol. The imbibed methanol has a gradient of T_1s ranging from 0.54 sec to 0.36 sec dropping rapidly approaching the glassy core. The T_2 image shows similar trends across the sample.

5. Discussion

5.1 Imaging Case II Diffusion

Determination of diffusion characteristics with NMR imaging depends on the ability to obtain a pure spin-density image. In a normal spin-echo imaging sequence, spin-spin and spin-lattice relaxations influence the image intensities so that accurate determination of concentration profiles is nearly impossible. To circumvent this problem two possible approaches are available. One is to take several images with different echo and repetition times and use them to create a spin-density image on a computer. The more direct approach is to use a pulse sequence that is independent of spin-spin and spin-lattice relaxations. The FLASH imaging sequence in Fig. 1 uses a short repetition time and a small tip angle to eliminate the relaxation dependencies. The combination of the short repetition and echo times produces a steady state between the spin-lattice relaxation and the timing of the rf pulses. With the short repetition time, data acquisition is much faster and signal averaging is more beneficial than with a standard spin-echo sequence. The images depend only on the spin-density distribution across the sample.

The FLASH images shown in Figs. 4a and 4b show the spin-density distributions within a 30 mm PMMA sphere. Fig. 4a is taken immediately after

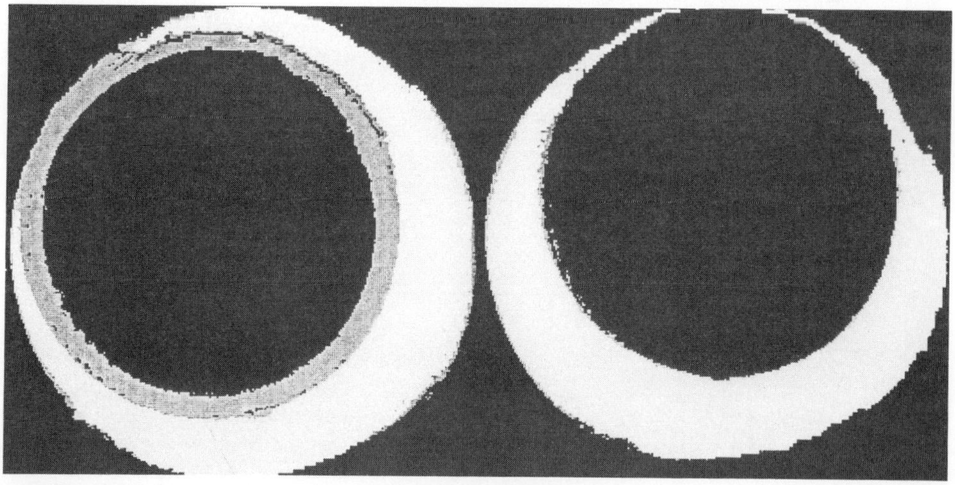

Figure 4. (a) The FLASH image of a 30 mm PMMA sphere immediately after submersion in methanol. (b) The Flash image of the same sphere after 48 hours submersion in methanol at 30°C. The brighter areas indicate higher methanol concentrations.

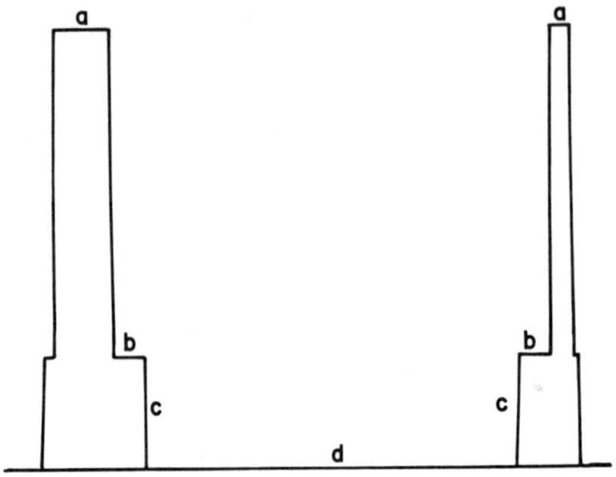

Figure 5. The intensity profile of Fig. 4b. (a) The bulk methanol, (b) the constant methanol concentration behind the front, (c) The sharp concentration gradient at the rubber-glass interface, (d) the glassy PMMA core.

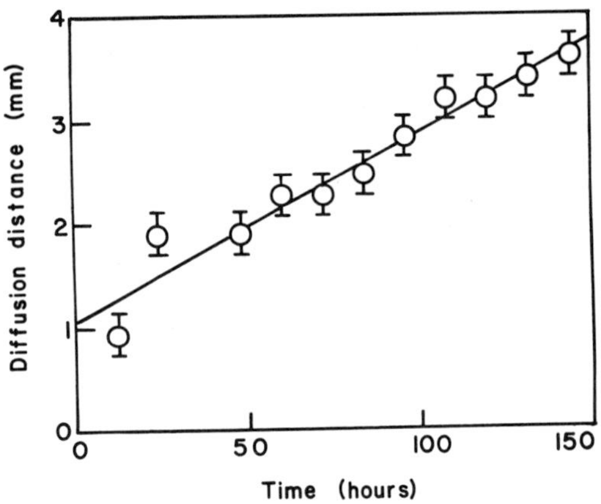

Figure 6. The plot of front position as a function of time for the sphere in Fig. 4.

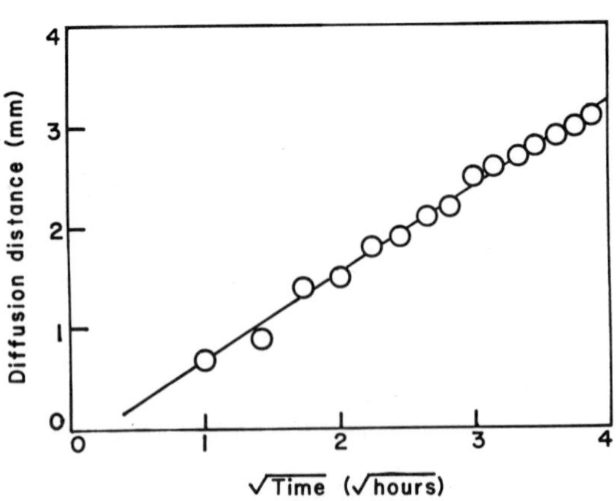

Figure 7. The plot of front position as a function of the square-root of time for a 19 mm sphere submersed in methanol at 60°C. The error in the distance measurement is ± 0.1 mm.

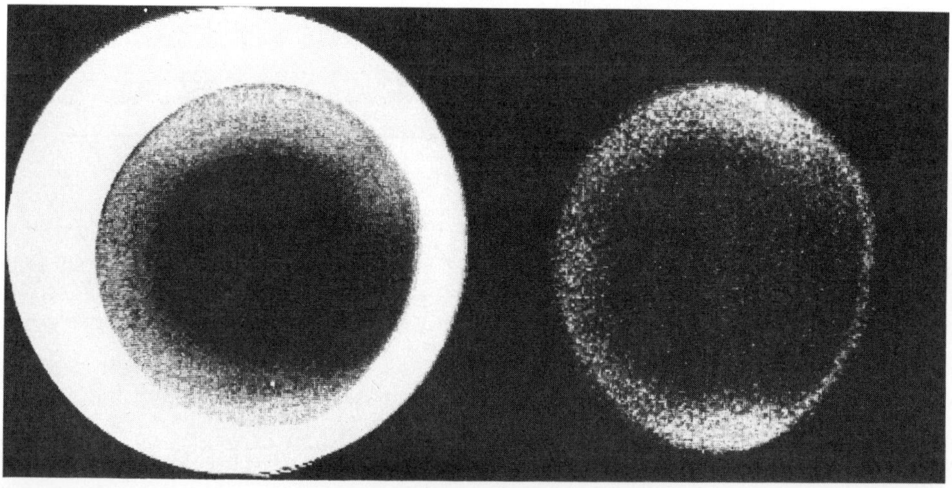

Figure 8. (a) The image from the diffusion pulse sequence in Fig. 2 of a 19 mm PMMA rod with an echo time of 62 msec. (b) The image from the spin-echo pulse sequence in Fig. 3 with an echo time of 12 msec. Both images were acquired at 30°C with a 2 sec. repitition time.

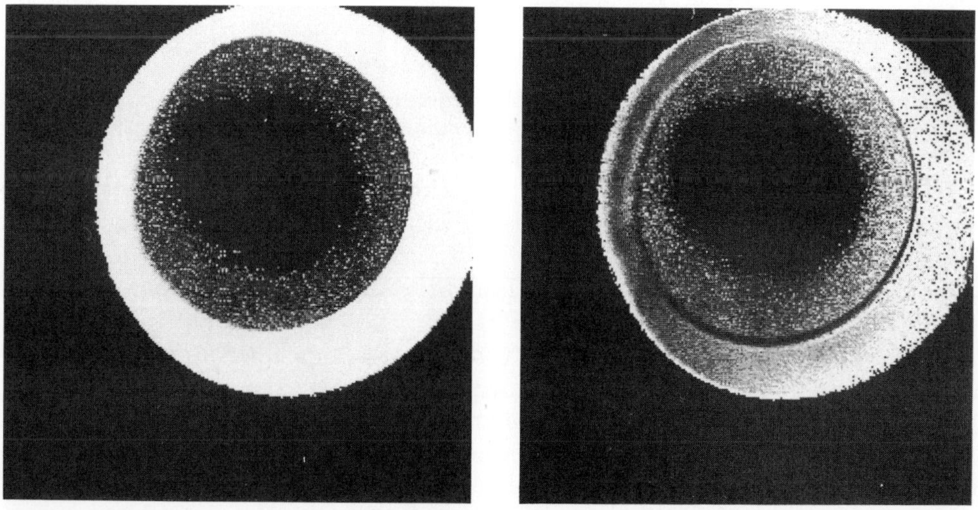

Figure 9. (a) The T_1 image of a 19 mm PMMA rod in methanol for 10 days reconstructed from 5 images with different repetition times and a 12 msec echo time. (b) The T_2 image of the same rod reconstructed from 4 images with different echo times and a 4 sec repetition time. In both cases brighter areas indicate longer relaxation times. All of the images were collected at 30°C.

Table I. SPIN-LATTICE AND SPIN-SPIN RELAXATION TIMES CALCULATED FROM IMAGES WITH DIFFERENT ECHO AND REPETITION TIMES

Region	T_1 (sec)	T_2 (msec)
Bulk	3.7 ± 0.3	50 ± 10
Bulk-Rubber Interface	0.54 ± 0.04	30 ± 10
Rubber-Glass Interface	0.36 ± 0.07	19 ± 7

submersion in methanol. The bulk methanol is the only area producing signal since no methanol has diffused into the sphere. Fig. 4b is the image after 48 hours exposure to methanol at 30°C. The methanol diffused 1.9 ± 0.2 mm into the sphere. Figure 5 is the profile of the image in Fig. 4b. The profile shows two of the aspects of Case II diffusion. The interface between the glassy polymer (d) and the swollen polymer has a steep methanol concentration gradient (c). Secondly, the methanol concentration (b) after the front is constant out to the bulk solvent. The concentration of methanol after the front is .26 ± 0.03 which is within experimental error of .275 from Thomas and Windle [1] data at 23°C.

The FLASH images also exhibit the third characteristic of Case II diffusion, the constant velocity. Several images are acquired over an extended period of time. The penetration depth is measured from the outer edge of the sphere to the edge of the sharp diffusion front. The penetration depth is plotted as a function of time in Fig. 6. The linear velocity of the front is 4.9 ± 0.8 nm/sec and agrees well with the velocities calculated from Thomas and Windle [8] data.

The Fickian concentration profiles preceding the front are not observed in the FLASH images. The signal from the Fickian precursor is probably below the noise level in the images but might be observed with increased signal averaging. The critical concentration for precursor formation increases with increasing temperature. Therefore, the Fickian precursor grows with increasing temperature. Near the boiling point of methanol, 60°C, the diffusion is nearly Fickian. The concentration profile is exponential at the front and the advance of the methanol goes as the square-root of time. Several images are acquired at 60°C over several time intervals using the spin-echo sequence in Fig. 3. The concentration profiles do not show the expected front profile. This is possibly due to relaxation effects which are discussed in a later section. Another contributing factor to the improper concentration profile is image smearing. Smearing occurs when the diffusion rate is faster than the time necessary to acquire an image. At 60°C, the concentration profile advances about 2 pixels over the 35 min imaging time. The diffusion distance does increase with the square-root of time as expected for Fickian diffusion. The diffusion coefficient of 2.0×10^{-8} cm^2/sec differs from 0.82×10^{-8} cm^2/sec calculated from Thomas and Windle [8] data that is averaged from data at 60 and 64°C. The difference possibly comes from the differences in the thermal history of the samples since Thomas and Windle annealed the thin PMMA slabs compared to our thick untreated samples. Experimental differences may also contribute to the difference.

5.2 Diffusion Imaged Directly

The pulse sequence in Fig. 2 uses gradients to probe the random motions in the solvent. The solvent diffusion causes random displacements of the spins. In the presence of a magnetic field gradient, these motions cause an attenuation of the spin-echo. The attenuation is given by

$$S(nTE) = S(O) \cdot \exp[(-n \bullet TE)/T_2] \cdot \exp(-b \bullet D) \tag{1}$$

where S(nTE) is the nth echo and D is the diffusion coefficient. The gradient attenuation factor, b, is given by

$$b_n = \gamma^2 \cdot \sum_{k=1}^{n} \sum_{l=x,y,z} G^2_{kl} \bullet d^2_{kl} \bullet \{[(d_{kl}+d'_{kl})/3] + I_{kl}\} \tag{2}$$

where n is the number of echoes, k is the interval between echoes, l is the axis, G and d are the amplitude and duration, respectively, of the first gradient of a gradient pair, G′ and d′ are the amplitude and duration, respectively, of the second gradient of a gradient pair, with the two being separated by the interval I [4].

The image produced from the diffusion pulse sequence is shown in Fig. 8. Only the imbibed methanol is present in the image. According to Equation 1, the intensities of two samples under the same echo time and motion probing gradients depend only on the diffusion coefficients of the samples, with the slower diffusing materials being more intense. Since the T_2 of the bulk methanol is larger than that of the imbibed methanol, the signal should be more intense for the bulk methanol without the special gradients. However the image shows only the imbibed methanol, indicating that the diffusion rate of the imbibed methanol is considerably slower than the self-diffusion of the bulk methanol as is expected.

5.3 Imaging Based on Relaxation Times

As discussed earlier, the relaxation times of the sample can inhibit imaging spin densities accurately. However, the distribution of relaxation times across the sample contains valuable structural information. The T_1 and T_2 images obtained with the reconstruction technique described previously are given in Fig. 9. The T_1 of the bulk methanol is larger than the imbibed methanol by a factor of 7. This reflects the faster motions of the bulk solvent and agrees with the diffusion pulse sequence data. The limited free volume within the polymer restricts the motions of the diffusing solvent. The difference in the T_1s of the imbibed methanol at the bulk-rubber interface and the rubber-glass interface about 33%. The T_1 at the bulk-rubber interface is the random motions limited by the free volume after the PMMA is swollen fully. The motions at the rubber-glass interface are more restricted since the material is in the process of swelling at this interface.

The T_1 and T_2 images both show an intense halo around the PMMA rod. Several sources may contribute to this feature. The magnetic susceptibility difference between the bulk methanol and the PMMA rod can increase the intensity at the interface. Data processing or experimental fluctuations that displace the information slightly may contribute to the halo. The halo may indicate that the relaxation is significantly different at the interface.

The T_2s are similar to the T_1 but are less dramatic. The T_2s show the difficulty of imaging imbibed fluids. With a standard spin-echo sequence and an echo time of 12 msec, the bulk and imbibed methanol signals are attenuated by 22% and 47%, respectively. A reduction of the T_2 attenuation difference to within 10% requires an echo time of 3 msec, within 5% requires an echo time of 1.5 msec. Even with these echo times imaging of diffusion is limited by the T_1s of the bulk methanol. A 20 sec repetition time is required for full relaxation of the spins. With 4 acquisitions of 256 echoes a 20 sec delay requires almost 6 hours which is long enough to cause motional artifacts with faster diffusion rates.

6. Conclusion

NMR imaging is useful for studying the diffusion of small organic molecules into glassy polymers. Three of the four aspects of Case II diffusion are readily available from imaging data. Increased signal-to-noise ratio may reveal the fourth characteristic. The special diffusion pulse sequence demonstrates that the motions of the imbibed methanol are much slower than the bulk methanol. The relaxation images also indicate slower motions in the swollen polymer. The gradient of relaxation times reflects the difference in free volume between the fully swollen and rubber and the rubber-glass interface. The relaxation times also indicate that FLASH is better suited for diffusion studies since it is less sensitive to relaxation processes.

7. References

1. Thomas, N. L. and Windle, A. H., *Polymer*, *23*, 529, 1982.
2. Hui, C. Y.; Wu, K. C.; Laskey, R. C. and Kramer, E. J., *J. Appl. Phys.*, *61*, 5137, 1987.
3. Haase, A.; Frahm, J.; Mattaei, D.; Hanicke, W. and Merboldt, K. D., *J. Magn. Reson.*, *67*, 258, 1986.
4. LeBihan, D.; Brenton, E.; Lallemand, D.; Grenier, P.; Cabanis, E. and Laval-Jeantet, M., *Radiology*, November, 401, 1986.
5. Stejskal, E. O. and Tanner, J. E., *Chem. Phys.*, *42*, 288, 1965.
6. Liu, J. and Koenig, J. L., submitted for publication.
7. In den Kleff, J. J. E. and Cuppen, J. J. M., *Magnetic Resonance in Medicine*, *5*, 513, 1987.
8. Thomas, N. L. and Windle, A. H., *Polymer*, *19*, 225, 1978.

Chapter 25

APPLICATION OF ONE-DIMENSIONAL NMR IMAGING TO SOLVENT DIFFUSION MEASUREMENTS IN POLYMERIC SAMPLES

STEPHEN PICKUP AND FRANK D. BLUM

1. Abstract

Small-scale, low-resolution NMR imaging has been performed on a standard NMR spectrometer equipped with a homogeneity spoiling unit without instrument modification. This technique uses the homogeneity spoiling unit to generate the magnetic field gradient needed in the imaging experiment. We demonstrate how this imaging technique can be used to observe a solvent diffuse into a polymer gel. By fitting experimental images to calculated ones it is possible to extract a diffusion coefficient from the imaging data. This experiment has been used to measure the diffusion of deuterium oxide in aqueous carboxymethyl-cellulose solutions. Two different samples were studied having concentrations of 4 and 6% cellulose by weight in water. Diffusion coefficients of 1.67×10^{-9} and 1.36×10^{-9} m^2/s were determined using the imaging experiment for the 4 and 6% samples, respectively. In samples of such low polymer concentration gradient the value of the solvent mutual-diffusion coefficient is nearly equal to that of the self-diffusion coefficient. Thus, for this set of samples, a comparison of self-diffusion measurements to the imaging results would serve as a check of the accuracy of the measurement. The pulsed-gradient spin-echo (PGSE) technique was used to determine the self-diffusion coefficients in the same set of samples. The results of the two diffusion measurements are found to be within experimental error of each other. These experiments demonstrate the feasibility of measuring mutual-diffusion coefficients for other systems in this way.

STEPHEN PICKUP - Department of Chemistry, University of Missouri-Columbia, Columbia, MO 65211 and FRANK D. BLUM - Department of Chemistry, University of Missouri-Rolla, Rolla, MO 65401.

Solid State NMR of Polymers, Edited by L. Mathias
Plenum Press, New York, 1991

2. Introduction

A wide variety of methods for studying molecular transport in polymeric systems are presently available. The results of such work generally yield a diffusion coefficient, however the exact meaning of the diffusion coefficient is often dependent on the measurement technique. Reviews of the transport processes, the methods of measurement and interpretation of results of such measurements have been published [1-3]. Of the several different types of diffusion coefficients that have been studied, the mutual-diffusion coefficient is most commonly probed. Mutual- or binary molecular-diffusion is a measure of the rate at which two components diffuse into each other in the presence of a concentration gradient. Mutual-diffusion is of importance and has been measured in such processes as drying and coloration of paints [4-6], curing of resins, and plasticization of polymers. Most of the techniques for determining mutual-diffusion coefficients monitor the changes in some physical property which can be related to the concentration (optical absorbance, index of refraction, etc.), as a function of time and position. Another method for determining diffusion rates is to attach a label which is easily monitored to the species of interest (e.g. radio-isotopes or fluorescence chromophores). The addition of such a label may dramatically effect the dynamics of the system although the use of isotope labeling is usually nonperturbing. In some cases secondary measurements may be used to monitor a property which is directly affected by solvent transport such as the rate of evaporation from the surface of a sample [2]. However, the more indirectly the property is related to the transport process, the more variables need to be controlled and the greater the probability for error.

NMR imaging has previously been used in transport studies [7-11]. Experiments have been reported which generate diffusion maps in which the contrast in a two dimensional image is related to the rate of diffusion [7,8]. These techniques have found use in examining vascular systems in living samples. Imaging has also been used to generate concentration profiles of a diffusing species in polymer matrices from which mutual-diffusion coefficients [9] and solvent transport through defects [10] have been determined. ESR imaging techniques which can be quite sensitive, have also been applied to diffusion measurements [11]. However these techniques usually require labeling of the diffusing species.

Typically, the resources required to do NMR imaging are unavailable for these types of applications due to the cost of high resolution imaging instrumentation. Much of the early imaging work has been performed on modified or home built instruments, and today several authors continue to modify existing high resolution NMR equipment in order to do microscopic imaging [12,13]. This too requires resources which are often unavailable to the small laboratory. It has been previously demonstrated that small-scale, low resolution, one-dimensional imaging can be performed on a standard NMR spectrometer equipped with a homogeneity-spoiling unit [14]. The magnetic field gradient generated by the homogeneity spoiling unit is used to convert spatial information into frequency information. This technique gives the spectroscopist one controllable field gradient which permits one-dimensional imaging only. Though this limits the use of this technique in many of the applications for which NMR imaging is commonly used, samples can be designed for diffusion measurements such that these constraints do not play a critical role. Thus this simple imaging technique is readily applicable to mutual-diffusion measurements.

We currently report the use of a NMR imaging technique to monitor solvent transport in a polymeric system. The advantages of this approach are in its selectivity, time scale and insensitivity to the optical properties of the sample. Most systems of interest contain NMR active nuclei. Selection of different molecules (even different moieties) is made possible by a choice of nucleus or resonance. Short relaxation times and overlapping signals can often be overcome by application of the proper pulse sequence. In addition, clarity or inhomogeneities in the sample provide no significant problem for the NMR experiment. These points have previously been addressed in several review articles [15-18]. In many cases, slowly diffusing materials may be studied by "running" a sample periodically over a long time scale (e.g. months). The technique is similar to pulsed-field gradient spin-echo studies [19-21] in that it tags nuclei in the sample with a magnetic-field gradient, but it is different in that it measures a mutual rather than a self-diffusion coefficient.

Here we make a comparison between diffusion coefficients from NMR imaging data and self-diffusion measurements made with the pulsed-gradient spin-echo (PGSE) technique. Though these are two distinctly different measurements, in the limit of zero polymer concentration gradient the two are equal. With different concentration gradients these imaging experiments may be used to determine a true mutual diffusion coefficient. The samples used in this study were designed to minimize differences between self and mutual diffusion rates. Thus the PGSE data can be used as a check of the accuracy of the imaging experiment. Since the technique will in general yield mutual diffusion coefficients, we will refer to these as mutual diffusion coefficients.

3. Experimental

In the present analysis we chose to study diffusion of deuterium oxide in aqueous solutions of carboxymethyl-cellulose (CMC) because it forms a very viscous solution at low polymer concentrations. The highly viscous system is necessary in order to minimize convection and other mechanical sources of transport. Also, the deuterated form of the solvent (D_2O) is readily available. Since we have only one controllable field gradient in the imaging experiment and it is perpendicular to the axis of the sample tube, there is no simple method available to limit the image to a particular plane perpendicular to the tube axis. Thus the image of the sample must be independent of location along this axis. This problem is readily overcome if one designs the sample such that the species of interest diffuses radially from the center of a 10 mm sample tube. Samples used in the diffusion experiment were designed to meet this requirement as discussed below.

Solutions of CMC (Hercules Chemical Co., Wilmington, DE) in water were made by adding weighed amounts of the constituents to a flask. The flask was then sealed and sonicated until homogeneous. A similar procedure was followed to make solutions of CMC in D_2O at the same concentrations. Portions of the aqueous solution were then placed in 10 mm NMR tubes. A column of the D_2O solution was injected into the center of the gel using a steel tube and rod as shown in Fig. 1. The tube was filled with the D_2O solution by drawing the rod up while the bottom was submerged in the solution. The tube containing the D_2O solution was then inserted into a 10 mm sample tube containing a sample of the undeuterated solution (Fig. 1b). Two vortex plugs were used to guide the apparatus and insure concentricity of the injected column. The D_2O solution was then injected into the aqueous solution by pulling up the steel tube while holding the rod steady (Fig. 1c). Injection of solutions containing a dye showed that the injected column was concentric and had the same diameter as the plunger rod. A timer was started immediately after injection. One-dimensional images were then taken of the deuterium nuclei at fixed time intervals until it became apparent that the deuterium had reached the tube wall.

The imaging technique uses a single pulse sequence modified such that the field gradient (generated by the homogeneity spoiling unit) is on during data

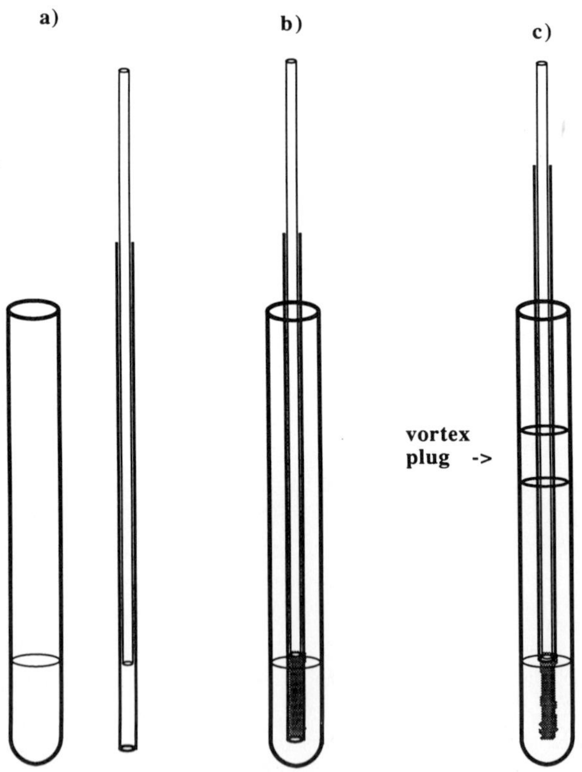

Figure 1. The sample preparation apparatus (a), insertion of the apparatus into the sample (b), and injection of the D_2O solution into the sample (c).

acquisition [14]. It is also recommended that the power to the homospoil coils be set at its maximum on the FX series spectrometers. The strength and linearity of the field gradient were calibrated by performing the PGSE experiment on a sample having several NMR active nuclei. The gradient strength was found to be ca. 0.05 T/m and to be linear within a few percent. This gives a spectral width of 1000 Hz for deuterium across a 10 mm sample.

All experiments reported here were performed on a JEOL FX90-Q operating at 13.7 MHz for deuterium. The one-dimensional images of deuterium in the gels were taken using 256 points, a 1300 Hz window, 128 scans and an equilibration delay of 50 ms, giving a total imaging time of 6 seconds. Several images were taken of each sample at an interval of ca. 50 seconds. The PGSE experiment was performed using 4096 points in a 1000 Hz window, 64 scans and a pulse delay of 1 second. Deuterium self-diffusion coefficients were determined from the slope of a linear least-square fit of the log of the peak height as a function of the gradient pulse length [20]. An external lithium lock operating at 34.8 MHz was used to correct for field drift during both experiments.

4. Data Interpretation

The radial concentration dependence on time and diffusion coefficient for radial diffusion in an infinite medium is given by the following equation [3];

$$C/C_o = (1/2Dt)\exp(-r^2/4Dt)\int_o^a \exp(-r'^2/4Dt)I_o(rr'/2Dt)r'dr' \qquad (1)$$

where C is the concentration, C_o is the initial concentration in the injected column, "a" is the width of the injected column, D is the mutual-diffusion coefficient, t is time, I_o is a Bessel function of the second kind of order zero, r is position along the radius and r' is the variable over which the integration is performed. The major difficulty in evaluating this equation is in performing the integration of the Bessel function (I_o). The integral has not been evaluated analytically and therefore numerical methods were applied. Two different polynomial expansions were used to approximate the value of the Bessel function for the operand ranges of 0 to 3.75 and 3.75 to infinity [22, equations 9.7.1 and 9.8.1]. These approximations were found to give at least four significant figures based on a comparison to tabulated values of the Bessel function [23, p. 428].

Radial concentration profiles were calculated using equation 1 for a wide variety of values of Dt (note that the value of Dt characterizes the curve). Calculated plots of the concentration as a function of radius are shown in Fig. 2a for five different values of Dt. The sharp step function corresponds to early times and the broader functions correspond to later times. Calculation of the one-dimensional images was performed by laying the cross section of the sample (perpendicular to the sample tube axis) out on a 250 x 250 grid. The radius was calculated at each point on the grid and the concentration at that radius was determined from the previously

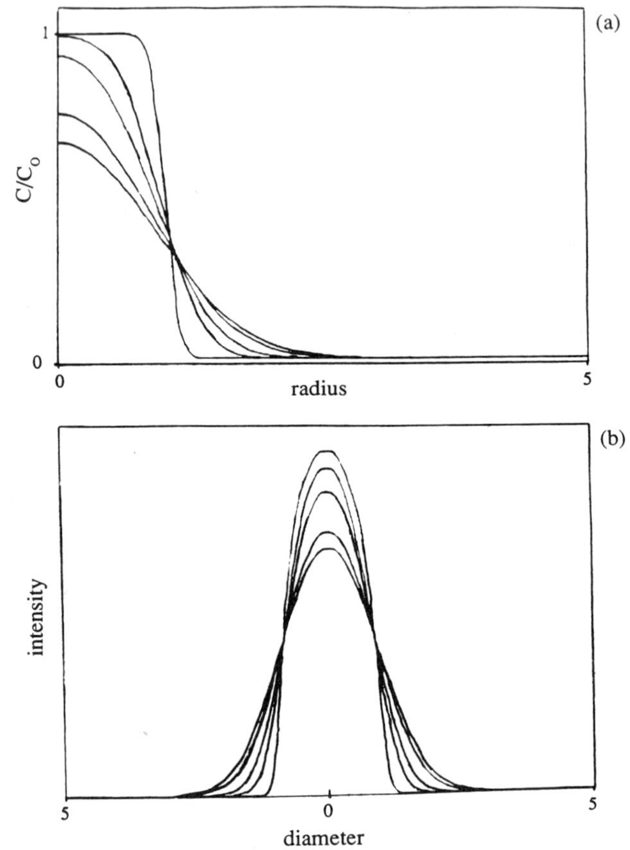

Figure 2. Concentration profiles calculated using equation 1 and Dt values of 0.12 x 10^{-7}, 1 x 10^{-7}, 2 x 10^{-7}, 4 x 10^{-7}, and 5 x 10^{-7} m^2 (a) and one-dimensional images that were calculated using these profiles (b). Note the two figures are on different scales.

calculated concentration profiles. The sum of the concentrations in the vertical dimension of the grid is determined for each position in the horizontal dimension. Plots of the concentration sums versus horizontal position are the calculated one dimensional images. The one-dimensional images calculated using the concentration profiles shown in Fig. 2a are shown in Fig. 2b. Note that the two figures are on different scales. In contrast to the radial plots, the total intensity in the calculated images should be independent of Dt. Summation of the image intensity over all positions for each image gave a total intensity which varied less than 0.7% for all of the calculated images. This is an indication that the errors in the numerical analysis of equation 1 are not significant.

The experimental images were fit to the calculated ones by digitizing the experimental images and overlaying them on the calculated ones. With a knowledge of the width of the experimental image at time infinity and the initial diameter of the injected column it is possible to scale the experimental images to the calculated ones. Since the shape of the sample tube is the same as that of the injected column, the image at infinite time should have the same shape as that of the calculated image at time zero. The relative widths of the images at time zero and time infinity should be equal to the ratio of the diameters of the sample tube to that of the injected column. Thus the x expansion of the experimental image is set equal to the product of the value that gave the best fit of the image at time infinity and the ratio of the tube diameter to that of the injected column. The first experimental image is then fit using this value of the x expansion, by comparing it to the various calculated images and adjusting the y expansion. The remaining experimental images are fit by comparing them to the calculated images using the same scaling factors that were used in the first image. A plot of Dt for the calculated images vs. experimental time should be linear with a slope equal to the diffusion coefficient and an intercept of zero.

5. Results

A typical set of one-dimensional images for the mutual-diffusion experiment is shown in Fig. 3 for the sample containing 4% CMC. The elapsed times between injection of the D_2O column and completion of the images are listed to the right of each image in the figure. No artificial line broadening was used in transformation of the data in order to minimize distortion of the peak shape. Also, a minimum number

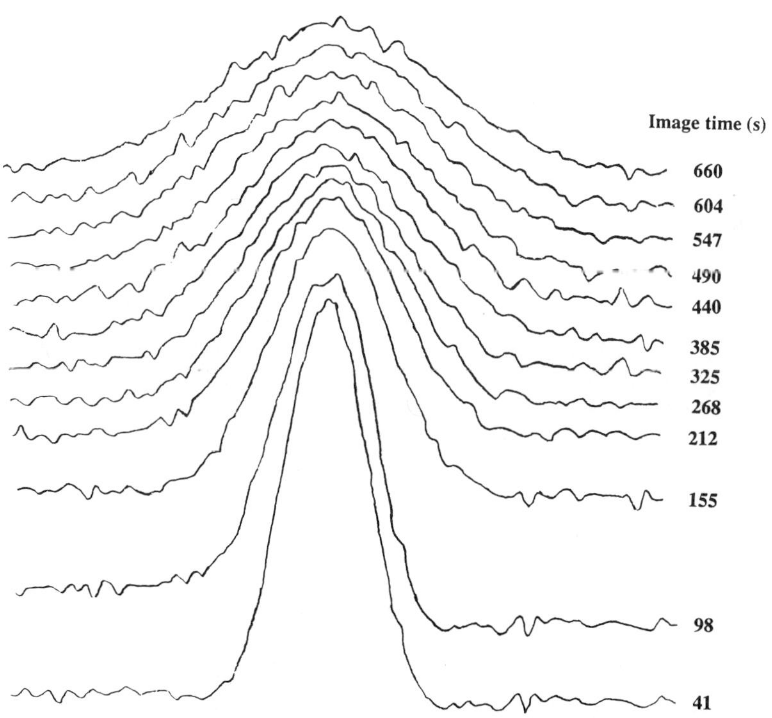

Figure 3. A typical set of images of deuterium oxide diffusing into a solution of 4% carboxymethylcellulose in water. The time at which each image was taken is listed to the right of the corresponding image.

of scans were used to acquire the images in order to minimize distortions due to the image changing during data acquisition because of diffusion. This resulted in a somewhat noisy image having little or no systematic distortion. The integrated intensity of the experimental images was found to vary by less than 5%. The diffusion of the D_2O into the gel is indicated by the increasing width and decreasing height of the image at later times. Equation 1 assumes radial diffusion in an infinite system and therefore only the images in which the deuterium has not reached the tube wall were fit to the calculated images. For example, the first four images in Fig. 3 fit this requirement. The fits of these experimental images to the calculated ones are shown in Fig. 4. The circles in the figures represent the experimental data and the lines are the calculated images. The quality of the fits is good for all of the images shown.

Also shown in Fig. 4 are the values of Dt which were used for the calculated images and the times at which the experimental images were taken. Plots of the Dt

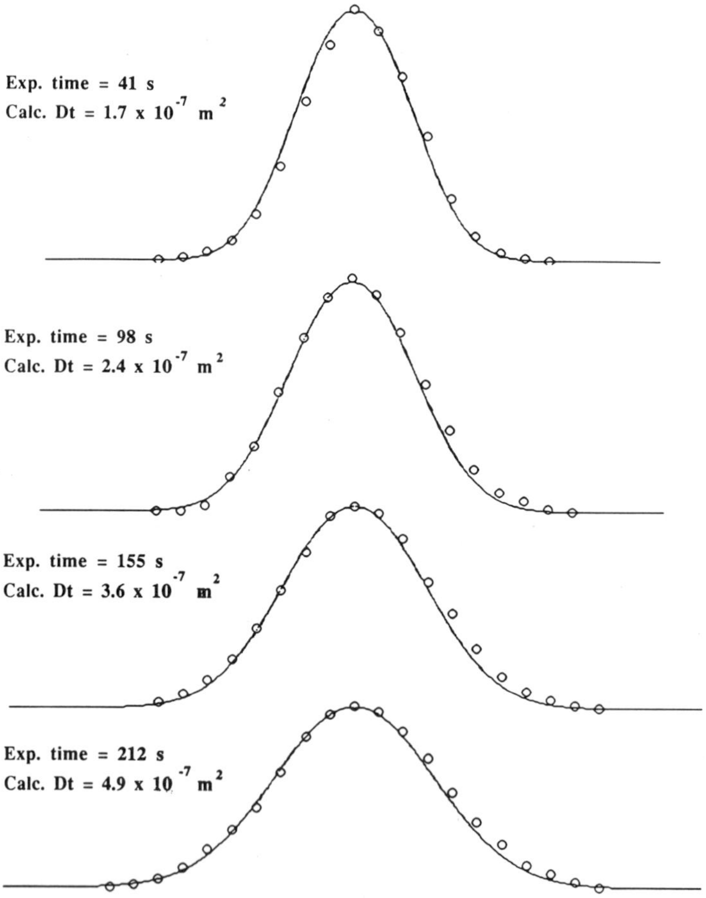

Figure 4. Fits of the first four experimental images shown in Figure 3 to the calculated images. The circles represent the experimental data and the line represents the calculated image.

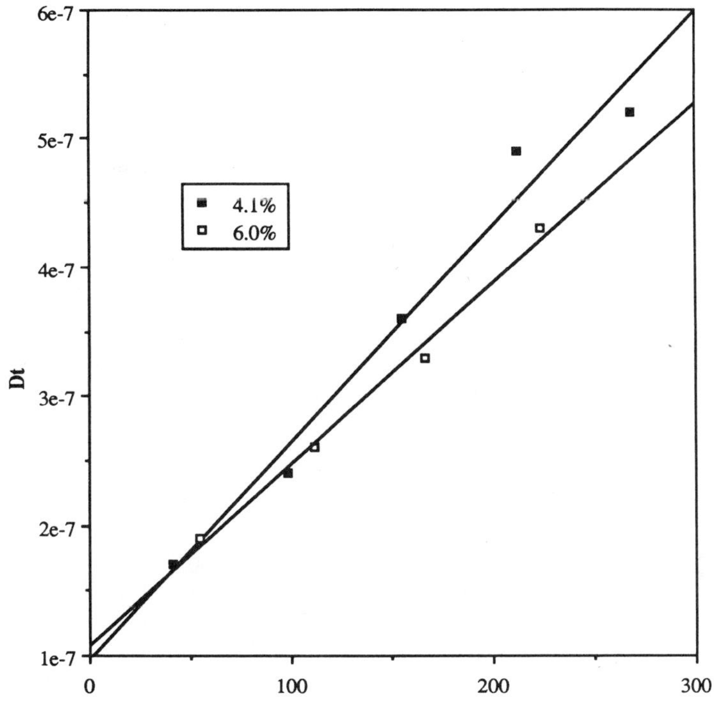

Figure 5. Plots of the Dt value used to calculate the best fitting image versus the experimental time at which the image was taken for both of the samples studied. The slope of the line is equal to the mutual-diffusion coefficient.

values verses experimental time are expected to be linear with a slope equal to the diffusion coefficient. Such plots are shown in Fig. 5 for both of the samples studied. Though only a small number of data points are available, the data are linear with little deviation. The mutual-diffusion coefficients are determined by a linear least-squares fit of the data in the figure and are given in Table I. Also shown in the table are the self-diffusion coefficients as determined by the PGSE technique for the same samples. The data are in close agreement however the mutual-diffusion measurement consistently gives a lower value.

6. Discussion

In this study we attempted to make the mutual-diffusion rate as similar to the self-diffusion rate as possible in order to facilitate a direct comparison between two experimental techniques. In the mutual-diffusion experiment, the only concentration gradient present is due to deuterated water although in general any concentration gradient is possible. The properties of D_2O are only slightly different from that of water and therefore the resulting diffusion coefficient should be very similar to the self-diffusion coefficient. The concentration of polymer in this system is so low that

Table I. MUTUAL AND SELF-DIFFUSION COEFFICIENTS OF DEUTERIUM OXIDE IN CARBOXYMETHYL-CELLULOSE SOLUTIONS

%CMC	mutual-diffusion coefficient (m^2/sec)	self-diffusion coefficient (m^2/sec)
4.1	1.67×10^{-9}	2.10×10^{-9}
6.0	1.36×10^{-9}	2.08×10^{-9}

polymer concentration gradients are not expected to effect the diffusion process. Our observations are consistent with this premise.

The mutual-diffusion measurement yields a value which is ca. 25% lower than the PGSE result. These differences are probably due to the assumption of radial diffusion in an infinite medium. Actually, the diffusion process is bounded by the sample tube. The later images are therefore not as broad as they should be which results in a lower apparent diffusion rate in the imaging experiment. Although the use of equation 1 (radial diffusion in an infinite medium) significantly limits our data set, this approach does offer some advantages. The most significant advantage is that the images only need to be calculated once without concern for the relative diameters of the sample tube and the injected column. In contrast, an approach which includes the boundary effects would require a knowledge of this relationship prior to the image calculation. Also, in the latter stages of the experiment, when the boundary effects play a major role, the images do not change significantly. Thus there is little to be gained by such a consideration.

7. Conclusion

We have shown that a rather simple NMR imaging technique can be readily applied to mutual-diffusion measurements in polymer solutions. Although we used a labeled solvent in our analysis, this may be avoided if selective pulse sequences are applied. The samples used in this study were designed to minimize any differences between mutual and self-diffusion rates. This allowed a direct comparison between the results of the imaging experiment and self-diffusion data. The two experiments were in close agreement though the imaging result was consistently lower than that of the PGSE experiment. These results show that the imaging technique for measuring mutual-diffusion coefficients has much potential due to its flexibility and simplicity.

8. References

1. Cussler, E. L., *Diffusion: Mass Transfer in Fluid Systems*, Cambridge University Press, Cambridge, 1984.

2. Crank, J. and Park, H. S., *Diffusion in Polymers*, Academic Press, New York, 1968.
3. Crank, J., *The Mathematics of Diffusion*, Oxford University Press, London, 1975.
4. Guerra, G. and Paolone, C., *J. Coatings Tech.*, *55(701)*, 53, 1983.
5. Ellis, W. H., *J. Coatings Tech.*, *55(696)*, 63, 1983.
6. Eaton, R. F. and Willeboordse, F. G., *J. Coatings Tech.*, *52(660)*, 63, 1980.
7. Merbolt, K. D.; Hanicke, W. and Fragm, J., *J. Magn. Reson.*, *64*, 479, 1985.
8. Merbolt, K. D.; Hanicke, W. and Fragm, J., *J. Magn. Reson.*, *67*, 336, 1986.
9. Blackband, S. and Manssfield, P., *J. Phys. C: Solid State Phys.*, *19*, L49, 1986.
10. Rothwell, W. P.; Holecek, D. R. and Kershaw, J. A., *J. Polym. Sci., Polym. Phys. Ed.*, *22*, 241, 1984.
11. Stemp, E. D. A.; Eaton, G. R.; Eaton, S. S. and Maltompo, M. M., *J. Phys. Chem.*, *91*, 6467, 1987.
12. Aguayo, J. B.; Blackband, S. J.; Schoeniger, J.; Mattingly, M. A. and Hintermann, M., *Nature*, *322(10)*, 190, 1973.
13. Hall, L. D.; Luck, S. and Raganayagam, V., *J. Magn. Reson.*, *66*, 349, 1986.
14. Pickup, S. and Blum, F. D., *Spectroscopy*, *2(6)*, 53, 1987.
15. Hoult, D. I., *Magnetic Resonance in Biology*, Vol. 1; John Wiley & Sons, New York, 1980, pp 70.
16. Andrew, R. E., *Acc. Chem. Res.*, *16*, 114, 1983.
17. Smith, S. L., *Anal. Chem.*, *57(4)*, 595, 1985.
18. Mansfield, P. and Morris, P. G., *Advances in Magnetic Resonance*, Suppliment 2; Waugh, J., Ed., Academic Press, New York, 1982.
19. Stejskal, E. O. and Tanner, J. E., *J. Chem. Phys.*, *42(1)*, 288, 1964.
20. Blum, F. D., *Spectroscopy*, *1(5)*, 32, 1986.
21. Stilbs, P., *Prog in NMR Spectros.*, *19*, 1, 1987.
22. Abramowitz, M. and Stegun, I. A., *Handbook of Mathematical Functions*, Dover Publications, Inc., New York, 1965.

CONTRIBUTORS

Mladen Andreis
Department of Macromolecular Science
Case Western Reserve University
Cleveland, OH 44106

Laurence A. Belfiore
Department of Chemical Engineering
Colorado State University
Fort Collins, CO 80523

Frank D. Blum
Department of Chemistry
and Materials Research Center
University of Missouri-Rolla
142 Schrenk Hall
Rolla, MO 65401-0249

F.A Bovey
AT&T Bell Laboratories
600 Mountain Avenue
Murray Hill, NJ 07974

C.C. Chang
Department of Radiology,
Pathology and Biochemistry
Univ. of Arkansas for Medical Sciences
4301 West Markham
Little Rock, AR 72205-7199

Chihmin Cheng
Department of Chemical Engineering
Colorado State University
Fort Collins, CO 80523

A.L. Cholli
B.O.C. Group
100 Mountain Avenue
Murray Hill, NJ 07974

Ronald F. Colletti
Monsanto Chemical
P.O. Box 12830
Pensacola, FL 32575

E.M. Conwell
Xerox Webster Research Center
800 Phillips Road 0114-39D
Webster, NY 14580

D.G. Cory
Laboratory of Physical Chemistry II
Faculty of Science
University of Nijmegan
Toernooiveld, 6525 ED Nijmegen
The Netherlands
CURR. ADD: Bruker Instruments, Inc.
Manning Park
Billerica, MA 01821

M.H. Cozine
AT&T Bell Laboratories
600 Mountain Avenue
Murray Hill, NJ 07974

Richard C. Crosby
Department of Chemistry
Texas A&M University
College Station, TX 77843

J.C. De Boer
Laboratory of Physical Chemistry II
Faculty of Science
University of Nijmegan
Toernooiveld, 6525 ED Nijmegen
The Netherlands

Hansotto Drotloff
Institut fur Makromolekulare Chemie
Universitat Freiburg
Stefan Meier Str. 31
7800 Freiburg
Federal Republic of Germany

Cecil Dybowski
Department of Chemistry
and Biochemistry
University of Delaware
Newark, DE 19716

Michael L. Froehlich
Energy and Minerals Research Center
and Department of Chemistry
University of North Dakota
Box 7185 University Station
Grand Forks, ND 58202

Robert B. Funchess
Department of Chemistry
and Materials Research Center
University of Missouri-Rolla
142 Schrenk Hall
Rolla, MO 65401-0249

M.A. Gomez
AT&T Bell Laboratories
600 Mountain Avenue
Murray Hill, NJ 07974

D. Granger
Hughes Research Laboratories
3011 Malibu Canyon Road, M/S RL70
Malibu, CA 90265-4737

R.A. Grinsted
Department of Macromolecular Science
Case Western Reserve University
Olin Bldg., Room 209
Cleveland, OH 44106

James F. Haw
Department of Chemistry
Texas A&M University
College Station, TX 77843

Martin B. Jones
Department of Chemistry
Adams State College
Alamosa, CO 81102

S. Kaplan
Xerox Webster Research Center
800 Phillips Road 0114-39D
Webster, NY 14580

Jack L. Koenig or J. L. Koenig
Department of Macromolecular Science
Case Western Reserve University
Cleveland, OH 44106

Gerhard Kögler
Institut fur Makromolekulare Chemie
Universitat Freiburg
Stefan Meier Str. 31
D-7800 Freiburg
Federal Republic of Germany

R.A. Komoroski
Department of Radiology,
Pathology and Biochemistry
Univ. of Arkansas for Medical Sciences
4301 West Markham, Slot #582
Little Rock, AR 72205-7199

G. Kothe
Institute of Physical Chemistry
University of Stuttgart
Pfaffenwaldring 55,
D-7000 Stuttgart 80
Federal Republic of Germany

Jonathan Lipowitz
Dow Corning Corporation
Mil Stop 540
Midland, MI 48686

R. Ludicky
Morton International
Specialty Chemicals Division
1275 Lake Avenue
Woodstock, IL 60098

Thomas J. Lutz
IBM Faculty Loan Program
Boulder, CO 80503

A.G. MacDiarmid
Department of Chemistry
University of Pennsylvania
Chem. Bldg., 3301 Spruce Street
Philadelphia, PA 19104

Lon J. Mathias
Department of Polymer Science
University of Southern Mississippi
Southern Station Box 10076
Hattiesburg, MS 39406-0076

C.N. Matthews
Department of Chemistry (M/C 111)
University of Illinois at Chicago
Box 4348
Chicago, IL 60680

Wiriya Meesiri
Department of Chemistry
and Materials Research Center
University of Missouri-Rolla
142 Schrenk Hall
Rolla, MO 65401-0249

Martin Möller
Universiteit Twente
Postbus 217
7500 AE Enschede

K. Müller
Institute of Physical Chemistry
University of Stuttgart
Pfaffenwaldring 55,
D-7000 Stuttgart 80
Federal Republic of Germany

Almera Natansohn
Department of Chemistry
Queen's University
Kingston, Ontario, K7L 3N6
Canada

Dieter Oelfin
Institut fur Makromolekulare Chemie
Universitat Freiburg
Stefan Meier Str. 31
7800 Freiburg
Federal Republic of Germany

B.C. Perry
Department of Macromolecular Science
Case Western Reserve University
Cleveland, OH 44106

Stephen B. Pickup
Department of Radiology
University of Missouri-Columbia
M201 Health Sciences Center
Columbia, MO 65212

A.M. Reichwein
Laboratory of Physical Chemistry II
Faculty of Science
University of Nijmegan
Toernooiveld, 6525 ED Nijmegen
The Netherlands

A.F. Richter
Department of Chemistry
University of Pennsylvania
Philadelphia, PA 19104

F. C. Schilling
AT&T Bell Laboratories
600 Mountain Avenue
Murray Hill, NJ 07974-2070

A. Schleicher
Institute of Physical Chemistry
University of Stuttgart
Pfaffenwaldring 55,
D-7000 Stuttgart 80
Federal Republic of Germany

Hajime Tanaka
AT&T Bell Laboratories
600 Mountain Avenue
Murray Hill, NJ 07974

A.E. Tonelli
AT&T Bell Laboratories
600 Mountain Avenue
Murray Hill, NJ 07974

Gary L. Turner
Spectral Data Services, Inc.
818 Pioneer
Champaign, IL 61820

J.W.M van Os
Laboratory of Physical Chemistry II
Faculty of Science
University of Nijmegan
Toernooiveld, 6525 ED Nijmegen
The Netherlands

W.S. Veeman
Laboratory of Physical Chemistry II
Faculty of Science
University of Nijmegan
Toernooiveld, 6525 ED Nijmegen
The Netherlands

R. Voelkel
Kunststofflaboratorium
Abt. Polymerphysik, Festkörperphysik
BASF Aktiengesellschaft
D-6700 Ludwigshafen

L.A. Weisenberger
Department of Macromolecular Science
Case Western Reserve University
Cleveland, OH 44106

INDEX

Acrylate copolymers, 233
Acrylates, 34
Acrylic dispersions, 233
Acrylonitrile, 34
Additives imaging, 72
Adhesive imaging, 74
n-Alkane, 89
Aminobutyltriethoxysilane, deuterated, 274
Aminopropyltriethoxysilane, 271, 272
  deuterated, 272, 274
Amorphous, 95, 117, 121, 146, 159, 161, 162, 187, 364
Asymmetric, 8
Axial symmetry, 8, 12, 14, 30

Back projection reconstruction, 346, 349
Biphenyl polymers, 112
Blends, 131, 139, 145, 146, 174, 176
Blends, zinc-promoted, 155
BR-24, 6, 357
Bridging methylenes, 110
Brill transition, 42
Broad-line ^{1}H NMR, 90

^{13}C NMR, 81, 92, 107, 117, 295, 322, 331
  blends, 131, 132, 145, 147, 151, 174, 176
  chemical shifts, 92, 98
  variable-temperature, 92
^{13}C CP/MAS, 42, 44, 108
  bound coupling agents, 274
  1,4-diazabicylco-[2.2.2]octane, 322
  N-ethylcarbazole, 172
  hydrogen cyanide polymers, 331
  polyanilines, 284, 286
  poly(isopropyl acrylate), 277, 280
  poly(methyl methacrylate), 279
  poly(N-vinylcarbazole), 169
  2,4,7-trinitro-9-fluorenone, 171
^{13}C MAS, 307
  organosilicon polymers, 305
  poly(di-n-hexylsilylene), 298, 300
  Si-C, 314
  Si-C-O fibers, 307
  Si-C-N-O fibers, 307
  Si-N, 314
^{13}C spin-echo MAS, 317
Carr-Purcell sequences, 349
Carrot cell protein, ^{15}N CP/MAS, 335
CDP/MAS, 285, 286
Chain scission, 117
Chain alignment, 14
Charge transfer complex, 167, 174, 176
Chemical shift, 27
  anisotropic contributions, 1
  scale factor, 7
  tensor, 8
Chloranil, 168, 176
Conformation, 81, 95, 117

Conformational disorder, 50, 245, 246, 364
Contaminants, 72
Contrast level, 68
Copolymers
  ethylene-methacrylic acid, blends, 146, 148, 152, 161
  styrene-butadiene, 148, 162
Correlation time, 252
Coupling agents, 271
CP/MAS, 119, 167, 184, 217, 233, 272, 273, 285
  variable temperature, 233, 321
CRAMPS, 10, 17, 18, 145, 147, 150, 156, 163, 355
Cross-polarization, 23
Crosslink density, 71
Crosslinked, 34, 35, 117
Crosslinking, 111, 124, 201, 364
  effect on T_1, 208
  effect on T_2, 209
  gamma-irradiation, 126
  sulfur vulcanization, 201
Crosslinks
  natural rubber, 201
  poly(methacrylic acid), 180, 181, 182, 188, 192
Crystalline, 14, 81, 95, 117, 121, 146, 159, 161, 162, 187, 245, 266
  disorder, 267
  fraction, 17
  hexagonal, 246
  monoclinic, 248
  orthorhombic, 248
  packing, 87
Cyanoethyltriethoxysilane, 272
Cyanopropyltriethoxysilane, 272
Cycloalkanes, 245, 247
Cyclodoeicosane, 249
Cycloheptacontane, 253

2D Fourier imaging method, 63
2D NMR, 33
2-Dimensional Fourier imaging, 346, 349, 350

Decoupling, 3
Degradation, 72
Delta, δ, 28
Deuterium NMR, 23
Deuterium lineshape, 30
Diamagnetic anisotropy, 54
Diaminomaleonitrile, ^{15}N CP/MAS, 335
Diffusion, radial concentration, 391
4,4′-Dimethylbiphenyl, 112
Dipolar coupling, 2
Dipolar decoupling, 23
Dipolar-dephased, 118, 120
  spectral editing, 120
  two-dimensional, 120
Dipolar dephasing, 161
Dipole-dipole coupling, 1, 16, 26
Double cross polarization, ^{1}H/^{15}N/^{13}C, 335
Double-resonance, 2
Dynamics, 81

Electric field gradient, 28
N-Ethylcarbazole, 169

^{19}F NMR, 1, 13, 14
Fickian diffusion, 71, 384
Fluorene polymers, 113
Fourier transformation, 2
Free decay, 2
Fumed silicas, 272

Gamma-irradiation, 117
γ-Gauche interaction, 86, 87
γ-Gauche effect, 81, 97
Gauche interactions, 12
Gauche conformations, 50
GHPD, natural rubber, 203
Glass transition, 34, 48, 120, 160, 169, 218, 223, 237, 265, 324, 363
Goldman-Shen experiment, 160

^{1}H NMR, 1
^{1}H MAS, imaging, 343
^{2}H labelling, 26
^{2}H NMR, 23, 245, 252, 265, 273, 283,

285,
computer simulation, 34
coupling agents, 276
2D, 41, 264
inversion recovery, 262
Jeener-Broekaert sequence, 263
nylon 6, 46
nylon 11, 47
nylons, 44
polyanilines, 287, 289
N-*n*-propyl stearamide, 48
quadrupole echo, 30, 263
simulations, 253, 255, 262, 265
spin alignment, 33, 37
spin-lattice relaxation, 30, 38
spin-spin relaxation, 38
T_2 minimum, 36
variable temperature, 247, 253, 257

Heteronuclear dipolar interactions, 120
Heteropolyamidines, 333
Hexachlorocyclotriphosphazene, polymerization, 321, 326
Hexafluoropropylene, 12
High resolution ^{13}C NMR, 81, 118, 124, 179, 184
N-ethylcarbazole, 172
poly(di-n-hexylsilylene), 298
poly(isopropyl acrylate), 277
poly(methacrylic anhydride), 179, 189
poly(methyl methacrylate), 279
2,4,7-trinitro-9-fluorenone, 171
poly(N-vinylcarbazole), 169
High spinning speed, 3
Homonuclear dipolar coupling, MAS, 343, 354
Hydrogen cyanide polymers, 331, 332
comets, 337
^{15}N CP/MAS, 334
origins of life, 337
Hydrogen bonded arrays, 48

Imaging
adhesive bonding, 74
boundary detection, 68
contrast level, 68
2D Fourier method, 63
3D method, 63
diffusion kinetics, 71, 377, 387, 391
FLASH pulse sequence, 379
gradient echo, 365, 366
interfacial flaws, 72
internal dimensions, 70
low density polyethylene, 355
microcracks, 72
multiple pulse methods, 354, 356
one dimensional, 388, 391
polybutadiene, 353, 363
polyisobutylene, 366
polyisoprene, 366
poly(methyl methacrylate), 380
resolution, 64, 68
relaxation time dependent, 385
rubber, 353, 363
solids, 343, 363, 377
spin echo, 66, 365, 366, 385, 387
spin density, 63, 381
theoretical basis, 61
transverse magnetization, 67
voxels, 67
Inhomogeneous resonance, 9
Inversion-recovery, 83, 108

Jeener-Broekaert pulse sequence, 33

KEL-F, 17

Larmor angular frequency, 61
Levenberg-Margdt curve fitting, 203
Line-narrowing techniques, 1, 76
Lineshapes, 24
Linewidth, 111
Liouville-von Neumann equation, 3
Lipid membranes, 57
Liquid crystals, 246, 261
conformational order, 267
Liquid crystalline, 49, 53, 215, 224, 230, 261, 266, 298
magnetic field aligned, 53

smectic, 231

Magic angle, 9
Magic-angle spinning, 17, 18, 23
Magnetic field gradient, 344, 345, 359
Magnetic moment, 28
Magnetic shielding, 26
Magnetogyric ratio, 61
MAS, 245, 298, 307
variable temperature, 298, 322
Mesomorphous, 245, 324
Microcrystallites, 8
Microstructures, 81
Molecular dynamics, 17
Molecular motion, 25, 120
Morphology, 123, 157, 158, 234, 322
Motional effects, 122
MREV, 5
MREV-8, 5, 6, 7, 12, 16, 18
MRI, 64
Multiple pulse methods, 1, 355, 379
BLEW-48, 357
BR-24, 357
^{1}H, 150
MREV, 5
MREV-8, 355, 357
TREV-4, 5
WAHUHA, 4
WAHUHA-4, 5, 6, 7, 12
Multipulse dynamic NMR, 262

^{15}N CP/MAS, 42, 45, 331, 334, 335, 336
hydrogen cyanide polymers, 331, 334
Natural rubber, 201, 363
NMR imaging, 61, 65, 69, 343, 363, 377, 378, 387
adhesive, 74
boundary detection algorithm, 68
computer processing, 68
diffusion, 71
eye, 65
internal dimensions, 70
internal imperfections, 72
line-narrowing techniques, 76
magic angle spinning, 76
polymers, 69
surface coil, 76
swelling, 71
tomato, 65
Non-uniform magnetic field, 61
Nylon 66, 42
Nylon 6, 42
Nylon 11, 42

Order parameter, 57
Orientation, 53

^{31}P MAS, variable temperature, 323
Packing effects, 89
Perfluorodecyl decane, 50
Perfluoromethylvinyl ether, 12, 13
Phase encoding, 352
Polyacrylamide, 239
Poly(acrylonitrile-d_{α}), 34
Polyaminocycanomethylene, 332
Polyaminomalononitrile, 333, 334
Polyanilines, 283, 284
Poly(arylene methylene)s, 107
Polybenzyls, 108
Poly[bis(ethoxy)phosphazene], 322
^{31}P MAS, 324
^{13}C CP/MAS, 324
Poly[bis(3-methylphenoxy)phosphazene], 322
^{31}P CP/MAS, 325
Polybutadiene
1,4-cis-, 17, 363, 367, 369, 372
1,4-trans-, 128
Poly(1-butene), 82, 128
Poly(butyl acrylate)
emulsions, 233
glass transition, 237
Poly(butyl acrylate-d_{α}), 34
Poly(t-butyl methacrylate)
syndiotactic, 182, 185
isotactic, 179, 182, 185
Poly(butylene terephthalate), 83, 90, 91
α-form, 90

β-form, 90
crystal-crystal transition, 90
model compounds, 91
Polycarbonate, 24
Poly(chlorotrifluoroethylene), 17
Poly-α-cyanoglycine, ^{15}N CP/MAS, 335
Poly(di-n-alkylsilylenes), 295
Poly(3,3-diethyloxetane), 83, 87
Poly(di-n-hexylsilylene), 297
Poly(3,3-dimethyloxetane), 86
Poly(di-n-pentylsilylene), 29
Poly(ethyl acrylate)
emulsions, 233
glass transition, 238
Poly(ethyl acrylate-d_α), 34
Polyethylene, 2, 16, 17, 49, 81, 120, 128, 245, 355
Poly(ethylene oxide), 117
blends, 146, 148, 151, 156
chain scission, 127
crystalline, 123, 146
radiation induced changes, 127
Polyisobutylene, 363, 367
Poly(isopropyl acrylate), 272, 277
Polymer dynamics, 24
Polymers, surface bound, 271
Poly(methacrylic acid), isotactic, 179
Poly(methacrylic anhydride), 179, 188
Poly(methyl acrylate-d_α), 34
Poly(methyl methacrylate), 18, 131, 136, 272, 377, 378
Poly(naphthalenyl methylene)s, 109
Polyoxetane, 86, 89
cyclic tetramer, 89
Polyoxymethylene, 83, 120, 128
Polypeptides, 56
Polyphosphazenes, 321
Polypropylene
α-form, 97
β-form, 97
isotactic, 82, 83, 97
smectic form, 97
syndiotactic, 82
Polysilylenes, 295
Polystyrene, Isotactic, 18
Poly(tetrafluoroethylene), 12, 13, 49
Polyurethane foam, 69
Poly(vinyl acetate), 74, 131, 139
Poly(N-vinylcarbazole), 167, 174, 176
Poly(vinylidene fluoride), 131, 136, 139
Poly(2-vinylpyridine) blends, 153
Poly(4-vinylpyridine) blends, 146, 148, 152
Powder patterns, 8
N-*n*-Propyl stearamide, 43
Protonated aromatic, 110, 115
Pseudoaffine deformation, 15
Pulsed-gradient spin-echo technique, 389, 395

Quadrupolar coupling, 24
Quadrupolar interactions, 26, 27
Quadrupole coupling constant, 28, 57
Quaternary aromatic, 110, 113

Radiation induced morphological changes, 117
Radiofrequency pulse averaging, 3
Rapid spinning, 3
Regiochemistry, 115
Relative detectability, 28
Relaxation times, 24
Resorcinol blends, 146, 148, 151, 156
Rotational isomeric states, 128
Rubber, 71

Scaler coupling, 27
Semicrystalline, 81, 117, 118, 121, 154, 261, 324
Semifluorinated alkanes, 49
^{29}Si CP/MAS
poly(di-n-butylsilylene), 302
poly(di-n-hexylsilylene), 301
poly(di-n-pentylsilylene), 302
^{29}Si MAS, 308
organosilicon polymers, 305
Si-C, 310
Si-C-N-O fibers, 309

Si-C-O fibers, 309
Si-N, 312
^{29}Si spin-lattice relaxation times, 301
Signal-to-noise, imaging, 68
Silicon carbides, 310
Silicon nitrides, 312
Silicon oxycarbides, 305
Single-crystal sample, 8
Solids imaging, 61, 343, 365, 377
Solid-solid transitions, 246, 322, 323
Solid state ^{13}C NMR, 119, 123, 131, 132, 145, 179, 184, 201, 215, 217, 233, 245, 261, 272, 283, 285, 295
liquid crystalline polymers, 215, 261
poly(methacrylic anhydride), 179, 190
rubber, 201
two-dimensional, 261, 264
Solvent absorption, 71
Solvent diffusion
PMMA, 377, 384
polymer solution, 388
Spatial resolution, 64, 68
Spherically symmetric, 8
Spin-diffusion, ^{1}H, 150, 157, 159, 161
interdomain, 159, 164, 241
intermolecular, 158, 234
intramolecular, 158
Spin dynamics, 1
Spin-echo techniques, 66
Spin-lattice relaxation times, 82, 86, 90, 92, 97, 98, 100, 120, 131, 300, 306, 384
Spin-spin coupling, 26
Spin-spin relaxation times, 131, 160, 263, 384

T_1, 51, 66, 82, 83, 100, 120, 131, 132, 169, 175, 205, 207, 215, 217, 220, 300, 306, 379, 384, 386
^{29}Si, 301, 306
variable temperature, 208, 210, 213, 215, 220, 225, 229, 300
$T_{1\rho}$, 122, 131, 132, 169, 218, 233
variable temperature, 233,237
$T_{1\rho C}$, 233, 234, 236
$T_{1\rho H}$, 131, 133, 135, 169, 175, 233, 234
blends, 131, 139, 142
copolymers, 235
T_{1H}, 109
T_2, 64, 120, 132, 346, 357, 363, 366, 372, 379, 384, 386
T_{CH}, 131, 133, 135
blends, 131, 136
Terephthaloyl polyesters, 96
Three dimensional imaging, 63
Toggling reference frame, 5
Torchia method, 83
Transverse magnetization, 67
TREV-4, 5
2,4,7-Trinitro-9-fluorenone, 167
Two-dimensional experiments, 10

Voxels, 67

WAHUHA, 4
WAHUHA-4, 5, 6, 7, 12

Zeeman interaction, 3, 26